Methods in
Reproductive Aquaculture

Marine and Freshwater Species

MARINE BIOLOGY

SERIES

The late Peter L. Lutz, Founding Editor
David H. Evans, Series Editor

PUBLISHED TITLES

Methods in
Reproductive Aquaculture

Marine and Freshwater Species

Edited by

Elsa Cabrita
Vanesa Robles
and Paz Herráez

CRC Press
Taylor & Francis Group
Boca Raton   London   New York

CRC Press is an imprint of the
Taylor & Francis Group, an **Informa** business

CRC Press
Taylor & Francis Group
6000 Broken Sound Parkway NW, Suite 300
Boca Raton, FL 33487-2742

© 2009 by Taylor & Francis Group, LLC
CRC Press is an imprint of Taylor & Francis Group, an Informa business

International Standard Book Number-13: 978-0-8493-8053-2 (Hardcover)

Library of Congress Cataloging-in-Publication Data

Methods in reproductive aquaculture : marine and freshwater species / editors, Elsa Cabrita, Vanesa Robles, Paz Herráez.
   p. cm. -- (Marine biology)
  Includes bibliographical references and index.
  ISBN 978-0-8493-8053-2 (hardback : alk. paper)
  1. Fishes--Artificial spawning. 2. Fishes--Germplasm resources--Cryopreservation. 3. Shellfish--Artificial spawning. 4. Shellfish--Germplasm resources--Cryopreservation. I. Cabrita, Elsa. II. Robles, Vanesa. III. Herráez, Paz. IV. Title. V. Series.

  SH155.6.M48 2008
  639.3--dc22
                                    2008013533

Visit the Taylor & Francis Web site at
http://www.taylorandfrancis.com

and the CRC Press Web site at
http://www.crcpress.com

THE EDITORS

The editors have been working together in the field of reproductive aquaculture for the last ten years. Numerous scientific publications have arisen as a direct result of their intensive work on sperm and egg quality evaluation and in developing sperm cryopreservation protocols for different fish species. The evaluation of cryodamage and the improvement of fish embryo cryoresistance have also been a priority in their research. Nowadays, *Dr. Elsa Cabrita* is a researcher associate in the Spanish National Research Council ICMAN-CSIC, Spain, where she works on fish reproductive physiology. *Dr. Vanesa Robles* is a researcher associate in the Centre of Regenerative Medicine in Barcelona CMR(B), Spain, where she works in fish germ cells and stem cell biology. Both have worked previously at the University of León, Spain, and then moved to the Center for Marine Sciences in Faro, Portugal, where they have continued to develop research in the cryopreservation area. *Dr. Paz Herráez* is a professor in the Department of Molecular Biology at the University of León, Spain, where she teaches biology of reproduction and cellular biology. She is leader of the Cryobiology Group and she has been responsible for the training of several doctorate and master's degree students.

METHODS IN REPRODUCTIVE AQUACULTURE:
MARINE AND FRESHWATER SPECIES

Gamete quality and management

PREFACE

The idea of writing this book came to us some years ago when we realized that we had the conditions and opportunity to be involved in such a project. I remember arriving at the lab and discussing this with my colleagues and we were all very enthusiastic about the idea of publishing a book on gamete quality and cryopreservation, our initial idea and objective at that time. The following months were of hard work, especially after the publisher encouraged us to write a wider overview of reproductive aquaculture. Actually, I must confess that at the time we were not aware of what that decision would imply, and today, after looking through these pages, we can say, with great satisfaction, that it was all worth the effort. The fact that over 80 authors contributed to this unique project encouraged us to carry on and overcome the difficulties we encountered. Some of you know the problems we have had to face during these long months, especially in coordinating all the work long distance.

Methods in Reproductive Aquaculture was written by several authors in response to all the questions raised concerning gamete quality and management. Reproduction has been a focus of research and interest in several marine and freshwater species. Many species have been reproduced in captivity, most of them with interest in aquaculture, fisheries or conservation. While the majority of these techniques have been applied with some success under field conditions, some aspects related to broodstock reproduction have been unattended. Gamete quality and management are two important aspects in fish reproduction that deserve special attention, particularly if we consider that both are requirements for the production of new individuals. This constitutes the main focus of the book and all our efforts were directed towards providing consistent and sound information, new objectives and areas of application and research, and principally new tools to deal with some aspects in reproduction of aquatic species.

Not all sectors have access to the considerable amount of information available on fish reproduction, which is also sometimes difficult to process. Although several books containing a wide range of subjects related to reproduction have been published, we have attempted to focus on those aspects that we believe have been neglected.

This is the aim of our book, to reflect some of the aspects considered important in the reproduction of marine and freshwater species, presenting all the available information in an easily understandable way. To accomplish these objectives, the book is divided into five sections. The first section is a review of the basic methods and techniques for gamete extraction, spawning stimulation and stripping. The second deals with sperm and egg quality and is focused on gamete characteristics and methods used to evaluate quality; a third section includes techniques used for artificial fertilization and the procedures for obtaining modified offspring (androgenesis, polyploids, etc.) and the fourth section is a review of methods and advancements in gamete and embryo preservation and storage. The last section includes several protocols for sperm cryopreservation in different aquatic species. As our background is mostly in the area of cryobiology, I must say that putting all this material from different authors together and coordinating this section gave us particular pleasure.

The first part provides basic knowledge of gamete extraction methods, mainly sperm collection and egg stripping in different species. Also, several techniques of spawning stimulation (in males and females) by photoperiod and/or temperature control and hormonal treatments are described. This section could be of special value for teachers and students, clarifying concepts and specific techniques. For aquaculturists and researchers it could be of great value in solving problems related to their broodstock management and research.

The second section is based on sperm and egg characteristics commonly used for quality assessment. Gamete quality is associated with larval quality and is a requirement not only in production but also in several fields of research such as embryology, genetics, biotechnology and cryopreservation, among others. Sperm cell morphology, physiology and metabolism have been a focus of research basically in cryopreservation studies. In this section, sperm analysis will also be considered from a practical point of view in the management of broodstock conditions or in the selection of appropriate breeders. Egg quality is a requirement for fertilization success and several biomarkers such as egg morphological characteristics, enzymatic and metabolic markers and novel molecular markers are discussed.

The third section of the book provides information on artificial fertilization from normal practice used in fertilization trials to more specific fertilization procedures allowing the manipulation of the normal process to produce more interesting progeny from the viewpoint of aquaculturists and researchers.

In the fourth section the different chapters try to present an overview of methods, research activities and directives in the management of gametes using biotechnology tools to store the breeders' reproductive potential for longer or shorter periods of time. An introduction to cryopreservation procedures both in sperm, oocytes and embryos is given as well as to materials and facilities required for gamete preparation and freezing-thawing procedures. During recent years, fish sperm cryopreservation has become an important tool in aquaculture facilities and conservation programs. It can be used by fish farmers to facilitate daily procedures and increase production as well as preserving genetic heritage, particularly of those individuals presenting desirable characteristics. From a conservation point of view it enables the preservation of genetic material from endangered species. For this reason we decided to incorporate a special section describing specific protocols for the cryopreservation of sperm from several species. Sperm cryopreservation protocols are detailed for fifty-six species, several marine, freshwater, anadromous and catadromous species of teleosts, chondrosts, molluscs, decapods or equinoderms. Each protocol includes three sub-parts. A general description of the relevance of the species and sperm particularities is made. A detailed description of the freezing-thawing protocol is given, including equipment and reagents used, sperm extraction method, sperm extender and cryoprotectants, dilution rates, sperm characteristics such as cell concentration, plasma osmolality and pH, sperm loading and freezing/thawing rates, and so on. This information will be very useful, particularly for people working in different fields and who are only interested in using sperm cryopreservation as a tool.

In conclusion, this book provides information on endangered, highly profitable species for aquaculture or fisheries and species with high potential in laboratory research, constituting a complete guideline in an important aspect of fish reproduction-gamete quality and management. This is the first practical guide in reproduction of aquatic species that includes cryopreservation as one of the techniques in gamete management. It establishes the basis for the reproduction of aquatic species, making it accessible to interested parts, from fish farmers to scientific researchers and from teachers to students, therefore having a wide range of potential readers.

We hope you enjoy reading this book as much as we have compiling it.

Elsa Cabrita

Vanesa Robles

Paz Herráez

ACKNOWLEDGMENTS

We are very grateful to our collaborators, relatives and friends who "suffered" this long process: Sonia Martínez-Páramo, Serafín Pérez-Cerezales, José Beirão, Paulo Gavaia, Javier Gallego, Julia Alvarez, Catherine Martin, Jeff Wallace, aquagroup and to the reproductive group from the University of León. We would also like to thank the collaboration of Adelina Jimenez, Ricardo Leite and Patricia Roberson for their assistance in editing, to all the authors for their comprehension and important contribution, to Fernando Romero, author of the cover photography, as well as to the aquaculturist who supplied the biological material in several projects and contributed to the development of our research. We also thank the confidence of the publishers who accepted this idea and encouraged us in publishing this book, suggesting we continue with a more ambitious project. Finally to Portuguese Science Foundation, Spanish Ministry of Science and Innovation and Junta de Castilla y León for their financial support.

CONTRIBUTING AUTHORS

Abad, Zoila, Center for Genetic Engineering and Biotechnology, P.O. Box 387, C.P. 70 100, Camagüey, Cuba.

Abelli, Luigi, Department of Environmental Sciences,Tuscia University, Viterbo, Italy.

Acker, Jason P., Canadian Blood Services, 8249-114 Street, Edmonton, Alberta, Canada, T6G 2R8.

Adams, Serean L., Aquaculture Scientist, Cawthron, Private Bag 2, Nelson, New Zealand.

Akarasanon, Khattiya, Center for Reproductive Biology of Economic Aquatic Animals, Institute of Science and Technology for Research and Development, Mahidol University, Nakornpatom 73170, Thailand.

Alvarez, Bárbaro, Center for Genetic Engineering and Biotechnology, P.O. Box 387, C.P. 70 100, Camagüey, Cuba.

Arenal, Amilcar, Center for Genetic Engineering and Biotechnology, P.O. Box 387, C.P. 70 100, Camagüey, Cuba.

Asturiano, Juan F., Grupo de Investigación en Recursos Acuícolas, Departamento de Ciencia Animal, Universidad Politécnica de Valencia, Camino de Vera, s/n, 46022 Valencia, Spain.

Babiak, Igor, Department of Fisheries and Natural Sciences, Bodø University College, 8049 Bodø, Norway.

Barbato, Fabio, Enea Biotec Amb, Centro Ricerche Casaccia s.p. 028, Via Anguillarese 301, 00123 Santa Maria di Galeria (Roma), Italy.

Berlinsky, David L., Department of Zoology, University of New Hampshire, Durham NH 03824, USA.

Bhavanishankar, S., Department of Biotechnology, KRMM College of Arts and Science, University of Madras, #4, Crescent Avenue Road, Gandhi Nagar, Chennai 600 020, India.

Bobe, Julien, Fish Reproduction, INRA,SCRIBE, UR 1037, Campus de Beaulieu, F-35000, Rennes, France.

Bokor, Zoltán, Department of Fish Culture, Szent István University, Páter Károly u. 1, Gödöllö H-2103, Hungary.

Butts, Ian A.E., Department of Biology, University of New Brunswick, PO Box 5050, Saint John, NB, E2L 4L5, Canada.

Canese, Stefano, Enea Biotec Amb, Centro Ricerche Casaccia s.p. 028, Via Anguillarese 301, 00060 S.M. di Galeria (Roma), Italy.

Chen, Song-Lin, Yellow Sea Fisheries Research Institute, Chinese Academy of Fisheries Sciences, Nanjing Road 106, 266071 Qingdao, China.

Chereguini, Olvido, Instituto Español de Oceanografía, Planta de Cultivos Marinos El Bocal, Barrio de Corbanera, Monte, Santander, B.O. 240, España.

Cloud, Joseph, Department of Biological Sciences, University of Idaho, Moscow, ID 83844-3051.

Damrongphol, Praneet, Department of Biology, Faculty of Science, Mahidol University, Rama VI. Rd., Bangkok 10400, Thailand.

Dinis, María Teresa, Centro de Ciências do Mar, Universidade do Algarve, Campus de Gambelas, 8005-139, Faro, Portugal.

Dong, Qiaoxiang, Aquaculture Research Station, Louisiana Agricultural Experiment Station, Louisiana State University Agricultural Center, Baton Rouge, Louisiana 70820 USA.

Duncan, Neil, IRTA, Crta. Poble Nou, km. 5,5, E-43540, Sant Carles de la Rapita, Tarragona, Spain.

Eudeline, Benoit, Taylor Resources Hatchery, Quilcene, Washington 98376 USA.

Fausto, Annamaria, Department of Environmentl Sciences, Tuscia University, Viterbo, Italy.

Fauvel, Christian, Ifremer LRPM Chemin de Maguelone, 34250 Palavas, France.

Fuentes, Roberto, Center for Genetic Engineering and Biotechnology, P.O. Box 387, C.P. 70 100, Camagüey, Cuba.

Granja, Clarissa, Centro de Investigaciones de la Acuacultura en Colombia CENIACUA, Carrera 8A # 96-60 Bogota, Colombia.

Gwo, Jin-Chywan, PhD, Department of Aquaculture, Taiwan National Ocean University, Keelung 20224, Taiwan.

Hagen, Andreas, Dana-Farber Cancer Institute, Department of Pediatric Oncology, Harvard Medical School, Boston, MA 02115, USA.

Hazlewood, Leona, Xiphophorus Genetic Stock Center, Texas State University, San Marcos, Texas 78666 USA.

He, Shuyang, Department of Animal and Avian Sciences, University of Maryland, College Park, MD, USA.

Hessian, Paul A., Departments of Physiology and Marine Science, University of Otago, P. O. Box 56, Dunedin, New Zealand.

Horváth, Ákos, PhD, Department of Fish Culture, Szent István University, Páter Károly u. 1., Gödöllõ H-2103, Hungary.

Huang, Changjiang, School of Marine Science and Technology, Zhejiang Ocean University, Zhoushan 316004, P.R. China.

Janke, Achim R., Aquaculture Scientist, Cawthron, Private Bag 2, Nelson, New Zealand.

Jenkins-Keeran, Karen, National Institutes of Health, Bethesda, MD, USA.

Ji, Xiang Shan, Yellow Sea Fisheries Research Institute, Chinese Academy of Fisheries Sciences, Qingdao 266071, China.

Kanki, John P., Dana-Farber Cancer Institute, Department of Pediatric Oncology, Harvard Medical School, Boston, MA 02115, USA.

Kaspar, Heinrich F., Aquaculture Scientist, Cawthron, Private Bag 2, Nelson, New Zealand.

King, Nick G., Aquaculture Scientist, Cawthron, Private Bag 2, Nelson, New Zealand.

Komen, Johannes, Animal Breeding and Genetics Group, Wageningen Institute of Animal Sciences, The Netherlands.

Kopeika, Eugeny, Institute for Problems of Cryobiology and Cryomedicine of the National Academy of Science of the Ukraine, 23 Pereyaslavskaya, Kharkov 61015, Ukraine.

Kopeika, Julia, Luton Institute for Research in the Applied Natural Sciences, University of Luton, The Spires, 2 Adelaide street, Luton, Bedfordshire, LU1 5DU, UK.

Labbé, Catherine, Fish Reproduction, INRA, UR 1037 SCRIBE, Campus de Beaulieu, F-35000 Rennes, France.

Lahnsteiner, Franz, Institute of Zoology, University of Salzburg, Hellbrunnerstrasse 34, A-5020 Salzburg, Austria.

Lezcano, Magda, Centro de Investigaciones de la Acuacultura en Colombia CENIACUA, Carrera 8A # 96-60 Bogota, Colombia.

Liao, Lin-Yen, Shei-Pa National Park, Taichung, Taiwan.

Lin, Young-Fa, Shei-Pa National Park, Taichung, Taiwan.

Litvak, Matthew K., Department of Biology, University of New Brunswick, PO Box 5050, Saint John, NB, E2L 4L5, Canada.

Lubzens, Esther, National Institute of Oceanography, Israel Oceanographic and Limnological Research, P. O. Box 8030, Haifa, 31080, Israel.

Maisse, Gerard, Fish Reproduction, INRA, UR 1037 SCRIBE, Campus de Beaulieu, F-35000 Rennes, France.

Mañanós, Evaristo, Consejo Superior de Investigaciones Científicas, Instituto de Acuicultura de Torre la Sal, 12595-Ribera de Cabanes, Castellón, Spain.

Mansour, Nabil, Institute of Zoology, University of Salzburg, Hellbrunnerstrasse 34, -5020 Salzburg, Austria.

Maria, Alexandre Nizio, Animal Sciences Department, Federal University of Lavras, DZO-UFLA, Lavras, MG, Brazil.

Martínez-Pastor, Felipe, National Wildlife Research Institute (IREC) (CSIC-UCLM-JCCM), and Game Research Institute (IDR), 02071, Albacete, Spain.

Mazzini, Massimo, Department of Environmental Sciences, Tuscia University, Viterbo, Italy

Mims, Steven D., Aquaculture Research Center, Kentucky State University, Frankfort KY 40601, USA.

Mladenov, Philip V., Departments of Physiology and Marine Science, University of Otago, P. O. Box 56, Dunedin, New Zealand.

Moretti, Filippo, Enea Biotec Amb, Centro Ricerche Casaccia s.p. 028, Via Anguillarese 301, 00060 S.M. di Galeria (Roma), Italy.

Morris IV, John P., Dana-Farber Cancer Institute, Department of Pediatric Oncology, Harvard Medical School, Boston, MA 02115, USA.

Mylonas, Constantinos, Institute of Aquaculture, Hellenic Center for Marine Research, Ex US Military Base of Gournes, P.O.Box 2214, Heraklion, Crete 71003, Greece.

Ogier de Baulny, Benedicte, Fish Reproduction, INRA, UR 1037 SCRIBE, Campus de Beaulieu, F-35000 Rennes, France.

Ohta, Hiromi, Department of Fisheries, Faculty of Agriculture, Kinki University, Nara 631-8505, Japan.

Okuzawa, Koichi, National Research Institute of Aquaculture, Nansei, Mie, Japan.

Paniagua-Chavez, Carmen G., Centro de investigación Científica y de Educación Superior de Ensenada-CICESE, Department of Aquaculture, Km 107 Carretera Tijuana-Ensenada, Apartado Postal 2732, 22800, Ensenada, Baja California, México.

Patton, S., Department of Biological Sciences, Life Science South, Rm 252, University of Idaho, Moscow, ID 83844-3051.

Pimentel, Eulogio, Center for Genetic Engineering and Biotechnology. P.O. Box 387, C.P. 70 100, Camagüey, Cuba.

Pimentel, Rafael, Center for Genetic Engineering and Biotechnology. P.O. Box 387, C.P. 70 100, Camagüey, Cuba.

Pugh, Patricia A., AgResearch, Ruakura Research Centre, Private Bag 3123, Hamilton, New Zealand.

Rana, Krishen J., Stirling University, Institute of Aquacukture, Stirling, Scotland FK9 4LA UK and Division of Aquaculture, Stellenbosch University, Stellenbosch, South Africa

Rawson, David, Luton Institute for Research in the Applied Natural Sciences, University of Luton, The Spires, 2 Adelaide street, Luton, Bedfordshire, LU1 5DU, UK.

Ribeiro, Laura, Centro de Ciências do Mar, Universidade do Algarve, Campus de Gambelas, 8005-139, Faro, Portugal.

Rideout, Rick M., Fisheries and Oceans Canada, Northwest Atlantic Fisheries Centre, PO Box 5667, St. John's, NL, A1C 5X1, Canada.

Ritar, Arthur Jeremy, Tasmanian Aquaculture & Fisheries Institute, University of Tasmania, Marine Research Laboratories, Nubeena Crescent, Taroona Tasmania 7053 Australia.

Roberts, Rodney D., Aquaculture Scientist, Cawthron, Private Bag 2, Nelson, New Zealand.

Salazar, Marcela, Centro de Investigaciones de la Acuacultura en Colombia CENIACUA, Carrera 8A # 96-60, Bogota, Colombia.

Sarasquete, Carmen, Center for Marine Sciences from Andalusia-ICMAN, Spanish Research Council, Poligono Industrial Rio San Pedro s/n, 11510 Puerto Real, Cadiz, Spain.

Smith, John F., AgResearch, Ruakura Research Centre, Private Bag 3123, Hamilton, New Zealand.

Soares, Florbela, Centro de Ciências do Mar, Universidade do Algarve, Campus de Gambelas, 8005-139, Faro, Portugal.

Subramoniam, T., Professor Emeritus, Department of Zoology, Life Sciences Building, University of Madras, Guindy Campus, Chennai 600 025, India.

Suquet, Marc, IFREMER, Laboratoire ARN, BP 70, 29280 Plouzané, France.

Taddei, Annarita, Department of Environmental Sciences, Tuscia University, Viterbo, Italy.

Tervit, Harry R., AgResearch, Ruakura Research Centre, Private Bag 3123, Hamilton, New Zealand.

Tian, Yong-Sheng, Yellow Sea Fisheries Research Institute, Chinese Academy of Fisheries Sciences, Qingdao 266071, China.

Tiersch, Terrence R., Aquaculture Research Station, Louisiana Agricultural Experiment Station, Louisiana State University Agricultural Center, Baton Rouge, Louisiana 70820 USA.

Urbányi, Béla, Department of Fish Culture, Szent István University, Páter Károly u. 1., Gödöllő H-2103, Hungary.

Viveiros, Ana T.M., Departamento de Zootecnia, Universidade Federal de Lavras, DZO-UFLA, Caixa postal 3037, 37200-000, Lavras, MG, Brazil.

Walter, Ron B., Xiphophorus Genetic Stock Center, Texas State University, San Marcos, Texas 78666 USA.

Webb, Steven C., Aquaculture Scientist, Cawthron, Private Bag 2, Nelson, New Zealand.

Woods III, L. Curry, Department of Animal and Avian Sciences, University of Maryland, College Park, MD, USA.

Zhang, Tiantian, LIRANS Institute of Research in the Applied Natural Sciences, University of Bedfordshire, The Spires, 2 Adelaide Street, Luton, Bedfordshire, LU1 5DU, UK.

CONTENTS

SECTION V – PROTOCOLS FOR SPERM CRYOPRESERVATION

SECTION I - GAMETE EXTRACTION TECHNIQUES

CHAPTER 1
REPRODUCTION AND CONTROL OF OVULATION, SPERMIATION AND SPAWNING IN CULTURED FISH

E. Mañanós, N. Duncan and C. Mylonas

1. INTRODUCTION

Fish are the largest phylum of living vertebrates, with around 30,000 fish species out of approximately 50,000 vertebrate species (www.fishbase.com) [1]. Fish inhabit almost every aquatic environment on the planet, which present an enormous variation in temperature, salinity, oxygen, and other chemical and physical water properties. These environments have exerted evolutionary pressures that have resulted in the evolution of the enormous number of fishes and an immense variety of reproductive strategies. Fish exhibit various types of sex determination, from genetic to environmental control, sex differentiation from hermaphroditism to gonochorism, age of puberty, from a few months to many years, fecundity, from a few to millions of eggs, internal or external fertilization, a wide range of egg sizes, some that float and others that sink or stick to substrates, uncared eggs scattered into the environment to parental care of eggs to live "birth" (ovoviviparity) [2-4]. The existence of these diverse reproductive strategies has important implications for finfish culture and broodstock management.

Finfish culture is the fastest growing food production industry in the world, and in 2005 a total of 28.3 million T of finfish were produced, which is around 20% of the world's fisheries production (aquaculture and capture fisheries) [5]. The number of species being cultured is also increasing rapidly, and 67 different cultured finfish species in 2005 had an annual production of over 10,000 T, which is twice the amount produced by the 28 species being cultured in 1980 [6]. One of the most important aspects at the basis of this continuing increase in the number of cultured species is our growing understanding of the complexities of the many different reproductive strategies of various fishes and how these behave in captivity. This is perhaps best demonstrated by the development of the culture of the catfishes (*Pangasius ssp*) in Vietnam. These catfishes do not mature in captivity, but during the 1990s hormone stimulation techniques were developed to induce ovulation/spermiation. This technology gave the control required to produce eggs, larvae and juveniles that formed the basis of an expanding aquaculture industry, producing 40,000 T in 1997 and 376,000 T in 2005 [6].

This chapter aims at giving a general vision of fish reproduction and the reproductive dysfunctions found in captive-reared fishes, and describe how this knowledge is used to develop treatments for the control of fish reproduction in aquaculture. This general view will be covered under four broad areas relevant to the control of gonad maturation and spawning of fish in captivity. First, it will be described the normal gonadal development as can be expected under optimal (natural) conditions, particularly the morphology of gonad development and its endocrine and environmental regulatory mechanisms. Second, the reproductive dysfunctions that are observed in fish held in captivity; often the captive environment does not provide the environmental conditions that a species requires to complete maturation. Third, the available techniques for the stimulation of ovulation and spermiation, both environmental manipulations and hormonal treatments of reproductively dysfunctional fish held in captivity. Finally, the last sections will describe the application of adequate broodstock management protocols and hormonal therapies for the stimulation of fish reproduction in captivity for some of the main food fish species under production, both as a reference and indication of the possible strategies available to stimulate ovulation and spermiation in species considered candidates for new aquaculture developments.

2. THE REPRODUCTIVE CYCLE OF FISH

The reproductive cycle is an ensemble of successive processes from immature germ cells to the production of mature gametes, with the final purpose of obtaining a fertilized egg after the insemination with a spermatozoon. The process of gamete growth and differentiation is called gametogenesis, and leads to the formation of the female oocyte (oogenesis) or the male spermatozoon (spermatogenesis). Both the female and male gametes have a common origin in the population of embryonic primordial germ cells (PGC) that migrate during embryonic development to the place of gonad formation, the germinal epithelium. The PGC proliferate through mitotic divisions to form the gonia, which differentiate into oogonia or spermatogonia depending on the sex of the individual. With the last mitotic division, gonia enter meiosis and become oocytes or spermatocytes, thus initiating gametogenesis in adult animals [7-8].

In both males and females, the reproductive cycle involves two major phases, the phase of gonadal growth and development (gametogenesis) and the phase of maturation, which culminates in ovulation/spermiation and spawning. The release of mature gametes to the external environment (spawning) is a highly synchronized event, leading to fertilization of the egg and development of the embryo. The success of reproduction depends on the successful progression through every process of the reproductive cycle, which leads to the production of good quality gametes. This section describes some general features of reproductive physiology and gonadal development in fish.

2.1 Ovarian development in females: oogenesis, maturation and ovulation

The ovary of female fish is a bilateral elongated organ, localized in the abdominal cavity. The ovarian lobules are surrounded by the mesovarium and project posterior through a pair of oviducts that connect to the genital papilla, which opens to the external environment. The ovaries are compartmentalized by folds of the germinal epithelium that project transversally to the ovarian lumen, the ovigerous lamellae. In these lamellae, the oocytes undergo the various phases of gametogenesis, until mature ova (i.e., eggs) are released into the ovarian cavity or abdominal cavity (e.g., salmonids) at ovulation and then to the external environment during spawning. Ovulated ova may remain in the ovarian/abdominal cavity for a period of time before spawning. There, they maintain their maturational competence (fertilizing capacity) for a certain period of time, but if not spawned, the ova become "over-ripe" through a process of degeneration. This is an important consideration in cultured fish whose reproduction is based on manual egg stripping and artificial insemination, because stripping should be performed before over-ripening occurs. The lapse of time between ovulation and over-ripening varies greatly among fish, from minutes (e.g., striped bass, Morone saxatilis) to days (e.g., salmonids) and depends greatly on water temperature. In salmonids, which do not have a complete mesovarium and the oocytes are ovulated directly into the abdominal cavity, the ovulated ova can remain for several days with no evident over-riping.

The germinal unit of the ovary consists of an oocyte surrounded by two layers of follicular cells. These follicular cells envelop the germ cell and offer structural and functional support to the developing oocyte, mediating the internalization of external molecules and synthesizing hormones and factors necessary for the differentiation, growth and survival of the oocyte. Each oocyte is surrounded by an inner mono-layer of granulosa cells and an outer mono-layer of theca cells [9]. Between the two follicular layers there is a thin basal membrane, which separates them. Also, a thick acellular envelop surrounds the oocyte (i.e., zona radiata), to which the granulosa cells are directly attached. The zona radiata develops progressively during gametogenesis, becoming increasingly thick and compact to constitute the egg chorion or egg shell.

The female reproductive cycle of fish is characterized by the specific process of vitellogenesis, the synthesis of vitellogenin (VTG), the precursor of the vitellin reserves of the egg [10-13]. The VTG is a lipophosphoglycoprotein synthesized in the liver under the stimulation of estradiol (E_2). It is released into the bloodstream and incorporated progressively into the growing oocytes, via receptor mediated endocytosis, were VTG is proteolytically cleaved into smaller components (phosvitin, lipovitellin and â-component), giving rise to the vitellin reserves of the egg, the yolk or vitellus [14]. The optimal accumulation and processing of VTG is of vital importance for the quality of the egg and further survival of the hatched larvae, as this constitutes the sole nutritional reserves of the larva, until the initiation of external feeding, several days after fertilization. The VTG is a female specific protein and its circulating blood levels correlate well with the initiation and progression of the gametogenic period.

These characteristics of the VTG molecule make the use of specific VTG immunoassays a very useful tool in aquaculture. Such VTG immunoassays are used for sexing broodstock in those species without external sexual dimorphism, because VTG detection in blood indicates clearly a female. The VTG immunoassays are also used to follow the stage of gonadal development, as VTG blood levels increase concomitantly with oogenesis [15-17].

The reproductive cycle of female fish can be divided into the period of oocyte growth (gametogenesis or vitellogenesis) and the period of oocyte maturation, along which the oocyte goes though different stages of development, before ovulation and spawning (Fig. 1) [10-13,18-19]. Oocyte development has been described in detail in a few species of fish [20-23]. Before intiation of the reproductive cycle, the immature ovary contains nests of oogonia along the ovigerous lamellae, which proliferate through mitotic divisions. At a certain moment, part of the oogonia population enters meiosis and become primary oocytes, which are arrested immediately at prophase I. This is the initiation of gametogenesis, and meiosis will not be resumed until final oocyte maturation (FOM). The primary oocytes go through a primary growth phase or previtellogenesis, which involves an increase in size, the appearance of pale material in the cytoplasm and the appearance of the granulosa and theca cellular layers (i.e., the follicle). This is a hormone-independent phase, before the period of E_2-induced VTG synthesis. The secondary growth phase, or vitellogenesis, is characterized by the synthesis and enormous accumulation of VTG and vitellin related proteins into the oocyte, resulting in a 10-fold increase in size. This period involves several transformation processes of the oocyte and associated follicular cells, which lead to the further classification of the vitellogenic period into different developmental stages. These transformations involve successive changes in both the oocyte cytoplasm and surrounding membrane and follicular cells. At early vitellogenesis, the oocyte is small (around 100 μm in diameter), with an opaque cytoplasm almost deprived of inclusions, except some oil droplets. With the progression of vitellogenesis, new inclusions appear in the cytoplasm, such as the cortical alveoli, lipid globules and yolk granules. The order of appearance of each type of inclusion is species-specific. These inclusions increase in size and number during vitellogenesis, promoting the increase of oocyte size. Also, the zona radiata, as well as the granulosa and theca layers become increasingly thick in order to support the rapid oocyte growth. At the end of the vitellogenic period (late vitellogenesis), the post-vitellogenic oocyte is characterized by a large transparent cytoplasm completely filled with yolk granules and lipid globules, a centrally located nucleus (or germinal vesicle, GV), and a thick, clearly striated zona radiata, enveloped by the granulosa and theca follicular layers.

After vitellogenesis, oocytes undergo maturation, with the resumption of meiosis and its advance to metaphase II, at which time the first polar body is released and the oocyte becomes a secondary oocyte [24]. At early maturation, lipid globules and yolk granules start coalescence and the germinal vesicle begins its migration to the animal pole (GV migration, GVM). As maturation advances, there is a massive coalescence of yolk inclusions and localization of the GV at a peripheral position.

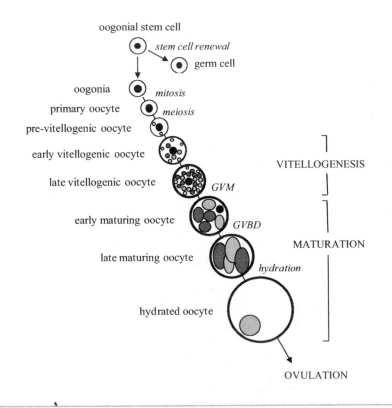

Figure 1 - The process of oocyte development and maturation in female fish. It initiates with the mitotic proliferation of the oogonia that undergo primary oocytes when entering to meiosis. The primary oocytes go through a primary growth phase (pre-vitellogenesis), which involves the appearance of pale material in the cytoplasm and formation of the two layers of surrounding granulosa and theca cells (i.e., follicular wall). The secondary growth phase (vitellogenesis) involves the synthesis and incorporation into the oocyte of vitellogenin (VTG), and is associated with a drastic increase in size. During vitellogenesis, new inclusions appear in the cytoplasm, such as the cortical alveoli (white circles), lipid globules (light grey circles) and yolk granules (dark grey circles) and the oocyte wall (i.e., zona radiata) and follicular wall become increasingly thick. At the end of vitellogenesis, the cytoplasm is filled completely with lipid globules and yolk granules at the onset of coalescence, the nucleus (germinal vesicle, GV) (black circle) is centrally located and a thick zona radiata surrounds the oocyte. At early maturation, lipid globules and yolk granules continue coalescence and the nucleus migrates to the animal pole (GV migration, GVM). As maturation advances, there is a massive coalescence of yolk granules and localization of the nucleus at a peripheral position. Final oocyte maturation (FOM) is characterized by the dissolution of the GV membrane (GV break down, GVBD) and hydration of the oocyte. Oocytes are finally ovulated into the ovarian or abdominal cavity, and are released in the water during spawning. In this diagram, the cell sizes are relative.

Final oocyte maturation is characterized by the dissolution of the nuclear membrane, a process called GV break down (GVBD). The transformation of the lipid and yolk inclusions modifies the ionic composition of the cytoplasm, causing a drastic incorporation of water inside the oocyte, through increases in osmotic pressure. This tremendous hydration is especially relevant in fishes producing pelagic eggs, and causes a rapid 2-3 fold increase in oocyte volume. After hydration, the follicular wall ruptures and the oocyte is ovulated into the ovarian/abdominal cavity and released to the water during spawning.

During the spawning season, post-ovulatory follicles (POF) can be found in the ovary. These are the empty follicular envelops that remain after the oocytes are released; under the microscope they have the aspect of folded structures, and disappear during ovarian reorganization a few days after ovulation. Atretic or apoptotic oocytes can also be found. These are oocytes that interrupted the process of vitellogenesis or FOM because of a failure in the hormonal regulation of the reproductive process. When the oocytes die, the vitelline envelop is fragmented and the hypertrophied follicular cells invade the ooplasm for phagocytosis. Follicular atresia appears under the microscope as a compact and well vascularized structure. The number of atretic oocytes increase along the pre-spawning and spawning season and it is a clear reflect of the success of reproduction of the female. It occurs in all species and relates to the appropriateness of the reproductive environment. Follicular atresia can occur at all stages of oogenesis and regulates the number of oocytes that advance through the reproductive process, affecting the fecundity of the species.

The determination of the stage of gonadal development in female breeders is an important tool in aquaculture. This can be determined by examination of biopsies of oocyte samples. The biopsy is performed in anesthetized females, by insertion of a cannula through the gonoduct and gentle aspiration of intra ovarian oocytes. The collected oocytes are observed under the binocular and classified according to their size, position of the GV (central, migrating or peripheral), degree of yolk granule coalescence, etc.; these classifications give a relative indication of the stage of gonad development of the females.

2.2 Testicular development in males: spermatogenesis, maturation and spermiation

The male gonad (testes) is also comprised of germinal and somatic tissue [25]. The germinal cells develop during spermatogenesis to give rise to the gametes, the spermatozoa. The somatic tissue of the testes forms the seminiferous tubules and supporting connective tissue, as well as specialized somatic cells, the Leydig and Sertoli cells [26]. These somatic cells offer structural and functional support to the germinal cells and play a crucial role in the production of hormones and other factors necessary for germ cell differentiation, development and survival. The Sertoli cells envelop the germ cells to form units called cysts or spermatocysts. The sum of all cysts constitutes the germinal epithelium of the testes. The Sertoli cells are attached to a basement membrane, which separates the germinal epithelium from the interstitial compartment. The interstitial compartment is formed by somatic tissue, in which the Leydig cells are located, between the seminiferous tubules. The Leydig cells, as the Sertoli cells, are also specialized endocrine cells, with an important role in the production of the necessary hormones for germ cell development.

The process of spermatogenesis can be divided in three major phases, 1) the mitotic proliferation of the spermatogonia, 2) the meiotic division of the spermatocytes, and 3) the transformation of the haploid spermatids into flagellated spermatozoa (spermiogenesis). Through these, the germ cells go through different developmental stages, spermatogonia A and B, primary (2n) and secondary (1n) spermatocytes, spermatids and spermatozoa (Fig. 2). The presence and relative abundance of each of these classes in the testes is used as an indication of the degree of testicular development [27-28]. Before initiation of gonadal development, the immature testes contain spermatogonia (spermatogonial stem cells) that proliferate by mitotic divisions, through a self-renewal process. During this phase the population of stem cells in the testes increases in number. At a certain moment, some spermatogonial stem cells enter the process of spermatogenesis and are committed to produce spermatogonia. During the phase of mitotic proliferation of the spermatogonia (phase 1), each spermatogonium goes through several cycles of mitotic divisions, ranging from five to fifteen, depending on the species. During division, cytokinesis is incomplete and daughter cells maintain direct cytoplasmic bridges between them, remaining together in a cluster, called spermatocyst. This cluster of daughter cells is thus formed by a clone of spermatogonia, as they all come from a single original cell. Each cyst is enveloped by a wall of somatic Sertoli cells that maintain different clones separated from each other. During this phase of mitotic proliferation, the spermatogonia go first through a phase of slow division rate (spermatogonia A) and then through a phase of rapid division rate (spermatogonia B). The last mitotic division of spermatogonia B gives rise to primary spermatocytes, which will enter to the process of meiosis (phase 2). During phase two, the primary spermatocytes proceed to the first meiotic division, which involves DNA duplication and recombination of the genetic information, leading to the formation of secondary spermatocytes. They go rapidly to the second meiotic division, without DNA duplication, leading to the formation of haploid germ cells, the spermatids.

The spermatids enter the process of spermiogenesis (phase 3), in which the haploid spermatids differentiate into flagellated spermatozoa. This process does not involve cellular proliferation, but only cell transformation, which includes a drastic reduction in size (>80%) due to nucleus condensation and extrusion of the cytoplasmic content to the surrounding Sertoli cells.

In addition, concentration of mitochondria, formation of a midpiece and formation of the flagellum take place at this time. In all but a few species of fish, there is no formation of the acrosome present in all other vertebrates, because fertilization is achieved through the micropyle of the fish egg. Once spermiogenesis is completed, the spermatocysts opens up by rupture of the Sertoli cell wall and the flagellated spermatozoa are released into the testicular lumen. Here, the spermatozoa will undergo the final process of maturation or capacitation, by which they acquire the fertilizing capacity (capacity of motility) [29].Maturation occurs during sperm migration along the efferent duct and involves only physiological changes. Simultaneous to sperm maturation, the efferent duct produce a high amount of fluid (sperm hydration), leading to the formation of the milt, the fluid containing suspended spermatozoa. The released spermatozoa are stored before spawning and, depending on the species, the storage place is the tubular lumen, the efferent duct system or the germinal vesicles.

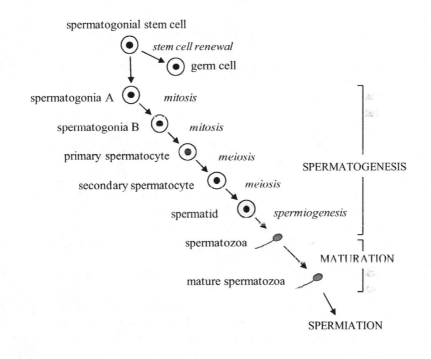

Figure 2 - The process of spermatozoa development and maturation in male fish. It initiates with the mitotic proliferation of the spermatogonia, first through slow division rates (spermatogonia A) and then through rapid division rates (spermatogonia B). Spermatogonia B emerge as primary spermatocytes, which enter to the first meiotic division; they become secondary spermatocytes that enter to the second meiotic division, leading to the formation of haploid spermatids. The spermatids differentiate into flagellated spermatozoa during spermiogenesis, which involves a drastic reduction in size (>80%) and formation of the flagellum. The flagellated spermatozoa are then released into the testicular lumen, were they undergo the process of maturation, by which they acquire the fertilizing capacity. The mature spermatozoa are stored in the testes until they are released in the water during spermiation. In this diagram, the cell sizes are relative.

Also, depending on the species, there are variations in the degree of maturation (fertilizing capacity) exhibited by the stored spermatozoa. For example, in salmonids, the intratesticular spermatozoa presents reduced fertilizing capacity as compared to spermatozoa present in the efferent duct. This is important to consider when collecting sperm from a given species for fertilization experiments, as the fertilizing capacity of the collected sperm can vary depending on the location of sperm collection.

Structurally, the testes of teleost fish can be classified in two major types, tubular or lobular. The tubular testes is the most common among fishes. In this type, the spermatocysts are distributed throughout the testes, in a tubular structure, and they do not move during the process of spermatogenesis. The lobular testes is found in some perciformes and atheriniformes and is characterized by the existence of lobules that blind-ended in the periphery of the testes; in these lobules the spermatogonia are restricted to the tips and the spermatocytes/spermatids move towards the efferent duct system during spermatogenesis.

2.3 Types of gonadal development

The high diversity of reproductive strategies of fish and that of inhabited environments is also reflected by the existence of a high variety in the types of gonadal development. This has important consequences to the fecundity and spawning characteristics of each species. The gonadal development in fish can be classified in three major types: synchronous, group-synchronous and asynchronous [11,19,30].

The synchronous type is exhibited by those species spawning only once in their life. This is the case of the lamprey (*Petromyzon spp*), the freshwater eel (*Anguilla spp*), some shad (*Alosa spp)* and Pacific salmons (*Oncorhynchus spp)*. In this type of ovary, all oocytes advance in synchrony through all phases of gametogenesis, FOM and ovulation. Thus, only one type of developing oocyte class is present in the ovary. The group-synchronous type is exhibited by the seasonal spawners, those species that spawn one or more times during the annual reproductive season. In this type, a cluster of vitellogenic oocytes is recruited and advance synchronously through further stages of development, whereas the rest of the oocyte population remains arrested. The cluster of recruited oocytes will undergo maturation, ovulation and spawning. This type of ovarian development can be divided in two subgroups: single-batch and multiple-batch spawners. In the single-batch group synchronous species, only one batch of oocytes undergoes maturation every season and thus, they produce one single spawn per year (e.g., rainbow trout, *Oncorhynchus mykiss*). The multiple-batch group synchronous species are able to repeat this process several times during the spawning season, with the recruitment of successive batches of oocytes and thus the production of several spawns per year. The number of spawning depends on the number of recruitments, e.g., the European sea bass (*Dicentrarchus labrax*) producing 2-4 spawns per season. The asynchronous type of ovarian development is exhibited by those species that produce multiple spawns through an expanded period of time (several months), normally on a daily basis. This is typical of some tropical species, but also many Mediterranean fishes of the Sparidae family [31-33]. The oocyte population develops in an asynchronic manner and all classes of oocytes (from early vitellogenesis to late maturation) can be found in the ovary at any moment of the reproductive cycle. There are no batches of oocyte growth. This would represent the extreme of a multiple-batch group synchronous type of ovarian development, making the classification in one of those categories difficult for some species. In fact, the classification of the ovarian development just represents a continuum and all possible strategies between the two extremes are found in fish.

In respect to male fish, the development of the testes is somehow more homogenous and could be described as an asynchronous type of development for all species. Male fish used to be fluent on a daily basis through a long period of time, normally overlapping the spawning period of the females. At every moment, several classes of cell development, from immature spermatogonia to spermatozoa, can be found in the testes. At the full spermiation period, the testes are mostly occupied by mature spermatozoa, ready for spermiation, while early in the season, a high percentage of less mature spermatocytes is present.

3. ENVIRONMENTAL REGULATION OF FISH REPRODUCTION

The aim of reproduction is to have offspring that survive. It has been recognized for a long time that food availability and environmental conditions are "ultimate" factors that determine survival and hence how a species evolves through natural selection [34]. Food availability or flow of energy (energetics) have been acknowledged as the ultimate factors central to reproduction [35-36]. Food availability and the ability to store energy determine when a fish proceeds to the completion of maturation (Fig. 3). Experimentally it has been demonstrated that groups of fish fed low rations exhibited a reduction in the percentage of fish that complete maturation [37-39]. Theories suggest that fish have the ability, through a genetically determined bio-chemical threshold, to ascertain what size and/or age conditions are optimal to complete maturation both during the first and subsequent maturation episodes [37,40]. Food availability for off-spring and hence off-spring survival determines the timing of reproduction. Food availability exhibits seasonal variation in the higher latitudes as well as lower latitudes in the tropics. Reproduction is timed to ensure that critical periods of feeding for the survival of the offspring, particularly larvae and juveniles, coincide with the seasonal periods of high food availability. For species in the high latitudes this is usually in the spring and for species in the lower latitudes this is often in relation to fluctuations in nutrient levels caused by changes in ocean currents, temperature cycles or weather cycles such as rainy seasons (Fig. 3). Therefore, maturation is a complex process that must be perfectly timed to ensure that spawning or critical off-spring feeding periods coincide with seasonal highs in food availability that are months or even years after maturation was initiated.

Fish have evolved to entrain maturational development with the predictable but constantly changing environmental parameters (e.g., photoperiod); these parameters cue the progress of maturational development and predict the approach of optimal conditions for offspring survival (Fig. 3). It is perhaps not surprising that the predictable parameters are often the same environmental factors that drive the seasonal change, weather systems and changes in ocean currents which result in cycles of food availability. These predictable parameters are termed proximate factors and examples are photoperiod, temperature, food availability, lunar cycle, rainfall, currents and pressure (Fig. 3). The degree to which these proximate factors have been studied is variable and therefore our true understanding of the importance of the factor is also variable ranging from a complete understanding of the factors role to circumstantial evidence that an aspect of maturation was repeatedly observed to coincide with a change in the proximate factor.

Perhaps the most important proximate factor is photoperiod. The role of photoperiod as a proximate factor has been comprehensively described for the salmonids and in particular the rainbow trout, through studies that examined the effect of natural and altered photoperiods on reproduction when other proximate factors such as temperature were maintained constant. From this type of studies it was demonstrated that photoperiod entrained an endogenous rhythm that controlled all aspects of maturational development, i.e., the entire brain-pituitary-gonad axis [41-42]. Therefore, in the rainbow trout the increasing spring photoperiod was shown to entrain the decision to proceed to the completion of maturation and in turn the start of vitellogenesis/spermatogenesis, the passage of photoperiod from spring to summer to autumn entrained the progress of vitellogenesis/spermatogenesis and the decreasing autumn photoperiod entrained final maturation, ovulation and spermiation. Photoperiod probably plays an important role in the timing of reproduction of most temperate fish species and has been shown to influence the timing of maturation in many species from a wide range of families that inhabit both temperate latitudes, such as the Atlantic salmon (*Salmo salar*, family: Salmonidae), European seabass (*Dicentrarchus labrax*, Percichthyidae) [43], gilthead bream, (*Sparus aurata*, Sparidae) [31], red drum (*Sciaenops ocellatus*, Sciaenidae) [44], Atlantic cod (*Gadus morhua*, Gadidae) [45], Atlantic halibut, (*Hippoglossus hippoglossus*, Pleuronectidae) [46], sole (*Solea solea*, Soleidae), turbot (*Psetta maxima*, Scophthalmidae) [47] and tropical latitudes, such as the Nile tilapia (*Oreochromus niloticus*) [48], grey mullet (*Mugil cephalus*, Mugilidae) [49], catfish (*Heteropneustes fossilis*, Heteropneustidae) [50] and common carp (*Cyprinus carpio*, Cyprinidae) [51-53].

However, the aspects of maturational development entrained by the different phases of the photoperiod cycle and the interaction with other proximate factors will depend on the reproductive strategy of each species.

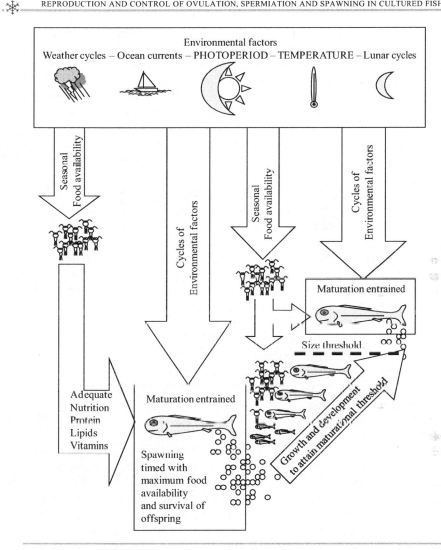

Figure 3 - Diagram of environmental factors affecting fish. The environmental factors such as photoperiod, temperature, lunar cycles, weather cycles and ocean currents, control the seasonality of food availability and entrain maturational development of fish. Food availability and the ability to store energy determine when a fish attains a genetical threshold and proceeds to the completion of maturation. Maturational development of the fish is entrained by environmental factors to ensure that critical off-spring feeding periods coincide with peaks in food availability which are months or years after maturation is initiated.

Temperature also plays an important role as a proximate factor and many species, particularly tropical and sub-tropical species appear to time spawning in relation to changes in temperature. However, the role of temperature is not clear and can be argued to act as a controlling or entraining factor such as photoperiod or a permissive factor that has a direct effect on the biological processes, but which is not actually used as a cue with which an organism times maturational development [42]. This unclear situation is partly caused by the lack of studies that have examined in isolation the interaction of temperature and maturation, with other proximate factors maintained constant, as studies have with photoperiod. However, despite this poor understanding, the importance of temperature in the maturational process cannot be disputed. Most species examined have an optimum temperature profile for the various stages of maturational development and generally temperatures below this optimal range will delay maturation and temperatures above accelerate maturation [54]. Temperatures that are extremely different from the optimal temperature range stop gametogenesis and induce atresia. In particular these temperature effects have been observed in relation to spawning. For example, the dab (*Limanda limanda*) in the North Sea has been observed to mature in relation to seasonally changing temperatures, gametogenesis from October to January and spawning as the temperature begins to rise from February to April [55]. Spawning time of different stocks of dab was progressively later the further north the stock was found and spawning time was correlated with progressively later latitude dependant rises in temperature [54]. These observations have given rise to findings that spawning can be predicted from the temperature profile. For example, Baynes et al. [56] demonstrated that for sole (*Solea solea*) there existed a positive correlation ($r = 0.9$) between winter temperature and the time that spawning started in the spring and Rothbard and Yaron [57] described how in Israel degree days after the temperature rises above 15°C are used to predict when carp are in spawning condition.

Other environmental parameters which have been observed to coincide with aspects of maturational development include food availability, lunar or tidal cycles, rainfall, currents and pressure. The knowledge of these sorts of parameters is based almost entirely on observations. Such observations offer little explanation as to the parameters role in the timing of maturation and the usefulness of such parameters in aquaculture is questionable, when the difficulty or impossibility of manipulating or reproducing these parameters is considered. One explanation could be that many of these parameters are final cues and fish mature to late stages of vitellogenesis/ spermatogenesis in relation to photoperiod and/or temperature and await a direct effect to cue final maturation and spawning. However, these parameters do highlight the diversity and complexity of reproductive strategies that have evolved. Some interesting examples include, timing of spawning of pelagic fish to coincide with plankton blooms [54], the spawning of Indian carps in relation to heavy monsoon rain and floods [58], timing of spawning with changes in currents on the Californian coast [59] and the association of captive seabream spawning with the lunar cycle [60].

4. HORMONAL REGULATION OF FISH REPRODUCTION

The reproductive cycle is regulated by a cascade of hormones along the brain-pituitary-gonad (BPG) axis, the so-called reproductive axis (Fig. 4). In this axis, the pituitary gonadotropins (GTHs), Follicle-Stimulating Hormone (FSH) and Luteinizing-Hormone (LH), are the key players in the endocrine control of reproduction. The secretion of the two GTHs is controlled by the brain via the stimulatory action of the Gonadotropin-Releasing Hormone (GnRH). This neuropeptide is the primary system regulating reproduction, acting as an integrator of external information (e.g., environment) and sending neuroendocrine inputs for the regulation of the reproductive axis. The GnRH acts directly at the pituitary gland to stimulate FSH and LH secretion that are released into the bloodstream to act on the gonad, where they stimulate the synthesis of gonad steroid hormones, which are the ultimate effectors of gonadal development.

At initial stages, GTH stimulation (mainly FSH) induces the secretion of androgens (e.g., testosterone (T) and 11-ketotestosterone (11KT)) in males and estrogens (e.g., estradiol (E_2)) in females, which act concomitantly with FSH in the control of gametogenesis. The E_2 plays an additional important role in female gametogenesis, with the stimulation of VTG synthesis from the liver. Thus, the period of gametogenesis is characterized by high blood levels of FSH and increasing levels of androgens in males, and increasing E_2 and VTG in the females. At the end of gametogenesis, secretion of LH from the pituitary induces a shift in the steroidogenic pathway of the gonad, stimulating the synthesis and secretion of progestin-like steroids, the maturation-inducing steroids (MIS). The concomitant action of LH with the MIS stimulates the process of gonadal maturation. This period is characterized by decreasing blood levels of FSH and androgens/estrogens and increasing blood levels of LH and MIS. Once gonadal maturation is completed, the brain GnRH system stimulates a high surge of LH secretion from the pituitary, which induces ovulation in the females, whereas in the males, relatively stable but elevated levels of LH induce spermiation (Fig. 5) [61-63]. The GnRH-induced pre-ovulatory LH surge in the plasma is essential for successful ovulation. In fact, the demonstration that this characteristic LH surge was absent in captive fish that failed to ovulate, but not in wild fish ovulating spontaneously, set up the basis for the development of hormone-based spawning induction therapies in aquaculture [64-65].

The success of reproductive maturation and viable gamete release depends on the correct functioning of all components of the reproductive axis throughout the entire reproductive cycle, from gametogenesis to spawning. The synchronized secretion of GnRH, GTHs and steroids through the reproductive cycle and their coordinated action is essential for successful spawning. The stress associated with captivity or the absence of appropriate environmental conditions in culture facilities may act on the brain-inhibiting neuroendocrine secretions, and thus blocking the reproductive axis, and inhibiting reproductive success.

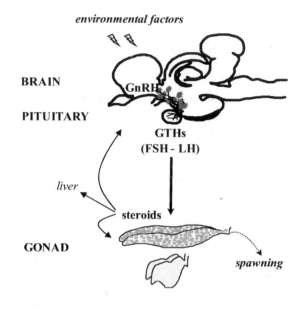

Figure 4 - The brain-pituitary-gonad (BPG) axis, showing the critical hormones involved in the regulation of fish reproduction. The reproductive process initiates in the brain, which integrates external information (e.g., environment) and respond with the activation of the Gonadotropin-Releasing Hormone (GnRH) system. The GnRH stimulate the synthesis and release of the pituitary gonadotropins (GTHs), Follicle-Stimulating Hormone (FSH) and Luteinizing-Hormone (LH), which act on the gonad (ovary or testes) to stimulate the synthesis of sex steroids, the ultimate effectors of gonadal development. The sex steroids play additional roles over nongonadal tissues, mainly feedback actions on the brain/pituitary and in females, the stimulation of vitellogenin (VTG) synthesis in the liver.

4.1 Brain Gonadotropin-Releasing Hormone (GnRH)

The brain is the highest level of the reproductive axis and acts as the director of reproduction, integrating external and internal information and responding with neuroendocrine signals. The primary neuronal system in the regulation of reproduction in all vertebrates is the GnRH. This is a neurohormone synthesized in specific areas of the brain, from where the GnRH cells project neuronal fibers directly into the pituitary. This system is unique in fish, since in higher vertebrates GnRH neurons do not project directly to the pituitary, but instead end into the median eminence and GnRH is released into a portal system, from where it reaches the pituitary gland. TheGnRH is released in close vicinity to the gonadotropic cells, binds to specific membrane receptors and stimulates the synthesis and release of both FSH and LH.

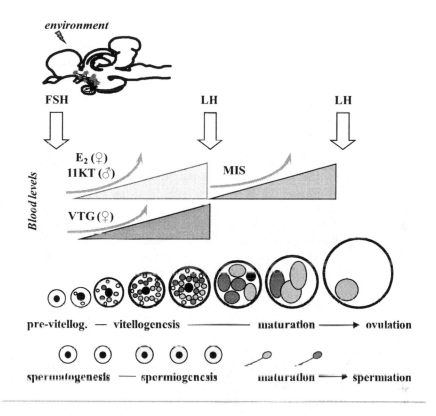

Figure 5 - Evolution of the endocrine and gonadal changes associated to the reproductive cycle of female and male fish. The top half of the diagram shows the pituitary hormone secretions and plasma hormone concentrations; the bottom half shows the correlated stages of oocyte and sperm development. At initial stages, pituitary FSH stimulation induces gonadal secretion of estrogens in females (estradiol (E_2)) and androgens in males (11-ketotestosterone (11KT)) that regulate gonad development. In females, E_2 plays an additional role on the liver, stimulating VTG synthesis (vitellogenesis). The period of gametogenesis is characterized by high blood levels of FSH and increasing levels of androgens in males, and E_2 and VTG in females. At the end of gametogenesis, pituitary LH secretion induces the synthesis of maturation-inducing steroids (MIS), which regulate the process of gonadal maturation; this is characterized by decreasing blood levels of FSH and androgens/estrogens and increasing blood levels of LH and MIS. At completion of maturation, a GnRH induced LH surge stimulates ovulation, spermiation and spawning.

Due to its crucial role in the integration and regulation of the neuroendocrine signaling governing reproduction, the GnRH system has been the focus of intensive research in reproductive biomedicine, for both basic research and applied uses of GnRH derived drugs for treatment of reproductive disorders [66-77].

The GnRH is a decapeptide that was first discovered in the brain of a mammalian species, and originally named Luteinizing-Hormone Releasing Hormone (LHRH), because of its LH-releasing activity [78-79]. It was also later named mammalian GnRH (mGnRH), a more appropriate nomenclature after the demonstration of its stimulatory action on both FSH and LH secretion. Since the characterization of the first GnRH, other GnRH forms have been isolated and characterized from the brain of other species, and up to date 24 different GnRH forms are known [77]. From them, 14 variants have been found in vertebrates, 9 in tunicates [80] and 1 in a cephalopod mollusk [81]. All GnRHs are decapeptides (except the octopus GnRH which is a dodecapeptide), with slight variations in their amino acid sequence. Each newly identified GnRH has been named after the species in which it was discovered first (Table 1).

In addition to the multiplicity of GnRH variants, an important finding was the demonstration that most vertebrate species express more than one GnRH form in the brain. As a general rule, two GnRHs are expressed simultaneously in the brain of a given species, localized in different regions of the brain and apparently exerting distinct biological functions. One GnRH system is directly involved in the regulation of pituitary secretion (i.e., the hypophysiotrohic system), whereas the other GnRH system does not. Fish are unique among vertebrates, because some teleost species have been found to express three GnRH forms in the brain [77]. In these species, the third GnRH system is related to the hypophysiotrophic GnRH system and probably both work coordinately in the regulation of pituitary secretion.

Due to the increasing number of GnRH forms and proposed names, the GnRH nomenclature has become somehow confusing. Recently, a new and more appropriate nomenclature for the GnRH family has been proposed, based on phylogenetic and neuroanatomical data. This nomenclature grouped all GnRHs in three main types, called GnRH-1, GnRH-2 and GnRH-3 [97-98]. The GnRH-1 would refer to the GnRH form directly involved in the regulation of pituitary GTH secretion, the classical hypophysiotrophic system. In the old nomenclature this GnRH-1 would correspond to one of eleven GnRH variants, depending on the species: mGnRH or gpGnRH in mammals and primitive fish (e.g., *Anguilla spp*), cGnRH-I in birds and reptiles, frGnRH in amphibians and sGnRH, ctGnRH, sbGnRH, hgGnRH, mdGnRH, whGnRH or dfGnRH in fish. These GnRHs have in common their biological function and brain distribution. They are synthesized by neurons located in the hypothalamus, from where they send numerous projections to the pituitary, where GnRH-1 is released to stimulate GTH secretion.

The GnRH-2 refers to the form synthesized by neurons located in the midbrain. These neurons do not send projections to the pituitary, where this form is absent, and therefore, it is believed that GnRH-2 does not have a direct role in the control of pituitary GTH secretion. In contrast to the previous system, the GnRH-2 corresponds always to the cGnRH-II variant of the old nomenclature. All the species studied, from fish to mammals, express cGnRH-II in this area of the brain. This ubiquitous system, highly conserved in the evolution, should have important functions, but to date there are no clear evidences of the specific biological functions of GnRH-2, although a potential role in the regulation of reproductive behavior has been claimed [99-100]. The GnRH-3 system is unique in teleost fishes. It is clearly related to the GnRH-1, considering phylogenetic and morphological data.

Table 1 - Primary structure of the 24 known native GnRH forms. The first discovered (mGnRH) is taken as the reference. Octopus GnRH is the only variant with 12 amino acids, presenting an Asn-Tyr insertion at the N-terminus. Medaka GnRH (mdGnRH [90]) is also known as pejerrey GnRH (pjGnRH [96]).

	1	2	3	4	5	6	7	8	9	10	ref.
Vertebrates:											
Mammalian (mGnRH)	pGlu	His	Trp	Ser	Tyr	Gly	Leu	Arg	Pro	Gly-NH$_2$	[78]
Guinea pig (gpGnRH)	-	Tyr	-	-	-	-	Val	-	-	-	[82]
Chicken-I (cGnRH-I)	-	-	-	-	-	-	-	Gln	-	-	[83]
Chicken-II (cGnRH-II)	-	-	-	-	His	-	Trp	Tyr	-	-	[84]
Frog (frGnRH)	-	-	-	-	-	-	-	Trp	-	-	[85]
Salmon (sGnRH)	-	-	-	-	-	-	Trp	Leu	-	-	[86]
Catfish (ctGnRH)	-	-	-	-	His	-	-	Asn	-	-	[87]
Seabream (sbGnRH)	-	-	-	-	-	-	-	Ser	-	-	[88]
Herring (hgGnRH)					His			Ser			[89]
Medaka (mdGnRH)	-				Phe	-	-	Ser	-	-	[90]
Whitefish (whGnRH)		-	-	-	-	-	Met	Am	-	-	[91]
Dogfish (dfGnRH)	-	-	-	-	His	-	Trp	Leu	-	-	[92]
Lamprey I (lGnRH-I)	-	-	-	-	His	Asp	Phe	Lys	-	-	[93]
Lamprey III (lGnRH-III)	=	=	Tyr	-	Leu	Glu	Trp	Lys	-	-	[94]
Invertebrates:											
Tunicate I (tGnRH-I)	-	-	-	-	Asp	Tyr	Phe	Lys	-	=	[95]
Tunicate II					Leu	Cys	His	Ala	=	=	[95]
Tunicate III	-	-	-	-	-	Glu	Phe	Met	-	-	[80]
Tunicate IV	-	-			Asn	Gln	-	Thr	-	-	[80]
Tunicate V	-	-	-	-	-	Glu	Tyr	Met	-	-	[80]
Tunicate VI	-	-	-	-	Lys	-	Tyr	Ser	-	-	[80]
Tunicate VII	-	-	-	-	-	Ala	-	Ser	-	-	[80]
Tunicate VIII	-	-	-	-	Leu	Ala	-	Ser	-	-	[80]
Tunicate IX	-	-	-	-	Asn	Lys	-	Ala	-	-	[80]
Octopus GnRH [Asn,Tyr]	-	-	Phe	-	Asn	-	Trp	His	-	-	[81]

    The localization of GnRH-3 neurons varies slightly between species, but normally overlaps with that of the GnRH-1 neurons. Nevertheless, GnRH-3 is always predominant in anterior regions (olfactory bulbs), whereas GnRH-1 predominates in the preoptic area (hypothalamus). The GnRH-3 neurons send a few projections to the pituitary, suggesting a potential involvement in the co-regulation of pituitary secretion [77,101-102].

    The multiplicity of GnRH variants and simultaneous presence of several GnRHs in the brain has raised important questions on their specific biological functions but also on the development of specific GnRH-derived analogues for therapeutic applications. It has been demonstrated that all GnRH forms stimulate LH secretion. Research on GnRH structure-activity has been directed towards the development of GnRH agonists (GnRHa), in which modifications of the GnRH structure could lead to increased bioactivity with respect to the native form. The highly conserved regions of the decapeptide structure, the NH_2-terminus (pGlu-His-Trp-Ser), the COOH-terminus (Pro-GlyNH$_2$) and the amino acid at position six, are indicative of the importance of these sequences in the bioactivity of the molecule, in regard to enzymatic resistance, receptor binding and activation. Based on these studies, thousands of GnRHa´s and antagonists have been developed for therapeutic applications in the control of reproductive disorders.

The stimulatory action of GnRH on GTH secretion is dependent on the presence of GnRH receptors (GnRH-R) in the membrane of the pituitary gonadotrops. Similarly to the situation for the GnRH ligands, multiple GnRH-Rs are expressed in a single species. In mammals, two types of GnRH-R, named type I and type II, have been identified, which display ligand specificity for each GnRH variant [71]. In fish, multiple GnRH-Rs have been identified and, in contrast to mammals, they do not show ligand specificity. All fish GnRH-Rs display higher affinity for cGmRH-II than for sGnRH or the hypophysiotrophic form [70,103]. Expression levels of the GnRH-R genes in the pituitary show a seasonal pattern, which is an important factor influencing the seasonal responsiveness of the pituitary to GnRH stimulation. Highest levels of GnRH-R and thus highest responsiveness of the pituitary occur at the pre-spawning period, whereas lowest GnRH-R levels are found during the resting period and early stages of gonadal development. This is critical not only for the natural development of the reproductive cycle, but also when applying hormonal therapies, as this affects greatly the efficiency of GnRHa-based hormonal treatments, depending on the moment of the cycle when the treatments are applied.

In addition to the primary GnRH stimulatory system, GTH secretion is under the influence of a brain inhibitory tone, the dopaminergic system [104]. Neurons secreting dopamine (DA) exert an inhibitory action on both the brain and pituitary. Over the GnRH system, DA inhibits GnRH synthesis and GnRH release. In the pituitary, DA down-regulates GnRH-R and interferes with the GnRH signal-transduction pathways, inhibiting GnRH-stimulated LH secretion from the pituitary [105]. A dopaminergic inhibition on LH release has been demonstrated in all vertebrates, including amphibians, birds, mammals and humans [106-108]. Its intensity and moment of action may differ greatly between species, depending mostly on different reproductive strategies. In fish, a dopaminergic inhibition of reproduction has been demonstrated in cyprinids, silurids, salmonids, tilapia (*Oreochromis spp.*), European eel (*Anguilla anguilla*) and grey mullet (*Mugil cephalus*) [109-114]. In these species, DA inhibits strongly the pre-ovulatory GnRH-stimulated LH surge and thus, ovulation and spawning; it also seems to be involved in the inhibition of puberty. In contrast, an active DA inhibitory system seems to be very weak or absent in most marine fishes.

Although GnRH is the primary regulator of reproduction, the brain synthesizes other neurohormones and neurotransmitters that have been shown to stimulate LH secretion and participate in the regulation of fish reproduction, being the most relevant the neuropeptide Y (NPY) and the neurotransmitter γ-amino-butiric acid (GABA) [61,115-118]. The NPY is involved in the regulation of the nutritional status of the fish; NPY neurons exert stimulatory actions on both GnRH and GTH and seem to play an important role in mediating interrelationships between nutrition and reproduction. The GABA is the most relevant neurotransmitter of the brain; in mammals and in fish it exerts a stimulatory action over LH secretion. It seems that the profusion of GABA neurons in the brain, plays an important role interconnecting different neuronal systems, synchronizing and fine tuning neuronal secretions from different systems. Other neuronal systems have also been shown to exert some LH stimulatory action, but they are of minor relevance. In general, all these neuronal systems act over both the GnRH neurons and/or the gonadotrops stimulating GnRH secretion, GnRH-R levels and FSH/LH synthesis and release. They can also act over the dopaminergic neurons, inhibiting DA secretion and thus exerting a stimulatory action on LH release. This neuronal network seems to fine tune the correct functioning of the primary GnRH-GTH endocrine system.

4.2 Pituitary Gonadotropins (GTH)

The pituitary or hypophysis is a major endocrine gland localized in the ventral part of the brain and is responsible for the release of the GTHs, in addition to several other hormones involved in growth, metabolism and stress adaptation. The two pituitary GTHs, FSH and LH, together with the Thyroid-Stimulating Hormone (TSH) and the placental chorionic gonadotropin (CG), constitute a family of structurally related molecules, the glycoprotein hormones [119]. They are heterodimeric proteins, constituted by a common α subunit, noncovalently linked to a hormone-specific β subunit, which confers the biological specificity to the hormone. Each glycoprotein subunit is encoded by a different gene. Early after synthesis, the peptide chain is folded, glycosilated and assembled to form the dimeric conformation, required for the biological activity of the hormone. The bioactivity of the GTHs depend on the duration of time that the hormone is present in the circulation (half-life), the binding to specific receptors and the activation of intra-cellular signal transduction mechanisms that leads to the biological response [120-121]. The half-life of the GTHs in the bloodstream is determined mainly by its degree of glycosilation. This is one of the main reasons for the use of human CG (hCG) in the hormonal treatment of several reproductive disorders, including spawning induction protocols in fish. The hCG is the highest glycosilated GTH and thus, it presents higher resistance to degradation than any other glycoprotein, thus having long acting effects. The stimulation of the target cells also depends on GTH binding and activation of specific membrane receptors. There are two types of GTH receptors, exhibiting ligand specificity for each gonadotropin. The hormone specific β subunit determines the specificity of the binding (FSH for the FSH-R, LH for the LH/CG-R), preventing interaction of a given GTH with the receptors of other glycoproteins. The human CG (hCG) binds to the same receptors as LH. This is an important reason that justify the use of hCG as a hormonal treatment for spawning induction in fish, as hCG can display LH related functions and thus induce ovulation and spermiation in captive fish (see following sections).

Research in the field of fish GTHs has followed that in mammals. For many years, it was believed that the fish pituitary produced a single GTH responsible for the control of all aspects of reproduction, in contrast to higher vertebrates. This single GTH in fish had characteristics similar to the LH of higher vertebrates. In 1988, two distinct GTHs were purified and identified for the first time from the pituitary of a fish species, and named GTH-I and GTH-II [122]. The similarity of these fish GTHs with tetrapod FSH and LH was further established through molecular, biochemical and immunological techniques [123]. Thus, when reviewing the bibliography referred to fish reproduction, one has to be aware of the nomenclature used for fish GTHs. The bibliography before 1988 cite only the name "GTH," which refer to a hormone that was later confirmed to correspond to LH. During the next decade, information regarding the fish GTHs, refer to the names "GTH-I and GTH-II" (or "GTH-1 and GTH-2"), GTH-II being the previously known GTH (LH-like hormone) and GTH-I the newly discovered hormone. This nomenclature has now been abandoned and recent fish bibliography refers to the names "FSH and LH," standardizing the nomenclature with that of all vertebrates, FSH being homologous to the previous GTH-I and LH homologous to the previous GTH-II.

Information on the structure-activity and biological functions of LH in fish reproduction is much more extensive than that of FSH [105,124]. This is because immunoassay methods for analyzing LH secretion in fish have been available for many decades, while FSH immunoassays were only available since 1988, and almost exclusively limited to salmonid species [124]. The initiation of the reproductive cycle is characterized by increased FSH levels, which are maintained high during gametogenesis, whereas LH levels remain undetectable. During gonadal maturation, FSH levels decline and LH increase, showing a sharp LH peak prior to ovulation. The recent development of molecular tools has allowed the analysis of FSH and LH gene expression levels in several fish species, obtaining information on the biological functions of both hormones in a broader range of fishes. In salmonid species, showing single-batch group synchronous ovarian development, mRNA levels of βFSH increase during early gametogenesis while βLH predominates during FOM [125]. Information in nonsalmonid species shows a slightly different picture. In the gilthead seabream (*Sparus aurata*), with asynchronous ovarian development, both βFSH and βLH are expressed throughout the year, increasing both during the reproductive season [126]. In other nonsalmonid species, exhibiting multiple-batch group synchronous or asynchronous ovarian development, such as the blue gourami (*Trichogaster trichopterus*), red seabream (*Pagrus major*), European seabass (*Dicentrarchus labrax*) and stickleback (*Gasterosteus aculeatus*), FSH and LH gene expression levels are found throughout the reproductive cycle, although in most cases FSH synthesis is advanced with respect to that of LH [127-130]. The general view is that FSH controls mainly early stages of gametogenesis, while LH regulates FOM, ovulation and spermiation. Nevertheless, it becomes clear that there are important differences between fish species, most probably related to different patterns of gonadal development and different reproductive strategies.

4.3 Gonad steroids

The gonad is the tissue for generation of gametes but also a major endocrine organ, specialized in the synthesis of sex steroid hormones. These steroid hormones are the final endocrine effectors of gonadal development, in coordination with the pituitary GTHs [2,9,131-132]. Steroidogenesis takes place in the somatic cells of the gonad, the granulosa and theca cells in the ovary and the interstitial Leydig cells and Sertoli cells in the testes. The major steroid hormones in the regulation of fish gametogenesis are the estrogen E_2 in females and the androgen 11KT in males. In mammals, the main estrogen in females is also E_2, but the main androgen in males is T, instead of 11KT, and to a lower extent dihydrotestosterone (DHT). The fish ovary also synthesizes T, which plays other reproductive related functions. Similarly, males also synthesize E_2, but this is found in much lower levels than in females. The testes of male fish produce other androgens than 11KT (e.g., T), which exert complementary functions during testicular development [26].

In addition to their role in regulating gonadal development, sex steroid hormones also exert both positive and negative feedback on the brain-pituitary axis and thus, regulating GTH release. A major positive action of the steroids is to enhance pituitary responsiveness to GnRH, probably by stimulating GnRH-R. A major negative action of these steroid hormones is exerted through the dopaminergic system, increasing DA turnover and thus enhancing the DA inhibitory tone over GTH secretion. In this way, the brain is constantly informed about the evolution of gonad development, through the action of the fluctuating circulating levels of steroids during the reproductive cycle [133].

4.3.1 Steroids regulating female oogenesis and maturation

In females, E_2 acts in coordination with the pituitary GTHs in the regulation of oocyte development. In the ovary, steroidogenesis is a two-cell biosynthetic process, in which the outer theca layer synthesizes steroid precursors that are transported into the granulosa cells, where they are transformed into derivates. During vitellogenesis, the theca cells synthesize T that is converted into E_2 in the granulosa cells, by the action of the enzyme aromatase. During vitellogenesis, E_2 exerts two main functions, one in the gonad regulating oocyte development and one in the liver stimulating the synthesis of VTG and other yolk related proteins.

Once vitellogenesis is completed, pituitary LH secretion induces a shift in the steroid biosynthetic activity of the ovary with a reduction in T and E_2 production and enhancement of the synthesis of MIS. This is caused by reduction of aromatase activity and increased activity of enzymes of the MIS pathway. There are two major

MIS identified in fish, $17\alpha,20\beta$,dihydroxy 4 pregnen-3-one ($17,20\beta$-P or DHP) and $17\alpha,20\beta,21$-trihydroxy-4-pregnen-3-one (20β-S). They both probably act as MIS in most fishes, but normally one of them is the predominant MIS for a given species. The $17,20\beta$-P is the major MIS in several salmonid and nonsalmonid species, while 20β-S is the major MIS in Atlantic croaker, spotted sea trout, striped bass and black porgy [134]. The synthesis of MIS is also a two-cell process, by which the precursor 17α-hydroxyprogesterone is synthesized in theca cells and converted into $17,20\beta$-P in the granulosa cells, by the enzyme 20β-hydosysteroid dehydrogenase. The MIS together with pituitary LH secretion regulate gonadal maturation. The action of MIS on FOM is not direct, but mediated by the complex interaction of different factors, including prostaglandins (PGE1, PGE2, PGF1α, PGF2α), insulin-like growth factors (IGF-I and IGF-II), activin B and other signal transduction pathways [135]. It is the rise of MIS rather than the reduction of other steroids which is responsible of inducing FOM. That is why in many fish species blood levels of estrogens remain high during gonadal maturation. In multiple spawners and some single spawners, plasma levels of MIS correlates well with the maturation cycle and E_2 levels are maintained high through the entire period of maturation.

During maturation, the oocyte goes first through a phase of oocyte maturational competence (OMC) before FOM can be achieved. During this phase, the oocyte acquires the competence to mature, including LH stimulation of follicle cells to produce necessary factors for MIS biosynthesis (enzymes, etc.) and the stimulation of germ cell capacity to respond to MIS (MIS receptors, etc.). During OMC, which is developed without MIS secretion, the first signs of FOM are evident (lipid globule coalescence and GVM). During FOM, LH dependent MIS secretion from follicular cells acts over membrane receptors in the oocyte to undergo final coalescence of yolk granules, GVBD and the resumption of meiosis.

4.3.2 Steroids regulating male spermatogenesis and maturation

Testicular spermatogenesis and maturation is also regulated by pituitary GTH secretion, but the action of the steroids secreted by the testes has a stronger influence. The androgen 11KT is the major regulator of spermatogenesis, while MIS regulates sperm maturation. Both steroids are synthesized by the somatic cells of the testes under GTH stimulation. The LH is mainly involved in the stimulation of androgen production in Leydig cells, whereas FSH seems to exert more complex functions in the male testes, stimulating androgen production in Leydig cells and regulating Sertoli cell activity during spermatogenesis. Although the regulatory mechanisms of FSH are mostly unknown, possible functions of FSH in the testes are the stimulation of Sertoli cell proliferation and differentiation and synthesis of growth factors, which act as autocrine and paracrine factors involved in Sertoli cell proliferation and differentiation and germ cell development.

Before initiation of spermatogenesis, spermatogonial stem cell renewal seems to be regulated by E_2 acting on Sertoli cells. At a certain moment, secretion of pituitary

GTHs (mainly FSH) induces a switch from spermatogonial self-renewal to spermatogonial proliferation, which represents the initiation of spermatogenesis. The FSH acts on Sertoli cells and stimulate 11KT biosynthesis through activation of specific enzymes (11β-hydroxylase and 11β-hydroxysteroid dehydrogenase). From then on, 11KT regulates the full process of spermatogenesis, an action that is mediated by growth factors secreted by the Sertoli cells. In males, FSH levels are high at early spermatogenesis, increase to maximum levels during the rapid testicular growth phase and then decline after spawning. On the other hand, LH is very low during early spermatogenesis, start increasing during the rapid testicular growth phase and peaks during spawning. As spermatogenesis advances, LH becomes important in supporting 11KT production. After completion of spermatogenesis, secretion of LH from the pituitary induces a shift in the steroiodogenic pathway of the testes leading to the production of MIS, which in turn regulate sperm maturation. During maturation, 17α-hydroxyprogesterone synthesized in Leydig cells is converted to MIS in the spermatozoa due to the activity of 20β-hydosysteroid dehydrogenase. The action of MIS on sperm maturation is not direct on the sperm, but via activation of specific enzymes that increase seminal plasma pH, which in turn induces spermatozoa capacitation. In males, androgen production remains high through the entire period of sexual maturation, even while MIS levels are high.

5. REPRODUCTIVE DYSFUNCTIONS IN CAPTIVE FISH

As mentioned in previous sections, there is a significant variation in fish reproductive strategies and types of gonad development. During their reproductive cycle, which can last for days, months or years depending on the species, fish experience a variety of external influences. In their natural habitats, the endocrine reproductive axis of the fish functions correctly and reproduction develops successfully, with spawning taking place at the moment when the fish detect that the external conditions are the most appropriate for the survival of the offspring and of course, its own survival. Unfortunately, the situation may change drastically when fish are reared in captivity and reproduction is somehow affected by the captive conditions. In fact, all fish species held in captivity exhibit some degree of reproductive dysfunction; and, it is normally the females who exhibit more serious reproductive problems. These dysfunctions depend on the species and can vary from a total absence of spawning to significant reductions in the quantity and quality of the eggs and sperm produced.

The reproductive problems detected in captive fish are derived from two causes: the stress associated with captivity and the absence of appropriate environmental signals permissive for reproduction [136]. The action of one of these factors or the combination of both underlies the total or partial inhibition of reproduction in captivity. Thus, the primary task of a broodstock manager will be to minimize the negative effects of these two parameters, in order to obtain the best reproductive performance of the cultured breeders. The negative influence of stress should be minimized by appropriate broodstock management (fish manipulation, fish care, prophylaxis, etc.)

and adequate culture conditions (tank design, water supply, light intensity, etc). This should be adapted to each species, considering that resistance and adaptiveness to stress varies greatly among species. The second parameter, the absence of appropriate environmental signals, is somehow much more difficult to solve. The broodstock manager should learn as much as possible about the reproductive biology of the species in its natural habitat and try to adapt the culture conditions to the natural situation. For many species, it is almost impossible to mimic the environmental conditions that the fish experience during their reproductive season (e.g., migration to the spawning grounds). The complexity of environmental factors to which the fish is exposed during the whole reproductive period is basically unknown for most species and would, anyway, most probably be difficult to reproduce in culture conditions. This is normally more feasible for nonmigratory species inhabiting stable habitats, but becomes more complex or unfeasible for long-distance migratory species (e.g., *Anguilla spp*, *Seriola spp*, and *Thunnus spp*). In any case, the better we provide the required environmental signals, the less reproductive problems will be exhibited by the breeders. If reproductive disorders persist even after taking maximum care to reduce the negative effects of these parameters, then the use of hormonal treatments can overcome reproductive problems, as demonstrated for many cultured fishes [62,64-65,137-139].

The sensory and endocrine system of fish has evolved to recognize when external and internal conditions are optimal for reproductive development and has the capacity to, 1) under optimal conditions maintain reproductive development to completion and spawning, 2) under sub-optimal conditions arrest development at a particular stage and postpone gamete production for improved conditions, or 3) under nonoptimal conditions abort reproductive development, reabsorb nutrients invested in the gonad and return the gonad to a resting stage. These pathways have developed to ensure survival of both offspring and the parents. Under optimal conditions the sensory and endocrine system of the parents has the ability to recognize that the possibility of offspring survival is high and, therefore, the parents risk personal survival to invest energies in reproduction and spawning. However, when conditions are not optimal for spawning the endocrine system has the ability to recognize that the risk to parental survival of investing energy in maturation may not be rewarded with survival of the offspring and maturation is arrested or in extreme conditions aborted.

In captive broodstock, the females normally exhibit more serious reproductive problems than the males; female dysfunctions can be classified in three main types (Fig. 6). The first type is the inhibition of vitellogenesis. In this species, reproduction is blocked at very early stages of development, e.g., *Anguilla spp* and sometime *Seriola spp* [140]. Physiologically, this is the most serious reproductive disorder, as the endocrine reproductive system of the fish has not functioned at any moment of the reproductive process.

The second type of reproductive dysfunction is the inhibition of the process of FOM. In species exhibiting this problem, vitellogenesis is completed correctly, but post-vitellogenic oocytes are unable to undergo FOM and become atretic. The degree of inhibition varies depending on the species and even on the environmental conditions of each specific reproductive season. Atresia can affect the whole population of post-vitellogenic oocytes of the gonad, causing the total absence of spawning, or may affect only part of the post-vitellogenic oocytes, which finally causes a reduction in the number of eggs released. The diminished egg production can be slight, or can be dramatic, causing only sporadic spawning of a few eggs. This second type is the most common reproductive dysfunction and is detected in the majority of fish species reared in captivity. The third type of reproductive dysfunction is the inhibition of spawning only. Fish exhibiting this dysfunction undergo all phases of the reproductive cycle correctly, with oocytes going through vitellogenesis, FOM and ovulation, but spawning is blocked and the ovulated oocytes remain in the ovarian or abdominal cavity. This is, physiologically, the least serious of all reproductive dysfunctions, as only the spawning event is inhibited from the whole reproductive process, although the end consequence is similar, the absence of spontaneous spawning. This dysfunction is observed in salmonids and some flatfishes (e.g., turbot *Psetta maxima*). In these species, manual removal of the eggs (i.e., stripping) is required. If stripping is not performed, then the eggs degenerate and are reabsorbed, but in such situations they may cause the death of the female.

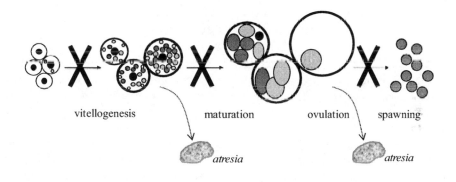

Figure 6 - Major reproductive dysfunctions observed in captive female fish. They are classified in three main types (indicated by X), 1) the inhibition of vitellogenesis, 2) the inhibition of oocyte maturation, which causes atresia of post-vitellogenic oocytes and, 3) the inhibition of spawning only, with ovulated oocytes retained in the ovarian or abdominal cavity. Each type is physiologically different, but the end consequence is similar for the aquaculturist, the absence of spontaneous spawning in the tank. The application of hormonal treatments has effectively resolved reproduction in many species exhibiting dysfunction type two. Solution of dysfunction type one is under investigation, whereas reproduction of species with dysfunction type three can be achieved through artificial fertilization, after manual stripping of eggs and sperm.

Although stripping and artificial fertilization is a common activity in hatcheries of these species, it still represents a serious management issue, as stripping is labor intensive and must be timed precisely in order to avoid over-ripening of the eggs and reduction in quality. It also requires repeated manipulation of the broodstock, as normally females are unshychronized and each breeder has to be checked to determine its ovulation stage, which is also stressful to the broodstock.

Although reproductive disorders are more common and serious in females, male fish also display some important problems. These are usually less important than in females and except in rare cases, males of all fish species usually are spermiating in captivity. The reproductive dysfunctions detected in captive male fish are diminished sperm volume and diminished milt fluidity, which can affect negatively the success of egg fertilization. Diminished sperm production in male broodstock represents a serious problem for those species in which hatchery production is based on artificial fertilization and the acquisition of gametes by manual stripping. The difficulties of obtaining enough sperm from male breeders can block fertilization programs and requires the handling of a higher number of male breeders. On the other hand, for species that spawn spontaneously in the tank, the production of highly viscous milt reduces the rapid dispersal of the spermatozoa and thus reduces the sperm fertilization capacity.

6. TECHNIQUES FOR ENVIRONMENTAL MANIPULATION OF FISH REPRODUCTION

As described in the environmental section and more fully in the reproductive dysfunction section, when environmental conditions are not optimal, as is often the case for fish held in captivity, maturation may be arrested until conditions are experienced that allow maturation to proceed or under extreme conditions reproduction may be aborted and the gonads regressed. The aim of environmental manipulations is to provide conditions that are sufficiently close to optimal or natural conditions to induce the majority of the fish to complete maturation and spawning. Environmental manipulations can be used to achieve two purposes, 1) adjust aspects of the naturally changing captive environment to ensure that fish proceed to maturation and spawning during the natural spawning season or, 2) to provide the complete cycle of environmental changes required to induce the full cycle of maturation outside of the natural spawning season and thereby obtain out-of-season egg production. During these manipulations the primary concern is to provide the correct or optimal conditions for spawning and close to optimal conditions for gametogenesis. Gametogenesis appears to be more flexible or have a wider optimal range for development, while spawning often requires quite precise conditions. Few studies have actually examined this environmental flexibility of gametogenesis. However, a number of studies that used environmental control to successfully obtain out-of-season spawns have also inavertedly altered the timing of gametogenesis to coincide with quite extreme temperatures compared to temperatures normally experienced during gametogenesis and apparently with no adverse effect on spawning or egg quality in rainbow tout (*Oncorhynchus mykiss*)

[42,141], Atlantic salmon (*Salmo salar*) [142] and European seabass (*Dicentrarchus labrax*) [43].

To ensure spawning during the natural spawning season the primary considerations are tank environment, water quality, space and social aspects and temperature (assuming that photoperiod is ambient). Simply put, water quality should be the best available, sex ratios need to be correct of the spawning species, stocking densities should be low (e.g., <5 kg m^{-3}) and sufficient space is required for the formation of social groups and courtship. Lastly and perhaps most importantly, the temperature should be in the range that the species requires for spawning. Examples described below in the cultured species section include selection and transfer of the correct proportion of male and female adult carp and catfish to specially prepared spawning ponds which provide the space, spawning substrate and the required temperature profile, and selection of Atlantic salmon broodstock and transfer to fresh water facilities with naturally decreasing temperatures. These kinds of manipulations can also be used to alter the timing or increase the length of the spawning period, advancing, delaying and extending the window of optimal temperatures has been observed to advance, delay or extend the spawning period (see below catfish).

To obtain out-of-season spawning, both photoperiod and temperature should be manipulated. In the rainbow trout it has been established that it is the change in photoperiod or daylength, rather than actual daylength that entrains rhythms controlling maturation [41-42]. Therefore, as maturational development in the rainbow trout is entrained by the spring-summer-autumn photoperiod, the application of such a photoperiod outside of the normal seasons will alter the timing of maturation and spawning [41-42] and it has been demonstrated that a phase shifted photoperiod (displacing the entire 12 month photoperiod by a number of months) will phase shift maturation, a compressed photoperiod (12 month cycle compressed to less than 12 months) will advance maturation and an expanded photoperiod (12 month cycle expanded to more than 12 months) will delay maturation. It has also been demonstrated that square wave photoperiods (direct increases or decreases in daylength) can have a similar effect as gradually changing photoperiods that resemble a natural photoperiod. However, a number of aspects need to be considered. First, spawning may not coincide in relation to the altered photoperiod as was observed under natural conditions, for example with a compressed photoperiod spawning can be expected to be later in relation to the compressed photoperiod (but earlier compared to controls under ambient conditions), i.e., winter or even spring spawning rather than autumn for trout [42]. Second, when applying a modified photoperiod to fish under ambient conditions, any change made to start the photoperiod will affect the endogenous rhythms, for example applying an increasing spring photoperiod (increasing from LD 12:12) in the summer (LD 18:6) would be the equivalent of applying a direct decrease (autumn) followed by a short winter photoperiod and finally an increasing spring photoperiod. As mentioned previously, much of the work with trout was completed under constant temperature conditions of 8-11°C, which are within the range of optimal temperatures for rainbow trout gametogenesis and spawning. Out-of-season spawning of other species can be achieved by adjusting the photoperiod and temperature cycle in the same way that photoperiod was used to manipulate spawning for the rainbow trout. The simplest and most secure way to manipulate spawning is to use phase shifted photothermal cycles, as described for gilthead seabream [31]. As understanding increases of a species response to environmental control, compressed or square wave photoperiods can be used, as described for rainbow trout [41-41], European seabass [43] and Sciaenidaes [44].

Lastly a few principally tropical or sub-tropical species appear not to require periods of gonadal regression, resting and recrudescence, which seasonal environmental cycles often stimulate (in addition to maturation). Species such as tilapia, carp and red drum when brought to spawning condition through environmental manipulation and held under constant optimal spawning conditions (photoperiod and temperature) have been observed to spawn for an entire year. In the case of red drum two females and two males were maintained spawning for 7 years under a constant photoperiod of LD 12:12 and temperatures close to 24°C [44].

7. TECHNIQUES FO R HORMONAL STIMULATION

The development and application of hormonal therapies for the treatment of reproductive disorders in cultured fish have permitted the reproduction in captivity of several fish species that did not do so spontaneously. Hormonal therapies have not only permitted reproduction, but also allowed the improvement of the reproductive performance of broodstock and the development of a technological and economically successful aquaculture industry for several fish species [64,139,143]. In captive females exhibiting inhibition of early gonadal development (dysfunction type I), hormonal treatments can stimulate gametogenesis and oocyte maturation, but this is under investigation and not feasible yet for egg production on a commercial scale. In the case of females with inhibition of spawning (dysfunction type III), hormonal treatments are not really needed to obtain eggs, as they can be stripped, but it is often used as a management tool to synchronize ovulation and thus accelerate egg acquisition activities. It is in females with inhibition of FOM (dysfunction type II) that the use of hormonal therapies has given the best results.

There are two principal situations in aquaculture for the use of hormonal stimulation for obtaining gametes, 1) to stimulate spawning in fish species that due to reproductive dysfunction do not complete maturation in captivity or spawning is unpredictable, and 2) to synchronize spawning of the broodstock and hence improve management. Naturally, the timing of spawning of individuals in a group of broodstock held under the same conditions will approximate that of a normal distribution. The majority of the broodstock spawn during the peak of the season with some early and some late spawners. This can mean that at the beginning and the end of the spawning period sufficient eggs are not obtained to form a cohort of juveniles to stock ongrowing facilities, and, conversely, in the peak of the spawning season too many eggs may be produced for the facilities available for incubation and larval rearing. This can be a particular problem in species that spawn once a year, such as trout and salmon. Hormone stimulation can be used to synchronize and advance spawning of a group of broodfish, giving the opportunity to manage a number of large batches of eggs over the spawning season.

The history of the hormone therapies for the treatment of reproductive dysfunctions in fish has been closely linked to research discoveries in the field of reproductive endocrinology and technical advances in the purification and study of reproductive hormones. A key discovery in the area was the determination of the major endocrine failure underlying the blockage of the reproductive process in captive fish. It was demonstrated in several fishes, that the inhibition of spontaneous spawning in captivity was clearly related to the inhibition of LH release from the pituitary [31,144]. For the species under investigation, it was shown that wild fish spawning in their natural spawning grounds exhibited high levels of LH in the bloodstream during oocyte maturation, with the typical LH surge preceding ovulation and spawning [145]. In contrast, congeners held in captivity that did not spawn spontaneously, exhibited highly reduced or absent LH levels in the blood without any LH surge, even in individuals that presented high concentrations of LH in the pituitary. Thus, the blockage of ovulation was specifically related to an inhibition of LH release from the pituitary, independently of the pituitary concentration of the hormone. This discovery was further corroborated by the application of hormonal treatments that simply induced the release of the LH stored in the pituitary of these fishes, which further stimulated the progression of FOM and spawning.

The hormonal therapies developed and applied for fish aquaculture can be grouped in two major types, "first generation" and "second generation" techniques [64,139]. The first generation are the pituitary hormone based preparations, and include the pituitary extracts and purified GTHs. The second generation are the brain hormone based treatments and includes the GnRH agonists (GnRHa) and dopamine (DA) antagonists. These two types of hormonal therapies act at different levels of the reproductive BPG axis. Drugs pertaining to the first type act directly on the gonads, while drugs of the second type act on the pituitary and thus indirectly on the gonad, through stimulation of endogenous pituitary GTH release. This is an important consideration when deciding the hormonal treatment to be applied for a specific species; the efficacy of the second generation drugs depends on the responsiveness of the pituitary of the treated fish. This means that these treatments can be totally ineffective in species in which the pituitary gland, for any reason, will not release GTHs under GnRHa treatment and thus, no effect on the gonad will be obtained. On the other hand, since first generation drugs act directly on the gonad, their efficacy does not depend on the functioning of the pituitary of the treated fish.

7.1 First generation: Gonadotropin preparations

The term "GTH preparations" refers to all those hormone preparations that display GTH activity and include pituitary homogenates, pituitary extracts (PE) and purified

GTHs. These are generally named the first generation because they were the first type of hormonal treatments developed and applied for the stimulation of reproduction. They all have in common their target organ, as they all act directly on the gonad of the treated fish, to stimulate gonad development and spawning. These different preparations differ in the degree of purity of the active component (GTH) and they have been developed, historically, in relation to technological advances in protein purification.

The first gonadotropin preparations used for spawning induction of captive fish were the fish pituitary homogenates (i.e., hypophysation). The basis of this treatment is simple and consists of the extraction of the pituitary gland from a fully mature fish, its homogenization in an appropriate buffer solution and the administration of the obtained homogenate in a recipient fish [146-147]. The homogenate induces ovulation and spawning in the treated fish. In China, hypophysation has been used extensively in the stimulation of carp reproduction, much before the understanding of the physiological bases of the treatment [139]. It was later understood that the effectiveness of the treatment is due to the high LH content of the pituitary of the donor fish. Although primitive, the application of pituitary homogenates represented for many years the only method that permitted the stimulation of reproduction in captive fishes and set up the basis for further and more sophisticated hormonal treatments. The hypophysation technique has several advantages. It is a custom-made preparation that can be easily obtained in the fish farm and does not require specialized people and instrumentation for its preparation. The major disadvantage is the inaccuracy of the method. A pituitary homogenate is not a calibrated preparation and it is not known exactly the dose of the active component (LH), because of the variable LH pituitary content of the donor fish. This makes it difficult to establish an accurate method. Another disadvantage is the risk of transmission of pathogens, when transferring biological material (pituitaries) from one fish to another. This is obviously a primitive technique, maybe appropriate for small fish farms localized in remote areas, but not for the development of an intensive aquaculture facility, though the method is still employed today.

An improvement of the hypophysation method was the use of pituitary extracts (PE), which are enriched preparations of the hormone component of the pituitary homogenate without the cellular parts. The PE require some technical expertise and specialized equipment and are, thus, prepared by qualified personnel. The physiological basis of the treatment is similar to the pituitary homogenates. A variety of PE have been used in aquaculture, and some of them, such as the salmon (sPE) and carp (cPE) pituitary extracts, are available commercially. The pituitary extracts are more effective than the pituitary homogenates as they are usually calibrated using bioassays. Nevertheless, they maintain the disadvantages of risk of pathogen transmission and high degree of species specificity [138,148-149].

The last and more sophisticated type of GTH preparations are the purified or recombinant GTHs. They require a more elaborate preparation, as only the GTH of the pituitary homogenate is used. These treatments became available with developments in GTH research and technological advances in protein purification, which allowed isolation of highly purified GTHs from a variety of species and biological sources. Purified GTHs obtained from human and mammalian biological material have been extensively used for clinical and veterinary uses, such as FSH and LH isolated from mammalian pituitaries, pregnant mare serum gonadotropin (PMSG) or human chorionic gonadotropin (hCG), isolated from the urine of pregnant women. The more common GTHs used for spawning induction of fish are hCG and purified fish GTHs. Functionally, the hCG display LH bioactivity, as it binds to gonadal LH receptors and stimulate ovarian and testicular development, gamete maturation and gamete release. The hCG, although from human origin, has been used extensively in aquaculture, because of its high availability in the market, low cost and standardized activity. In contrast, the technological difficulties in isolating fish GTHs and smaller aquaculture market compared to human and veterinary applications, have limited the use of fish GTHs, which would be physiologically more appropriate for fish than hCG. Currently, only salmon and carp GTH preparations are available in the market for aquaculture. The use of purified GTHs has important advantages over pituitary homogenates and PE, mostly the calibration of the preparation, which allows accurate dosing, the repetitiveness of the treatment and reduced risk of pathogen transmission of the highly purified preparations. The treatment with hCG has an important disadvantage, which relates to the complex structure of the molecule. The hCG is a large and species-specific protein, which may cause immune response when administered to non-mammalian species [139]. Such immune response in the treated fish may render them unresponsive or less responsive to successive treatments. Obviously, the best approach would be the use of purified LH from the same species to which the treatment will be applied, which is the case of spawning induction of carp using purified carp GTH (cGTH). However, this is in most cases not feasible, resulting in the use of hCG as a general GTH treatment in many farmed fish.

7.2 Second generation: GnRH agonists (GnRHa)

The second generation of hormonal therapies was developed after the discovery of the brain hormone responsible for the regulation of pituitary GTH secretion, the GnRH [78]. The application of GnRH-based therapies has important advantages over the previous GTH preparations, due to the possibility of acting at a higher level of the reproductive axis and thus promoting a more general and physiological stimulation of the whole reproductive process.

As seen in previous sections, there are several native GnRHs produced in the brain of vertebrate species, all having GTH releasing activity. The GnRH is a short decapeptide synthesized in specific areas of the brain, transported through neuronal fibers to the pituitary and released by the terminal nerves in close proximity to the gonadotrops, were it acts immediately on GnRH–R stimulating GTH release. This role is correlated with a very short half-life in circulation. The GnRHs are degraded rapidly by specific proteolytic enzymes in circulation [150-151], making exogenous administration of native GnRHs an ineffective treatment. Studies on the structure-activity relation of native GnRHs have prompted the development of GnRHs, which are structurally modified GnRHs exhibiting increased GTH releasing potency when administered exogenously. The most important characteristics for a GnRHa are (a) high resistance to enzymatic degradation, (b) high binding to the GnRH-R and (c) high activation of the gonadotropes, which together determine the agonist's GTH releasing potency. Accordingly, structure modifications are mostly focused on protecting amino acids at positions 6 and 9, which are the regions of enzymatic recognition and degradation, and enhancing amino acids of the NH_2- (pGlu-His-Trp-Ser) and COOH- (Pro-GlyNH$_2$) terminus, which are the regions responsible for receptor binding. The native GnRHs have a half-life in circulation of ~5 min, whereas the most active GnRHa's have a half-life of around ~20 min [152]. Therefore, an injection of even the most potent GnRHa stimulates a GTH elevation in circulation for 24-72 h, although this is highly variable depending on the species and temperature [153-155].

Administration of GnRHa induces release of LH from the pituitary, which in turn stimulates gonadal maturation. In females, GnRHa treatment induces FOM, ovulation and spawning, and in males increases sperm volume and, sometimes, spermatozoa density. The GnRHa based hormonal therapies have important advantages over GTH preparations. The GnRHa's are easy to prepare and result in a lower cost preparation than a purified GTH. They are generic and thus useful for a wide range of species, from fish to mammals, making them widely available commercially. The small size of the GnRHa does not induce immune response in treated animals and thus repeated treatments can be applied with no desensitization problems. Also, GnRHa act on the pituitary inducing the endogenous release of the fish's own GTHs and a more appropriate gonadal stimulation than obtained with the administration of high levels of exogenous GTHs. These advantages have made the use of GnRHa the best choice for spawning induction in fish.

The GnRHa based treatments have a limitation, which is the short half-life of the decapeptides. The classical mode of administration is the intra-peritoneal or intra-muscular injection of saline-dissolved GnRHa, at the required dose. Depending on the GnRHa type, fish species and water temperature, a single GnRHa injection induces a LH surge that lasts around 12-72 hours, before the effect disappears. In some cases, this short-lived effect of a single GnRHa injection is enough to induce spawning 2-3 days after treatment. But, in many cases, further injections are necessary to induce prolonged LH release and stimulate complete gonad maturation and spawning. In females, multiple GnRHa injections are normally required for asynchronous or multiple-batch group synchronous species, and also in species whose gonadal development is inhibited at early stages and need prolonged stimulation to affect the whole maturation process. In males, multiple injections are normally recommended, because of the asynchronous type of development of the testes. However, multiple injections are hazardous and stressful to broodstock, and in the long term can cause inhibition of reproduction by itself, appearance of stress-associated pathologies and even the death of some breeders.

For these reason, GnRHa treatments have been also administered in the form of sustained-release delivery systems [64-65]. The single administration of GnRHa via delivery systems causes sustained delivery of GnRHa to the bloodstream and thus a prolonged stimulation of pituitary LH release, which can last for several weeks, depending on the species and temperature [156-157]. A single administration of a delivery system replaces efficiently the effect caused by 4-5 injections. Many different types of delivery systems have been developed for GnRHa administration, including pellets of cholesterol-cellulose, Ethylene-Vinyl Acetate (EVAc) implants, biodegradable microspheres, osmotic pumps, etc. One of the most convenient and efficient preparation is the EVAc implants, which can be easily applied into the dorsal musculature with a syringe type applicator [140,159-160]. These implants have been proved to induce ovulation, spermiation and spawning in a wide range of fish species, being also highly efficient in the stimulation of male spermiation.

Other than GnRHa, there is another type of second generation hormone treatments, the dopamine antagonists (DAant), which are drugs that block the dopamine (DA) system of the brain, something necessary for the stimulation of reproduction of several fish species. As described previously, the endocrine regulation of reproduction is under a dual control from the brain, the stimulatory action of GnRH and the inhibitory action of DA. The activity rate of both systems determines finally the endogenous stimulation or inhibition of LH release. Not all fish species posses an active DA system in the brain. It seems that the DA inhibitory system is strong in fresh water species, but weak or absent in most marine species. The activity of the DA inhibitory tone also varies depending on the season and physiological steroid levels of the fish. In fish species with an active DA system, the inhibition of ovulation and spermiation in captivity is caused by both increased DA activity and decreased GnRH activity, whereas in fish species with a weak DA system, reproductive disorders are due almost exclusively by decreased GnRH activity. A strong DA inhibitory tone has been demonstrated in cyprinids, silurids, salmonids and some perciformes. In these species, treatment with GnRHa only stimulates LH release, similar to the treatment with only DAant, but it is the combined treatment (GnRHa + DAant) which provides best results. Normally, only the co-treatment is efficient in the stimulation of oocyte maturation, spermiation and spawning. There are several DAants available in the market that proved to be useful for hormone treatments in aquaculture, the most common being pimozide, domperidone and metoclopramide. Normally, they are administered as a liquid solution injected intraperitoneally or intramuscularly, on a weekly basis.

8. GENERAL PROTOCOL FOR SPAWNING INDUCTION AND STRIPPING

The basis for developing a protocol to control reproductive maturation in a fish species is the understanding of its reproductive strategy and reproductive cycle in relation to environmental conditions, particularly photoperiod and temperature. Knowledge of sexual differentiation, size at first maturity, maturational development in relation to environmental changes, reproductive endocrinology, spawning behaviour and egg parameters, enable the design of a protocol for maintaining the fish in conditions that allow maturation to advance to late stages of gametogenesis or spawning (Table 2). The protocol describes the nutrition, environmental conditions, light, photoperiod, temperature, space, substrate and social conditions required during maturation and, when necessary, the hormonal therapy required to stimulate final maturation. Generally, for a species with a long history in aquaculture such information is known under captive conditions and protocols, which may not have actually been described, are routinely followed and not necessarily fully understood. Where a protocol is followed but not fully understood, it can be difficult to explain an unexpected change in reproductive development caused by, for example, abnormal environmental conditions, whereas a full understanding of a protocol and hence the conditions required, may enable anticipation of possible problems.

For a species with no established aquaculture history, the following sources of information can be used to propose protocols that have good opportunities for success:

- Studies on reproductive strategies, maturational development and egg parameters from wild populations,
- Anecdotal information from fishermen, whom often have a very sound knowledge of the seasonal changes in a species gonad,
- Information on environmental parameters from areas where mature individuals of the particular species are caught,
- Extrapolation from reproductive strategies, maturational development and egg parameters from as many closely related species as possible.

In particular, the protocol should identify critical points, such as broodfish selection, nutrition before and during vitellogenesis, environmental conditions during vitellogenesis and spawning, space and social conditions for spawning and hormone therapies required. The most common critical points that cause problems with maturational development and, therefore, the definition of a protocol to obtain gametes can be ordered as nutrition > environmental factors that control or entrain the initiation and progress of vitellogenesis or spermatogenesis > environmental factors that control or entrain final maturation and ovulation or spermiation > hormone induction therapies (Fig. 7). Nutrition has implications on both the decision to proceed with maturation in a given year (based on energy reserves and energetics, see environment section) and whether all the necessary nutritional components are available for the developing gametes. When an adequate nutrition is not provided, profound effects have been observed on fecundity and egg quality. Amongst the most important nutritional aspects for reproduction are protein quality, lipid/fatty acid composition, and vitamins. Often the only solution to poor spawning due to inadequate nutrition is to improve the nutrition for the next maturational episode [31]. No environmental or hormonal control will solve a nutritionally-based problem, and before any attempts are made to environmentally or hormonally control maturation, it is essential that the broodstock nutrition is adequate (Fig. 7).

Table 2 - Reproductive information of interest for designing a spawning protocol and how the information may be used in practice.

Reproductive information of interest	Practical use of information
Sexual differentiation strategy, gonochoristic (separate sexes), hermaphrodite, changing sex.	Indicates proportion of sexes required and when selection should be made in species that change sex.
Size at first maturation.	Gives minimum sizes and indication of optimal sizes for brood fish. Indication of tank design.
Timing of start of vitellogenesis in relation to environmental changes.	Preferably, months before this date brood fish should be selected and care should be taken to ensure that fish are nutritionally prepared for reproduction.
Timing and environmental parameters naturally experienced during vitellogenesis.	Captive environmental conditions during vitellogenesis should be controlled to ensure a close approximation to natural conditions.
Timing and environmental parameters naturally experienced during the spawning season, particularly: · When sperm can easily be expressed from males. · When females present late stages of vitellogenesis. · Timing of the peak or middle of the spawning season. · Timing of the end of the season when gonads return to a resting stage.	· Captive environmental conditions during the spawning season should be controlled to ensure a close approximation to natural conditions. · Spawning can be anticipated and brood fish can be checked for maturational status. · When no spawning is observed, a revision of brood fish with the objective of hormonal induction can be timed to coincide with the peak of the spawning season.
· Endocrinology of reproduction. · Hormone induction reports. · In particular dopamine activity.	Selection of hormone and doses to hormonally induce spawning.
Spawning behaviour. · Polygamy, monogamy. · Number of spawns per season: Daily batch spawner, Group synchronous batch spawner or Single spawn per season.	Indication of: · Tank design. · Proportion of sexes required. · Method for collection of eggs, stripping, substrate spawning or pelagic egg collection.
Egg size.	Indication of: · Fecundity (generally larger eggs = lower fecundity) · Oocyte size required for hormone induction · Care required for larvae
Fecundity.	Indicates the number of brood fish required.
Type of eggs: pelagic, demersal, adhesive.	Indication of: · Method for collection of eggs, stripping, substrate spawning or pelagic egg collection. · Incubation methods.
Strategy of parental care for eggs.	Indication of: · Method for collection of eggs, stripping, substrate spawning or pelagic egg collection. · Incubation methods. · Fecundity (generally more parental care = lower fecundity).

Incorrect or sub-optimal environmental parameters during gametogenesis and/ or final maturation will result in respective primary or secondary reproductive dysfunctions (see reproductive dysfunctions). Solutions include improving the captive environment or applying hormonal therapies, when it is technically not possible to improve conditions or the optimal conditions are not known. Primary dysfunctions require a prolonged hormone therapy to stimulate the entire gametogenic period, similar to that used to spawn the freshwater eels (see protocol below).

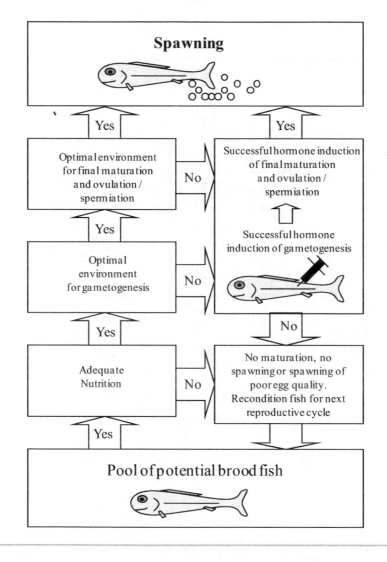

Figure 7 - Flow diagram of critical points in the induction of spawning in captivity.

Secondary dysfunctions, which are the most common dysfunctions encountered in aquaculture, require a short-term hormone therapy to induce final maturation, ovulation or spermiation. These types of therapies have been used in various fish, including carps, basses, bream, drums, croakers, grouper, snapper, salmonids, catfish, mullet, flatfish and puffer fish. When developing a hormone therapy for a secondary dysfunction, two critical factors should be considered, 1) the stage of ovarian development, often measured as oocyte size, and 2) the hormone dose. Generally, spawning with hormone induction therapies is compromised when fish are treated outside of an optimal range of oocyte size or hormone doses [161]. Fish with smaller than the minimum oocyte diameter either do not spawn or exhibit a poor spawning response [161-164]. Higher than optimum doses result in reduced egg quality [161,165] and lower doses result in reduced spawning frequency [161,165-166]. There appears to be a relationship between minimum oocyte size for successful induced spawning and egg size (Fig. 8) and the equation, Minimum oocyte size = -95.98 + 0.624 x egg size, can be used to give an indication or starting point to identify the required oocyte size for developing a hormone induction therapy.

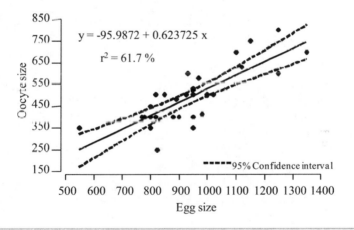

Figure 8 - Regression of minimum oocyte size (μm) for successful hormone induced spawning against egg size (μm). Data plotted from review Table 7-4 in [167].

Optimal hormone doses should be determined for each species. A starting point is to examine doses that have been used in closely related species. Some optimal doses for several fish species are presented in Table 3. It should also be considered that oocyte size and hormone dose appear to interact [161].

When ovulation has been achieved but the gametes must be stripped, a final consideration is the timing of stripping in relation to ovulation. After ovulation, unfertilized eggs left in the ovarian or abdominal cavity have been observed to over-ripen, a process by which the eggs lose viability, as measured by percentage of fertilization and embryo development [39]. Over-ripening appears to be a process in which the eggs undergo a series of morphological and chemical changes, which are either the result of or the cause of the eggs losing viability. A number of species undergo a ripening period before over-ripening begins, i.e., after ovulation the viability of the unfertilized eggs may increase before decreasing due to over-ripening, resulting in an optimum period for stripping and fertilization. Care should be taken to strip eggs during the period of optimal viability [39,167].

Maturation in fish is a very complex process and its manipulation and achievement of the spawning of good quality eggs requires a complete understanding and careful consideration of all the different factors that influence the outcome of the manipulations. Many studies have demonstrated that fish can be environmentally and/or hormonally manipulated to spawn good quality eggs, equal in quality to eggs from naturally spawning fish [31,39,42]. It should, therefore, be considered that poor egg quality from manipulated fish indicates that some aspect of the spawning protocol was not correct and all parameters should be reconsidered.

9. ESTABLISHED PROTOCOLS FOR CULTURED FISH

As mentioned in the introduction, the production of finfish from aquaculture is the fastest growing agriculture industry in the world; our understanding and control of reproduction forms the basis for this growing production. In this section, the knowledge on spawning protocols for a selected group of important species is compiled and summarized. Summaries have been focused on the top seven (by volume of production) orders of fish that are being cultured, that is the Cypriniformes, Perciformes, Salmoniformes, Gonorynchiformes, Siluriformes, Anguilliformes and Mugiliformes [6] and a further three orders, Tetraodontiformes, Acipenseriformes and Pleuronectiformes, which are relatively important in terms of production or relevant for aquaculture diversification. For Perciformes, due to the diversity of families and the presence of many promising aquaculture candidates, species from ten different fish families or "groups" are described, such as tilapias, basses, snakeheads, breams, amberjacks, drums, groupers, cobia, tunas and snapper. This information can be consulted as a guide to how spawning has been achieved for the species described and as an indication of the problems and solutions that may be encountered when working with either an established aquaculture species or a candidate aquaculture species that is related to the described species.

Table 3 - Selected hormonal therapies that have successfully induced ovulation or spawning in a range of species from different families. Similar therapies have been grouped as an indication that the therapies could be considered for related species. Inj. = injection, imp. = implant, PIM = pimozide, MET = metaclopramide.

Family	Species	Similar hormone therapies	Hormone therapy (μg kg^{-1} body weight)
Percichthyidae	*Dicentrarchus labrax*		10 μg kg^{-1} GnRHa repeated inj. [168]
Moronidae	*Morone saxatilis* x *M. chrysops*		62 μg kg^{-1} GnRHa EVAc imp. [169]
Centropomidae	*Lates calcarifer*		37.5–75 μg kg^{-1} GnRHa imp. [165]
Sparidae	*Sparus aurata*		100 μg kg^{-1} GnRHa EVAc imp. [31]
Sciaenidae	*Sciaenops ocellatus*	GnRHa effective for inducing spawning. Doses required are 25-100 μg kg^{-1} for GnRHa imp. and 10-40 μg kg^{-1} for GnRHa inj.	20 μg kg^{-1} GnRHa inj. [44]
Carangidae	*Seriola dumerili*		2 x 40 μg kg^{-1} GnRHa EVAc imp. [140]
Scombridae	*Thunnus thynnus*		40-100 μg kg^{-1} GnRHa EVAc imp. [170]
Paralichthyidae	*Paralichthys lethostigma*		100 μg kg^{-1} GnRHa imp. [171]
Scophthalmidae	*Psetta maxima*		25 μg kg^{-1} GnRHa imp. [164]
Tetraodontidae	*Sphoeroides annulatus*		40 μg kg^{-1} GnRHa inj. [172]
Chanidae	*Chanos chanos*		10-33 μg kg^{-1} GnRHa inj. [173-174]
Salmonidae	*Salmo salar*		50 μg kg^{-1} GnRHa EVAc imp. [175]
Acipenseridae	*Acipenser baeri*		10 μg kg^{-1} GnRHa inj. [176]
Lutjanidae	*Lutjanus argentimaculatus*	GTH preparations or high doses of GnRHa. Doses required are 1000-1760 IU kg^{-1} for hCG inj. and 100 μg kg^{-1} for GnRHa inj.	1000 IU kg^{-1} hCG inj. 100 μg kg^{-1} GnRHa inj. [177]
Ictaluridae	*Ictalurus punctatus*		1000-1760 IU kg^{-1} hCG inj.; 1° = 10;, 2° = 90 μg kg^{-1} GnRHa inj. [178]
Clariidae	*Clarias gariepinus*	Dopamine inhibition described. GTH or GnRHa + dompamine antagonist. Doses required are 44-4000 IU kg^{-1} for hCG, 300-600 μg kg^{-1} for cGTH and 20-50 μg kg^{-1} for GnRHa + DA antagonist (PIM, MET).	4000 IU kg^{-1} hCG inj. 50 μg kg^{-1} GnRHa + 500 μg kg^{-1} PIM inj. [179]
Mugilidae	*Mugil Cephalus*		1° = 14-21, 2°= 30-48 IU g^{-1} of hCG [281] 1° = 10 μg kg^{-1} GnRHa + 15 mg kg^{-1} of MET 2° = 20 μg kg^{-1} GnRHa + 15 mg kg^{-1} MET [180]
Cyprinidae	*Cyprinus carpio*		1° = 50-100 μg kg^{-1} cGtH; 2° = 250-500 μg kg^{-1} cGtH [57] 1° = 2 μg kg^{-1} GnRHa + 0.5 μg kg^{-1} PIM; 2° = 18 μg kg^{-1} GnRHa + 4.5 μg kg^{-1} PIM [181]

9.1 Cypriniformes - Carps (Cyprinidae)

Generally, in captivity carps will reach late stages of vitellogenesis and spermiation, and can either be hormonally induced to spawn or left to spawn naturally in specially prepared ponds or natural waterbodies. Dopamine inhibition of GnRH action has been demonstrated. In 2005, species from the family Cyprinidae had the highest aquaculture production (9,428,518 T) of any taxonomic family [6]. The highest producing countries were China (4,999,307 T) and India (2,558,599 T). The dominant species being produced were 3,043,712 T of common carp (*Cyprinus carpio*), 2,086,311 T of crucian carp (*Carassius carassius*), of the Chinese carps were 3,904,799 T of grass carp (*Ctenopharyngodon idellus*), 4,152,506 T of silver carp (*Hypophthalmichthys molitrix*) and 2,208,678 T of bighead carp (*Hypophthalmichthys nobilis*), and of the Indian carps were 1,235,992 T of catla (*Catla catla*) and 1,195,965 T of roho labeo (*Labeo rohita*). The carps are gonochoristic and have been reported to mature in the first year, but larger 2- or 3-year-old common carp (1-5 kg) are considered the most adequate for broodstock [57]. In general, natural spawning requires warm spring or summer conditions either in ponds or flooded areas [58]. Carps are batch spawners and species such as the common carp will spawn all-year-round (5 spawns) in the tropics and just once a year in northern latitudes. Spawning is polygamous and eggs are scattered onto substrate, vegetation (common carp), river beds (Chinese carps) or flooded areas (Indian carps). Eggs or fry for aquaculture are obtained from natural spawning in natural water bodies (particularly in Asia), in specially prepared ponds (common carp) or by hormonally induced spawning in hatcheries (all carps). Common carp are both spawned naturally in ponds or hormone induced spawning with carp pituitary extract (cPE) or carp gonadogropin (cGtH). In both natural and induced spawning, the broodstock must first be environmentally induced to reach the late stages of gametogenesis. In ponds on the coastal plain of Israel, vitellogenesis begins in August-September and the peak of spawning is in May, as temperatures rise above 18°C to optimal temperatures of 22-24°C [57,182]. The timing of spawning can be estimated by counting 1,000-1,200 (2 to 3-year-old fish) or 2,000 (>3-year-old fish) degree days from when spring temperatures rise above 15°C. Therefore, vitellogenesis takes place under a natural photoperiod (LD 10:14 - 14:10) and thermal cycle (13-30°C). Spermiating males can be encountered all year round. Broodfish are carefully selected males by the presence of flowing milt and females by the presence of migrating germinal vesicle in 65% of the cleared oocyte sample. Selected broodfish were transferred to ponds for natural spawning (male:female ratio from 1.5:1 to 1:2) or to the hatchery for hormonal induction (male:female ratio of 1:2) [57,182]. Spawning ponds were carefully prepared [57-58,182] and pond spawning was initiated by adding vegetation or fibrous spawning mats, or by introducing broodstock to a pond with substrate for spawning. In the hatchery, female fish were induced with a priming dose of 50-100 μg kg^{-1} of cGtH in calibrated carp pituitary extract or the extract of 0.1-0.2 of a carp pituitary and 10-12 h later a resolving dose of 250-500 μg kg^{-1} cGtH, or the extract of 1-1.4 pituitaries [57,182]. Eggs were hand-stripped after a latency period of around 6-14 h that varies with temperature (240-260 degree hours). The optimum period for stripping carps was reported to be 2 h after ovulation [183]. Fecundities were 100,000-300,000 eggs kg⁻¹ [57-58,182] and egg size varies from 2 to 2.5 mm, depending on the size of the female [182]. Males are given a single dose of 100-250 µg kg⁻¹ cGtH, or the extract of 0.5-0.8 pituitaries. Rothbard and Yaron [57] compared the two spawning methods; induction with pituitary extract can produce 1 million fry using 15 breeders (10 females and 5 males) and highly technical knowledge and infrastructure, while pond spawning can produce 1 million fry using 400-500 breeders (200 females and 200-300 males) and low technical knowledge and infrastructure. Recently, interest has focused on using GnRHa to induce spawning in carps. Alone, GnRHa will not induce ovulation in carps; it should be co-administered with a dopamine antagonist to successfully induce ovulation [62]. Several GnRHas have been used with the dopamine antagonists pimozide, metoclopramide or haloperidol to successfully induce ovulation [148-149,181]. However, care should be taken as different strains of common carp gave slightly different results, particularly in the synchronization of spawning [149]. Mikolajczyk et al. [181] obtained 95% ovulation of eggs with 49-54% hatching after administering 20 µg kg⁻¹ of GnRHa in conjunction with 5 µg kg⁻¹ of pimozide, in two doses that were administered as a priming dose of 10% followed 6-12 h later by a resolving dose of 90%. The latency time from the resolving injection to spawning was 9-13 h.

9.2 Perciformes

9.2.1 Tilapias (Cichlidae)

In captivity, tilapias complete maturation and spawn naturally. In 2005, species from the family Cichlidae had an aquaculture production of 2,025,560 T of which 1,703,125 T were the Nile tilapia (*Oreochromis niloticus*) and 43,369 T were the Mozambique tilapia (*Oreochromis mossambicus*) [6]. Most of the production originated from China (978,135 T), Egypt (217,019 T), Indonesia (189,570 T), Thailand (109,742 T), Taiwan (83,435 t) and Brazil (67,851 T). Nile tilapia is gonochoristic and mature in the first year at the earliest opportunity (>40g). The most adequate broodstock for aquaculture are 1+ year-old and a body weight of 150-250 g, as relative fecundities, spawning frequencies and larval survival were higher compared to younger or older fish [184-185]. The Nile tilapia reproduces in almost any container, and have been spawned all year round [184]. Optimal temperatures for spawning were 25-30°C, whereas outside of this range spawning frequency decreases and spawning stops below 20°C [186]. Vitellogenesis had a duration of approximately 14-30 days, to give monthly spawns of batches of eggs [184,187]. Dominant male broodstock defend an area or nest to which females visit to spawn, but spawning is polygamous [184]. Broodstock groups can be biased to females in ratios of 2:1. Eggs are 2.4 mm in diameter, demersal and are incubated orally by the females. Fecundities are approximately 2,000 eggs per fish [184] or 3,000-8,000 eggs kg⁻¹ [188]. A major problem for tilapia culture is the low fecundity and asynchronous spawning of broodstock groups. MacIntosh and Little

[184] compared different management practices to obtain large batches of eggs for stocking facilities. For research purposes mature tilapias have been strip-spawned to obtain gametes; females are maintained separately and observed for signs of ovulation (swollen belly and genital papilla) before being stripped [189]. Nile tilapia is at an optimum for stripping 1 h after ovulation.

9.2.2 Basses (Percichthyidae, Moronidae, Centropomidae)

Basses of the families Percichthyidae, Moronidae and Centropomidae will reach late stages of vitellogenesis and spermiation, and some will spawn naturally in captivity. However, spawning frequency can be low and unpredictable, in which case hormone induced spawning may be applied. In 2005, species from these families had an aquaculture production of 562,269 T, of which 427,487 T were from the family Percichthyidae, 100,769 T from Moronidae and 34,545 t from Centropomidae [6]. The producing countries were China with 424,857 T, Turkey with 37,490 T and Greece with 30,959 T. The dominant species being produced were 251,770 T of Japanese seabass (*Lateolabrax japonicus*), 175,687 T of mandarin fish (*Siniperca chuatis*), 95,040 T of European seabass (*Dicentrarchus labrax*), 30,970 T of barramundi (*Lates calcarifer*) and 5,729 T of hybrid stripped bass (cross between *Morone chrysops* and *M. saxatilis*). European seabass is gonochoristic and males mature normally at their second year (approx. 300 g), while females mature the following year (>500 g) [43]. Recommended size for broodstock are males of 700 g (3-4 years) and females of 1.5-2 kg (6-8 years) [190]. Vitellogenesis begins in September-October, around 4 months before spawning, which takes place from January to March [17,43]. Therefore, environmental conditions for vitellogenesis are a declining photoperiod (LD 12:12 to LD 9:15, in Spain) and thermal cycle (25-13°C), and environmental conditions for spawning are an increasing photoperiod (LD 9:15 to 12:12) and thermocycle (10-15°C). Spawning has been observed over the temperature range of 9-18°C, while optimal temperatures are 13-15°C and spawning stops at temperatures above 18°C [190]. An influence of the temperature on gametogenesis has also been described in photothermal manipulated sea bass broodstock, showing that a reduction of water temperature, from the highest summer to mild autumn temperatures, at least 1-2 months before the spawning period is necessary for initiation of vitellogenesis and further successful egg spawning [17]. European seabass are group-synchronous batch spawners that spawn several batches of eggs over the spawning season [43,191]. Through hormonal induction, 2-4 spawns have been detected in each female breeder, during the spawning season [168,192]. Fecundities in the wild have been estimated at 2 million per fish over the spawning season [191], and in captivity 300,000-600,000 eggs kg[-1] have been reported [17,168,190]. Spawning behaviour is polygamous and the spawned eggs are pelagic with a diameter of 1.15-1.2 mm. European seabass spawn naturally in captivity, but hormonal induction is often used to synchronize or ensure spawns, when required. Carrillo et al. [43] reviewed the development of induced spawning protocols and concluded that the most adequate protocol was the administration of two injections of 5 and 10 μg kg[-1] of

GnRHa 6 h apart, to females that had oocyte diameters > 650 μm. A latency period of 72 h from the first injection was observed before spawning, although this can vary slightly depending on water temperature and the time of the day when the treatment is applied [193-194]. Multiple spawnings (2-4 during the season) have been obtained using repeated GnRHa injections of 10 μg kg^{-1}, that were spaced 7-14 days apart [168]. Different GnRHa delivery systems have been tested in the European seabass, including fast and slow release implants, liquid solution of microspheric implants and saline dissolved injections. These studies have shown that administration of GnRHa via sustained release delivery systems does not improve spawning performance of females, compared to the classical saline dissolved GnRHa injection [192], but increase significantly sperm production in males [158]. Environmental manipulations have been used to obtain all-year-round spawning [17,43,190]. Displaced spawning periods were obtained by altering both the photoperiod and the thermal cycle, either by phase shifting the cycles, using square wave cycles or using compressed or expanded cycles. One or two month square wave period of long daylength (LD 15:9) in an otherwise constant short daylength (LD 9:15) regime could be applied each month from March through to September to obtain spawning from October through to May [43].

9.2.3 Snakeheads (Channidae)

Snakeheads complete maturation and spawn naturally in captivity, however, much of the aquaculture production is from juveniles collected from natural sources or culture ponds. In 2005, species from the family Channidae had an aquaculture production of 308,938 T of which 277,763 T were snakehead (*Channa argus*), 13,036 T Indonesian snakehead (*Channa micropeltes*), and 8,593 T striped snakehead (*Channa striata*) [6]. Most of the production originated from China (277,511 T), Indonesia (11,525 T), India (8,213 T), Thailand (7,508 T) and Nigeria (1,333 T). The striped snakehead is gonochoristic and first maturity is at about 11 months and a size of 25-30 cm [195. The striped snakehead can spawn all year round in India, but peak spawning is observed during the rainy season, which also results in a cooler period with temperatures below 30°C. In Southern India two peak periods of spawning coincide with two rainy seasons (June-July and November) and in Northern India one peak spawning period during April to September. Snakeheads appear to form spawning pairs, which build a nest for courtship, spawning and egg incubation. The eggs float, but are spread in a kind of film over the centre of the nest. Striped snakehead eggs are 1.15-1.46 mm in diameter and water temperatures for egg incubation range from 16 to 33°C. Fecundities for striped snakehead were reported between few hundreds and few thousands, depending on the size of the female; for other species, fecundities range from 2,214 eggs for a 50 cm fish (*C. marulius*) to 33,873 for a 22.2 cm fish (*C. punctata*) [195]. Striped snakehead has been reported to spawn naturally in captivity just two months after being air freighted from Thailand to Hawaii [196]. The broodstock were 1-2 kg and were held at 28-32°C; the eggs were collected from the bottom of the tank. Striped snakehead has been spawned with carp and catfish pituitary extract. A dose in the range of 40-80 mg of pituitary per female given in two injections induced spawning 11-12 h after the second injection in *C. marulius*, *C. puntatus* and *C. gachua* [193].

9.2.4 Breams (Sparidae)

Generally, in captivity, the Sparidae will reach late stages of vitellogenesis and spermiation and spawn naturally, but hormone induced spawning has also been applied successfully when required. In 2005, species from the family Sparidae had an aquaculture production of 245,217 T, of which 110,705 T were gilthead seabream (*Sparus aurata*) and 82,083 T silver seabream (*Pagrus auratus*) [6]. Most of the production originated from Japan (76,082 T), China (44,222 T), Greece (44,124 T), Turkey (28,334 T) and Spain (15,552 T). Gilthead seabream is a protandrous hermaphrodite, with individuals maturing first as males and then as females. The change from males to females occurs from one year to the next and appears to be controlled socially when the ratio of large females to smaller males decreases [31]. All gilthead seabream mature as males in the first year (approx. 200-300 g) and can, under the correct social conditions, change to females in the second or third year (approx. >600 g). Recommended size for broodstock for a hatchery are males of 300-500 g (2-3 years) and females of 1-1.5 kg (4-6 years) [190]. The gonads, from May to September, may exhibit both morphology of testes and ovary. From September onwards vitellogenesis begins in maturing females and spermatogenesis in maturing males. Environmental conditions for vitellogenesis are a declining photoperiod (LD 12:12 to LD 9:15, in Spain) and thermal cycle (25-13°C). The spawning season extends from January to May, under an increasing photoperiod (LD 9:15 to 14:10) and thermal cycle [31]. Spawning has been observed over the temperature range of 14-20°C, while optimal temperatures were 15-17°C and spawning stops at temperatures above 24°C [190]. Gilthead seabream is a batch spawner and can spawn daily for 3-4 months; fecundities can reach 2-3 million eggs kg^{-1} over the spawning season [31]. Spawning behaviour is polygamous and the spawned eggs are pelagic with a diameter of 0.94-0.99 mm. Although hormonal induction is now not needed for seabream, well established hormonal induction protocols were necessary when the industry started, with the use of wild broodstock. Zohar et al. [31] reviewed the development of induced spawning protocols and concluded that the most adequate protocol was the use of slow release implants containing 100 μg kg^{-1} of GnRHa, in females that had oocyte diameters >530 μm. A latency period of 48-72 hours was observed before spawning began in 80% of the fish and daily spawning continued for 4 months. Often a proportion of females within a pool of broodstock do not spawn naturally, and hormone induction protocols have been applied to spawn these females [166]. Environmental control can be used to obtain all-year-round spawning [31,190]. The 12-month photoperiod and thermal cycle is phase-shifted 3, 6 and 9 months to give 4 different spawning groups (including the ambient cycle). A 3-4 month spawning period is achieved soon after the period of short winter days, December, March, June and September under respective photoperiods ambient and phase shifted 3, 6 and 9 months, to give an all-year-round production of eggs. No reduction in egg quality was observed from fish spawned with hormonal or environmental control.

9.2.5 Amberjacks (Carangidae)

In 2005 species from the family Carangidae had an aquaculture production of 178,270 T, of which 159,798 T were Japanese amberjack yellowtail (*Seriola quinqueradiata*), 2,392 T Japanese jack (*Trachurus japonicus*) and 15,534 T unidentified *Seriola spp* and *Trachurus spp* [6]. Most of the production originated from Japan (164,808 T) and China (11,973 T), but more recently kinfish (*Seriola lalandi*) has also been produced commercially in Australia [197]. In Europe, the species of interest is the greater amberjack (*Seriola dumerili*), but commercial production has been hindered by substantial problems in reproduction, as cultured fish often fail to reach advanced stages of gametogenesis [140,198-199]. Depending on the species, amberjacks reach reproductive maturation at 2 years (*Seriola rivoliana*) [200] or may require more than 4 years to mature (greater amberjack), reaching sizes of >5 kg [201-202]. Amberjacks in the temperate zone initiate their reproductive cycle in the spring and spawn in the early summer months (June-July), at water temperatures ranging between 21 and 25°C [140,203].

All amberjacks are either group-synchronous multiple-batch spawners or asynchronous spawners, and depending on the species, they can spawn a few times during the reproductive season, as in greater amberjack and kingfish [140,197,201,203], or can spawn on a daily basis for many weeks, as is the case in the Japanese yellowtail [204-205]. However, the response to hormonal therapies is not the same in all amberjacks. For example, in the greater amberjack hormonal treatment with either a single GnRHa injection or a GnRHa delivery system may not always induce more than a single spawning event [206], whereas in the Japanese yellowtail, a GnRHa delivery system induced multiple spawning [204]. Spawning induction in amberjacks has been achieved with the use of hCG injections [205,207-208] and GnRHa delivery system (40 µg kg⁻¹) [140].

Egg fecundity can be extremely variable in response to hormonal treatment, which is probably related to the stage of reproductive development at the time of hormone treatment (see earlier section). In the greater amberjack, fecundity may range between 3,000 and 43,000 eggs kg⁻¹ [140,208], whereas in Japanese yellowtail it may range between 18,000 and 172,000 eggs kg⁻¹ [205,207].

9.2.6 Drums, croakers (Sciaenidae)

Generally in captivity, the Sciaenidae reach late stages of vitellogenesis and spermiation and spawn naturally, however, spawning frequency can be low and unpredictable, in which case hormone induction of spawning was used. In 2005 species from the family Sciaenidae had an aquaculture production of 117,141 T, of which 69,641 T was large yellow croaker (*Larimichthys croceus*), 46,632 T red drum

(*Sciaenops ocellatus*) and 800 T meagre (*Argyrosomus regius*) [6]. Most of the production originated from China (115,383 T). Red drum is gonochoristic and matures at a size of 4-5 kg or 4-5 years in wild stock, however, the age has been reduced to 2 years in cultured fish [44]. Vitellogenesis has been observed to begin in July-August indicating a period of 1-3 months before the peak of spawning in September-October [209]. Therefore, environmental conditions for vitellogenesis are decreasing photoperiod and temperatures of 28-30°C [209]. The spawning season extends from August to January with a peak in September-October, i.e., under the declining autumn photoperiod and thermal cycle. Spawning has been observed at 24-26°C, but stops below 20°C [44]. The red drum was described as group synchronous batch spawners [210] and two females have been recorded to spawn 10-20 times during one month [44]. The eggs are pelagic, with a diameter of 0.9-1.0 mm and fecundities over the spawning season have been estimated at 3×10^7 eggs per 9-14 kg female [211]. Although hormonal induction is not needed for red drum, studies have indicated that the sciaenids, red drum, spotted sea trout (*Cynoscion nebulosus*), orangemouth corvine (*Cynoscion xanthulus*) and Atlantic croaker (*Micropogonias undulatus*), with mean respective oocyte diameters of 600, 400, 440 and 550 µm, can be spawned with a single injection of 20-100 µg kg^{-1} of GnRHa [44].

An amazing aspect of the red drum and other sciaenids has been the environmental control of spawning using phase-shifted and compressed photoperiod and thermal cycles. Thomas et al. [44] reviewed how two females and two males were exposed to an abbreviated photoperiod and thermal cycle until the autumn spawning environment was reached and held at LD 12:12 and 24°C. Under these conditions the fish spawned continually for seven years, producing approximately 250 million fertilized eggs from 360 spawns. During the seven year period variations in the temperature were simulated to control the spawning, optimal spawning was obtained at 24-26°C, spawning slowed at below 23°C and stopped at below 20°C. Meagre is a promising candidate for aquaculture in the Mediterranean, and females with oocyte sizes >500 µm were successfully induced to spawn with 50 µg kg^{-1} GnRHa implants (Duncan et al., unpublished).

9.2.7 Groupers (Serranidae)

In 2005, species from the family Serranidae had an aquaculture production of 65,815 T, most of which (61,815 T) was not reported by species [6]. Most of the production originated from China (38,915 T), Taiwan (13,582 T), Indonesia (6,883 T), Malaysia (2,572 T) and Thailand (2,280 T). There are various species of groupers with interest for aquaculture [212], some of them being the dusky grouper (*Epinephelus marginatus*) [212], the white grouper (*E. aeneus*) [214], the Nassau grouper (*E. striatus*) [215], the spotted grouper (*E. akaara*) [216], the honeycomb grouper (*E. merra*) [217] and the sevenband grouper (*E. Septemfasciatus*) [218]. Groupers are protogynous hermaphrodites, with sex inversion taking place at a rather large size and age [219], and in the wild a single male usually fertilizes the eggs of many females [220]. The reproductive season is in the summer, and the spawned eggs are pelagic in nature, but are smaller than other marine fishes (< 800 μm in diameter). Groupers have a group-synchronous multiple-batch ovarian development and spawn over an extended reproductive season.

Spontaneous spawning in captivity is difficult in groupers [221-222], and it usually requires the employment of very large tanks or ponds, and low stocking densities [215,223-224]. This may be due to an elaborate breeding behaviour and pairing requirement of most groupers [216,221,225-226]. In addition, the large age-size that fish undergo sex inversion results in a scarcity of males and the need to artificially sex invert females to functional males [227]. Females often do not complete vitellogenesis and the diameter of the oocytes is too small for hormonal induction. In cases that oocytes reach the end of vitellogenesis, induction can be done using GnRHa [228] or hCG [229] in injectable form, or using GnRHa delivery systems [159,214]. The use of GnRHa delivery systems for the induction of ovulation in groupers is advantageous to injectable forms, as it may stimulate multiple ovulation events [159]. Still, no available system can induce spawning for the whole duration of the natural reproductive season, and re-administration of the hormonal therapy may be necessary. As a result, total fecundity of captive females after GnRHa implantation is lower than reported for wild groupers from the natural environment.

9.2.8 Cobia (Rachycentridae)

Generally in captivity, cobia complete maturation and spawn naturally. In 2005 cobia (*Rachycentron canadum*) had an aquaculture production of 22,751 T; no other species from the Rachycentridae family has a reported aquaculture production. Most of the production originated from China (18,882 T) and Taiwan (3,863 T) [6]. Cobia is a rapidly emerging aquaculture species; in 1999 world production was just 820 T from Taiwan. Cobia is gonochoristic and mature at 1-2 years old, with a size of 10 kg [230-231]. Under natural conditions in ponds in Taiwan, cobia have been spawned all-year-round with two peak spawning periods, one in the spring (February-May) and a second in the autumn (October) [230-231]. Suitable temperatures for spawning were 24-29°C [230], and peak spawning was at tempertaures of 24-26°C [230] and 23-27°C [231]. Liao et al. [231] described how cage reared mature fish were selected and 100 breeder (ratio males:females of 1:1) were transferred to each pond with an area of 400–600 m² and 1.5 m depth. Spawned eggs were collected from the ponds; eggs were pelagic and 1.35-1.40 mm in diameter.

9.2.9 Tunas (Scombridae)

In 2005, species from the family Scombridae had an aquaculture production of 22,995 T, of which 7,869 T was Pacific bluefin tuna (*Thunnus orientalis*), 7,583 T Atlantic bluefin tuna (*Thunnus thynnus*) and 7,458 T Southern bluefin tuna (*Thunnus maccoyii*) [6]. Most of the production originated from Mexico (7,869 T), Australia

(7,458 T), Croatia (3,425 t) and Spain (3,364 T). The tuna aquaculture is a "capture-based" industry [232] and involves the capture of migrating wild fish and their fattening in floating cages, for periods ranging from 2 months to 2 years [232-236]. Tunas are asynchronous spawners with a frequency of every 1-2 days [237] and spawn during the late spring, early summer at water temperatures >23°C. Being pelagic fishes, tunas migrate great distances to reach their spawning grounds, and spawning takes place on the water surface after dusk. The eggs are positively buoyant with a diameter of about 1 mm.

Efforts at developing a captive tuna broodstock were initiated in Japan with the Pacific bluefin tuna [236,239]. Broodstock weighing >100 kg are maintained in large cages or enclosures and are allowed to spawn naturally. Fish reared from eggs obtained during the 1990s at Kinki University reached reproductive maturation in 2004 and spawned in captivity [239]. Currently, a small number of fully farmed-raised Pacific bluefin tuna are sent to the market on a regular basis. However, spawning is not consistent, since broodstock maintained in sea cages cannot be exposed to the optimal thermal conditions for reproductive maturation and spawning, and it has been observed that lower temperatures may delay or abolish the spawning season in a given year [241]. On the other hand, yellowfin tuna (*Thunnus albacares*) broodstock have been maintained in land-based tanks, and have spawned naturally over many years [241-242].

Recently, a GnRHa delivery system based method for the induction of spawning in captive Atlantic bluefin tuna has been developed [160]. The GnRHa implants were prepared by loading GnRHa into a matrix of poly [Ethylene-Vinyl Acetate] and the implants were attached to a p[ethylene] arrowhead using a 0.5 mm nylon monofilament. Administration of the GnRHa delivery system was done underwater using a spear gun fitted with a specially designed spearhead, since tunas cannot be anaesthetized and treated with the necessary hormones. After GnRHa treatment, FOM and post-ovulatory follicles occurred in 63% and 88%, respectively, of the GnRHa implanted females, compared to 0% and 21%, respectively, of the control females. In addition, eggs were obtained from GnRHa-implanted females and were fertilized in vitro with sperm from spermiating males, which resulted in viable embryos and larvae. Finally, fertilized eggs were collected from the cages after three days from the GnRHa delivery system administration.

9.2.10 Snappers (Lutjanidae)

Generally in captivity, the Lutjanidaes will mature to the late stages of vitellogenesis and spermiation and different species will either spawn naturally or be hormonally induced to spawn. In 2005 species from the family Lutjanidae had an aquaculture production of 3,911 T, of which the dominant species was the mangrove red snapper (*Lutjanus argentimaculatus*) with a production of 3,699 T. Most of the production originated from Malaysia (3,452 T) [6]. Snapper are gonochoristic and the red snapper (*Lutjanus guttatus*), a species with good aquaculture potential from the eastern coast of the Pacific (Mexico to Peru) mature at a size of 250g from the second year (personal observation). The snapper tend to have long spawning seasons that cover most of the year, but which have peak periods. Grimes [243] observed that the peak periods were associated with peaks in the productivity of the ecosystem; generally, species in stable island environments spawned throughout the year while species with continental environments, with seasonal ecosystem productivity, present long spawning periods with peaks. The red snapper exhibit the gonadal morphology of an active batch spawner during the whole year, with peak periods during March-April and August-November on the Mexican coast [244]. Vitellogenesis is short, 1-2 months, from February to April [161,244] under an increasing photoperiod (LD 11:13 to 12:12) and temperature (22 to 24°C). Spawning is polygamous, with the snapper forming spawning aggregations [243]. Red snapper eggs are pelagic, with a diameter of 0.8 mm and scattered into the environment [161]. Fecundities were 70,000-100,000 eggs kg^{-1} for red snapper that were induced to spawn. Many snapper species have been observed to spawn naturally in captivity, e.g., *L. campechanus* [245], *L. kasmira* [246], *L. stellatus* [247], *L. argentiventris* [248] and *L. argentimaculatus* [177]. However, other species do not spawn naturally in captivity, but hormonal induction has been used successfully, both hCG and GnRHa treatments. Red snapper with an oocyte size >440 μm were successfully spawned with 240–280 μg kg^{-1} GnRHa implants [161]. The mangrove red snapper, *Lutjanus argentimaculatus*, the principal species being produced, have been induced to spawn with a single injection of 100 μg kg^{-1} GnRHa or a single injection of 500 IU kg^{-1} of hCG when females had an oocyte size >400 μm [177].

9.3 Salmoniformes - Salmon, trout (Salmonidae)

Generally in captivity, the Salmonidae complete maturation through to ovi-postion and the gametes must be manually stripped and fertilized. In 2005 species from the family Salmonidae had an aquaculture production of 1,950,578 T, of which 1,235,972 T were Atlantic salmon (*Salmo salar*) and 486,928 T rainbow trout (*Oncorhynchus mykiss*) [6]. Most of the production originated from Norway (641,174 T), Chile (598,251 T), Scotland (142,613 T) and Canada (103,164 T). Atlantic salmon is gonochoristic and exhibits a plasticity of maturational strategies, which results in a proportion of individuals from a cohort maturing each year. This plasticity of maturation strategies depends on environmental and genetic influences that determine when maturation can proceed to completion [37,40,249]. The most adequate broodstock for aquaculture are the late maturing genetic strains with a minimum first maturation at 2 sea winters (3 kg) (3 years old, 1 year in fresh water + 2 years in sea water). Vitellogenesis last for approximately 10 months, beginning in the spring-summer months [250-251]. It is clear that the majority of the oocyte growth attributed to vitellogenesis takes place during spring to autumn under a spring-summer-autumn photoperiod (in Scotland LD 18:6 to 6:18) with appropriate thermocycle (in Scotland 2-14°C). The spawning period is during autumn-winter. Different genetic strains and different environmental conditions due to different latitudes result in different spawning periods, but the average peak of spawning in Scotland is in November [252]. Female broodstock spawn once a year. Optimal spawning conditions in Scotland are short daylength (6-8 h) and temperatures bellow 8°C. No ovulation and poor spermiation were observed at temperatures above 13°C [175,253] or 16°C [254]. GnRHa has been successfully used to induce ovulation and spermiation in fish maintained at 14-16°C [171]. Wild broodstock migrate from sea water to fresh water to spawn and generally culture practice is to transfer broodstock to fresh water for ovulation and stripping. A reduced proportion of females left in sea water were observed to ovulate [255]. GnRHa has been successfully used to induce ovulation in females left in sea water (personal observation). During courtship, in the natural spawning areas, the females excavates a reed (gravel nest) into which the eggs are spawned, fertilized and buried with gravel. Spawning is polygamous, a female spawns with a dominant male and precocious male sneakers. Eggs are 5-6 mm in diameter and demersal [256]. In culture, in the absence of substrate, the females will not release the eggs and gametes must be manually stripped and fertilized. In rainbow trout the optimum time to strip eggs was 4-6 days after ovulation [257]. Fecundities are approximately 2,000-4,000 eggs kg^{-1}. Hormone stimulation is not necessary to obtain ovulation or spermiation, but is a useful management tool to enhance ovulation or spermiation when conditions, such as temperature and salinity, are not correct (see above) or to synchronize and advance ovulation or spermiation. Sustained release of GnRHa using implants, microspheres or FIA-emulsion has been used to synchronize and advance ovulation in Atlantic salmon [156,175,258] and rainbow trout [259-260]. Successful doses were 20-50 µg kg^{-1} for rainbow trout [259-260] and 50 µg kg^{-1} for Atlantic salmon [175]. It was recommended to apply the sustained GnRHa delivery systems up to 6 weeks before the anticipated date of spawning, to obtain ovulation of 80-100% of the fish within 2 weeks of treatment [261-261]. Rainbow trout has been environmentally manipulated with altered photoperiods under constant temperatures to spawn all-year-round [41-42]. Although not as thoroughly studied, the Atlantic salmon appears to respond similarly to photoperiod manipulation [175,263-264]. It is important to ensure that optimal temperatures are provided during the displaced spawning period [42,175].

9.4 Gonorynchiformes - Milkfish (Chanidae)

Generally a large percentage of milkfish are cultured from wild juveniles, however, the proportion of hatchery produced juveniles is increasing (10% of production in the Philippines and the bulk of production in Indonesia) [265] and the majority of eggs for culture appear to be obtained from natural spawns in cages, ponds and tanks. In 2005 milkfish (*Chanos chanos*) had an aquaculture production of 594,783 T [6]; aquaculture production was not reported for any other species from the Chanidae family. Most of the production originated from the Philippines (289,153 T), Indonesia (254,067 T) and Taiwan (50,050 T). Milkfish are gonochoristic and mature at a size of >4 kg and 5 years, however, fish as old as 9 years are required for optimum egg quality [58]. Vitellogenesis or gonadal maturation occurs during February and March before the 6-7 month spawning season, the timing of which varies across the species geographic distribution. Temperatures of 26-29°C and 29-30°C were indicted for spawning and egg development [173-174,266]. Natural spawning has been achieved in cages and ponds with 1:1 sex ratios [58,266]. Spawning occurred in relation to the lunar cycle between midnight and 06.00 h and appeared to be polygamous with males chasing spawning females. The eggs were pelagic with a diameter of 1.1-1.25 mm. Fecundities have been estimated at 25×104 eggs kg^{-1}. Hormonal induction therapies have been developed, but due to difficulties related to handling large delicate broodstock, research has focused on natural spawning. Marte et al. [173-174] successfully induced spawning using either hCG or GnRHa. Spawning was induced with a single injection of 1,000 IU kg^{-1} of hCG and either an injection of 10-33 µg kg^{-1} of GnRHa or an implant of 19-36 µg kg^{-1}. The latency period between hormone administration and spawning was 16-49 h. Marte et al. [174] recommended an oocyte size of 730-780 µm for hormone induced spawning.

9.5 Siluriformes - Catfish, pangasius (Pangasiidae, Ictaluridae, Siluridae, Clariidae, Bagridae)

Generally in captivity, the catfish species will mature to the late stages of vitellogenesis and spermiation (particularly the families Ictaluridae, Siluridae, Clariidae) and can either be hormonally induced to spawn or left to spawn naturally in ponds, depending on the species. However, some species particularly in the Pangasiidae family must be hormonally induced to spawn. Dopamine inhibition of GnRH action has been demonstrated in the North African catfish (*Clarias gariepinus*) [267]. In 2005 species from the order Siluriformes had an aquaculture production of 1,468,357 T of which 440,611 T were from the family Pangasiidae, 382,112 T from Ictaluridae, 287,588 T from Siluridae, 264,723 T from Clariidae and 92,900 T from Bagridae [6]. The dominate species being produced were 440,611 T of pangas catfish (*Pangasius spp*), 379,707 T channel catfish (*Ictalurus punctatus*), 286,330 T amur catfish (*Silurus asotus*), 114,311 T of a catfish hybrid (cross between *Clarias gariepinus* and *C. macrocephalus*), 110,876 T of torpedo-shaped catfish (*Clarias spp.*), 84,565 T of yellow catfish (*Pelteobagrus fulvidraco*) and 28,746 T of North African catfish (*Clarias gariepinus*). Most of the production originated from China (478,004 T), Vietnam (376,000 T), USA (275,754 T), Thailand (130,784 T) and Indonesia (102,090 T). The channel catfish is gonochoristic and mature after 2-3 years at sizes greater than 1.5 kg, but for culture, broodstock of 3-4 years are recommended [268]. Vitellogenesis has been observed to begin in November and last for 6 months, before the spawning season in May-July [269-270]. Environmental conditions for vitellogenesis were a winter photoperiod and temperatures of 15-25°C [269]. The spawning season extended from May to July under the spring photoperiod and temperatures of 24-30°C [269] and spawning stopped at temperatures above 30°C. The channel catfish form pairs for spawning, the pair excavates a hole or nest where the egg mass is laid and the male guards the eggs during incubation. The females spawn once a year, with a fecundity of approximately 8,000,000 eggs kg^{-1}. The eggs are 3 mm in diameter, demersal and adhesive. Under culture conditions channel catfish spawn naturally in spawning ponds, utilizing 20-40 L spawning containers from which the egg masses can be collected [268]. Broodstock are stocked at 500-1,000 kg ha^{-1}, at a ratio of 1:1 to 3:5 males to females. Although hormonal induction is not needed and little used for channel catfish, studies have shown that successful spawning was possible with a latency period of 24-72 h after injection with respective priming and resolving doses of 2 and 9 mg kg^{-1} of cPE, a single injection of 1,000-1,760 IU kg^{-1} of hCG or respective priming and resolving doses of 10 and 90 µg kg^{-1} of GnRHa [178]. Hormone induction is used for the culture of the North African catfish and successful spawning was possible after injection with either 4,000 IU kg^{-1} of hCG or 50 µg kg^{-1} of GnRHa, combined with 500 µg kg^{-1} of pimozide [178-179]. The temperature dependant latency period was 12.5 and 16 h respectively, for hCG and GnRHa at 25°C. The inhibition of GnRH by dopamine has been described in the North African catfish, suggesting that a dopamine antagonist should be used with GnRHa [267]. The pangas catfish *spp* must be hormonally induced to spawn in captivity. *Pangasius bocourti* were observed to mature to advanced stages of vitellogenesis, but not to complete vitellogenesis. Cacot et al. [271] selected fish with a mean oocyte size of 1.1 mm and induced gametogenesis with a mean of 4 (range 1-10) daily injections of 500 IU kg^{-1} of hCG. This preparatory step induced the development to an average ovarian content of 52% of oocytes >1.6 mm and final maturation and ovulation was induced with either a single injection of 2,000-2,500 IU kg^{-1} of hCG or respective priming and resolving doses of 1,500 and 2,500 IU kg^{-1} of hCG. After ovulation, the eggs may be stripped and fertilized; in *Clarias macrocephalus* the optimum time for stripping was 10 h after ovulation [272]. Temperature has been used to both advance and delay spawning in the channel catfish. Hall et al. [269] used geothermal temperature control to heat pond water from February to April and advanced spawning by 2 months, compared to control fish at ambient temperatures. Similarly, Brauhn [273] obtained both natural and cPE induced spawns in August and November after maintaining channel catfish, from April, at 18°C for 109 days before increasing the temperature to 26°C over an 8 day period.

9.6 Anguilliformes - Eels (Anguilidae)

Generally eels are cultured from wild juveniles; eels do not mature in captivity. Studies have successfully induced gametogenesis and ovulation, but egg and larval quality has been poor and larval survival has not yet been sufficient for mass culture. In 2005 species from the family Anguilidae had an aquaculture production of 242,067 T of which 233,045 T were the Japanese eel (*Anguilla japonica*) and 8,329 T were the European eel (*Anguilla anguilla*) [6]. The highest producing countries were China (179,245 T), Taiwan (28,481 T), Japan (19,744 T), Republic of Korea (5,575 T) and the Netherlands (4,000 T). The life cycle of the European eel begins in the Sargasso sea where eels are believed to spawn, based on the presence of larval stages of eels, the larvae migrate with currents towards Europe and enter river systems as small juveniles [274]. In the river systems the eels grow to >40cm before leaving the rivers to migrate to the Sargasso sea. The eels prepare for migration by changing color, from yellow to silver, but at the time that the eels leave the river systems no or little gonadal development has taken place and GSI was 1-2%. The male eels leave the river systems in August with a size of 40 cm and females in September-October with a size of >50 cm. It was suggested that different swimming speeds resulted in the larger females meeting the smaller females in the spawning areas. It is not known what environment is experienced by migrating eels or what the environmental controls are for eel maturation. Despite many experiments to provide appropriate environmental stimulus for maturation, including temperature, light, salinity, pressure and swimming, little or no maturational development was observed [274]. However, both pressure (by submersion in a cage) and swimming did increase levels of maturational hormones and a slight increase in GSI. Maturational development and ovulation/spermiation have been hormonally induced in both the Japanese and European eel. The Japanese eel was first induced [275]. Males (200-300 g) were treated with weekly injections of 1 IU g^{-1} of hCG for 10-14 weeks to obtain spermiation. Females were treated with weekly injections of 20 mg fish^{-1} of salmon pituitary extract (sPE) until oocytes with migrating germinal vesicle were obtained (8-13 injections). A single injection of 17,20ß-dihydroxy-4-pregnen-3-one (DHP) (2 µg^{-1}) was given 24 h after the last injection of sPE (20 mg fish^{-1}) and most fish ovulated 15-18h later. Very similar protocols have been successfully applied to the European eel [276-277]. Observations indicated that the hormonally induced maturation was temperature dependant and the European eel takes about 20 days to mature at 25°C, 60 days at 15°C, but does not mature at temperatures below 10°C [274].

9.7 Mugiliformes - Mullets (Mugilidae)

Generally mullets are cultured from wild juveniles. Mullet do not spawn in captivity and only a variable proportion mature to the late stages of vitellogenesis. However, studies have successfully induced spawning and demonstrated a dopamine inhibition of GnRH. In 2005, species from the family Mugilidae had an aquaculture production of 167,946 T which was almost entirely flathead grey mullet (*Mugil Cephalus*) and unidentified mullet (*Mugilidae spp*) [6]. Most of the production originated from Egypt (156,441 T), Indonesia (11,668 T) and the Republic of Korea (5,501 T). Mullet are gonochoristic with males maturing at a size of 251-375 mm or 1-3 years and females at a size of 291-400 mm or 2-4 years [278]. The spawning season varies considerably across the species geographic distribution; in the Atlantic coast of the USA the spawning season is from October to April [279], in the Mediterranean from July to December [180] and in the Philippines peak spawning is from June to August [280]. Vitellogenesis appears to take place during the 2 months before the spawning season, from August to October [279], as temperatures and photoperiod begin to decrease. Mullets migrate from brackish lagoons to spawn in full strength sea water. Spawning appears to be polygamous and the eggs are pelagic with a diameter of 0.7-0.8 mm [180]. Fecundities have been estimated at 1×10^6 to 1.6×10^6 eggs kg^{-1} [58,180]. Natural spawning has not been achieved in captivity. A range of hormonal induction therapies have been developed [281]. It was recommended to use mullet with oocyte diameters >0.6 mm. Two therapies that were 100% successful included a priming dose of 4.3-8.4 µg g^{-1} of sGtH or 14.1-21.4 IU g^{-1} of hCG and 24 h later a resolving dose of 6-16.8 µg g^{-1} sGtH or 30.4-47.6 IU g^{-1} of hCG. The latency period was 10-15 h and percentage fertilization was 87-98%. Treatments using GnRHa were not as successful; doses of 101-455 µg kg^{-1} were 68% successful with fertilization of 0-95%. However, the poor efficacy of GnRHa has been shown to be due to the inhibitory effect of dopamine [180]. Aizen et al. [180] obtained induced spawns applying a priming dose of 10 µg kg^{-1} of GnRHa combined with 15 mg kg^{-1} of metaclopramide (MET, dopamine antagonist) followed 22.5 h later with a resolving dose of 20 µg kg^{-1} of GnRHa combined with 15 mg kg^{-1} of MET. The spawning success was 83% and the latency period was 21 h. Male spermiation was stimulated with EVAc slow release implants loaded with 4 mg kg^{-1} of 17á-methyltestosterone [180]. The timing of gonadal development has been altered through environmental manipulation. Generally, short photoperiod (LD 8:16) and cool temperatures (21°C) stimulated gonadal development and long photoperiod (LD 16:8) and high temperatures (31°C) inhibited development [49]. A short photoperiod (LD 8:16) with either low 21°C or high 31°C temperatures appeared to stimulate the cortical vesicle developmental stage, long or short photoperiod with low temperatures stimulated vitellogenesis and either long or short photoperiod with high temperatures induced atresia [49,281].

9.8 Tetraodontiformes - Puffer fish (Tetraodontidae)

Generally in captivity, the Tetraodontiformes will mature to the late stages of vitellogenesis and spermiation; hormonal treatment is used for spawning induction. In 2005 species from the family Tetraodontidae had an aquaculture production of 24,572 none of which was reported by species [6]. Most of the production originated from China (15,407 T) and Japan (4,582 T). The botete diana (*Sphoeroides annulatus*) is a species with good aquaculture potential for the Pacific coast of America. Botete diana is gonochoristic and mature at a size of 400-500 g (3 years) (personal observation). Vitellogenesis was observed to begin in March-April, 1-2 months before spawning [172]. Environmental conditions for vitellogenesis were a spring photoperiod (LD 12:12) and increasing temperatures (20-24°C). The spawning season extended from April to June with a peak in May under the increasing spring photoperiod (LD 12:12 to 14:10) and thermocycle (24-28°C). The botete appear to be group synchronous batch spawners from histology [172] and because a few individual fish in captivity have been spawned twice approx. a month apart in the same year (personal observation). Spawning appears to be polygamous and eggs are scattered into the environment. The eggs were demersal and adhesive with a diameter of 0.7 mm. Fecundities were 1×10^6 eggs kg^{-1} [172,282]. In captivity gametes should be stripped and artificially fertilized, as spawned eggs will adhere to most surfaces in the tank complicating egg collection. Few broodstock spawn naturally in captivity and hormone induction can be used to induce spawning. Duncan et al. [282] induced 82% of females to ovulate using GnRHa implants and injections, compared to 18% of control fish that spawned naturally. Females with oocyte sizes greater than 500 μm were implanted with 135±32 μg kg⁻¹ GnRHa (82% spawned) or injections of a priming dose of 20 μg kg⁻¹ and resolving doses of 40 μg kg⁻¹ (73% spawned), with a latency period of 16-40 h. Eggs were stripped and fertilized; fertilization rates were high (90-97%). Optimal time for stripping in tiger puffer (*Takifugu rubripes*) was close to ovulation (time 0), mean fertilization of eggs stripped within 4 h of ovulation was 70% [283]. To date, no environmental manipulation has been reported with the botete diana. However, spawning has been observed to start earlier when spring temperatures rise above 24°C early and the season was extended when temperatures did not rise above 28°C until late June early July (personal observation).

9.9 Acipenseriformes - Sturgeon (Acipenseridae)

Generally in captivity, the Acipenseriformes will mature to the late stages of vitellogenesis and spermiation and hormonally induced spawning has been applied. In 2005 species from the family Acipenseridae had an aquaculture production of 19,648 T, most of which was not reported by species. Most of the production originated from China (15,407 T), Russian Federation (2,470 T) and Italy (1,158 T) [6]. The sturgeons are gonochoristic and mature at a late age, the Siberian sturgeon (*Acipenser baeri*) matures after 10 years when the females reach a size of 7.9 kg and the males 5.3 kg [176]. The sturgeons are migratory fish which go to sea and return to the rivers to spawn. Spawning can be every year, every second year or even every third year and the most common strategy for the Siberian sturgeon was to spawn every two years. The spawning season for the Siberian sturgeon extended from April to May and optimal temperatures were 15°C for females and 12°C for males, but spawning could take place over the range 11-20°C [176]. It would appear that sturgeon spawn in captivity, but for management reasons ovulation is induced and males and females are maintained apart to ensure tank spawning does not take place. The eggs were demersal and adhesive with a diameter of 3-3.9 mm. Spawning has been induced using single injections of carp pituitary extract (5 mg kg⁻¹), sturgeon pituitary extract (2.5 mg kg⁻¹) and GnRHa (10 μg kg⁻¹) [176], over females having an oocyte diameter greater than 2.8 mm. Male Siberian sturgeon have been stimulated with a single injection of 2 mg kg⁻¹ of carp pituitary extract or 5 μg kg⁻¹ of GnRHa [176].

9.10 Pleuronectiformes - Flatfish (Bothidae, Paralichthidae, Scophthalmidae, Pleuonectidae)

The flatfishes normally undergo gametogenesis, ovulation and spermiation spontaneously in captivity, but egg spawning in females is often inhibited. Environmental manipulations are used to obtain all-year-round supply of eggs, whereas hormonal treatments are normally used to synchronize ovulations before stripping and to stimulate sperm production in males. In 2005, species from the order

Pleuronectiformes had an aquaculture production of 135,512 T of which 76,884 T were from the family Bothidae, 44,666 T from Paralichthyidae, 7,124 T from Pleuonectidae, and 6,838 T from Scophthalmidae [6]. Most of the production originated from China (82,560 T), Republic of Korea (40,075 T), Spain (5,572 T) and Japan (4,591 T). The dominate species being produced were 44,666 T of the Japanese flounder (*Paralichthys olivaceus*), 6,838 T of turbot (*Psetta maxima*) and 1,445 T of Atlantic halibut (*Hippoglossus hippoglossus*).

The reproductive biology of the most relevant aquaculture species, e.g., Japanese flounder, Atlantic halibut and turbot, has been studied for more than 20 years, which allowed the development of efficient broodstock management and reproductive control technologies and further establishment of a successful aquaculture industry. Environmental or hormonal manipulations are normally not necessary for egg production, but they have been developed and used to improve reproductive performance and alleviate some of the encountered reproductive problems. The major reproductive disorder found in females is the inhibition of spawning, after completion of oocyte maturation and ovulation. In males, the major reproductive problems are diminished sperm volume production and production of abnormally viscous milt, which has negative consequences for both natural and artificial fertilization of the eggs [284-285]. Small milt volumes (<0.3 ml) are difficult to collect and handle, mainly if the milt is highly viscous and also, it is difficult to avoid urine contamination, which can have deleterious effects on sperm quality and affect further fertilization rates. For a given species, these reproductive problems are found especially relevant in captive reared broodstocks in comparison to a more normal and efficient reproductive performance of wild broodstocks. For example, spontaneous tank-spawning of captive adapted wild broodstock has been described for Japanese flounder [286], turbot [287], Atlantic halibut [288] and sole [289], but further establishment of captive reared broodstock generations (F1 and successive) have shown that in most cases hand stripping of gametes and artificial fertilization is required.

Environmental manipulations are routinely used for out-of-season spawning. Normally, 12-month simulated natural photothermal cycles adequately shifted by several months in various broodstock batches is successfully used to have an all-year-round supply of gametes. Hormonal treatments are generally used to induce and synchronize ovulation in females and to enhance spermiation and milt fluidity in males. The most effective hormonal therapy for flatfishes is the single administration of GnRHa sustained release delivery systems, which has demonstrated to be highly effective in both females and males. The GnRHa implants are especially effective in female fish exhibiting group-synchronous multiple batch ovarian development, which is the case of most flatfishes [64].

From the family Paralichthyidae, several *Paralichthys spp* are of interest for aquaculture, e.g., the Japanese flounder in Asia and southern flounder (*Paralichthys lethostigma*) and summer flounder (*Paralichthys dentatus*) in North America. The Japanese flounder is a consolidated aquaculture species and hatchery and grow out technologies have been accomplished [286,290]. Females of this species can show spontaneous tank-spawning with adequate natural and artificial photothermal regimes, without hormonal treatment [286]. The southern and summer flounders are developing aquaculture species [291]; for both species, all-year-round egg production, usually after stripping, can be obtained from wild broodstock using adequate photothermal conditioning and when required, hormonal treatment with slow release GnRHa pellets (100 µg kg⁻¹) [162,292-293]. Female southern flounder have been induced to ovulate with either multiple hCG injections [294] or single treatment with GnRHa cholesterol pellets, at doses of 4, 20 and 100 µg kg⁻¹ [295]; in some cases, spontaneous tank-spawning has been described [171]. In the summer flounder, treatment with cPE injections, hCG injections and GnRHa implants were all effective in inducing ovulation in females, being the cPE the most effective in terms of number of ovulated females and egg fertilization rates [163].

From the family Pleuronectidae, the Atlantic halibut (*Hippoglossus hippoglossus*) is the most important aquaculture species, although some knowledge and interest exist on other species, such as, the starry flounder (*Platichthys stellatus*) and the yellowtail flounder (*Limanda ferruginea*). The starry flounder is a Pacific species and its reproduction in captivity is mainly based on artificial fertilization after stripping, which is sometimes limited by the males, because of diminished volume of abnormally viscous milt. Treatment of males with GnRHa cholesterol pellets (50, 100 or 200 µg kg⁻¹), during the spawning season, increase milt volume in a dose-dependent manner, mainly by increasing milt hydration [296-297]. The yellowtail flounder is a cold ocean flatfish with aquaculture potential for the Northwest Atlantic. It is a batch-spawner, with daily egg spawns during the summer-spring spawning period. Reproduction in captivity is based on artificial fertilization of the eggs after stripping. Treatment of females with GnRHa cholesterol pellets (224 µg kg⁻¹) or biodegradable microspheres (75 µg kg⁻¹) is effective in inducing multiple ovulations, causing increased egg production, increased egg quality, synchronization of females, advancement of spawning and shortening of inter-ovulatory periods [156,298]. Similar treatments applied on males, increased sperm production and milt volume, as well as sperm motility and seminal plasma pH, while having no negative effects on sperm fertilizing ability [156,285]. The Atlantic halibut is a consolidated aquaculture species, important for North Atlantic countries. Reproduction in captivity is mainly based on artificial fertilization after stripping. Major reproductive dysfunctions are poor egg quality and unpredictable timing of ovulation in females [299] and diminished sperm production and high viscosity of the milt in males, mainly towards the end of the spawning season [300]. Another problem is the desynchronization of the broodstock towards the end of the reproductive season, when females still produce ovulations of good quality eggs but spermiation of males is drastically reduced, which makes difficult the continuation of artificial fertilization trials. Environmental manipulations are used to obtain ovulation and spermiation throughout the year, by having broodstock batches under simulated 12-month photothermal regimes shifted in time several months; although successful, environmental manipulated broodstock may exhibit diminished egg and sperm quality as compared to natural broodstock [301]. Hormonal treatment of males with GnRHa implants, at doses of 5, 25, 30 and 50 µg kg^{-1}, advance the initiation of spermiation in 4 weeks and increase milt fluidity and sperm motility [284,302].

From the family Scophthalmidae, turbot (*Psetta maxima*) is the most important aquaculture species. It is naturally distributed in the Mediterranean and Atlantic coast of Europe and spawning occurs in spring-summer. Females are multiple batch spawners, each female producing up to 10 spawns per season. Simulated photothermal regimes are successfully used for all-year-round production of eggs [287,303]. Egg production in captivity is based on stripping and artificial fertilization. Egg stripping should be performed within 10 h after ovulation (at 12-13°C) to avoid over-ripening [304]. In captivity, turbot broodstock exhibit the common reproductive problems of other flatfishes, such as, inhibition of spawning, desynchronization of ovulation and spermiation, lack of ovulation in some female breeders and unpredictable timing of ovulation. Hormonal treatment with GnRHa pellets (25 µg kg^{-1}) is effective in inducing ovulation of 100% of the females, compared to around 50% ovulations in control broodstock and reduce by half the duration of the spawning period [164].

The family Soleidae includes several species of soles with potential aquaculture interest, mainly the common sole (*Solea solea*) and the Senegalese sole (*Solea senegalensis*). Their reproductive biology and culture techniques have been studied for more than 20 years, but their aquaculture industry has not been yet consolidated, mainly due to reproductive and pathological problems [305-307]. They are distributed in the Atlantic and Mediterranean coast of Europe, with the common sole having a more northern distribution than the Senegalese sole. The natural spawning period occurs during spring-summer, but depends on the latitude and water temperature [307]. Both species reproduce spontaneously in captivity and in contrast to other flatfishes, egg spawning is obtained spontaneously in the tank, without egg stripping [56]. Spawning is highly dependent on water temperatures. The optimal temperature range for spawning of common sole is 8-12°C [289], whereas the Senegalese sole spawn at temperatures of 15-20°C [306,308]. Environmental manipulations (photoperiod and temperature) are used for all-year-round egg production [289,307]. Hormonal treatments have been used to induce spawning in females of both species, mainly with the purpose of advancing and synchronizing the spawning time. Spawning of common sole have been successfully induced with GnRHa injections, at a dose of 10 µg kg^{-1} [309] and with hCG injections, at doses of 250, 500 and 1,000 IU kg^{-1} [310]. All studies on the natural and environment/hormone manipulated reproduction of sole in captivity, have shown that this is quite easily accomplished, with high productions of good quality eggs obtained through spontaneous tank spawning [306-307]. Nevertheless, almost all information is obtained from wild broodstock adapted to captivity and as mentioned before for other flatfishes, captive reared generations (F1 and successive) may display reproductive disorders that were not detected in wild broodstock. In fact, captive sole broodstock present important reproductive disorders, mainly the absence of spawning in females and diminished sperm production in the males, which has limited to date the development of a consolidated aquaculture industry for this species.

10. CONCLUSIONS

As mentioned in the introduction and made more evident from the previous sections, fish exhibit a great variety of reproductive strategies, which must be recognized, studied and taken into account when a species is brought into captivity to function as a broodstock for aquaculture production. Characteristics such as hermaphroditism or gonochorism, age-at-puberty, fecundity, internal or external fertilization, egg size and buoyancy, oviparity and ovo-viviparity, as well as parental care have important implications for finfish culture and broodstock management. Once the previous information is identified for each species of interest, the first prerequisite for the development of a sustainable aquaculture industry is the ability of the fish to undergo gametogenesis, maturation and spawning under captive conditions. It is very common for wild fish to fail to reproduce reliably when reared in captivity, but in some species this "dysfunction" is reduced or abolished with subsequent generations, as fish are inadvertently selected for the conditions prevailing in captive environment. However, other fishes never become fully "domesticated" and there is a need for environmental or pharmacological interventions in order to control reproductive processes and induce gamete maturation and spawning. This chapter gave a general description of the fish reproductive system, its endocrine control and the reproductive dysfunctions exhibited by captive-reared fishes, and described how this knowledge was used to develop treatments for the control of fish reproduction in aquaculture. The information presented is not exhaustive, and the reader was directed to important reviews available, but it was meant to provide the reader with the necessary background to evaluate the reproductive biology of a species of interest and to experiment with the development of reproductive control protocols.

11. REFERENCES

1-Nelson, R.J., *Fishes of the World*, 4th edition, John Wiley & Sons Inc, New York, 2006, 601.
2-Devlin, R.H. and Nagahama, Y., Sex determination and sex differentiation in fish: another view of genetic, physiological, and environmental influences, *Aquaculture*, 208, 191, 2002.
3-Kjesbu, O.S. and Whitthames, P.R., Evolutionary pressure on reproductive strategies in flatfish and groundfish: Relevant concepts and methodological advancements, *J Sea Res*, 58, 23, 2007.
4-Murua, H. and Soborido-Rey, R, Female reproductive strategies of marine fish species of the North Atlantic, *J Northwest Atl Fish Sci*, 33, 23, 2003.
5-FAO, *The State of World Fisheries and Aquaculture 2006*, Food and Agriculture Organization of the United Nations, Fisheries and Aquaculture Department, Rome, (www.fao.org/docrep/fao/009/a0699e/a0699e.pdf), 2007.
6-FAO, *Aquaculture production: quantities 1970-2005*, Fisheries Department, Fishery Information, Data and Statistics Unit, FISHSTAT Plus: Universal software for fishery statistical time series, Version 2.3. 2000, 2005.
7-Mellinger, J., *Sexualité et reproduction des poissons*, CNRS editions, Paris, 2002.
8-Jalabert, B., Particularities of reproduction and oogenesis in teleost fish compared to mammals, *Reprod Nutr Dev*, 45, 261, 2005.

9-Nagahama, Y., Endocrine regulation of gametogenesis in fish, *Int J Develop Biol*, 38, 217, 1994.

10-Wallace, R.A., Vitellogenesis and oocyte growth in nonmammalian vertebrales, in: *Developmental Biology. Vol. I: Oogenesis*, Browder, L.W., Ed., Plenum Press, New York, 1985, 127.

11-Wallace, R. and Selman, K., Cellular and dynamic aspects of oocyte growth in teleosts, *Am Zool*, 21, 325, 1981.

12-Wallace, R. and Selman, K., Ultrastructural aspects of oogenesis and oocyte growth in fish and amphibians, *J Electron Micr Tech*, 16, 175, 1990.

13-LaFleur, G.J., Vitellogenins and vitellogenesis, in: *Encyclopedia of Reproduction*, Vol. 4, Academic Press, San Diego, CA, 1999, 985.

14-Carnevali, O., Ciarletta, R., Cambi, A., Vita, A., Broamge, B., and Yola, N., Formation and degradation during oocyte maturation in seabream *Sparus aurata*: Involvement of two lysosomial proteinase, *Biol Reprod*, 60, 140, 1999.

15-Mañanós, E., Zanuy, S., Le Menn, F., Carrillo, M., and Núñez, J., Sea bass (*Dicentrarchus labrax* L.) vitellogenin. I: Induction, purification and partial characterization, *Comp Biochem Physiol*, 107B, 205, 1994.

16-Mañanós, E., Núñez, J., Zanuy, S., Carrillo, M., and Le Menn F., Sea bass (*Dicentrarchus labrax* L.) vitellogenin. II: Validation of an enzyme-linked immunosorbent assay (ELISA), *Comp Biochem Physiol*, 107B, 217, 1994.

17-Mañanós, E.L., Zanuy, S., and Carrillo, M., Photoperiodic manipulations of the reproductive cycle of sea bass (*Dicentrarchus labrax*) and their effects on gonadal development, and plasma 17â-estradiol and vitellogenin levels, *Fish Physiol Biochem*, 16, 211, 1997.

18-Guraya, S.S., *The cell and molecular biology of fish oogenesis*, Sauer, H.W., Ed., Karger, London, 1986.

19-Tyler, J.R. and Sumpter, J.P., Oocyte growth and development in teleosts, *Rev Fish Biol Fish*, 6, 287, 1996.

20-Selman, K. and Wallace, R.A., Gametogenesis in *Fundulus heteroclitus*, *Am Zool*, 24, 173, 1986.

21-Iwamatsu, T., Ohta, T., Oshima, E., and Sakai, N., Oogenesis in the medada *Oryzias latipes* - stages of oocyte development, *Zool Sci*, 5, 353, 1988.

22-Mayer, I., Shackley, S.E., and Ryland, J.S., Aspects of the reproductive biology of the bass, *Dicentrarchus labrax* L. I. An histological and histochemical study of oocyte development, *J Fish Biol*, 33, 609, 1988.

23-Grau, A., Crespo, S., Riera, F., Pou, S., and Sarasquete, M.C., Oogenesis in the amberjack *Seriola dumerili* Risso, 1810. An histological, histochemical and ultrastructural study of oocyte development, *Sci Mar*, 60, 391, 1996.

24-Nagahama, Y., Yoshikuni, M., Yamashita, M., and Tanaka, M., Regulation of oocyte maturation in fish, in: *Fish Physiology, vol. XIII: Molecular Endocrinology of Fish*, Sherwood, N.M. and Hew, C.L., Eds., Academic Press, San Diego, CA, 1994, 393.

25-Billard, R., Spermatogenesis and spermatology of some teleost fish species, *Reprod Nutr Develop*, 26, 877, 1986.

26-Schulz, R.W. and Miura, T., Spermatogenesis and its endocrinology, *Fish Physiol Biochem*, 26, 43, 2002.

27-Loir, M., Sourdaine, P., Mendis-Handagama, S.M.L.C., and Jegou, B., Cell-cell interactions in the testis of teleosts and elasmobranches, *Mic Res Tech*, 32(6), 533, 1995.

28-Miura, T. and Miura, C.I., Molecular control mechanisms of fish spermatogenesis, *Fish Physiol Biochem*, 28, 181, 2003.

29-Miura, T., Kasugai, T., Nagahama, Y., and Yamauchi, K., Acquisition of potential for sperm motility *in vitro* in Japanese eel *Anguilla japonica*, *Fish Sci*, 61, 533, 1995.

30-Babin, P.J., Cerdá, J., and Lubzens, E., *The fish oocyte: From Basic Studies to Biotechnological Applications*, Springer, The Netherlands, 2007, 508.

31-Zohar, Y., Harel, M., Hassin, S., and Tandler, A., Gilt-head sea bream (*Sparus aurata*), in *Broodstock Management and Egg and Larval Quality*, Bromage, N.R. and Roberts R.J., Eds., Blackwell Science, Oxford, UK, 1995, 94.

32-Pavlidis, M., Keravec, L., Greenwood, L., Mourot, B., and Scott, A.P., Reproductive performance of common dentex, *Dentex dentex*, broodstock held under different photoperiod and constant temperature conditions, *Fish Physiol Biochem*, 25, 171, 2001.

33-Mylonas, C.C., Papadaki, M., Pavlidis, M., and Divanach, P., Evaluation of egg production and quality in the Mediterranean red porgy (*Pagrus pagrus*) during two consecutive spawning seasons, *Aquaculture*, 232, 637, 2004.

34-Baker, R.H., The reaction of esters with aluminum isopropoxide, *J Am Chem Soc*, 60(11), 2673, 1938.

35-Bronson, F.H., Mammalian reproduction: An ecological perspective. *Biol Reprod*, 32, 1, 1985.

36-Sumpter, J.P., General concepts of seasonal reproduction, in: *Reproductive seasonality in teleosts: environmental influences*, Munro, A., Scott, A., and Lam, T., Eds., CRC Press, London, UK, 1990, 13.

37-Thorpe, J.E., Talbot, C., Miles, M.S., and Keay, D.S., Control of maturation in cultured Atlantic salmon, *Salmo salar*, in pumped seawater tanks, by restricting food intake, *Aquaculture*, 86, 315, 1990.

38-Cerda, J., Carrillo, M., Zanuy, S., and Ramos, J., Effect of food ration on estrogen and vitellogenin plasma levels, fecundity and larval survival in captive sea bass, *Dicentrarchus labrax*: preliminary observations, *Aquat Living Resour*, 7, 255, 1994.

39-Bromage, N., Broodstock management and seed quality - general considerations, in: *Broodstock Management and Egg and Larval Quality*, Bromage, N.R. and Roberts R.J., Eds., Blackwell Science, Oxford, UK, 1995, 1.

40-Thorpe, J.E., Age at first maturity in Atlantic salmon, *Salmo salar*: Freshwater period influences and conflicts with smelting, in *Salmonid age at maturity*, Meerburg, D.J., Ed., Can. Spec. Publ. Fish. Aquat. Sci., 89, 1986, 7.

41-Bromage, N R , Randall, C.R., Thrush, M., and Duston, J., The control of spawning in salmonids, in *Recent Advances in Aquaculture*, Roberts, R.J. and Muir, J., Eds., Vol. 4, Blackwell, Oxford, 1993, 55.

42-Bromage, N.R., Porter, M.J.R., and Randall, C.F., The environmental regulation of maturation in farmed finfish with special reference to the role of photoperiod and melatonin, *Aquaculture*, 197, 63, 2001.

43-Carrillo, M., Zanuy, S., Prat, F., Cerda, J., Ramos, J., Mañanos, E., and Bromage, N., Sea Bass (*Dicentrarchus labrax*), in: *Broodstock Management and Egg and Larval Quality*, Bromage, N.R. and Roberts R.J., Eds., Blackwell Science, Oxford, UK, 1995, 138.

44-Thomas, P., Arnold, C.R., and Holt, G.J., Red drum and other Sciaenids, in: *Broodstock Management and Egg and Larval Quality*, Bromage, N.R. and Roberts R.J., Eds., Blackwell Science, Oxford, UK, 1995, 118.

45-Davie, A., Porter, M.J.R., Bromage, N.R., and Migaud, H., The role of seasonally altering photoperiod in regulating physiology in Atlantic cod (*Gadus morhua*). Part I. Sexual maturation, *Can J Fish Aquat Sci*, 64, 84, 2007.

46-Smith, P., Bromage, N.R., Shields, R., Gamble, J., Gillespie, M., Dye, J., Young, C., and Bruce, M., Photoperiod controls spawning time in the Atlantic halibut (*Hippoglossus hippoglossus*), presented at *Proc IV Int Symp Reproductive Physiology of Fish,* Scott, A.P., Sumpter J.P., Kime D.E., and Rolfe M.S., Eds., Norwich, UK, July 7"12, 1991, 172.

47-Girin, M. and Devauchelle, N., Decalage de la periode de reproduction par raccourcissement des cycles photoperiodique et thermique chez des poissons marin, *Ann Biol Anim Biochim Biophys*, 18, 1059, 1978.

48-Campos-Mendoza, A., McAndrew, B.J., Coward, K., and Bromage, N., Reproductive response of Nile tilapia (*Oreochromis niloticus*) to photoperiodic manipulation; effects on spawning periodicity, fecundity and egg size, *Aquaculture*, 231, 299, 2004.

49-Kelly, C.D., Tamaru, C.S., Lee, C.S., Moriwake, A., and Miyamota, G., Effects of photoperiod and temperature on the annual ovarian cycle of the striped mullet, *Mugil cephalus*, in: *Reproductive Physiology of Fish*, Scott, A.P., Sumpter J.P, Kime D.E., and Rolfe M.S., Eds., Norwich, 1991, 142.

50-Sundararaj, B.I. and Sehgal, A., Effects of a long or an increasing photoperiod on the initiation of ovarian recrudescence during the preparatory period in the catfish *Heteropneusfes fossilis* (bloch), *Biol Reprod*, 2, 413, 1970.

51-Davies, P.R. and Hanyu, I., Effect of temperature and photoperiod on sexual maturation and spawning of the common carp. 1. Under conditions of high temperature, *Aquaculture*, 51, 277, 1986.

52-Davies, P.R., Hanyu, I., Furukawa, K,. and Nomura, M., Effect of temperature and photoperiod on sexual maturation and spawning of the common carp. 2. Under conditions of low temperature, *Aquaculture*, 51, 51, 1986.

53-Davies, P.R., Hanyu, I., Furukawa, K., and Nomura, M., Effect of temperature and photoperiod on sexual maturation and spawning of the common carp. 3. Induction of spawning by manipulating photoperiod and temperature, *Aquaculture*, 52, 137, 1986.

54-Bye, V.J., Temperate marine teleosts, in: *Reproductive seasonality in teleosts: environmental influences,* Munro, A., Scott., A., and Lam, T., Eds., CRC Press, London, 1990, 125.

55-Htun-Han M., The reproductive biology of the dab *Limanda limanda* (L.) in the North Sea: Seasonal changes in the ovary, *J Fish Biol*, 13, 351, 1978.

56-Baynes, S.M., Howell, B.R., and Beard, T.W, A review of egg production by captive sole, *Solea solea* (L.), *Aquac Fish Manag*, 24, 171, 1993.

57-Rothbard, S. and Yaron, Z., Carps (Cyprinidae), in: *Broodstock Management and Egg and Larval Quality*, Bromage, N.R. and Roberts R.J., Eds., Blackwell Science, Oxford, 1995, 321.

58-Beveridge, M.C.M. and Haylor, G.S., Warm-water farmed species, in *Biology of Farmed Fish*, Black, K.D., and Pickering, A.D., Eds., Academic Press, Sheffield, 1998, 415.

59-Parrish, R.H., Nelson, C.S., and Bakun, A., Transport mechanisms and reproductive success of fishes in the Californian current, *Biol Oceanog*, 1, 175, 1981.

60-Saavedra, M. and Pousao-Ferreira, P., A preliminary study on the effect of lunar cycles on the spawning behaviour of the gilt-head sea bream, *Sparus aurata, J Mar Biol Assoc UK*, 86, 899, 2006.

61-Trudeau, V., Neuroendocrine regulation of gonadotrophin II release and gonadal growth in the goldfish, *Carassius auratus, Rev Reprod*, 2, 55, 1997.

62-Peter, R.E. and Yu, K.L., Neuroendocrine regulation of ovulation in fishes: basic and applied aspects, *Rev Fish Biol Fish*, 7, 173, 1997.

63-Weltzien, F.-A., Andersson, E., Andersen, O., Shalchian-Tabrizi, K., and Norberg, B., The brain-pituitary-gonad axis in male teleosts, with special emphasis in flatfish (Pleuronectiformes), *Comp Biochem Physiol*, 137A, 447, 2004.

64-Mylonas, C.C. and Zohar, Y., Use of GnRHa-delivery systems for the control of reproduction in fish, *Rev Fish Biol Fish*, 10, 463, 2001.

65-Mylonas, C.C. and Zohar, Y., Promoting oocyte maturation, ovulation and spawning in farmed fish, in: *The Fish Oocyte: from Basic Studies to Biotechnological Applications*, Babin, P.J., Cerdá, J., and Lubzens, E., Eds., Kluwer Academic Publishers, 2007, 433.

66-Dubois, E.A., Zandberge, M.A., Peute, J., and Goos, H.J.Th., Evolutionary development of three gonadotropin-releasing hormone (GnRH) systems in vertebrates, *Brain Res Bull*, 57, 413, 2002.

67-Somoza, G.M., Miranda, L.A., Strobl-Mazzulla, P., and Guilgur, L.G., Gonadotropin-Releasing hormone (GnRH): from fish to mammalian brains, *Cell Mol Neurobiol*, 22, 589, 2002.

68-Yamamoto, N., Three gonadotropin-releasing hormone neuronal groups with special reference to teleosts, *Anatom Sci Int*, 78, 139, 2003.

69-Morgan, K. and Millar, R.P., Evolution of GnRH ligand precursor and GnRH receptors in protochordate and vertebrate species, *Gen Comp Endocrinol*, 139, 191, 2004.

70-Lethimonier, C., Madigou, T., Muñoz-Cueto, J.A., Lareyre, J.J., and Kah, O., Evolutionary aspects of GnRHs, GnRH neuronal systems and GnRH receptors in teleost fish, *Gen Comp Endocrinol*, 135, 1, 2004.

71-Millar, R.P., GnRHs and GnRH receptors, *An Reprod Sci*, 88, 5, 2005.

72-Ando, H. and Urano, A., Molecular regulation of gonadotropin secretion by gonadotropin-releasing hormone in salmonid fishes, *Zool Sci*, 22, 379, 2005.

73-Pawson, A.J. and McNeilly, A.S., The pituitary effects of GnRH, *Anim Reprod Sci*, 88, 75, 2005.

74-Clarke, I.J. and Pompolo, S., Synthesis and secretion of GnRH, *Anim Reprod Sci*, 88, 29, 2005.

75-Belsham, D.D. and Lovejoy, D.A., Gonadotropin-releasing hormone: gene evolution, expression, and regulation, *Vitam Horm*, 71, 59, 2005.

76-Guilgur, L.G., Moncaut, N.P., Canario, A.V.M., and Somoza, G.M., Evolution of GnRH ligands and receptors in gnathostomata, *Comp Biochem Physiol*, 144A, 272, 2006.

77-Kah, O., Lethimonier, C., Somoza, G., Guilgur, L.G., Vaillant, C., and Lareyre, J.J., GnRH and GnRH receptors in metazoan: A historical, comparative, and evolutive perspective, *Gen Comp Endocrinol*, 153, 346, 2007.

78-Matsuo, H., Baba, Y., Nair, R.M., Arimura, A., and Schally, A.V., Structure of the porcine LH- and FSH-releasing hormone. I. The proposed amino acid sequence, *Biochem Biophys Res Commun*, 43, 1334, 1971.

79-Burgus, R., Butcher, M., Amoss, M., Ling, N., Monahan, M., Rivier, J., Fellows, R., and Guillemin, R., Primary structure of the ovine hypothalamic luteinizing hormone releasing factor (LRF) (LH-hypothalamus-LRF-gas chromatography-mass spectrometry-decapeptide-Edman degradation), *Proc Natl Ac Sci USA*, 69(1), 278, 1972.

80-Adams, B.A., Vickers, E.D., Warby, C., Park, M., Fischer, W.H., Grey Craig, A., Rivier, J.E,. and Sherwood, N.M., Three forms of gonadotropin-releasing hormone, including a novel form, in a basal salmonid, *Coregonus clupeaformis*, *Biol Reprod*, 67, 232, 2002.

81-Iwakoshi, E., Takuwa-Kuroda, K., Fujisawa, Y., Iisada, M., Ukena, K., Tsutsui, K., and Minakata, H., Isolation and characterization of a GnRH-like peptide from Octopus vulgaris, *Biochem Biophys Res Comun*, 291, 1187, 2002.

82-Jiménez-Liñán, M., Rubin, B.S., and King, J.C., Examination of guinea pig luteinizing hormone-releasing hormone gene reveals a unique decapeptide and existence of two transcripts in the brain, *Endocrinol*, 138, 4123, 1997.

83-Miyamoto, K., Hasegawa, Y., Minegishi, T., Nomura, M., Takahashi, Y., Igarashi, M., Kangawa, K., and Matsuo, H., Isolation and characterization of chicken hypothalamic luteinizing hormone-releasing hormone, *Biochem Biophys Res Commun*, 107, 820, 1982.

84-Miyamoto, K., Hasegawa, Y., Nomura, M., Igarashi, M., Kangawa, K., and Matsuo, H., Identification of the second gonadotropin-releasing hormone in chicken hypothalamus: evidence that gonadotropin secretion is probably controlled by two distinct gonadotropin-releasing hormones in avian species, *Proc Natl Acad Sci USA*, 81, 3874, 1984.

85-Yoo, M.S., Kang, H.M., Choi, H.S., Kim, J.W., Troskie, B.E., Millar, R.P., and Kwon, H.B., Molecular cloning, distribution and pharmacological characterization of a novel gonadotropin-releasing hormone ([Trp8] GnRH) in frog brain, *Mol Cell Endocrinol*, 164, 197, 2000.

86-Sherwood, N., Eiden, L., Brownstein, M., Spiess, J., Rivier, J., and Vale, W., Characterization of a teleost gonadotropin-releasing hormone, *Proc Natl Acad Sci USA*, 80, 2794, 1983.

87-Ngamvongchon, S., Sherwood, N M , Warby, (M , and Rivier, J.E., Gonadotropin-releasing hormone from thai catfish: chromatographic and physiological studies, *Gen Comp Endocrinol*, 87, 266, 1992.
88-Powell, J.F., Zohar, Y., Elizur, A., Park, M., Fischer, W.H., Craig, A.G., Rivier, J.E., Lovejoy, D.A., and Sherwood, N.M., Three forms of gonadotropin-releasing hormone characterized from brains of one species, *Proc Natl Acad Sci USA*, 91, 12081, 1994.
89-Carolsfeld, J., Powell, J.F., Park, M., Fischer, W.H., Craig, A.G., Chang, J.P., Rivier, J.E., and Sherwood, N.M., Primary structure and function of three gonadotropin-releasing hormones, including a novel form, from an ancient teleost, herring, *Endocrinol*, 141, 505, 2000.
90-Okubo, K., Amano, M., Yoshiura, Y., Suetake, H., and Aida, K., A novel form of gonadotropin-releasing hormone in the medaka, *Oryzias latipes*, *Biochem Biophys Res Commun*, 276, 298, 2000.
91-Adams, B.A., Tello, J.A., Erchegyi, J., Warby, C., Hong, D.J., Akinsanya, K.O., Mackie, G.O., Vale, W., Rivier, J.E., and Sherwood, N.M., Six novel gonadotropin-releasing hormones are encoded as triplets on each of two genes in the protochordate, *Ciona intestinalis*, *Endocrinol*, 144, 1907, 2003.
92-Lovejoy, D.A., Fischer, W.H., Ngamvongchon, S., Craig, A.G., Nahorniak, C.S., Peter, R.E., Rivier, J.E., and Sherwood, N.M., Distinct sequence of gonadotropin-releasing hormone (GnRH) in dogfish brain provides insight into GnRH evolution, *Proc Natl Acad Sci USA*, 89, 6373, 1992.
93-Sherwood, N.M., Sower, S.A., Marshak, D.R., Fraser, B.A., and Brownstein, M.J., Primary structure of gonadotropin-releasing hormone from lamprey brain, *J Biol Chem*, 261, 4812, 1986.
94-Sower, S.A., Chiang, Y.C., Lovas, S., and Conlon, J.M., Primary structure and biological activity of a third gonadotropin-releasing hormone from lamprey brain, *Endocrinol*, 132, 1125, 1993.
95-Powell, J.F., Reska-Skinner, S.M., Prakash, M.O., Fischer, W.H., Park, M., Rivier, J.E., Craig, A.G., Mackie, G.O., and Sherwood, N.M., Two new forms of gonadotropin-releasing hormone in a protochordate and the evolutionary implications, *Proc Natl Acad Sci USA*, 93, 10461, 1996.
96-Montaner, A.D., Min Kyu Park, Fischer, W.H., Craig, A.G., Chang, J.P., Somoza, G.M., Rivier, J.E., and Sherwood, N.M., Primary structure of a novel gonadotropin-releasing hormone in the brain of a teleost, pejerrey, *Endocrinol*, 142(4), 1453, 2001.
97-White, R.B., Eisen, J.A., Kasten, T.L., and Fernald, R.D., Second gene for gonadotropin-releasing hormone in humans, *Proc Natl Acad Sci USA*, 95, 305, 1998.
98-Fernald, R.D. and White, R.B., Gonadotropin-releasing hormone genes: phylogeny, structure and functions, *Front Neuroendocrinol*, 20, 224, 1999.
99-Millar, R.P., GnRH II and type II GnRH receptors, *Trends Endocrinol Metab*, 14, 35, 2003.
100-Kauffman, A.S., Emerging functions of gonadotropin-releasing hormone II in mammalian physiology and behaviour, *J Neuroendocrinol*, 16, 794, 2004.
101-Rodríguez, L., Carrillo, M., Sorbera, L.A., Soubrier, M.A., Mañanós, E., Holland, M.C.H., Zohar, Y., and Zanuy, S., Pituitary levels of three forms of GnRH in the European male sea bass (*Dicentrarchus labrax*, L.) during sex differentiation and puberty, *Gen Comp Endocrinol*, 120, 67, 2000.
102-González, D., Zamora, N., Mañanós, E., Saligaut, D., Zanuy, S., Zohar, Y., Elizur, A., Kah, O., and Muñóz-Cueto, J.A., Immunohistochemical localization of three different prepro-GnRHs in the brain and pituitary of the European sea bass (*Dicentrarchus labrax*) using antibodies to the corresponding GnRH-associated peptides, *J Comp Neurol*, 446, 95, 2002.
103-Jodo, A., Ando, H., and Urano, A., Five different types of putative GnRH receptor gene are expressed in the brain of masu salmon (*Oncorhynchus masou*), *Zool Sci*, 20, 1117, 2003.
104-Dufour, S, Weltzien, F.A., Sebert, M.E., Le Belle, N., Vidal, B., Vernier, P., and Pasqualini, C., Dopaminergic inhibition of reproduction in teleost fishes: ecophysiological and evolutionary implications, *Ann NY Acad Sci*, 1040, 9, 2005.

105-Yaron, Z., Gur, G., Melamed, P., Rosenfeld, H., Levavi-Sivan, B., and Elizur, A., Regulation of gonadotropin subunit genes in tilapia, Comp Biochem Physiol, 129B, 489, 2001.

106-Sotowska-Brochocka, J., Marty, L., and Licht, P., Dopaminergic inhibition of gonadotropic release in hibernating frogs, Rana temporaria, Gen Comp Endocrinol, 93(2), 192, 1994.

107-Sharp, P. J., Talbot, R.T., and Macnamee, M.C., Evidence for the involvement of dopamine and 5-hydroxytryptamine in the regulation of the preovulatory release of luteinizing hormone in the domestic hen, Gen Comp Endocrinol, 76(2), 205-213, 1989.

108-Huseman, C.A., Kugler, J.A., and Schneider, I.G., Mechanism of dopaminergic suppression of gonadotropin secretion in men, J Clin Endocrinol Metab, 51(2), 209, 1980.

109-Peter, R.E., Lin, H.R., and Van Der Kraak, G., Induced ovulation and spawning of cultured freshwater fish in China: advances in application of GnRII analogues and dopamine antagonists, Aquaculture, 74, 1, 1988.

110-Kah, O., Dulka, J.G., Dubourg, P., Thibault, J., and Peter, RE., Neuroanatomical substrate for the inhibition of gonadotrophin secretion in goldfish: Existence of a dopaminergic preoptico-hypophyseal pathway, Neuroendocrinol, 45(6), 451, 1987.

111-Saligaut, C., Linard, B., Mañanós, E.L., Kah, O., Breton, B., and Govoroun, M. Release of pituitary gonadotrophins GtH I and GtH II in the rainbow trout (Oncorhynchus mykiss): modulation by estradiol and catecholamines, Gen Comp Endocrinol, 109(3), 302, 1998.

112-Vacher, C., Mañanós, E.L., Breton, B., Marmignon, M.II., and Saligaut, C., Modulation of pituitary dopamine D1 or D2 receptors and secretion of follicle stimulating hormone and luteinizing hormone during the annual reproductive cycle of female rainbow trout, J Neuroendocrinol, 12(12), 1219, 2000.

113-Vidal, B., Pasqualini, C., Le Belle, N., Holland, M.C.H., Sbaihi, M., Vernier, P., Zohar, Y., and Dufour, S., Dopamine inhibits luteinizing hormone synthesis and release in the juvenile European eel: A neuroendocrine lock for the onset of puberty, Biol Reprod, 71(5), 1491, 2004.

114-Nocillado, J.N., Levavi-Sivan, B., Carrick, F., and Elizur, A., Temporal expression of G-protein-coupled receptor 54 (GPR54), gonadotropin-releasing hormones (GnRH), and dopamine receptor D2 (drd2) in pubertal female grey mullet, Mugil cephalus, Gen Comp Endocrinol, 150(2), 278, 2007.

115-Trudeau, V.L., Spanswick, D., Fraser, E.J., Larivière, K., Crump, D., Chiu, S., MacMillan, M., and Schulz, R.W., The role of amino acid neurotransmitters in the regulation of pituitary gonadotropin release in fish, Biochem Cell Biol, 78(3), 241, 2000.

116-Kah, O., Trudeau, V.L., Sloley, B.D., Chang, J.P., Dubourg, P., Yu, K.L., and Peter, R.E., Influence of GABA on gonadotrophin release in the goldfish, Neuroendocrinol, 55(4), 396, 1992.

117-Anglade, I., Mazurais, D., Douard, V., Le Jossic-Corcos, C., Mañanos, E.L., Michel, M., and Kah, O., Distribution of glutamic acid decarboxylase mRNA in the forebrain of the rainbow trout as studied by in situ hybridization, J Comp Neurol, 410, 277, 1999.

118-Mañanós, E.L., Anglade, I., Chyb, J., Saligaut, C., Breton, B., and Kah, O., Involvement of ã-Aminobutyric Acid in the Control of GTH-1 and GTH-2 Secretion in Male and Female Rainbow Trout, Neuroendocrinol, 69, 269, 1999.

119-Pierce, J.G. and Parsons, T.F., Glycoprotein hormones: structure and function, Annu Rev Biochem, 50, 465, 1981.

120-Lustbader, J.W., Lobel, L., Wu, H., and Elliott, M.M., Structural and molecular studies of human chorionic gonadotropin and its receptor, Rec Prog Horm Res, 53, 395, 1998.

121-Hearn, M.T.W. and Gomme, P.T., Molecular architecture and biorecognition processes of the cystine knot protein superfamily: part I. The glycoprotein hormones, J Mol Recog, 13, 223, 2000.

122-Suzuki, K., Kawauchi, H., and Nagahama, Y., Isolation and characterization of two distinct gonadotropins from chum salmon pituitary glands, Gen Comp Endocrinol, 71(2), 292, 1988.

123-Quérat, B., Tonnerre-Doncarli, C., Genies, F., and Salmon, C., Duality of gonadotropins in gnathostomes, *Gen Comp Endocrinol*, 124, 308, 2001.

124-Swanson, P., Dickey, J.T., and Campbell, B., Biochemistry and physiology of fish gonadotropins, *Fish Physiol Biochem*, 28, 53, 2003.

125-Weil, C., Bougoussa-Houadec, M., Gallais, C., Itoh, S., Sekine, S., and Valotaire, Y., Preliminary evidence suggesting variations of GtH1 and GtH2 mRNA levels at different stages of gonadal development in rainbow trout, *Oncorhynchus mykiss*, *Gen Comp Endocrinol*, 100(3), 327, 1995.

126- Meiri, I., Knibb, W.R., Zohar, Y., and Elizur, A., Temporal profile of â follicle-stimulating hormone, â luteinizing hormone, and growth hormone gene expression in the protandrous hermaphrodite, gilthead seabream, *Sparus aurata*, *Gen Comp Endocrinol*, 137, 288, 2004.

127-Jackson, K., Goldberg, D., Ofir, M., Abraham, M., and Degani, G., Blue gourami (*Trichogaster trichopterus*) gonadotropic beta subunits (I and II) cDNA sequences and expression during oogenesis, *J Mol Endocrinol*, 23(2), 177, 1999.

128-Gen, K., Okuzawa, K., Senthilkumaran, B., Tanaka, H., Moriyama, S., and Kagawa, H., Unique expression of gonadotropin-I and -II subunit genes in male and female red seabream (*Pagrus major*) during sexual maturation, *Biol Reprod*, 63(1), 308, 2000.

129-Mateos, J., Mañanós, E., Martínez, G., Carrillo, M., Querat, B., and Zanuy, S., Molecular characterization of sea bass gonadotropin subunits (á, FSHâ and LH-â) and their expression during the reproductive cycle, *Gen Comp Endocrinol*, 133, 216, 2003.

130-Hellqvist, A., Schmitz, M., Mayer, I., and Borg, B., Seasonal changes in expression of LH-beta and FSH-beta in male and female three-spined stickleback, *Gasterosteus aculeatus*, *Gen Comp Endocrinol*, 145(3), 263, 2006.

131-Borg, B., Androgens in teleost fishes, *Comp Biochem Physiol*, 109C, 219, 1994.

132-Patiño, R., Yoshizaki, G., Thomas, P., and Kagawa, H., Gonadotropic control of ovarian follicle maturation: the two-stage concept and its mechanisms, *Comp Biochem Physiol*, 129B, 427, 2001.

133-Kah, O., Anglade, I., Linard, B., Pakdel, F., Salvert, G., Bailhache, T., Ducouret, B., Saligaut, C., Le Goff., P., Valotaire, Y., and Jégo, P., Estrogen receptors in the brain-pituitary comples and the neuroendocrine regulation of gonadotropin release in rainbow trout, *Fish Physiol Biochem*, 17, 53, 1997.

134-Yueh, W.S., Thomas, P., and Chang, C.F., Identification of 17,20beta,21-trihydroxy-4-pregnen-3-one as an oocyte maturation-inducing steroid in black porgy, *Acanthopagrus schlegeli*, *Gen Comp Endocrinol*, 140(3), 184, 2005.

135-Epping, J.J., Growth and development of mammalian oocytes *in vitro*, *Arch Pathol Lab Med*, 116, 379, 1992.

136-Schreck, C.B., Contreras-Sanchez, W., and Fitzpatrick, M.S., Effects of stress in fish reproduction, gamete quality, and progeny, *Aquaculture*, 197, 3, 2001.

137-Tucker, J.W., Spawning of captive serranid fishes: a review, *J World Aquac Soc*, 25, 345, 1994.

138-Yaron, Z., Endocrine control of gametogenesis and spawning induction in the carp, *Aquaculture*, 129, 49, 1995.

139-Zohar, Y. and Mylonas, C.C., Endocrine manipulations of spawning in cultured fish: from hormones to genes, *Aquaculture*, 197, 99, 2001.

140-Mylonas, C.C., Papandroulakis, N., Smboukis, A., Papadaki, M., and Divanach, P., Induction of spawning of cultured greater amberjack (*Seriola dumerili*) using GnRHa implants, *Aquaculture*, 237, 141, 2004.

141-Davis, B. and Bromage, N., The effects of fluctuating seasonal and constant water temperatures on the photoperiod advancement of reproduction in female rainbow trout, *Oncorhynchus mykiss*, *Aquaculture*, 205, 183, 2002.

142-Taranger, G.L., Stefansson, S.O., Oppedal, F., Andersson, E., Hansen, T., and Norberg, B., Photoperiod and temperature affects gonadal development and spawning time in: Atlantic salmon (*Salmo salar*), in *Reproductive Physiology of Fish*, Taranger, G.L., Norberg, B., Stefansson, S.O., Hansen, T., Kjesbu, O., and Andersson, E., Eds., Bergen, 1999, 345.

143-Donaldson, E.M., Manipulation of reproduction in farmed fish, *Anim Reprod Sci*, 42, 381, 1996.

144-Mylonas, C.C., Woods, L.C., III, Thomas, P., and Zohar, Y., Endocrine profiles of female striped bass (*Morone saxatilis*) during post-vitellogenesis, and induction of final oocyte maturation via controlled-release GnRHa-delivery systems, *Gen Comp Endocrinol*, 110, 276, 1998.

145-Mylonas, C.C., Scott, A.P., and Zohar, Y., Plasma gonadotropin II, sex steroids, and thyroid hormones in wild striped bass (*Morone saxatilis*) during spermiation and final oocyte maturation, *Gen Comp Endocrinol*, 108, 223, 1997.

146-Houssay, B.A., Accion sexual de la hipofisis en los peces y reptiles, *Revista de la Sociedad Argentina de Biología*, 106, 686, 1930.

147-Fontenele, O., Injecting pituitary (hypophyseal) hormones into fish to induce spawning, *Prog Fish Cult*, 18, 71, 1955.

148-Arabaci, M., Cagirgan, H., and Sari, M., Induction of ovulation in ornamental common carp (Koi, *Cyprinus carpio* L.) using LHRHa ([D-Ser(tBu)6, Pro9-NEt]-LHRH) combined with haloperidol and carp pituitary extract, *Aquac Res*, 35, 10, 2004.

149-Brzuska, E., Artificial spawning of carp (*Cyprinus carpio* L.); differences between the effects of reproduction in females of Hungarian, Polish and French origin treated with carp pituitary homogenate or [D-TLC6, Pro9LEt9] GnRH (Lecirellin), *Aquac Res*, 35, 1318, 2004.

150-Goren, A., Zohar, Y., Fridkin, M., Elhanati, E., and Koch, Y., Degradation of gonadotropin-releasing hormones in the gilthead seabream, *Sparus aurata* I. Cleavage of native salmon GnRH and mammalian LHRH in the pituitary, *Gen Comp Endocrinol*, 79, 291, 1990.

151-Zohar, Y., Pagelson, G., Gothilf, Y., Dickhoff, W.W., Swanson, P., Duguay, S., Gombotz,W., Kost, J., and Langer, R., Controlled release of gonadotropin releasing hormones for the manipulation of spawning in farmed fish, *Control Release Bioact Mater*, 17, 51, 1990.

152-Gothilf, Y. and Zohar, Y., Clearance of different forms of GnRH from the circulation of the gilthead seabream, *Sparus aurata*, in relation to their degradation and bioactivities, in: *Reproductive Physiology of Fish*, Scott, A.P., Sumpter, J.P, Kime, D.E., and Rolfe, M.S., Eds., Norwich, 1991, 35.

153-Crim, L.W., Sherwood, N.M., and Wilson, C.E., Sustained hormone release. II. Effectiveness of LHRH analog (LHRHa) administration by either single time injection or cholesterol pellet implantation on plasma gonadotropin levels in a bioassay model fish, the juvenile rainbow trout, *Aquaculture*,74, 87, 1988.

154-Zohar, Y., Gonadotropin releasing hormone in spawning induction in teleosts: basic and applied considerations, in: *Reproduction in Fish: Basic and Applied Aspects in Endocrinology and Genetics*, Zohar, Y., and Breton, B., Eds., INRA Press, Paris, 1988, 47.

155-Harmin, S.A. and Crim, L.W., Influence of gonadotropin hormone-releasing hormone analog (GnRH-A) on plasma sex steroid profiles and milt production in male winter flounder, *Pseudopleuronectes americanus* (Walbaum), *Fish Physiol Biochem*, 10, 399, 1993.

156-Mylonas, C.C., Tabata, Y., Langer, R., and Zohar, Y., Preparation and evaluation of polyanhydride microspheres containing gonadotropin-releasing hormone GnRH, for inducing ovulation and spermiation in fish, *J Control Release*, 35, 23, 1995.

157-Mylonas, C.C., Woods, L.C., III, Thomas, P., Schulz, R.W., and Zohar, Y., Hormone profiles of captive striped bass (*Morone saxatilis*) during spermiation, and long-term enhancement of milt production, *J World Aquac Soc*, 29, 379, 1998.

158-Mañanós, F., Carrillo, M., Sorbera, L.A., Mylonas, C.C., Asturiano, J.F., Bayarri, M.J., Zohar, Y., and Zanuy, S., Luteinizing hormone and sexual steroid plasma levels after treatment of European sea bass with sustained-release delivery systems for gonadotropin-releasing hormone analogue, *J Fish Biol*, 60, 328, 2002.

159-Marino, G., Panini, E., Longobardi, A., Mandich, A., Finoia, M.G., Zohar, Y., and Mylonas, C.C., Induction of ovulation in captive-reared dusky grouper, *Epinephelus marginatus* (Lowe, 1834) with a sustained-release GnRHa implant, *Aquaculture*, 219, 841, 2003.

160-Mylonas, C.C., Bridges, C.R., Gordin, H., Belmonte Ríos, A., García, A., De la Gándara, F., Fauvel, C., Suquet, M., Medina, A., Papadaki, M., Heinisch, G., De Metrio, G., Corriero, A., Vassallo-Agius, R., Guzmán, J.M., Mañanos, E., and Zohar, Y., Preparation and administration of gonadotropin-releasing hormone agonist (GnRHa) implants for the artificial control of reproductive maturation in captive-reared Atlantic bluefin tuna (*Thunnus thynnus thynnus*), *Rev Fish Sci*, 15, 183, 2007.

161-Ibarra-Castro, L. and Duncan, N.J., GnRHa-induced spawning of wild-caught spotted rose snapper *Lutjanus guttatus*, *Aquaculture*, 272, 737, 2007.

162-Berlinsky, D.L., King, V.W., Smith, T.I.J., Hamilton, R.D.II, Holloway, J.Jr., and Sullivan, C.V., Induced ovulation of southern flounder *Paralichthys lethostigma* using gonadotropin releasing hormone analogue implants, *J World Aquac Soc* 27, 143, 1996.

163-Berlinsky, D.L., King, W., Hodson, R.G., and Sullivan, C.V., Hormone induced spawning of summer flounder *Paralichthys dentatus*, *J World Aquac Soc*, 28, 79, 1997.

164-Mugnier, C., Guennoc, M., Lebegue, E., Fostier, A., and Breton, B., Induction and synchronisation of spawning in cultivated turbot (*Scophthalmus maximus* L.) broodstock by implantation of a sustained-release GnRH-a pellet, *Aquaculture*, 181, 241, 2000.

165-Garcia, L.M.B., Dose-dependent spawning response of mature female sea bass, *Lates calcarifer* (Bloch), to pelleted luteinizing hormone-releasing hormone analogue (LHRHa), *Aquaculture*, 77, 85, 1989.

166-Barbaro, A., Francescon, A., Bozzato, G., Merlin, A., Belvedere, P., and Colombo, L., Induction of spawning in gilthead seabream, *Sparus aurata* L., by long-acting GnRH agonist and its effects on egg quality and daily timing of spawning, *Aquaculture*, 154, 349, 1997.

167-Tucker, J.W., *Marine Fish Culture*, Kluwer Academic Publishers, Massachusetts, 1998, 750.

168-Mylonas, C.C., Irini Sigelaki, I., Divanach, P., Mananos, E., Carrillo, M., and Afonso-Polyviou, A., Multiple spawning and egg quality of individual European sea bass (*Dicentrarchus labrax*) females after repeated injections of GnRHa, *Aquaculture*, 221, 605, 2003.

169-Mylonas, C.C., Magnus, Y., Gissis, A., Klebanov, Y. and Zohar, Y., Application of controlled-release, GnRHa-delivery systems in commercial production of *white bass × striped bass hybrids* (sunshine bass), using captive broodstocks, *Aquaculture*, 140, 265, 1996.

170-Corriero, A., Medina, A., Mylonas, C.C., Abascal, F.J., Deflorio, M., Aragón, L., Bridges, C.R., Santamaría, N., Heinisch, G., Vassallo-Agius, R., Belmonte, A., Fauvel, C., Garcia, A., Gordón, H., and De Metrio, G., Histological study of the effects of treatment with gonadotropin-releasing hormone agonist (GnRHa) on the reproductive maturation of captive-reared Atlantic bluefin tuna (*Thunnus thynnus* L.), *Aquaculture*, 272, 675, 2007.

171-Smith, T.I.J., McVey, D.C., Jenkins, W.E., Denson, M.R., Heyward, L.D., Sullivan C.V., and Berlinsky, D.L., Broodstock management and spawning of southern flounder, *Paralichthys lethostigma*, *Aquaculture*, 176, 87, 1999.

172-Duncan, N.J., Garcia-Aguilar, N., Rodriguez-M. de O., G., Bernadet, M., Martinez-Chavez, C., Komar, C., Estoñol, P. and Garcia-Gasca, A., Reproductive biology of captive bullseye puffer (*Sphoeroides annulatus*), LHRHa induced spawning and egg quality, *Fish Physiol Biochem*, 28, 505, 2003.

173-Marte, C.L., Sherwood, N.M., Crim, L.W., and Harvey, B., Induced Spawning of Maturing Milkfish (*Chanos chanos* Forsskal) with Gonadotropin-Releasing Hormone (GnRH) Analogues Administered in Various Ways, *Aquaculture*, 60, 303, 1987.

174-Marte, C.L., Sherwood, N., Grim, L., and Tan, J., Induced spawning of maturing milkfish (*Chanos chanos*) using human chorionic gonadotropin and mammalian and salmon gonadotropin releasing hormone analogues, *Aquaculture*,73, 333, 1988.

175-Taranger, G.L., Vikingstad, E., Klenke, U., Mayer, I., Stefansson, S.O., Norberg, B., Hansen, T., Zohar, Y., and Andersson, E., Effects of photoperiod, temperature and GnRHa treatment on the reproductive physiology of Atlantic salmon (*Salmo salar* L.) broodstock, *Fish Physiol Biochem*, 28, 403, 2003.

176-Williot, P., Reproduction, in: *Esturgeons et Caviar,* Billard, R., Ed., Editions TEC and DOC, London, 2002, 63.

177-Emata, A.C., Reproductive performance in induced and spontaneous spawning of the mangrove red snapper, *Lutjanus argentimaculatus*: a potential candidate species for sustainable *Aquaculture*, *Aquac Res*, 34, 849, 2003.

178-Legendre, M., Linhart, O., and Billard, R., Spawning and management of gametes, fertilized eggs and embryos in Siluroidei, *Aquat Living Res*, 9, 59, 1996.

179-Richter, C.J.J., Eding, E.H., Verreth, J.A.J., and Fleuren, W.L.G., African catfish (*Clarias gariepinus*), in: *Broodstock Management and Egg and Larval Quality*, Bromage, N.R. and Roberts R.J., Eds., Blackwell Science, Oxford, 1995, 242.

180-Aizen, J., Meiri, I., Tzchori, I., Levavi-Sivan, B., and Rosenfeld, H., Enhancing spawning in the grey mullet (*Mugil cephalus*) by removal of dopaminergic inhibition, *Gen Comp Endocrinol*, 142, 212, 2005.

181-Mikolajczyk,T., Chyb, J., Szczerbik, P., Sokolowska-Mikolajczyk, M., Epler, P., Enright, W., Filipiak, M., and Breton, B., Evaluation of the potency of azagly-nafarelin (GnRH analogue), administered in combination with different formulations of pimozide, on LH secretion, ovulation and egg quality in common carp (*Cyprinus carpio* L.) under laboratory, commercial hatchery and natural conditions, *Aquaculture*, 234, 447, 2004.

182-FAO, Mass production of eggs and early fry. Part 1. Common Carp, Collection: Capacitation - No. 8, Rome, Italy, 1985.

183-Susuki, R., Duration of developmental capability of carp eggs in the overian cavity after ovulation, *Bull Nat Res Inst Aquac*, 1, 1, 1980.

184-MacIntosh, D.J. and Little, D.C., Nile tilapia (*Oreochromis niloticus*), in: *Broodstock Management and Egg and Larval Quality*, Bromage, N.R. and Roberts R.J., Eds., Blackwell Science, Oxford, 1995, 277.

185-Watanabe, W., Wicklund, R., Olla, B. and Head, W., Saltwater culture of the Florida red tilapia and another saline-tolerant tilapias: A review, in: *Tilapia Aquaculture in the Americas,* Costa-Pierce, B. and Rakocy, E., Eds., Vol. 1, The World Aquaculture Society, Baton Rouge, Louisiana, 1997, 54.

186-Bhujel, R.C., A review of strategies for the management of Nile tilapia (*Oreochromis niloticus*) broodfish in seed production systems, especially hapa-based systems, *Aquaculture*, 181, 37, 2000.

187-Tacon, P., Ndiaye, P., Cauty, C., Le Menn, F., and Jalabert, B., Relationships between the expression of maternal behaviour and ovarian development in the mouthbrooding cichlid fish *Oreochromis Niloticus, Aquaculture*, 146, 261, 1996.

188-Rana, K.J., Reproductive Biology and the hatchery rearing of tilapia eggs and fry, in: *Recent Advances in Aquaculture*, Muir ,J.F. and Roberts, R.J., Eds., Vol. 3,406. Croom Helm Ltd, London, 1990, 343.

189-Poleo, G.A., Lutz, C.G., Cheuk, G., and Tiersch, T.R., Fertilization by intracytoplasmic sperm injection in Nile tilapia (*Oreochromis niloticus*) eggs, *Aquaculture,* 250, 82, 2005.

190-Moretti, A., Pedini Fernandez-Criado, M., Cittolin, G., and Guidastri, R., *Manual on hatchery production of seabass and gilthead seabream*, Vol. 1, FAO, Rome, 1999, 194.

191-Mayer, I., Shackley, S.E., and Witthames, P.R., Aspects of the reproductive biology of the bass, *Dicentrarchus labrax* L. II. Fecundity and pattern of oocyte development, *J Fish Biol*, 36, 141, 1990.

192-Forniés, M.A., Mañanós, E., Carrillo, M., Rocha, A., Laureau, S., Mylonas, C.C., Zohar, Y., and Zanuy S., Spawning induction of individual European sea bass females (*Dicentrarchus labrax*) using different GnRHa-delivery systems, *Aquaculture*, 202, 221, 2001.

193-Alvariño, J.M.R., Zanuy, S., Prat, F., Carrillo, M., and Mañanós, E., Stimulation of ovulation and steroid secretion by LH-RHa injection in sea bass (*Dicentrarchus labrax*): effect of the time of the day, *Aquaculture*, 102, 177, 1992.

194-Alvariño, J.M.R., Carrillo, M., Zanuy, S., Prat, F., and Mañanós, E., Pattern of sea bass oocyte development after ovarian stimulation by LHRH, *J Fish Biol*, 41, 965, 1992.

195-Leong-Wee, K., Snakeheads – their biology and culture, in: *Recent Advances in Aquaculture*, Muir, J.F. and Roberts, R.J., Eds., 1982, 179.

196-Qin, J. and Fast, A.W., Size and feed dependent cannibalism with juvenile snakehead *Channa striatus*, Aquaculture, 144, 313, 1996.

197-Poortenaar, C.W., Hooker, S.H., and Sharp, N., Assessment of yellowtail kingfish (*Seriola lalandi lalandi*) reproductive physiology, as a basis for *Aquaculture* development, *Aquaculture*, 201, 271, 2001.

198-Micale, V., Maricchiolo, G., and Genovese, L., The reproductive biology of the amberjack, *Seriola dumerilii* (Risso 1810). I. Oocyte development in captivity, *Aquac Res*, 30, 349, 1999.

199-Pastor, E., Grau, A., Riera, F., Pou, S., Massuti, E., and Grau, A.M., Experiences in *The culture of new species in the 'Estacion de Acuicultura' of the Balearic Government (1980-1998)*, Basurco, B., Ed., Cahiers Options Méditerranéennes, Vol. 47, CIHEAM, Zaragoza, 2000, 371.

200-Laidley, C.W. and Shields, R.J., Domestication of greater amberjack (*Seriola dumerili*) at the Oceanic institute in Hawaii, presented at *Proc 7th Int Symp Reproductive Physiology of Fish*, Mie, Japan, 2003.

201-Marino, G., Mandich, A., Massari, A., Andaloro, F., and Porrello, S., Aspects of reproductive biology of the Mediterranean amberjack (*Seriola dumerilii* Risso) during the spawning period, *J App Icht*, 11, 9, 1995.

202-Mandich, A., Massari, A., Bottero, S., Pizzicori, P., Goos, H., and Marino, G., Plasma sex steroid and vitellogenin profiles during gonad development in wild Mediterranean amberjack (*Seriola dumerilii*, Risso), *Mar Biol*, 144, 127, 2003.

203-Jerez, S., Samper, M., Santamaría, F.J., Villamados, J.E., Cejas, J.R., and Felipe, B.C., Natural spawning of greater amberjack (*Seriola dumerili*) kept in captivity in the Canary Islands, *Aquaculture*, 252, 199, 2005.

204-Kagawa, H., Reproductive physiology and induced spawning in yellowtail (*Seriola quinqueradiata*), *Mar Ranch*, 18, 15, 1989.

205-Watanabe, T., Verakunpiriya, V., Mushiake, K., Kawano, K., and Hasegawa, I., The first spawn-taking from broodstock yellowtail cultured with extruded dry pellets, *Fish Sci*, 62, 388, 1996.

206- Lazzari, A., Fusari, A., Boglione, A., Marino, G., and Di Francesco, M., Recent advances in reproductional and rearing aspects of *Seriola dumerilii*, in *Mediterranean Marine Aquaculture Finfish Species Diversification*, Basurco, B., Ed., Cahiers Options Méditerranéennes, Vol. 47, CIHEAM, Zaragoza, 2000, 241.

207-Mushiake, K., Kawano, K., Sakamoto, W., and Hasegawa, I., Effects of extended daylength on ovarian maturation and HCG-induced spawning in yellowtail fed moist pellets, *Fish Sci*, 60, 647, 1994.

208-García-Gómez, A. and De la Gándara, F., Observations on the embryonic and larval development of Mediterranean yellowtail (*Seriola dumerilii*), in *Sea farming today and tomorrow*, Basurco, B. and Saroglia, M., Eds., Special Publication No. 32, The European Aquaculture Society, Belgium, 2003, 246.

209-Craig, S.R., MacKenzie, D.S., Jones, G., and Gatlin, D.M., Seasonal changes in the reproductive condition and body composition of free-ranging red drum, *Sciaenops ocellatus*, *Aquaculture*, 190, 89, 2000.

210-Wilson, C.A. and Nieland, D.L., Reproductive biology of red drum, *Sciaenops ocellatus* from the neritic waters of the northern Gulf of Mexico, *Fish Bull*, 92, 841, 1994.

211-Gold, J.R., Burridge, C.P., and Turner, T.F., A modified stepping-stone model of population structure in red drum, *Sciaenops ocellatus* (Sciaenidae), from the northern Gulf of Mexico, *Genetica*, 111, 305, 2001.

212-Lutz, C.G., Groupers: emerging species, A*q Magaz*, 27, 57, 2002.

213-Marino, G., Azzurro, E., Massari, A., Finoia, M.G., and Mandich, A., Reproduction in the dusky grouper from the southern Mediterranean, *J Fish Biol*, 58, 909, 2001.

214-Hassin, S., de Monbrison, D., Hanin, Y., Elizur, A., Zohar, Y., and Popper, D.M., Domestication of the white grouper, *Epinephelus aeneus* 1. growth and reproduction, *Aquaculture*, 156, 305, 1997.

215-Tucker, J.W., Woodward, P.N., and Sennett, D.G., Voluntary spawning of captive Nassau groupers *Epinephelus striatus* in a concrete raceway, *J World Aquac Soc*, 27, 373, 1996.

216-Okumura, S., Okamoto, K., Oomori, R., and Nakazono, A., Spawning behavior and artificial fertilization in captive reared red spotted grouper, *Epinephelus akaara*, *Aquaculture*, 206, 165, 2002.

217-Bhandari, R.K., Hika, M., Nakamura, S., and Nakamura, M., Aromatase inhibitor induces complete sex change in the protogynous honeycomb grouper (*Epinephelus merra*), *Mol Reprod Develop*, 67, 303, 2004.

218-Shein, N.L., Chuda, H., Arakawa, T., Mizuno, K., and Soyano, K., Ovarian development and final oocyte maturation in cultured sevenband grouper, *Fish Sci*, 70, 360, 2004.

219-Abu-Hakima, R., Aspects of the reproductive biology of the grouper, *Epinephelus tauvina* (Forskal), in Kuwaiti waters, *J Fish Biol*, 30, 213, 1987.

220-Zabala, M., Garcia-Rubies, A., Louisy, P., and Sala, E., Spawning behaviour of the Mediterranean dusky grouper *Epinephelus marginatus* (Lowe, 1834) (Pisces, Serranidae) in the Medes Islands Marine Reserve (NW Mediterranean, Spain), *Sci Mar*, 61, 65, 1997.

221-James, C.M., Al-Thobaiti, S.A., Rasem, B.M., and Carlos, M.H., Breeding and larval rearing of the camouflage grouper *Epinephelus polyphekadion* (Bleeker) in the hypersaline waters of the Red Sea coast of Saudi Arabia, *Aquac Res*, 28, 671, 1997.

222-Spedicato, M.T. and Boglione, C., Main constraints in the artificial propagation of the dusky grouper *Epinephelus marginatus* (Lowe, 1834): three years experimental trials on induced spawning and larval rearing, in *Mediterranean Marine Aquaculture Finfish Species Diversification*, Basurco, B., Ed., Cahiers Options Méditerranéennes, Vol. 47, CIHEAM, Zaragoza, 2000, 227.

223-Tucker, J.W., Spawning of captive serranid fishes: a review, *J World Aquac Soc*, 25, 345, 1994.

224-Okumura, S., Okamoto, K., Oomori, R., H., S., and Nakazono, A., Improved fertilization rates by using a large volume spawning tank in red spotted grouper (*Epinephelus akaara*), *Fish Physiol Biochem*, 28, 515, 2003.

225-Aquilar-Perera, A. and Aquilar-Dávila, W., A spawning aggregation of Nassau grouper *Epinephelus striatus* (Pisces: Serranidae) in the Mexican Caribbean, *Environ Biol Fish*, 45, 351, 1996.

226-Zabala, M., Louisy, P., Garcia-Rubies, A., and Gracia, V., Socio-behavioural context of reproduction in the Mediterranean dusky grouper *Epinephelus marginatus* (Lowe, 1834) (Pisces, Serranidae) in the Medes Islands Marine Reserve (NW Mediterranean, Spain), *Sci Mar*, 61, 79, 1997.

227 Marino, G., Azzurro, E., Finoia, M.G., Messina, M.T., Massari, A., and Mandich, A., Recent advances in induced breeding of the dusky grouper *Epinephelus marginatus* (Lowe, 1834), in: *Mediterranean Marine Aquaculture Finfish Species Diversification*, Basurco, B., Ed., Cahiers Options Méditerranéennes, Vol. 47, CIHEAM, Zaragoza, 2000, 215.

228-Watanabe, W.O., Ellis, S.C., Ellis, E.P., Head, W.D., Kelley, C.D., Moriwake, A., Lee, C.S., and Bienfang, P.K., Progress in controlled breeding of Nassau grouper (*Epinephelus striatus*) broodstock by hormone induction, *Aquaculture*, 138, 205, 1995.

229-Tamaru, C.S., Carlstrom Trick, C., FitzGerald, W.J., and Ako, H., Induced final maturation and spawning of the marbled grouper *Epinephelus microdon* captured from spawning aggregations in the Republic of Palau, Micronesia, *J World Aquac Soc*, 27, 363, 1996.

230-Liao, I.C., Su, H.M., and Chang, E.Y., Techniques in finfish larviculture in Taiwan, *Aquaculture*, 200, 1, 2001.

231-Liao, I.C., Huang, T.S., Tsai, W.S., Hsueh, C.M., Chang, S.L., and Leaño, E.M., Cobia culture in Taiwan: current status and problems, *Aquaculture*, 237, 155, 2004.

232-Ottolenghi, F., Silvestri, C., Giordano, P., Lovatelli, A., and New, M.B., The fattening of eels, groupers, tunas and yellowtails, in *Capture-based aquaculture*, FAO, Rome, 2004, 308 pp.

233-Doumenge, F., L'Aquaculture des thons rouges, *Biol Mar Med*, 3, 258, 1996.

234-Miyake, P.M., De la Serna, J.M., Di Natale, A., Farrugia, A., Katavic, I., Miyabe, N., and Ticina, V., General view of bluefin tuna farming in the Mediterranean area, *ICCAT Collective Volume of Scientific Papers*, 55, 114, 2003.

235-FAO, Report of the third meeting of the Ad Hoc GFCM/ICCAT working group on sustainable bluefin tuna farming/fattening practices in the Mediterranean, in *FAO Fisheries Reports*, 779, FAO, March 16-18, Rome, 2005, 108.

236-Directorate General for Fisheries, E.U., Tuna: a global fishing activity, Fishing in Europe, 23, 1, 2004.

237-Schaefer, K.M., Reproductive biology of tunas, in: *Tuna: Physiology, Ecology, and Evolution*, Block, B.A. and Stevens, E.D., Eds., Vol. 19, Academic Press, San Diego, 2001, 225.

238-Lioka, C., Kani, K., and Nhhala, H., Present status and prospects of technical development of tuna sea-farming, in: *Mediterranean Marine Aquaculture Finfish Species Diversification*, Basurco, B., Ed., Cahiers Options Méditerranéennes, Vol. 47, CIHEAM, Zaragoza, 2000, 275.

239-Sawada, Y., Okada, T., Miyashita, S., Murata, O., and Kumai, H., Completion of the Pacific bluefin tuna *Thunnus orientalis* (Temmich et Schlegel) life cycle, *Aquac Res*, 36, 413, 2005.

240-Masuma, S., Maturation and spawning of bluefin tuna in captivity, presented at *Kinki Univ Int Symp Ecology and Aquaculture of Bluefin Tuna*, Amami Ohshima, November 10-11, 2006, 27.

241-Wexler, J.B., Scholey, V.P., Olson, R.J., Margulies, D., Nakazawa, A., and Suter, J.M., Tank culture of yellowfin tuna, *Thunnus albacares*: developing a spawning population for research purposes, *Aquaculture*, 220, 327, 2003.

242-Margulies, D., Suter, J.M., Hunt, R., Olson, R.J., Scholey, V.P., Wexler, J.B., and Nakazawa, A., Spawning and early development of captive yellowfin tuna *Thunnus albacares*, *Fish Bull*, 105, 249, 2007.

243-Grimes, C.B., Reproductive biology of the Lutjanidae: a review, in: *Tropical snappers and groupers. Biology and fisheries management*, Polovina, J.J. and Ralston, S., Eds., Westview Press, Boulder, 1987, 239.

244-Arellano-Martinez, M., Rojas-Herrera, A., Garcia-Domínguez, F., Caballos-Vazquez, B.P., and Villarejo-Fuerte, M., Ciclo reproductivo del pargo lunarejo *Lutjanus guttatus* (Steindachner, 1869) en las costas de Guerrero, Mexico, *Rev Biol Mar Oceanog*, 36(1), 1, 2001.

245-Arnold, C.R., Wakeman, J.M., Williams, T.D., and Treece, G.D., Spawning of red snapper (*Lutjanus campechanus*) in captivity, *Aquaculture*, 15, 301, 1978.

246-Suzuki, K. and Hioki, S., Spawning behavior, eggs and larvae of the lutjanid fish, *Lutjanus kasmira*, in an aquarium, *Japan J Ichtyol*, 26(2), 161, 1979.

247-Hamamoto, S., Kumagai, S., Nosaka, K., Manabe, S., Kasuga, A., and Iwatsuki, Y., Reproductive behaviour, eggs and larvae of a lutjanid fish, *Lutjanus stellatus*, observed in an aquarium, *Jap J Ichth*, 39, 219, 1992.

248-Muhlia-Melo, A., Guerrero-Tortolero, D.A., Perez-Urbiola, J.C., and Campos-Ramos, R., Results of spontaneous spawning of yellow snapper (*Lutjanus argentiventris* Peters, 1869) reared in inland ponds in La Paz, Baja California Sur, Mexico, *Fish Physiol Biochem*, 28, 511, 2003.

249-Herbinger, C.M. and Newkirk, G.F., Sources of family variability for maturation incidence in cultivated Atlantic salmon (*Salmo salar*), *Aquaculture*, 85, 153, 1990.

250-Bromage, N. and Cumaranatunga R., Egg Production in the rainbow trout, in *Recent Advances in Aquaculture*, Vol. 3, Muir, J.F. and Roberts, R.J., Eds., 1988, 64.

251-Scott, A.P., Salmonids, in: *Reproductive seasonality in teleosts: environmental influences*, Munro, A., Scott., A., and Lam, T., Eds., CRC Press, London, 1990, 323.

252-Duncan, N.J., Thrush, M.A., Elliott, J.A.K., and Bromage, N.R., Seawater growth and maturation of Atlantic salmon (*Salmo salar*) transferred to sea at different times during the year, *Aquaculture*, 213, 293, 2002.

253-Taranger, G.L. and Hansen, T., Ovulation and egg survival following exposure of Atlantic salmon, *Salmo salar* L., broodstock to different water temperatures, *Aquac Fish Manag*, 24, 191, 1993.

254-King, H. and Pankhurst, N.W., Ovulation of Tasmanian Atlantic salmon maintained at elevated temperatures: implications of climate changes for sustainable industry development, in *Reproductive Physiology of Fish*, Taranger, G.L., Norberg, B., Stefansson, S., Hansen, T., Kjesbu, O., and Andersson, E., Eds., Bergen, 2000, 396.

255-Magwood, S.J., Bromage, N., Duncan, N.J., and Porter, M., The influence of salinity on reproductive success in female Atlantic salmon (*Salmo salar*) grilse, in *Reproductive Physiology of Fish*, Taranger, G.L., Norberg, B., Stefansson, S., Hansen, T., Kjesbu, O., and Andersson, E., Eds., Bergen 2000, 346.

256-Thorpe, J.E., Miles, M.S., and Keay, D.S., Developmental rate, fecundity and egg size in Atlantic salmon, *Salmo salar* L., *Aquaculture*, 43, 289, 1984.

257-Springate, J., Bromage, N., Elliot, J.A.K., and Hudson, D.L., The timing of ovulation and stripping and the effects on the rates of fertilisation and survival to eying, hatch and swim-up in the rainbow trout (*Salmo gairdneri*, L.), *Aquaculture*, 43, 313, 1984.

258 Crim, L.W. and Glebe, B.D., Advancement and Synchrony of ovulation in Atlantic salmon with pelleted LHRH analog, *Aquaculture*, 43, 47, 1984.

259-Breton, B.,Weil, C., Sambroni, E.,and Zohar, Y.,. Effects of acute versus sustained administration of GnRHa on GtH release and ovulation in the rainbow trout, Oncorhynchus mykiss. *Aquaculture* 91, 373, 1990.

260-Arabaci, M., Diler, I., and Sari, M., Induction and synchronisation of ovulation in rainbow trout, *Oncorhynchus mykiss*, by administration of emulsified buserelin (GnRHa) and its effects on egg quality, *Aquaculture*, 237, 475, 2004.

261-Zohar, Y., Goren, A., Fridkin, M., Elhanati, E., and Koch, Y., Degradation of gonadotropin-releasing hormones in the gilthead seabream *Sparus aurata* II. Cleavage of native salmon GnRH, mammalian LHRH and their analogs in the pituitary, kidney and liver, *Gen Comp Endocrinol*, 79, 306, 1990.

262-Goren, A., Gustafson, H., and Doering, D., Field trials demonstrate the efficacy and commercial benefit of a GnRHa implant to control ovulation and spermiation in salmonids, in *Reproductive Physiology of Fish*, Goetz, F.W. and Thomas, P. Eds., Austin, 1995, 99.

263-Taranger, G.L., Haux, C., Stefansson, S.O., Bjornsson, B.Th., Walther, B.Th., and Hansen, T., Abrupt changes in photoperiod affect age at maturity, timing of ovulation and plasma testosterone and oestradiol-17b profiles in Atlantic salmon, *Salmo salar, Aquaculture*, 162, 85, 1998.

264-Taranger, G.L., Haux, C., Hansen, T., Stefansson, S.O., Bjornsson, B.J., Walther, B.Th., and Kryvi, H., Mechanisms underlying photoperiodic effects on age at sexual maturity in Atlantic salmon, *Salmo salar*, *Aquaculture*, 177, 47, 1999.

265-Marte, C.L., Larviculture of marine species in Southeast Asia: current research and industry prospects, *Aquaculture*, 227, 293, 2003.

266-SEAFDEC, Milkfish breeding and hatchery fry production, SEAFDEC Aquaculture Department, Tigbauan, Iloilo, Philippines, 6, 1999.

267-Goos, H.J.T. and Kichter, C.J.J., Internal and external factors controlling reproduction in the African catfish, *Clarias gariepinus*, *Aquat Living Resour*, 9, 45, 1996.

268-Dupree, H., Channel catfish (*Ictalurus punctatus*), in: *Broodstock Management and Egg and Larval Quality*, Bromage, N.R. and Roberts R.J., Eds., Blackwell Science, Oxford, 1995, 220.

269-Hall S.G., Finney, J., Lang, R.P., and Tiersch, T.R., Design and development of a geothermal temperature control system for broodstock management of channel catfish *Ictalurus punctatus*, *Aquac Eng*, 26, 277, 2002.

270-Barrero, M., Small, B.C., D'Abramo, L.R., Hanson, L.A., and Nelly, A.M., Comparison of estradiol, testosterone, vitellogenin and cathepsin profiles among young adult channel catfish (*Ictalurus punctatus*) females from four selectively bred strains, *Aquaculture*, 264, 390, 2007.

271-Cacot, P., Legendre, M., Dan, T.Q., Tung, L.T., Liem, P.T., Mariojouls, C., and Lazard, J., Induced ovulation of *Pangasius bocourti* (Sauvage, 1880) with a progressive hCG treatment, *Aquaculture*, 213, 199, 2002.

272-Mollah, M.F.A. and Tan, E.S.P., Viability of catfish (*Clarias macrocephallus*, Gunther) eggs fertilizad at varying post-ovultation times, *J Fish Biol*, 22, 563, 1983.

273-Brauhn, J.L., Fall spawning of channel catfish, *Prog Fish Cult*, 33, 150, 1971.

274-van Ginnekin, V., Vianen, G., Muusze, B., Palstra, A., Verschoor, L.V., Lugten, O., Onderwater, M., van Schie, S., Niemantsverdriet, P., van Heeswijk, R., Eding, E.P., and van den Thillart, G., Gonad development and spawning behaviour of artificially-matured European eel (*Anguilla anguilla* L.), *Anim Biol*, 55, 203, 2005.

275-Ohta, H., Kagawa, H., Tanaka, H., Okuzawa, K., Iinuma, N., and Hirose, K., Artificial induction of maturation and fertilization in the Japanese eel, *Anguilla japonica*, *Fish Physiol Biochem*, 17, 163, 1997.

276-Pedersen, B.H., Induced sexual maturation of the European eel *Anguilla anguilla* and fertilisation of the eggs, *Aquaculture*, 224, 323, 2003.

277-van Ginneken, V.J.T. and Maes, G.E., The European eel (*Anguilla anguilla*, Linnaeus), its lifecycle, evolution and reproduction: a literature review, *Rev Fish Biol Fish*, 15, 367, 2005.

278-McDonough, C.J., Roumillat, W.A., and Wenneret C.A., Fecundity and spawning season of striped mullet (*Mugil cephalus* L.) in South Carolina estuaries, *Fish Bull*, 101, 822, 2005.

279-McDonough, C.J., Roumillat, W.A., and Wenneret C.A., Sexual differentiation and gonad development in striped mullet (*Mugil cephalus* L.) from South Carolina estuaries, *Fish Bull*, 103, 601, 2003.

280-Tamaru, C.S., Lee, C.S., Kelley, C.D., and Banno, J.E., Effectiveness of chronic LHRH-analogue and 17á-methyltestosterone therapy, administered at different times prior to the spawning season, on the maturation of milkfish (*Chanos chanos*), *Aquaculture*, 70, 159, 1988.

281-Kuo, C.M., Manipulation of ovarian development and spawning in grey mullet, *Mugil cephalus* L, *Israeli J Aquac*, 47, 43, 1995.

282-Duncan, N.J., Rodriguez M. de O., G.A., Alok, D. and Zohar, Y., Effects of controlled delivery and acute injections of LHRHa on bullseye puffer fish (*Sphoeroides annulatus*) spawning, *Aquaculture*, 218, 625, 2003.

283-Matsuyama, M. and Chuda, H., Hormone induced propagation of tiger puffer (*Takifugu rubripes*), in *Reproductive Physiology of Fish*, Taranger, G.L., Norberg, B., Stefansson, S., Hansen, T., Kjesbu, O., and Andersson, E., Eds., Bergen, 2000, 431.

284-Vermeirssen, E.L.M., Shields, R.J., Mazorra de Quero, C., and Scott, A.P., Gonadotrophin-releasing hormone agonist raises plasma concentrations of progestogens and enhances milt fluidity in male Atlantic halibut (*Hippoglossus hippoglossus*), Fish Physiol Biochem, 22, 77, 2000.

285-Clearwater, S.J. and Crim, L.W., Gonadotropin releasing hormone-analogue treatment increases sperm motility, seminal plasma pH and sperm production in yellowtail flounder *Pleuronectes ferrugineus*, Fish Physiol Biochem, 19, 349, 1998.

286-Tsujigado, A., Yamakawa, T., Matsuda, H., and Kamiya, N., Advanced spawning of the flounder, *Paralichthys olivaceus*, in an indoor tank with combined manipulation of water temperatura and photoperiod, Int J Aquac Fish Technol, 1, 351, 1989.

287-Devauchelle, N., Alexandre, J.C., Le Corre, N., and Letty, Y., Spawning of turbot (*Scophthalmus maximus*) in captivity, Aquaculture, 69, 159, 1988.

288-Holmefjord, I., Gulbrandsen, J., Lein, I. Refstie, T., Leger, P., Harboe, T., Huse, I., Sorgeloos, P., Bolla, S., Olsen Y., Reitan, K., Vadstein, O., and Oie, G., An intensive approach to Atlantic halibut fry production, J World Aquacult Soc, 24, 275, 1993.

289-Devauchelle, N., Alexandre, J.C., Le Corre, N., and Letty, Y., Spawning of sole (*Solea solea*) in captivity, Aquaculture, 66, 125, 1987.

290-Tanaka, M., Ohkawa, T., Maeda, T., Kinoshita, I., Seikai, T., and Nishida, M. Ecological diversities and stock structure of the flounder in the Sea of Japan in relation to stock enhancement, Bull Natl Res Inst Aquacult, 3, 77, 1997.

291-Bengtson, D.A., Aquaculture of summer flounder (*Paralichthys dentatus*): status of knowledge, current research and future research priorities, Aquaculture, 176, 39, 1999.

292-Watanabe, W.O. and Carroll, P.M., Progress in controlled breeding of summer flounder, *Paralichthys dentatus*, and southern flounder, P lethostigma, J Appl Aquacult, 11, 89, 2001.

293-Watanabe, W.O., Woolridge, C.A., and Daniels, H.V., Progress toward year-round spawning of southern flounder broodstock by manipulation of photoperiod and temperature, J World Aquac Soc, 37, 256, 2006.

294-Benetti, D.D., Grabe, S.W., Feeley, M.W., Stevens, O.M., Powell, T.M., Leingang, A.J., and Main, K.L., Development of Aquaculture methods for southern flounder, Paralichthys lethostigma: I. Spawning and larval culture, J Appl Aquac, 11, 113, 2001.

295-Luckenbach, J.A. and Sullivan, C.V., Effective GnRHa dose and gamete ratio for reproduction of southern flounder, Paralichthys lethostigma (Jordan and Gilbert 1884), Aquac Res, 35, 1482, 2003.

296-Moon, S.H., Lim, H.K., Kwon, J.Y., Lee, J.K., and Chang, Y.J., Increased plasma 17-hydroxyprogesterone and milt production in response to gonadotropin-releasing hormone agonist in captive male starry flounder, Platichthys stellatus, Aquaculture, 218, 703, 2003.

297-Lim, H.K., Han, H.S., and Chang, Y.J., Effects of gonadotropin-releasing hormone analog on milt production enhancement in starry flounder Platichthys stellatus, Fish Sci, 68, 1197, 2002.

298-Larson, D.G.J., Mylonas, C.C., Zohar, Y., and Crim, L.W., Gonadotropin-releasing hormona analogue (GnRH-A) induces multiple ovulations of high-quality eggs in a cold-water, batch-spawning teleost, the yellowtail flounder (*Pleuronectes ferrugineus*), Can J Fish Aquat Sci, 54, 1957, 1997.

299-Norberg, B., Valkner, V., Huse, J., Karlsen, I., and Grung, G.L., Ovulatory rhythms and egg viability in the Atlantic halibut (*Hippoglossus hippoglossus*), Aquaculture, 97, 365, 1991.

300-Shields, R.J., Gara, B., and Gillespie, M.J.S., A UK perspective on intensive hatchery rearing methods for Atlantic halibut (*Hippoglossus hippoglossus* L.), Aquaculture, 176, 15, 1999.

301-Babiak, I., Ottesen, O., Rudolfsen, G., and Johnsen, S., Chilled storage of semen from Atlantic halibut, *Hippoglossus hippoglossus* L. II. Effect of spermiation advancement, catheterization of semen, and production-scale application, *Theriogen*, 66, 2036, 2006.

302-Vermeirssen, E.L.M., Mazorra de Quero, C., Shields, R.J., Norberg, B., Kime, D.E., and Scott, A.P., Fertility and motility of sperm from Atlantic halibut (*Hippoglossus hippoglossus*) in relation to dose and timing of gonadotrophin-releasing hormone agonist implant, *Aquaculture*, 230, 547, 2004.

303-Forés, R., Iglesias, J., Olmedo, M., Sanchez, F.J., and Peleteiro, J.B., Induction of spawning in turbot (*Scophthalmus maximus* L.) by a sudden change in the photoperiod, *Aquac Eng*, 9, 357, 1990.

304-Howell, B.R. and Scott, A.P., Ovulation cycles and post-ovulatory deterioration of eggs of the turbot, *Scophthalmus maximus* L., *Rap CIEM*, 191, 21, 1989.

305-Howell, B.R., A re-appraisal of the potential of the sole, *Solea solea* (L.), for commercial cultivation, *Aquaculture*, 155, 355, 1997.

306-Dinis, M.T., Ribeiro, L., Soares, F., and Sarasquete, C., A review on the cultivation potential of *Solea senegalensis* in Spain and in Portugal, *Aquaculture*, 176, 27, 1999.

307-Imsland, A.K., Foss, A, Conceiçao, LEC, Dinis, MT, Delbare, D., Schram, E., Kamstra, A., Rema, P, and White, P., A review of the culture potential of *Solea solea* and *S. senegalensis*, *Rev Fish Biol Fisher*, 13, 379, 2003.

308-Anguis, V. and Cañavate, J.P., Spawning of captive Senegal sole (*Solea senegalensis*) under a naturally fluctuating temperature regime, *Aquaculture*, 243(1-4), 133, 2005.

309-Ramos, J., Luteinizing hormone-releasing hormone analogue (LH-RHa) induces precocious ovulation in common sole (*Solea solea* L.), *Aquaculture*, 54, 185, 1985.

310-Ramos, J., Induction of spawning in common sole (*Solea solea* L.) with human chorionic gonadotropin (HCG), *Aquaculture*, 56, 239, 1986.

Complete affiliation:

Evaristo Mañanós, Institute of Aquaculture of Torre la Sal (C.S.I.C.), 12595-Cabanes, Castellón, Spain. evaristo@iats.csic.es

Phone: 0034964319500 Fax: 0034964319509

CHAPTER 2
METHODS FOR SPERM COLLECTION

J-C. GWO

1. INTRODUCTION

Successful cryopreservation is closely related to the quality of pre-frozen semen. The structure and cryoresistance of spermatozoa, the physicochemical composition of seminal plasma, and the quality and volume of semen vary considerably among species, strains, and even individuals of the same stock. In addition, the fitness of spermatozoa also varies according to the animal's health status, social hierarchy, diet, storage time, storage conditions, as well as collection techniques and season. Therefore the quality of the pre-frozen semen from the selected individual should be carefully examined. Incision is an exceptional way of removing testes and extracting sperm.

To maximize the quality of collected semen, the bloodstock should be kept under stress-free conditions. This includes collecting semen during the peak of the spawning season, choosing sexually mature and healthy bloodstock, anaesthetizing (2% 2-phenoxyethanol, benzocaine, Quinaldine, MS-222, or any suitable anaesthetic) animals during handling, and minimizing possible human damage to the animals. It is advisable to prevent semen from mixing with any contaminates and keep the sperm (of finfish and invertebrates) quiescent after collection to prevent the depletion of their energy reserves before freezing. Semen samples should be kept at 4-5°C under aerobic conditions and frozen as soon as possible after being collected. Fish sperm deteriorate rapidly once they are stripped from fish.

2. FINFISH

To collect sperm from finfish, the fish should be removed from water, with any traces of mucus and water wiped off from its body and, in particular, its urogenital opening area (genital papilla) kept absolutely clean and dry. Prior to collecting semen, the abdomen of the fish should be gently massaged to expel urine and fecal contaminates. Brood males (with oozing) semen should be selected first. When collecting semen by gentle abdominal stripping, it is important to avoid contamination with water, slime, scales, blood, bile, or peritoneal fluid. Contamination can be avoided by using a catheter for sperm collection [1,2]. For certain fishes, like the ocean pout (*Macrozoarces americanus*), semen can be obtained by inserting a polyethylene tube directly into the sperm duct via the urinogenital pore [3]. Centrifugation could be necessary when the semen sample is contaminated. In tilapia semen is centrifuged at 5000 rpm for 5 min and then supernatant is decanted [4]. Centrifuging at 1000 g has no obvious adverse effect on sperm motility [5,6]. Finally, store semen from individual fish separately on crushed ice until it is assessed for motility and quality, pooled if necessary. Avoid exposing semen to direct sunlight.

2.1 Semen samples collected from testes

Stripping semen from some fish is difficult because of their anatomy, reproductive physiology, body size, and limited semen volume. The process of releasing sperm involves excising the testes from the male, dissecting away adherent tissue, and macerating (catfish) or crushing (loach and small aquarium fish) the testes in a proper diluent [7-10]. Catfish males can survive surgical removal of testes and are capable of partially regenerating their testicular tissue [11,12]. To successfully collect vital spermatozoa from high sea black marlin (*Makaira indica*), the process involves removing the testes from live specimens, slicing the testes into small sections, and gently massaging the testicular semen from the spermatic duct by hand [13]. Care must be taken in the handling of testes from *post mortem* to minimize the risk of environmental fluids, water contamination and dehydration damaging sperm quality [2,7,14-21]. Autolysis rapidly destroys the internal organs of fish after death. Testes should be kept at 4-5°C under aerobic and moist conditions. The time of *post mortem* storage significantly affects sperm quality. Fish sperm ageing is much more rapid *in vitro* than *in vivo* [22,23].

2.2 Semen collected by hormone treatment

Hormone treatment is required for some species (eel and grouper) to induce maturation and spermiation so as to obtain a sufficient volume of semen for experiments [20,24-34]. Sperm can also be acquired from sex-reversed females produced by hormonal treatment with androgens for breeding unisex progeny [35-37]. Sex-reversed female rainbow trout lack sperm ducts. The scalpel technique enables dripping semen to be collected from the surgically extracted testes of sacrificed animals, reduces contamination in samples and retains the immature sperm in testicular cysts [37]. Hormonal treatments do not affect the percentage of motile sperm compared to that of untreated fish. Sperm concentration, compositions of semen, and osmolality of seminal plasma are also similar in the hormonally treated group and control group.

The fertilizing capacities of spermatozoa obtained from hormone-injected fish are similar, if not better than, the controlled ones, indicating that the sperm quality is not adversely affected by hormonal treatment.

The volume and quality of collectable semen vary with the quantity of available milt and size of the males. Semen should be directly collected into sterile syringes, Eppendorfs, pipette, test tube, plastic bag, or Petri dishes. For scarce semen, microhaematocrit tubes (10-50 μl) should be used [38,39]. Strip milt by abdominal massage into clean, dry, pre-labelled containers or crush testis in ringer solution. Inspect a semen sample under a microscope (100 x). Semen contaminated with mucus, water, blood and faeces will usually activate spermatozoa. Maintain semen samples at 4-5°C under aerobic conditions and freeze semen immediately after collection [40-51]. Use only samples showing >90% vigorous motility following activation, for cryopreservation or further experiments.

3. INVERTEBRATE MOLLUSC [BIVALVE (OYSTER), GASTROPOD (ABALONE)] AND SEA URCHIN

Purchase adult broodstocks in a mature, gravid condition from hatcheries during the spawning season. Determine the sex of abalone (*Haliotis diversicolor*) by external examination on the gonad index (male gonad milky white, female gonad dark green) since it is difficult to determine the sex of oysters and sea urchins by observing external appearance [10,52-55]. Drill a hole or remove the top valve of the Pacific oyster (*Crassostrea gigas*) shell or insert a syringe through the soft portion around the mouth of the sea urchin to sample a small piece of the gonad. Determine the sex and sexual maturity of individual animals by observing a smear of gonad material under a light microscope at 200 x magnification.

Maintain and condition animals in separate aquaria or tanks with aerated seawater at constant and natural water temperature, salinity, and photoperiod to prevent sudden spontaneous spawning and to avoid possible gamete cross-contamination. Scrub clean and rinse the outer surface of all adult molluscs (oysters and small abalone) in fresh seawater before collecting gametes. Cut off the spines on the sea urchin shell and wash the denuded shell with seawater and dry the sea urchin. This will reduce contamination by water, faeces, mucus and the risk of infection by ciliates and other protozoa often located on the shell.

Sperm can be obtained either by dissection or induced spawning [53]. Scrape the gonad after removing the oyster shell, insert the tip of the pipette or the needle of a syringe into the incision, and extract the gametes (dry testicular sperm) directly from the gonad. For larger oysters, semen can be collected using a syringe inserted into the gonad, through a hole drilled into the shell, which would not kill the animals. Remove the "Lantern of Aristotle" and body fluid. Place the sea urchin on paper, aboral side down. Inject about 0.5 ml of an isotonic KCl solution (0.5M) into the body cavity and allow dry sperm to get into a Petri dish through a central hole made on the paper. A Gilson pipette can prevent the thick, viscous semen from adhering to the tips.

In vitro fertilization of surgically removed oyster gametes is a routine method of obtaining functional gametes for experiments. But, the physiology of testicular sperm and ejaculated sperm of small abalone is different. Fresh testicular sperm collected by extracting the testes of mature abalone broodstocks cannot fertilize eggs successfully; hence sperm from ejaculated sperm-seawater suspension is often used [54]. To induce the spawning of abalone, expose mature abalones to air (out of water) for 100 min before placing them back into the aquaria and sequentially repeated temperature shock (raise seawater temperature 4~5°C above ambient room temperature for 3~4 h) two to three times. Usually this practice will result in successful spawning. As soon as spawning is observed (cloudy milky white milt are released from the genital pore), collect the gametes immediately. Filter the sperm-seawater suspension through a 20 μm filter to remove any gross debris, and centrifuge the sperm-seawater suspension at 3000 g for 10 min at 20°C to obtain concentrated sperm. Temperature induction is also adopted to induce the release of gametes in hard clam (*Meretrix lusoria*) [56].

Sperm motility is usually used as an indicator of viability. Mix sperm (stripped oyster semen or concentrated ejaculated abalone sperm) and fresh seawater thoroughly in a dilution ratio of 100 at room temperature. Examine the sperm motility under a light microscope at 200 x magnification. Only sperm samples containing over 80% motile sperm will be used in the study.

Dissect oyster eggs directly into filtered (1 μm) seawater and leave for 60 minutes to synchronize development. Eggs from the abalone and the hard clam (*Meretrix lusoria*) are not viable when removed from the adult by dissection. Both abalone and hard clam must be spawned to obtain viable eggs.

3.1 Decapod crustacean (Penaeoidea shrimp)

Male shrimp gametes can be prepared as complete spermatophores (sperm sac; some species do not form typical structured spermatophores such as *Sicyonia ingentis*), sperm mass, and sperm suspension [57-62]. Acquire mature shrimps (males and females) caught offshore from wild sources or suppliers. Select males based on their health and apparent good physical conditions and stock in tanks for 2-3 days to acclimatise to laboratory conditions. Choose males with a clear white swelling around the coxae at the base of the fifth walking leg (pereopods) from tanks. Healthy spermatophores show no signs of melanization. Spermatophores and sperm mass can be obtained either: 1) manually, by applying gentle pressure with the thumb and index finger between the abdomen and the base of the fifth walking leg; or 2) electrically, using electrostimulation (25 mV, 0.1-0.5 second pulse duration at 100 cps frequency) to the base of the fifth pair of pereopods, without sacrificing the shrimp [53,63-66]. Grasp the partially protruding spermatophores with tweezers and remove them from the terminal ampoules of the genital pore. Avoid damaging animals and hampering their ability to form new spermatophores. Successive spermatophore stripping (15 days apart) has no negative effect on sperm quality and quantity.

Sperm can also be obtained from the spermatophore or sperm mass stored inside the seminal receptacle of the female shrimp after mating. Penaeoidean female shrimps store sperm either on the outside of the thelycum (open thelycum shrimp) or within exoskeletal invagination (seminal receptacle) of the thelycum (closed thelycum shrimp). Open thelycum penaeoideans (e.g., *Litopenaeus setiferus, L. japonicus*) mate within a few hours of a spawn and their sperms do not need post-testicular modification to acquire fertilization ability after transfer to the female. Their sperm is capable of fertilization after a broadcasting release from males into the surrounding seawater. For closed thelycum penaeoideans (e.g., *L. monodom*), spermatophore or sperm mass received at mating is stored for extended periods and sperm inside the spermatophore or sperm mass undergo capacitation after translocation to the female. Anaesthetize closed thelycum shrimps with ice and cut through their thoracoabdominal segment just below the 5th pair of walking legs where the thelycum is situated. Isolate the whitish gelatinous sperm masses from the dissected thelyca.

Weigh each complete spermatophore. Press the posterior region of a spermatophore, rupture it and liberate the glutinous sperm mass that forms a droplet. Mechanically homogenize the spermatophore (by using a tissue grinder) or sperm mass (vigorous pipetting) in seawater to release sperm. Filter the sperm suspension through a 20 μm filter to remove any gross debris. The released sperm are then pelleted from seawater by centrifugation at 200 to 500 g for 5 min. Sperm will survive several rounds of resuspension and centrifugation (washing).

As mature crustacean sperms are nonmotile, several methods (vital dye stain exclusion, stain response, sperm acrosomal reaction assay, and flow cytometry) were compared to characterize male reproductive potential and sperm cell viability in penaeid shrimp [53,57-59,61 63,67,68]. Evaluate the sperm quality, through sperm count, sperm gross morphology, and dye exclusion using 1% Trypan blue or eosin-nigrosin (5% eosin and 10% nigrosin) under a phase contrast microscope, which is still utilized by most researchers to describe male viability. The sperm counts can be estimated using a standard haemocytometer. Sperm cells with irregularly (malformed, bent or missing) shaped heads or spikes are considered abnormal. Dead sperm cells are stained pink and live sperm cells remain unstained against the red background. An intact cell wall can be clearly visible in live sperm cells, while a dead one is irregular in shape.

Viability assessed by induction of *in vitro* acrosome reaction has increased recently. Induction of the acrosome reaction is the only definitive criterion for determining sperm viability. Egg water, the jelly precursor released by eggs during spawning in seawater, induces the acrosome reaction of sperm. Therefore, egg water is a practical tool to explore sperm maturation in both open and closed thelycum shrimps.

4. REFERENCES

1-Glogowski, J., Kwasnik, M., Piros, B., Dabrowski, K., Goryczko, K., Dobosz, S., Kuzminski, H., and Ciereszko, A., Characterization of rainbow trout milt collected with a catheter: semen parameters and cryopreservation success, *Aquac Res*, 31, 289, 2000.

2-Babiak, I., Ottesen, O., Rudolfsen, G., and Johnsen, S., Chilled storage of semen from Atlantic halibut, *Hippoglossus hippoglossus* L. II: effect of spermiation advancement, catheterization of semen, and production-scale application, *Theriogenology*, 66, 2036, 2006.

3-Yao, Z. and Crim, L.W., Spawning of ocean pout (*Macrozoarces americanus* L.): evidence in favour of internal fertilization of eggs, *Aquaculture*, 130, 361, 1995.

4-Rana, K.J. and McAndrew, B.J., The viability of cryopressed tilapia spermatoza, *Aquaculture*, 76, 335, 1989.

5-Yao, Z., Crim, L.W., Richardson, G.F., and Emerson, C.J., Motility, fertility and ultrastructural changes of ocean pout (*Macrozoarces americanus* L.): sperm after cryopreservation, *Aquaculture*, 181, 361, 2000.

6-Dong, Q., Huang, C., and Tiersch, T.R., Post-thaw amendment of cryopreserved sperm for use in artificial insemination of a viviparous fish, the green swordtail *Xiphophorus helleri*, *Aquaculture*, 259, 403, 2006.

7-Mongkonpunya, K., Chairak, N., Pupipat, T., and Tiersch, T.R., Cryopreservation of Mekong giant catfish sperm, *Asian Fish Sci*, 8, 211, 1995.

8-Aoki, K., Okamoto, K., Tatsumi, K., and Ishikawa, Y., Cryopreservation of medaka spermatozoa, *Zool. Sci*, 14, 641, 1997.

9-Horvath, A. and Urbanyi, B., The effect of cryoprectants on the motility and fertilizing capacity of cryopreserved African catfish *Clarias gariepinus* (Burchell 1822) sperm, *Aquac Res*, 31, 317, 2000.

10-Dong, Q., Huang, C., Eudeline, B., Allen Jr., S.K., and Tiersch, T.R., Systematic factor optimization for sperm cryopreservation of tetraploid Pacific oyster, *Crassostrea gigas*, *Theriogenology*, 66, 387, 2006.

11-Bart, A.N., Wolfe, D.F., and Dunham, R.A., Cryopreservation of blue catfish spermatozoa and subsequent fertilization of channel catfish eggs, *Trans. Am. Fish. Soc.*, 127, 819, 1998.

12-Kwantong, S. and Bart, A.N., Effect of cryoprotectants, extenders and freezing rates on the fertilization rate of frozen striped catfish, *Pangasiu.s hypophthalmus* (Sauvage), sperm, *Aquac Res*, 34, 887, 2003.

13-Van der Straten, K.M., Leung, L.K-P., Rossini, R., and Johnston, S.D., Cryopreservation of spermatozoa of black marlin, *Makaira indica* (Teleostei: istiophoridae), *CryoLetters*, 27, 203, 2006.

14-Okumura, S. and Hirose, K., Artificial fertilization in red tilefish, *Branchiostegus japonicus* with cryopreserved spermatozoa, *Suisanzoshoku*, 39, 441, 1991.

15-Steyn, G.J., The effect of freezing rate on the survival of cryopreserved African sharptooth catfish (*Clarias gariepinus*) spermatozoa, *Cryobiology*, 30, 581, 1993.

16-Caylor, R.E., Biesiot, P.M., and Franks, J.S., 1994. Culture of cobia (*Rachycentron canadum*) cryopreservation of sperm and induced spawning, *Aquaculture*, 125, 81, 1994.

17-Pillai, M.C., Yanagimachi, R., and Cherr, G.N., *In vivo* and *in vitro* initiation of sperm motility using fresh and cryopreserved gametes from the Pacific herring, *Clupea pallasi*, *J Exp Zool*, 269, 62, 1994.

18-Wayman, W.R., Thomas, R.G., and Tiersch, T.R., Refrigerated storage and cryopreservation of black drum (*Pogonias cromis*) spermatozoa, *Theriogenology*, 47, 1519, 1997.

19-Gwo, J.-C., Ohta, H., Okuzawa, K., and Wu, H.-C., Cryopreservation of sperm from the endangered formosan landlocked salmon (*Oncorhynchus masou formosanus*), *Theriogenology*, 51, 569, 1999.

20-Gwo, J.-C., Cryopreservation of sperm of some marine fishes, in: *Cryopreservation in Aquatic Species*, Tiersch, T.R. and P.M. Mazik, Eds., *Advances in World Aquaculture*, 7,138, 2000, World Aquaculture Society, Baton Rouge, LA, USA.

21-Lahnsteiner, F., Semen cryopreservation in the salmonidae and in the Northern pike, *Aquac Res*, 31, 245, 2000.

22-Billard, R. and Cosson, M.P., Some problems related to the assessment of sperm motility in freshwater fish, *J. Exp. Zool.* 261, 122, 1992.

23-Billard, R., Cosson, J., Crim, L.W., and Suquet, M., Sperm physiology and quality, in: *Broodstock Management and Egg and Larval Quality,* Bromage, N. and Roberts, R., Eds., Cambridge University Press, Cambridge, 1995, 25.

24-Hara, S., Canto, J.T., and Almendras, J.M.E., A comparative study of various extenders for milkfish, *Chanos chanos* (Forsskal), sperm preservation, *Aquaculture,* 28, 339, 1982.

25-Withler, F.C. and Lim, L.C., Preliminary observations of chilled and deep-frozen storage of grouper *(Epinephelus lauvina)* sperm, *Aquaculture,* 27, 389, 1982.

26-Khan, A.L., Lopez, E., and Leloup-Hatey, Induction of spermatogenesis and spermiation by a single injection of human chorion gonadotropin in intact and hypophysectomized immature European eel *(Anguilla anguilla* L.), *Gen Comp Endocr,* 68, 91, 1987.

27-Gwo, J.-C., Strawn, K., Longnecker, M.T., and Arnold, C.R., Cryopreservation of Atlantic croaker spermatozoa, *Aquaculture,* 94, 355, 1991.

28-Gwo, J.-C. and Arnold, C.R., Cryopreservation of Atlantic croaker spermatozoa: evaluation of morphological changes, *J. Exp. Zool,* 264, 444, 1992.

29-Gwo, J.-C., Strawn, K., and Arnold, C.R., Induced ovulation in Atlantic croaker (Sciaenidae) using hCG and an LHRH Analog: a preliminary study, *Theriogenology,* 39, 353, 1992.

30-Gwo, J.-C., Cryopreservation of black grouper *(Epinephelus malabaricus)* spermatozoa, *Theriogenology,* 39, 1331, 1993.

31-Gwo, J.-C., Kurokura, H., and Hirano, R., Cryopreservation of Spermatozoa from Rainbow Trout, Carp and Marine Puffer, *Nippon Suisan Gakk,* 59, 777, 1993.

32-Gwo, J.-C., Cryopreservation of yellowfin seabream *(Acanthopagrus latus)* spermatozoa (Teleost, Perciformes, Sparidae), *Theriogenology,* 41, 989,1994.

33-Ohta, H., Kagawa, H., Tanaka, H., Okuzawa, K., and Hirose, K., Milt production in the Japanese eel *Anguilla japonica* induced by repeated injections of human chorionic gonadotropin, *Fish Sci,* 62, 44, 1996.

34-Tanaka, S., Zhang, H., Horie, N., Yamada, Y., Okamura, A., Utoh, T., Mikawa, N., Oka, H.P., and Kurokura, H., Long-term cryopreservation of sperm of Japanese eel, *J Fish Biol,* 60, 139, 2002.

35-Tabata, K. and Mizuta, A., Cryopreservation of sex reversed gynogenetic female sperm hirame, *Fish Sci,* 63, 3, 482, 1997.

36-Geffen, A.J. and Evans, J.P., Sperm traits and fertilization success of male and sex-reversed female rainbow trout *(Oncorhynchus mykiss), Aquaculture,* 182, 61, 2000.

37-Robles, V., Cabrita, E., Cuñado, S., and Herráez, M.P., Sperm cryopreservation of sex-reversed rainbow trout *(Oncorhynchus mykiss):* parameters that affect its ability for freezing, *Aquaculture,* 224, 203, 2003.

38-Harvey, B., Kelley, R.N., and Ashwood-Smith, M.J., Cryopreservation of zebra fish spermatozoa using methanol, *Can J Zool,* 60, 1867, 1982.

39-Tiersch, T.R., Cryopreservation in aquarium fishes, *Mar Biotechnol,* 3, 212, 2001.

40-Stoss, J., Fish Gamete Preservation and Spermatozoan Physiology, in: *Fish Physiology,* Hoar, W.S., Randall, D.J. and Donaldson, E.M., Eds, Academic Press: New York, 1983, 305.

41-Magyary, I., Urbanyi, B., and Horvath, L., Cryopreservation of common carp *(Cyprinus carpio* L.) sperm 1. The importance of oxygen supply, *Journal Appl Ichthyol,* 12, 113, 1996.

42-Fabbrocini, A., Lavadera, L., Rispoli, A S., and Sansone, G., Cryopreservation of sea bream *(Sparus aurata)* spermatozoa, *Cryobiology,* 40, 46, 2000.

43-Rita, A.J. and Campet, M., Sperm survival during short-term storage and after cryopreservation of semen from striped trumpeter *(Latris lineate), Theriogenology,* 54, 467, 2000.

44-Suquet, M., Dreanno, C., Fauvel, C., Cosson, J., and Billard, R., Cryopreservation of sperm in marine fish, *Aquac Res,* 31, 231, 2000.

45-Lahnsteiner, F., Mansour, N., and Weismann, T., The cryopreservation of spermatozoa of the burbot, *Lola lota* (Gadidae, Teleostei), *Cryobiology,* 45, 195, 2002.

46-Sansone, G., Fabbrocini, A., Leropoli, S., Langellotti, A.L., Occidente, M., and Matassino, D., Effects of extender composition, cooling rate, and freezing on the motility of sea bass (*Dicentrarchus labrax*, L.) spermatozoa after thawing, *Cryobiology*, 44, 229, 2002.

47-Zilli, L., Schiavone, R., Zonno, V., Storelli, C., and Vilella, S., Evaluation of DNA damage in *Dicentrarchus labrax* sperm following cryopreservation, *Cryobiology*, 47, 227, 2003.

48-He, S. and Woods, C., Changes in motility, ultrastructure, and fertilization capacity of striped bass *Morone saxatilis* spermatozoa following cryopreservation, *Aquaculture*, 236, 677, 2004.

49-Ji, X.S., Chen, S.L., Tian, Y.S., Yu, G.C., Sha, Z.X., Xu, M.Y., and Zhang, S.C., Cryopreservation of sea perch (*Lateolabrax japonicus*) spermatozoa and feasibility for production-scale fertilization, *Aquaculture*, 241, 517, 2004.

50-Cabrita, E., Robles, V., Rebordinos, L., Sarasquete, C, and Herraez, M.P., Evaluation of DNA damage in rainbow trout (*Oncorhynchus mykiss*) and gilthead sea bream (*Sparus aurata*) cryopreserved sperm, *Cryobiology*, 50, 144, 2005.

51-Liu, Q., Li, J., Zhang, S., Ding, F., Xu, X., Xiao, Z., and Xu, S., An efficient methodology for cryopreservation of spermatozoa of red seabream, *Pagrus major*, with 2-ml cryovials, *J World Aquacult*, 37, 289, 2006.

52-Gwo, J.-C., Cryopreservation of eggs and embryos from aquatic organisms, in: *Cryopreservation in Aquatic Species*, Tiersch, T.R. and Mazik P.M., Eds., *Advances in World Aquaculture*, 7. 211, 2000, World Aquaculture Society, Baton Rouge, LA, USA.

53-Gwo, J.-C., Cryopreservation of aquatic invertebrate semen: a review, *Aquac Res*, 31, 259, 2000c.

54-Gwo, J.-C., Wu, C.-Y., Chang, W.-P., and Cheng, H.-Y., Evaluation of DNA damage in Pacific oyster (*Crassostrea gigas*) spermatozoa before and after cryopreservation using comet assay, *Cryo-Letter*, 24, 171, 2003.

55-Kawamoto, T., Narita, T., Isowa, K., Aoki, H., Hayashi, M., Komaru, A., and Ohta, H., Effects of cryopreservation methods on post-thaw motility of spermatozoa from the Japanese pearl oyster, *Pinctada fucata martensii*, *Cryobiology*, 54, 19, 2007.

56-Chao, N.H. and Liao, I.C., Cryopreservation or finfish and shellfish gametes and embryos, *Aquaculture*, 197, 161, 2001.

57-Anchordoguy, T., Crowe, J.H., Griffin, F.J., and Clark Jr., W.H., Cryopreservation of sperm from the marine shrimp *Sicyonia ingentis*, *Cryobiology*, 25, 238, 1988.

58-Lin, M.-N., Ting, Y.-Y., and Hanyu, I., Hybridization of two closed-thelycum penaeid species *Penaeus monodon* X *P. penicllatus* and *P. penicillatus* X *P. monodon*, by means of spermatophore transplantation, *Bull Taiwan Fish Res last*, 45, 83, 1988.

59-Bray, W.A. and Lawrence, A.L., Reproduction of Penaeus species in captivity, in:, *Marine Shrimp Culture: Principles and Practices*, Fast, A. and Lester, L.J., Eds., Elsevier, Amsterdam, 1992, 93.

60-Gwo, J.-C. and Lin, C.-H., Preliminary experiments on the cryopreservation of penaeid shrimp (*Penaeus japonicus*) embryos, nauplii and zoea, *Theriogenology*, 49, 1289, 1998.

61-Ceballos-Vazquez, B.P., Sperm quality over consecutive spermatophore regeneration in the Pacific white shrimp *Litopenaeus vannamei*, *J World Aquacult Soc*, 35, 178, 2004.

62-Bart, A.N., Choosuk, S., and Thakur, D.S., Spermatophore cryopreservation and artificial insemination of black tiger shrimp, *Penaeus monodon* (Faricius), *Aquac Res*, 37, 523, 2006.

63-Wang, Q., Misamore, M., Jiang, C.Q., and Browdy, C.L., Egg water induced reaction and biostain assay of sperm from marine shrimp *Penaeus vannamei*: dietary effects on sperm quality, *J World Aquac Soc*, 26, 261, 1995.

64-Misamore, M. and Browdy, C., Evaluating hybridization potential between *Penaeus setiferus* and *Penaeus vannamei* through natural rearing, artificial insemination and in vitro fertilization, *Aquaculture*, 150, 1, 1997.

65-Pratoomchart, B., Spermatophore regeneration in pond-reared *Penaeus monodon* Fabricius, *Aquaculture in the Tropics*, 14, 255, 1999.

66-Sampath, K. and Ramachandran, T., Spermatophore collection in *Penaeus monodon* by electrical stimulation, *Aquaculture in theTropics*, 14, 193, 1999.

67-Alfaro, J., Reproductive quality evaluation of male *Penaeus stylirostris* from a grow-out pond, *J World Aquacult Soc*, 24, 6, 1993.

68-Benzie, J.A.H., Kenway, M., and Bailment, E., Growth of *Penaeus monodon x Penaeus esculentus* tiger prawn hybrids relative to the parental species, *Aquaculture*, 193, 227, 2001.

Complete affiliation:

Jin-Chywan Gwo, Department of Aquaculture, Taiwan National Ocean University, Keelung 20224, Taiwan; e-mail: gwonet@ms.16.hinet.net.

SECTION II - SPERM AND EGG QUALITY EVALUATION

CHAPTER 3
SPERM QUALITY ASSESSMENT

E. Cabrita, V. Robles and P. Herráez

1. BASIC ASPECTS OF SPERM PHYSIOLOGY

1.1 General considerations

Aquacultured species belong to very different groups from echinoderms to teleosts with the common feature of spending their life spawn in water. Sperm from these groups have particular physiological characteristics, related to the mechanism of fertilization (external or internal) and the type of eggs that strongly conditioned spermatozoa characteristics and sperm physiology.

Most aquacultured species are external fertilizers: sperm is released into the water and their spermatozoa reach and fertilize the eggs. Many strategies have been developed to facilitate sperm-egg contact, from the simultaneous spawning of many individuals (males and females) in the same area (bivalve, herring), to the release of milt over the spawn, previously liberated by the female in a small "nest" (trout). Other groups are internal fertilizers, the spermatozoa entering the female body. Within crustacea decapoda, such as shrimps, crabs or lobsters, fertilization occurs internally: immotile spermatozoa are packed into a spermatophoric container, the spermatheca, which is transferred to the female's seminal receptacle or thelycum, where, as previously demonstrated in shrimp, sperm cells experience physiological changes (maturation or capacitation) required for further fertilization [1]. Some teleosts, including live-bearing fish such as swordtails and platyfish of the genus *Xiphophorus*, which produce ornamental fish as well as a model for developmental research, are also internal fertilizers and use a similar strategy: males transfer the sperm packed in spermatophores to females, which store the sperm in the *receptaculum seminis* of the genital tract for as long as several months after insemination [2].

The eggs from aquacultured species are also very different in structure. Almost all teleosts have relatively large eggs (usually more than 1 mm Ø) surrounded by the chorion, a rigid and impermeable envelope. The chorion is perforated by one or more channels (micropyles) according to the species, which allows the spermatozoa to reach the plasma membrane and fertilize the eggs. Taking into account the comparative size of sperm and egg and the scarce number of micropyles, many spermatozoa will never reach them and have no chance of fertilizing the egg. On the other hand, because spermatozoa enter through a channel, direct contact is produced between egg and sperm plasma membranes and the spermatozoa heads do not require specific enzymes to lyse the egg envelopes, as do those contained in the acrosome of mammalian spermatozoa. On the contrary, eggs from invertebrates, Agnatha (lamprey) or live-bearing fish, display jelly coats that are penetrated by the spermatozoa before reaching the plasma membrane and proceeding with the sperm-egg fusion.

Taking these facts into account, Jamieson and Leung [3] classified spermatozoa as "aquasperm," released into water and fertilizing externally and "introsperm" or internally fertilizing sperm, and as acrosomal and anacrosomal according to the presence of acrosome in the sperm head. Teleosts, except live-bearing fish, possess structurally simplified spermatozoa: a small head that lacks acrosome, probably as a result of secondary simplification during evolution because the development of the egg envelopes [3], a small mid-piede with a few mitochondria and a uniflagellar tail; whereas viviparous teleosts have a more complex structure, more similar to that of mammals [4]. Sturgeons may represent a transitional stage of gamete simplification: spermatozoa have an active acrosome that undergoes an acrosome reaction before fertilization, in spite of the eggs containing multiple micropyles. The acrosome function of sturgeon is not well understood. Recently, Sarosiek et al. [5] identified arilsulfatase activity, which is characteristic of mammalian spermatozoa and seminal plasma, but its role still remains unknown. From Agnatha (lampreys) to all invertebrates, spermatozoa are acrosomal and there are no signs of structure simplification.

When the spermatozoa structure is compared, the head is the cellular compartment showing more variability in morphology and size amongst aquatic organisms, even within the same group [6]. The mid-piece is frequently poorly developed, because the presence of a small number of rounded mitochondria is common in external fertilizers, from sea urchin to bivalves or teleosts. With regard to the tail, some teleosts of the genus *Ictalurus* [3] have biflagellate spermatozoa, and some crustacean aflagellate spermatozoa, but the most common gametes are uniflagellate.

Sperm (spermatozoa and seminal plasma) is produced in the testes. During their trajectory through the efferent duct system, at least in teleost, spermatic ducts synthesize and secrete enzymes, monosaccharide, lipids and proteins into the seminal fluid, regulate ionic composition of the plasma and have fagocytic activity, allowing the final maturation of spermatozoa and keeping them viable until spawning [7]. Milt release is dependent on different stimuli, and according to the species, can be limited to a specific breeding season or take place all year round. Cells from the type of aquasperm are non-motile in the seminal plasma, and motility is activated upon release into the water or, as in the case of some ascidians, even later, upon contact with egg secretions [8]. Motility activation is a complex process that deserves a wider explanation. Motility duration is, in most of the species, very short, no longer than a few minutes, but exceptionally, the spermatozoa of some teleosts such as herring, or chondrostean such as sturgeon, retain their motility and their fertilization capacity for several hours after activation [9]. Aquasperm are very simple cells, with low energy production efficiency and are released into an aggressive medium, which does not allow long-term survival. Under such conditions mechanisms providing long-term movement are not needed and fertilization success depends on the release of a high number of spermatozoa very close to the egg, as well as on possible mechanisms facilitating their contact. In viviparous and ovoviviparous species, motility starts after sperm discharge into the genital tract. Semen from viviparous teleosts possesses atypical features such as well-developed mitochondria and glycolytic activity comparable to that of mammalian sperm, which enables long-term movement and survival in the female reproductive tract, being more mammalian-like in morphology and behaviour [4].

1.2 Physiological aspects of spermatozoa motility triggering

Sperm motility has been extensively studied in most species of interest in aquaculture and fisheries, endangered species and in model species such as sea urchin. The trigger factors involved in motility activation have been well studied, although some aspects of the mechanism are not well understood. Two major factors have been identified as motility triggers, environmental factors involving sperm after its release and factors present in the egg, coelomic fluid or seminal plasma. As these factors depolarize the cell membrane, they may initiate a signal cascade that stimulates flagellar movement [10].

1.2.1 Environmental factors: the role of ions, osmolality, temperature and pH

In most fish, motility is dependent on the osmolality of the external media. Changes in intracellular ion concentration caused by swelling or shrinkage promoted by external osmolality, appear to regulate sperm motility in most species, particularly in marine teleosts and cyprinids. In Chondrostean, such as sturgeon, and salmonids, K^+ concentration, but not osmolality, seems to be the principal factor regulating motility [9]. The initiation of motility in marine teleosts also appears to involve an increase in the internal concentration of Ca^{2+} probably derived from intracellular stores [8]. Intracellular parameters such as cAMP, ATP, concentration of Ca^{2+}, temperature and pH, affect the capacity and duration of motility [11]. The role of ions, osmolality and pH seems to be crucial and often interrelated.

Osmolality- In several species osmolality has been identified as a motility trigger regardless of temperature, pH or the composition of the activation solution. In freshwater and marine species activation occurs when spermatozoa come into contact with hypotonic or hypertonic medium, respectively. In some species, such as gilthead seabream (*Sparus aurata*) [12], Senegalese sole (*Solea senegalensis*) [13,14], puffer fish (*Takifugu niphobles*) [15-17] and flounder [15], motility was activated using hyperosmotic solutions made from sugars or other nonionic compounds, demonstrating that the main triggering factor is in this case osmolality, probably mediated by osmotic-pressure receptors present in the plasma membrane. In euryhaline species it will depend on the period of adaptation to the environment, but usually, in well-adapted freshwater individuals motility activation will occur in a large range of osmolarities tending to the hypotonicity, and in individuals adapted to seawater sperm will be activated in osmolarities near 1000 mOsm/Kg (similar to seawater) [18]. A good example of this performance can be found in tilapia spermatozoa (*Oreochromis mossambicus*), which can spawn in both seawater and freshwater.

Krasznai et al. [19] proposed a cell-signalling cascade for the activation of motility in carp sperm. According to this author, the hypoosmotic shock produced upon release into freshwater causes K^+ channels to open and, due to the low K^+ concentration in the surrounding medium, a K^+ efflux is induced through the membrane. This hyperpolarization is followed by depolarization of the plasma membrane and a Ca^{2+} influx into the cell, flagellar beating being activated in a Ca^{2+} dependent cascade. In several studies the role of cyclic AMP has been discussed as a prerequisite for the phosphorilation of specific protein subunits involved in the triggering of axonema motility [20], but according to the Krasznai et al. model [19], at least in carp, cAMP/ phosphorilation signaling would not be necessary.

In anadromous and catadromous species, motility can be activated in hypo-, iso- and hypertonic solutions (up to 600 mOsm/kg for striped bass) [21], suggesting that in these species, other mechanisms must be involved in controlling sperm motility (for details see He et al. [22]).

Potassium- A recent review by Alavi and Cosson [11] describes the interaction and influence of Ca^{2+} and K^+, two major ions present in seminal plasma that have been considered the key ions for sperm motility activation and duration in marine fish as well as in Salmonidae and Acipenseridae. In Salmonids, seminal plasma contains a high K^+ concentration (from 20 to 60 mM as summarized by Alavi and Cosson [11]), related to the immotility of spermtozoa in the milt. Motility is triggered by a decrease in the concentration of K^+ in the external medium that produces an efflux of K^+ through the plasma membrane [23,24], with a consequent hyperpolarization. However, it has been suggested that other factors must be involved since sperm is activated in ovarian fluid which has a high concentration of this ion. Several studies confirm that divalent cations such as Ca^{2+} and Mg^{2+} in the fertilization media, usually antagonize K^+ inhibition. Alavi and Cosson [11] hypothesize that the inhibitory effects of high concentrations of K^+ could depend on the sensitivity of the spermatozoa to this ion, which is highly variable between males and breeding season, probably due to the seasonal changes in concentrations of K^+ and Ca^{2+} in the seminal plasma. There is less information about mechanisms regulating motility in sturgeon, but all the studies suggest significant similarities with salmonids.

Spermatozoa from cyprinids are less sensitive to K^+, but the motility of common carp spermatozoa seems to be recovered after incubation in K^+-rich media, in which the cells are immotile [25].

Calcium- Extracellular calcium has been considered a prerequisite for the initiation of motility in several fish species [11,19,26]. The influx of the external Ca^{2+} is promoted by the stimulus causing the activation, and the Ca^{2+} entering the sperm cytoplasm participates in the activation of certain enzymes or other proteins. However, recent studies pointed out the possible effect or contribution of internal calcium stores in mitochondria [27] or other structures similar to the endoplasmatic reticulumm in the increase in calcium inside the cell upon spermatozoa "stimulation." In carp sperm, Krasznai et al. [19] suggested that the influx of extracellular calcium produces the release of Ca^{2+} from internal stores, but in the absence of this influx of calcium from outside, the release of calcium from internal stores induced by other mechanisms does not activate sperm motility, suggesting that entry of external Ca^{2+} is a critical event.

Similar findings in which complete sperm motility inhibition was observed when sperm was incubated with desmethoxyverapamil, a calcium channel blocker, were reported by Alavi and Cosson [11] for salmonid sperm. These authors report changes in internal Ca^{2+} concentration from 30 nM to 180 nM before and after swimming, respectively. In other fish species such as pufferfish, *Takifugu niphobles* [15] or freshwater-acclimated tilapia, *Oreochromis mossambicus* [18] strong evidence suggests that in the absence of external Ca^{2+}, intracellular Ca^{2+} concentration increases during motility activation. Morita and coworkers [18] pointed out the role of sleeve structures in the flagella with similar characteristics to the endoplasmatic reticulum, known to accumulate this ion, in its release to the cytoplasm. Other structures that might be involved are mitochondria since spermatozoa from teleost fish lack an endoplasmatic reticulum. In those species able to activate motility in nonionic hypertonic solutions, it is possible that the increase in calcium inside the cell required to activate sperm or to produce the cascade of events, is produced by release from internal stores, together with the calcium contained in the seminal plasma.

Sodium- Na^+ ions are known to have a secondary role in the activation and regulation of fish and invertebrate sperm motility. Indeed, not much literature can be found on this issue and only recently was the role of Na^+ in motility initiation and regulatory process of herring spermatozoa [28] demonstrated. Herring spermatozoa are immotile both in the testes (seminal plasma) and in contact with seawater when released [29]. However, spermatozoa acquire motility upon contact with the eggs, as will be explained below. Recently, Vines and co-workers [28] detected a Na^+/Ca^{2+} exchanger in sperm from this species and postulated that motility initiation and regulation occur by reverse Na^+/Ca^{2+} exchange. This regulatory mechanism would allow the spermatozoa to change from forward to reverse operating channel under the control of ion concentration inside and outside the cell [30]. Since motility in this species could be longer than 60 minutes, it seems clear that the cells could regulate the increase in calcium inside by operating in a reverse mode, thus calcium ions would be balanced by the incoming sodium ions. This would permit motility for extended periods as well as the maintenance of sperm in contact with seawater in a quiescent mode prior to contact with the specific egg factor.

pH and temperature- Changes in pH upon sperm liberation trigger motility activation in some species. Sea urchin spermatozoa cannot swim in the gonads because high CO_2 tension in semen maintains pH acid (7.2) with respect to sea water, and dinein, the engine for axoneme beating, is inactive below pH 7.3. Spawning produces an increase in pH and dinein activation. Then, ADP production activates mitochondrial respiration 50-fold and initiates motility [8]. Another recent study in starfish, *Asterina pectinifera*, demonstrated that an increase in intracellular pH induced phosphorylation of the axonemal proteins required for flagellar activation in this species [31]. It is not clear how or whether pH from external solutions affects motility initiation in fish. In some teleost, mainly marine species, an increase in intracellular pH has been found during motility activation. In gilthead seabream, artificial seawater with pH 9.3 produced the highest percentage of motile spermatozoa as well as the fastest sperm motility [12]. This fact was also observed in puffer fish and flounders, in which motility could be initiated under isotonic conditions if the intracellular pH increased [32]. Nevertheless, in 1991, Boitano and Omoto [33] demonstrated that pH has no effect on trout sperm. In almost all species the range of pH for motility activation is quite large, ranging from acid (pH = 5) to alkaline (pH = 10) (seabass, carp, mullet) [25,34,35]. Thus, it might speculate the role of external solution pH in regulating motility, although some interference was described for some species. This is the case of sea urchin and other species with an acrosomal reaction and chemotactic behaviour, in which internal pH increases upon acrosomal reaction or chemoattractant exposure, but in some of them, it is not a prerequisite for calcium entry into the cell, meaning that the regulatory effect of pH on sperm activation was not demonstrated at this step [36].

Temperature from the external media changes the motility pattern as well as motility duration in several species. Low temperatures result in prolonged duration of motility with a reduction in the velocity of cells and an increase in the frequency of flagellar beat (reviewed in Cosson et al. [37], and Alavi and Cosson [38]). In rainbow trout the duration of forward movement is about 140 s at 5°C and drops to 70 s at 10°C [39] This phenomenon is mainly due to the energetic resources spent much faster at higher temperatures and spermatozoa do not compensate by producing more energy. The optimum temperature of the motility solution has been described in detail for salmonids, cyprinids and acipenserids (see Alavi and Cosson [38]).

1.2.2 Biological factors: egg secretions and seminal plasma

Sperm motility is sometimes increased in response to secretions released from the eggs or the female reproductive organs. Sperm chemotaxis towards the egg occurs under the influence of chemoattractants released from the egg to complete fertilization, guiding spermatozoa to it. In some cases, these secretions may act not only by increasing motility or regulating the process but also by triggering the initiation of motility which is not initiated by a change in environmental conditions upon sperm release. In other cases, this secretion may be contained in the seminal plasma and enhances spermatozoa motility and metabolism (chemokinesis). This phenomenon has been described in ascidians and echinoderms, as well as in some fish such as the Japanese bitterling (*Acheilognathus lanceolata*), the herring (*Clupea pallasii*), the rosy-barb (*Barbus conchonis*) and the sea lamprey (*Petromyzon marinus*) [28,40,41]. In all cases, spermatozoa from these species behave differently when activated and exposed to these substances.

Sea urchin sperm can be released as close as 1 cm away from an egg and must swim around 50-fold their length to reach it, but chemotaxis is probably effective at distances shorter than 0.2-0.5 mm. The jelly surrounding the eggs contains small peptides (speract, resact) that species-specifically change the metabolic state and motility, and possibly facilitates the triggering of acrosome reaction (reviewed by

Darszon et al. [8]). In the ascidian *Ciona*, the role of egg components seems to be even more crucial: spermatozoa are immotile after release into the water and become active and attracted to the egg under the influence of an egg factor (called SAAF), that increases K^+ permeability [42]. In the lamprey, a small accumulation of spermatozoa can be seen after motility activation near the jelly-substance covering the eggs, suggesting the presence of some motility enhancer. In some fish spermatozoa, such as the bitterling or rosy-barb, movement changes near the micropyle, leading all spermatozoa in their way. In herring, spermatozoa motility is initiated in the presence of an initiation factor (glycoprotein) located in the micropyle region of the egg [43].

Secretions from the seminal vesicle present in some teleosts can also play a role in the "protection" of spermatozoa motility by reducing sperm dispersion in the nature, such as in the case of toad-fish, *Halobatrachus didactylus*. However, Mansour et al. [44] found no evidence of the effect of seminal vesicle secretion on the motility of sperm from African catfish, *Clarias gariepinus*. More recently, several proteins present in the milt of European seabass, *Dicentrarchus labrax* [45] and trout [46], were identified as possibly being involved in the regulation of sperm motility, as was also described in human sperm [47].

2. QUALITY EVALUATION

2.1 General considerations

The quality of sperm is related to their capacity to produce viable embryos when in contact with good quality eggs in an appropriate environment. This includes several events, each with different requirements: i) the ability of the spermatozoa to reach the egg, ii) the capacity to cross the egg envelopes or enter through the micropyle, iii) recognizing the oolema and the fusion of both plasma membranes, iv) correct activation of the egg metabolic pathways and v) contribution to the future embryo with an undamaged genome. The definition of reliable and objective biomarkers of sperm quality is, subsequently, very difficult, because quality does not rely on a particular characteristic of the milt, but on its overall "fitness." Fertilization trials could be considered as the best unequivocal measurement, but results of fertilization tests are also dependent on egg quality, sperm/egg ratio, as well as incubation conditions, being very difficult to be absolutely objective and standardized. Another concern for the use of fertilization trials in the assessment of semen quality is related to long-lasting results. Quality assessment must be as reliable and fast as possible to be useful. Fertilization rates could be calculated some hours or minutes after fertilization but, in this case, only the activation of the cleavage is really taken into account. The rate of hatching embryos is a more reliable result, but development could take too long (from days to weeks depending on the species).

Traditionally, much more effort has been put into analyzing egg quality than sperm quality in aquaculture. Usually females undergo more difficulties in spawning than males and, except in those species in which there are evident difficulties in spermiation or very poor sperm quality in captivity, problems in breeding are mainly attributed to egg failure. The use of a large quantity of milt in artificial fertilization, or the use of pooled milt from several different males, apparently solved problems in fertilization, because bad quality samples are compensated with better ones. Nevertheless, this procedure represents a waste of resources, and could even impair fertilization results because of the use of an inadequate sperm/egg ratio. On the other hand, the use of several males does not guarantee the contribution of all of them to offspring development, because good milt producers will successfully fertilize most of the eggs and could result in undesirable and unpredicted inbreeding. Consequently, a good evaluation of the sperm quality is of considerable interest in commercial aquaculture. In order to increase the effectiveness of artificial fertilization, checking the quality of milt makes it possible to discard low quality samples. The repeated analysis of sperm produced by the broodstock enables the identification of those males with better reproductive success. This is an important characteristic, which together with other genetic considerations, must be taken into account to select good quality males in the broodstock (Martinez Pastor, personal communication). Sample quality, as we will further explain, varies between ejaculations and throughout the year, so the identification of "good males" cannot be based on simple evaluation. Apart from the evaluation of males as potential breeders, quality analyses are of great help in the evaluation of artificial fertilization procedures. Any extender for fertilization, or variation in the environmental conditions, should be first tested for their effect on the sperm quality.

On the other hand, the management of sperm in aquafarms often requires that sperm be stored for hours, days or even years (sperm banking). The development of methods for fresh storage at 4°C, but especially for cryopreservation in liquid nitrogen at -196°C, has tremendously potentiated the analysis of sperm quality. Cryopreservation is a very complex process and sperm quality must be evaluated after each step of the protocol in order to formulate the least damaging procedure. Moreover, only those semen samples of good quality are susceptible to be successfully cryopreserved, as has been pointed out by several authors [48,49], and this fact has encouraged the analysis of sperm quality. Studies on sperm cryopreservation have been performed in many species of fish, molluscs or crustaceans and the field of sperm quality assessment, as well as procedures for sperm management and artificial fertilization have benefited from this fact. Ecotoxicology has also contributed to the evaluation of sperm quality, since sperm from aquatic species is very sensitive to pollutants, and the quality of milt from different species (sea urchin, zebrafish, goldfish, etc.) is often used as biomarker of water contamination.

2.2 How to check sperm quality

Despite the great deal of research carried out in this field there is no consensus about which parameter better correlates with "global" quality. Correlations have been found between motility and fertilization ability in different species [50-52], but there arc also reports on fertilization with immotile sperm, generally after cryopreservation. Fertility has also been correlated to some seminal plasma characteristics, such as osmolality or ion content [53], sperm density [54], or other parameters in different species. Nevertheless these correlations could be different if applied in different species or experimental conditions. Lahnsteiner and co-workers [55] performed a complete characterization of rainbow trout sperm, analyzing different aspects of motility (motility rate, swimming patterns, etc.), plasma composition (including inorganic compounds, lytic enzymes and organic compounds related to energy metabolism) and markers of sperm metabolism (enzymes and metabolites) and then described three multiple regression models, which, according to the results, defined the fertilization rate. The same concern for repeatability in other experimental conditions should also be taken in this case.

Parameters related to motility, especially the rate of progressive cells, are widely used for routine milt evaluation. Nevertheless, an accurate assessment should always be based on the combination of several traits and should include general parameters of the sperm (cell density, pH, plasma composition, etc.), some motility characteristics and an evaluation of the spermatozoa status. The choice of the combination of parameters depends on the objective of the evaluation as well as on the available equipment and experience. Quality assessment can be relatively simple for routine sperm handling in a farm, but should be much more precise for experimental purposes. In this field it is also important to consider that semen is not a homogeneous mixture of cells and plasma, but a pool of cells originating from different spermatogonia, whose spermatogenesis could be asynchronous giving place to a heterogeneous population of cells with different genotype, maturation stage and characteristics. This is the reason why new trends in sperm evaluation emphasize the analysis of spermatozoa subpopulations for some determinations, finding more concerns for the use of average values of the sperm sample.

3. PARAMETERS USED FOR QUALITY EVALUATION

3.1 General semen characteristics

Volume, pH, osmolality, cell density, spermatocrit and plasma composition, are general characteristics of a semen sample which are traditionally analyzed. Results of these analyses are important in order to determine the general status of the sperm, as well as to detect possible contamination with water, urine or faeces during extraction. Most of them are simply and quickly determined, and could help us to rapidly discard bad quality samples.

Milt volume is a characteristic of each species and ranges from microliters to tens of milliliters. It is easily measured and can be expressed as an absolute value, or in volume units per kilogram of body weight. The extraction of a very fair quantity of sperm could be due to the male being too recently stripped, or previously spawned by itself but could also indicate that the male is not at the peak of the breeding season, and the quality will probably not be optimal.

pH and osmolality must be assessed using specific previously calibrated equipment: a pH meter endowed with an electrode for small volumes and an osmometer. A cryosmometer or a vapor osmometer are recommended for this purpose. Both parameters strongly affect cell motility, as has been previously explained. Deviations from the characteristic range of the species could result in the activation of the sperm before it is used to fertilize the eggs, making the sample unviable for further use. Usually, low osmolalities reflect urine contamination, and the sample should be discarded.

Cell concentration inside the sperm has often been used for the evaluation of fish milt, and some studies have correlated it to fertilization [53]. Evaluation is frequently done by counting the cells in a known volume of milt, using a haemocytometer, a Neubauer chamber or any other counter chamber and a phase contrast microscope [48,56,57]. Before counting, the sperm must be diluted at a pre-established rate to give an appropriate concentration of cells in the observation field. The viscosity of the semen often makes it difficult to obtain a homogeneous dilution of the sample, in such way that depending on the aliquots pipetted for observation, scores can be very different in the same sample. Three readings or more must be made on each sample, very extreme values must be discarded and the average of several measures must be taken as the result. This method is time-consuming, and can be replaced by an electronic device to count particles, the Coulter counter [58]. A faster and more reliable method is the evaluation of the optical density of milt using a spectrophotometer, as described by Ciereszko and Dabrowski [59]. Optical density varies according to the cell concentration and is characteristic of each species. A standard curve of cell concentration can be done by centrifuging the sperm to separate cells from plasma and thereafter mixing them at different rates. After checking cell density with a haemocytometer, optical density is determined with a spectophotometer in each sample with a different concentration. Once the standard curve is determined, cell density evaluation is very fast. The evaluation of spermatocrit is carried out by filling special glass capillaries for haematocrite with the sample and then centrifuging to separate the cells from plasma. Spermatocrit is expressed as the ratio of cell volume per total sperm volume x 100. Other small tubes can be used for determination if specific equipment is not available. This parameter is also a direct index of cell density, and also correlates with the counting using counting chambers [60].

3.1.1 Analysis of plasma composition

Seminal plasma contains substances capable of withstanding spermatozoa metabolism but also substances reflecting functions of the reproductive system. Milt quality will depend on the conditions in which spermatogenesis occurs as well as on sperm intratesticular storage and the analysis of plasma content can reflect both scenarios. Sperm plasma composition is a very good trait to evaluate sperm quality and the metabolic activity of the spermatozoa that has been specially analyzed in salmonids [55,61,62] and cyprinids [63]. Seminal fluid is a secretory product of the testis and of the sperm ducts [64,65] and disturbances in its composition will lead to changes in spermatozoa quality since its main role is to create an optimal environment for spermatozoa function [66]. Thus, the study of seminal plasma constituents could give some clues about sperm quality and inform us about possible ageing processes, metabolic alterations, contaminations or any other factor affecting sperm quality.

The mineral constituents of seminal plasma have been identified as principal constituents affecting sperm quality. Their presence or absence interferes with osmolality, pH and some spermatozoa functions such as motility activation [10]. In most of the studied species five ions predominate in seminal plasma: sodium, potassium, chloride, calcium and magnesium. The determination of their concentrations will change from species to species, but there is a suitable range for each ion which provides the best conditions for spermatozoa. Changes in this concentration could indicate several disruptions that will affect spermatozoa. Contamination with blood, faeces or urine could affect the composition of these ions and a low or high concentration of a determined ion could also indicate the beginning or ending of the reproductive season, indicating sperm of a worse quality [67]. However, ion composition and concentration can vary widely for the same species, thus making it difficult to interpret the meaning of the results obtained.

Organic compounds of seminal plasma could be more indicative of spermatozoa quality or of the factors affecting it. Some of them, including triglycerides, glycerol, fatty acids, glucose, malate or lactate, are indicative of energy metabolism [68]. Others have a high anti-peroxidation effect, for example, vitamins (E and C), selenium and citric acid [69], which avoid oxidative processes and ageing during the storage in the reproductive tract. In the case of vitamins, and specially vitamin C, which cannot be synthetized by the animal and depends on their uptake from food, their absence or low concentration in the seminal plasma can indicate a nutritional dysfunction of the broodstock, normally associated with the administration of frozen food. Ciereszko and co-workers [70] found that low levels of vitamin C in rainbow trout seminal plasma promoted reduced sperm concentration, motility and fertilizing ability. Ascorbic acid has also been associated with its anti-oxidative action in protecting spermatozoa plasma membrane integrity. Other constituents of seminal plasma such as sugars, lipids, free amino acids and proteins may also characterize the quality of sperm (see the review by Ciereszko and his colleages [66]).

Lytic enzymes are also present in plasma. Some of them: acid phosphatase, alcaline phosphatase,βD-glucuronidase or proteases, are responsible for eliminating degenerated spermatozoa at the end of spawning [65] and others are leaking out of the spermatozoa, as will be explained in section 3.3.1, indicating phenomena such as cell lysis or ageing. Lactate dehydrogenase (LDH), aspartate aminotransferase (AspAT) and acid phosphatase (AcP) are the enzymes most commonly determined as associated with lythic processes and correlated with cell viability. Ciereszko and Dabroski [54] found a significant correlation between AspAT activity in fresh milt and the percentage of hatched fry, suggesting the application of biochemical methods in the evaluation of the quality of fish sperm. The study of these enzymatic activities has often been applied to the analysis of the effects of cryopreservation on spermatozoa: Glogowski et al. [71] used the AspAT and AcP assay to determine cryopreservation efficacy in trout milt. The same analysis was successfully performed to characterize post-thawed pike sperm by Babiak and co-workers [72] and bream (*Abramis brama*) sperm by Glogowski et al. [73], among others.

For the analysis of seminal plasma, milt must be centrifuged (350 g for 10 min at 4°C for rainbow trout) and the supernatant centrifuged a second time. Then, plasma can be stored at -20°C until further analysis. Metabolites and enzymes are determined by routine spectophotometric assays or commercial test-kits, but adaptation of protocols to rainbow trout have been described by Lahnsteiner et al. [55]. These authors analyzed a high number of plasma constituents and carried out fertility tests, concluding that plasma pH, â-D-glucuronidase activity, total lipid levels and calcium levels, were predictors of trout fertilization ability.

3.2 Spermatozoa morphology

Structural and ultrastructural studies provide very interesting information in the field of seminology. Spermatozoa structural pattern, shape or size could be indicative of very different features, from the evolutive position of the species, to the maturation stage of the cell or the injuries caused by environmental factors. In mammals, there exist complete descriptions of abnormal morphological patterns, and structural features, such as the presence of cytoplasmic drops in the head base, are related to incomplete maturation in the epididyme. Abnormal spermatozoa morphometry has also been associated to low fertility in different species such as the bull, boar or stallion. In aquatic organisms, several structural studies are descriptive [3,6,74-76] and focus on the understanding of taxonomic and evolutionary relationships (for a good review see Jamieson and Leung [3]). The evaluation of damage promoted by cryopreservation or by the effect of pollutants has also aimed the analysis of sperm morphology and structural modifications of spermatozoa [50,77-82]. Sperm deformities promoted by pollutants such as Hg^{2+} or tributyltin, have been associated with reduced motility and fertilizing capacity [60], and alterations in plasma membrane, mitochondria swelling and tail morphology with changes in motility patterns and decrease in fertilization ability after freezing/thawing [78]. However, information about the correlation between structural or morphometrical characteristics and fertilization ability is very scarce compared to that for mammals.

As we have seen, there exist different morphological patterns of sperm among aquacultured species, from the mammalian-like cells of the live-bearing fish, with a head containing the acrosome and nucleus, a well developed mitochondrial sheath in the mid-piece and one or two flagella, to the most primitively-shaped aquasperm of the majority of teleosts, in which structure is very simple (small rounded head with nucleus, very few rounded mitochondria, and flagella), or the aflagellate spermatozoa of crabs which, in turn, display a well-developed acrosome. The information provided by structural analysis is, subsequently, of different value. The more complex the spermatozoa, the more cellular structures need to maintain their organization, the more information provided by structural studies.

The simpler morphological evaluation is performed by visual assessment using light microscopy (phase contrast), but, considering the small size of most aquatic spermatozoa, the information available is very limited. This method can be used to evaluate acrosome status, or acrosome reacted cells, as well as very evident morphoanomalies or spermatozoa subtypes. Transmission and scanning electron microscopy (TEM and SEM) are frequently used [1,3,75,77,78,81-83]. Fine structure gives more information, and alterations such as chromatin decondensation, mitochondria swelling, formation of blebs in the plasma membrane, changes in the shape of the tail or insertion into the head or modifications of the axonemal pattern, can be detected (Fig. 1). Special care must be taken with sample preparation to avoid structural artifacts that make it difficult to interpret the results, especially for TEM. According to some authors the information provided by electron microscopy is partial, and the methods are expensive and time-consuming [84]. Other options are laser light-scattering spectroscopy and stroboscopic illumination, used by VanLook and Kime [79] to evaluate the effect of mercury on goldfish sperm. With the aim of providing methods that are more accurate and objective and less time-consuming, an automated system for spermatozoa head morphology analysis (ASMA) was developed and validated for mammals, has been used in goldfish [79] and is now being tried in eel, (Fig. 2) rainbow trout and cod to establish correlations between morphometry and other quality parameters [75,85,86].

Considering the particular information provided by morphological studies and given the development of more functional evaluation methods that we will see hereafter, this type of study is now frequently used in combination with other methods of analyzing cell viability [75,82]. Nevertheless, in species with immotile spermatozoa, for example, decapod crustaceans, it is not possible to use motility as an indicator of cell viability and quality evaluation is mainly done by morphotype analysis under light microscopy. Therefore, peneid cells are classified into three morphological groups: spiked cells, non-spiked cells and everted cells, and viability is expressed as the percentage of cells with spike over the total number of evaluated cells, as explained by Salazar et al. (protocol 32 of this volume).

In all acrosomal spermatozoa, the evaluation of acrosome status is crucial for a complete evaluation of quality, and morphological evaluation is frequently the only method of choice. Moreover, the knowledge of structural damage promoted by some treatments, such as dilution in cryoprotectant extenders, is very useful in order to modify the design of the protocols according to the observed alterations [82].

3.3 Viability and cell membrane resistance

The term viability is sometimes confusing: viability *sensu stricto* is the ability of the cell to proceed with its specific functions; in the particular case of spermatozoa, to fertilize the egg. Lahnsteiner defined viability as "sperm motility which could be activated" [46]. Nevertheless, this term is usually used as a synonym of plasma membrane integrity. Plasma membrane is a key structure responsible, among other functions, for the exchanges between the cell and the environment or for the reception of stimulus and the triggering of responses, such as activation of motility after dilution in water. The membrane is also considered the most sensitive structure to environmental stress, and alterations of their architecture have consequences in their permeability and functionality. For these reasons the evaluation of membrane permeability and integrity is considered an index of cell viability: any cell with an injured membrane should be unable to develop its functions, and any non-functional cell should reveal alterations in membrane structure and/or permeability.

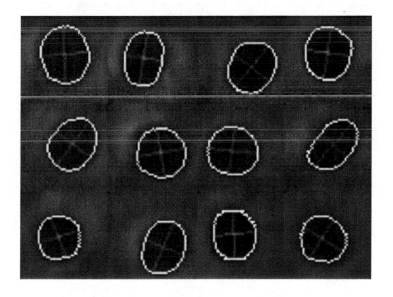

Figure 1 - Transversal section of a rainbow trout flageda showing anormal axonema structure.

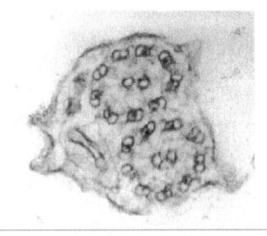

Figure 2 - Morphometric analysis performed in eel sperm using ASMA. Each cell is registred and the cell contour is marked. Length and weight are measured automatically by the software. (*Source*: J. Asturiano)

3.3.1 Plasma membrane functionality

The use of selective dyes or fluorescent probes, which specifically label viable or non-viable cells, is a simple method for evaluation. There are many different options ranging from the traditional stain with eosin-nigrosin or trypan blue to the use of more reliable and evolved fluorescent probes. Trypan blue is unable to cross the plasma membrane. Only those cells with a damaged membrane incorporate the stain, allowing the proportion of living and dead cells to be estimated by light microscopy: viable spermatozoa remain unstained and non-viable ones appear stained in blue. This is a fast, cheap and simple method which has been often used [87], and is useful for fast determinations when a well-equipped laboratory is not available. More accurate is the wide range of fluorescent probes, commonly used to determine many different cell types. Their advantages are related to (i) specificity, (ii) higher label intensity which provides unambiguous identification of cell status, (iii) the possibility of being used either alone or in combination with other fluorescent dyes for the evaluation of other cell characteristics (i.e., mitochondrial activity) and (iv) the possibility of scoring by either fluorescent microscopy or flow cytometry.

Among the different available probes, the choices for sperm evaluation are those related to DNA staining: most of them including Hoescht 33258, Propidium Iodide (PI) or Rhodamine 123 (R123) are non-permeable substances able to penetrate only damaged or dead cells that have lost control of membrane permeability, and then intercalate to DNA and provide the nucleur with fluorescence, whereas others, such as SYBR 14, are permeable to the plasma membrane, staining the nucleus of viable cells. Non-permeable probes are used solely or in combination with permeable ones to facilitate identification. Double labeling with SYBR14 (green) / IP (red) is frequently used for assessing sperm quality in mammals [88] and has been used more and more frequently with fish [89-91] and shellfish sperm [92]: live spermatozoa with intact membranes fluoresce green and damaged cells fluoresce red. Scoring must be carried out immediately after labeling to avoid changes in labeling patterns, because permeation of the probes increases when the incubation time is exceeded.

The use of fluorescence microscopy requires simpler equipment than flow cytometry, but the latter allows the identification of more than 10000 cells in a few seconds, making it possible to accurately evaluate a high number of samples in a short time. Fluorescence microscopy has been used for the evaluation of spermatozoa viability in different aquacultured species using Hoetch 33258, PI or SYBR-14/PI [90,93,94] and according to Flajshans [90] is as accurate as flow cytometry when used in combination with phase contrast microscopy evaluation of the same objects, which facilitates cell status identification. Nevertheless, the use of flow cytometry is at present the method of choice for most researchers working with fish [89,91,95], bivalves [92,96], sea urchins [96] or shrimps [97], in order to evaluate the quality of samples before and after different treatments. Using this method with the oyster *Crassostrea virginica,* membrane integrity was correlated with motility of frozen/thawed sperm samples [92]. Flow cytometry enables the evaluation of different parameters in each particular particle of any suspension, such as size, shape complexity, or light emission at different wave lengths. Analysis of the data is performed with the help of specific software (Fig. 3).

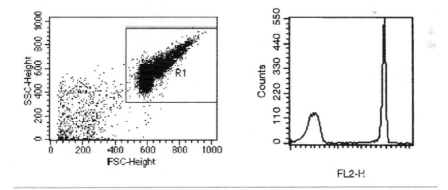

Figure 3 - Acquisition and analysis graphics provided by the Cell Quest program corresponding to the analysis of an IP labeled sperm sample by flow cytometry. A) Acquisition dot plot of the forward–scattered light (FSC) and the size-scattered light allowing the identification of the particles corresponding to spermatozoa (R1). For further analysis this region is considered and other particles and cellular debris discarded. B) Emission of red fluorescence (FL2-H) in the particles from R1. Two populations are defined, viable cells with a low FL2 signal and non-viable cells, displaying a high FL2 signal.

Differences exist between species in the permeability of spermatozoa to the different stains and probes. Spermatozoa from species with spermatophores, such as those from decapod crustaceans, are particularly difficult to analyze due to their packaging into sperm sacs or spermatophores, and the presence of the spermatophore wall as an additional envelope [98]. To avoid inaccurate measurements, the generation of standard curves is recommended: aliquots of the sperm are submitted to a lethal treatment (i.e., heating to 50°C for 10 min for oyster sperm [92]), and then mixed in different ratios with untreated sperm samples (0:100, 25:75, 50:50, 75:25, 100:0). Suitable methods for the identification of viable cells (choice of probes, concentration of dyes, time of staining, etc.) should provide scores which correlated with the predicted viable/ non-viable ratios.

Other methods used to study the percentage of viable cells in a sperm sample are based on the analysis of cytoplasm components delivered in the seminal plasma. The presence of cytoplasmic enzymes such as aspartate aminotransferase (AspAT), lactate dehydrogenase (LDH) or acid phosphatase (AcP) in the seminal plasma has been used by Ciereszko and Dabrowski [99] and Lahnsteiner et al. [50] to evaluate sperm quality in different fish species. The increase in the concentration of these enzymes in plasma has been correlated to the loss of fertility in sperm samples submitted to cryopreservation [83]. Other authors have analyzed the activity of LDH, AspAT and AcP which is retained inside the cells as a quality biomarker (the more activity detected, the higher the percentage of viable cells) [71,99,100]. This second approach is considered less accurate than the analysis of seminal plasma because the incomplete extraction of the intracellular content could provide unreliable results [100]. The main concern for the use of biochemical determinations is related to the fact that only membrane breakdown or sperm lysis is detected, but information about alterations in membrane permeability in dysfunctional cells is not provided. From a practical point of view, large quantities of sperm are required and determination is time consuming, being inappropriate for immediate quality assessment.

3.3.2 Plasma membrane composition

Plasma membrane content in phospholipids, cholesterol and lipoproteins has also been investigated in an attempt to characterize spermatozoa membrane and relate this information to sperm quality [101,102]. These components are responsible for membrane fluidity and have a significant influence in the resistance of the membrane to environmental conditions, such as those occurring during fertilization, exposure to seawater or freshwater, to activation mediums or other extenders [103,104]. There are many papers related to the characterization of plasma membrane components of spermatozoa in mammals and birds [105-108], but little effort has been made in fish [101,102]. Isolation of the membrane is done by cell lysis and ultracentrifugation or by nitrogen cavitation, and lipid determination is performed by biochemical procedures or using commercial kits. The information available for fish demonstrated that the analysis of lipid constituents in the plasma membrane could be a good index of sperm quality. This information reflects the status of broodstock and the effect of environmental conditions (temperature, salinity) on gametogenesis and spermatozoa during the reproductive period and even the fertilizing ability of spermatozoa [103]. During cryopreservation it was demonstrated that the plasma membrane undergoes loss of components and lipid rearrangements. Alterations in the cholesterol/ phospholipids ratio may be responsible for the lower or higher resistance of spermatozoa to further damage and rearrangements of membrane components could impair the sperm egg recognizing mechanisms and plasma membrane fusion, both of which are essential steps in the fertilization process.

3.3.3 Resistance to osmotic shock

The information provided by the previously described methods for checking cell viability, refers to the status of the plasma membrane at a particular moment, when the analysis is done, but does not inform us about the behaviour of the cell under physiological conditions. The plasma membrane is the first structure to suffer stress caused by changes in the environment. External fertilization requires the milt to be diluted in fresh or salt water, and the spermatozoa to be exposed to intense osmotic stress. Under such conditions, viable spermatozoa are those whose membrane remains undamaged until reaching the eggs, to be able to fertilize them. This means that spermatozoa must react to osmotic variations, suffering a change in volume, but must be able to regain their original shape when the medium is replaced by the previous osmolality.

Hyper and hypo-osmotic sensitivity tests are often performed in mammals, which are never exposed to dramatic changes in osmolality, as are external fertilizers. In aquatic species hypo-osmotic sensitivity tests frequently promote membrane leakage. Sperm resistance to the osmotic shock caused by dilution in the fertilization media could depend on many factors, mainly the osmolality of both seminal plasma and fertilization media, but also, as has been explained, on the membrane composition. In freshwater species hypo-osmotic shock caused by water (~4mOsm/Kg) is often very aggressive, and cells lose their viability in a very short time (most of them, as in rainbow trout, in less than 1 min, [93]) (Fig. 4). Consequently a high spermatozoa/egg rate is needed for fertilization: only those spermatozoa delivered close to the micropyle will have some chance of reaching it before suffering lyses. Marine species suffer a lower degree of damage and the survival time in salt water is not such a determining factor for fertilization success. The quality of milt obtained in the middle of the breeding season, which will be used for fertilization immediately after stripping, can be accurately evaluated without performing the analysis of resistance to osmotic shock. Nevertheless, low membrane resistance to this stress could indicate suboptimal membrane functionality, which could affect the ability of the sample to be submitted to further handling procedures, such as short- or long-term storage at low temperatures. Resistance of fresh spermatozoa to hypo-osmotic shock has been correlated to fertilization ability after freezing/thawing in rainbow trout sperm [56,109], milt samples with a higher percentage of very resistant cells providing higher fertility rates after cryopreservation. So, it is considered an appropriate parameter for selecting samples for cryopreservation.

Determinations of the osmotic behaviour and fragility of spermatozoa have been performed in trout after dilution in different hypo-osmotic solutions (from 10 to 300mOsm/Kg) and analyzing through a time course (from 30 sec to 15 min) by direct observation using phase contrast microscopy [110], or the use of a coulter-counter, which informs about the number of cells in the sample as well as about the size of each particle [109]. These experiments allowed the identification of different subpopulations of spermatozoa according to their resistance, but changes of volume are not great enough to be easily identified and quantified. A more accurate method is the evaluation of cell viability using fluorescent probes at different times after dilution in the solution to be tested. This kind of time course determinations has been used with carp [111] and trout sperm [56,58].

Different factors affecting plasma membrane composition, such as breeders feeding or temperature during gametogenesis and spawning [103], can also modify membrane fluidity and resistance to osmotic shock and, as a consequence, the future freezability or resistance to further treatments of the sperm samples.

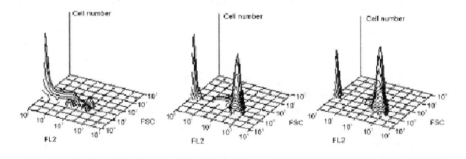

Figure 4 - Flow cytometry analysis of an IP labeled trout sperm sample 30 sec (a), 5 min (b) and 15 min (c) after dilution in a 100mOsm/Kg solution. Red fluorescence emission (FL2) and forward scattered signals are represented. Population of non viable-cells (FL2 emission about 10^3 units) clearly increases in time despite the relatively high osmolality of the tested solution.

3.3.4 Acrosome status

Sperm competence of species having acrosomal spermatozoa, such as shellfish, crustacea, echinoderm, eels, sturgeons or viviparous fish, requires that not only plasma membrane, but also the acrosome membrane, remains intact. Fluorescence labeled lectins from *Pisum sativum* aglutinin (PSA) are commonly used for the evaluation of acrosome integrity in mammalian sperm, but have not yet been adapted for their use in marine invertebrates such as oyster [92]. Evaluation of acrosome alterations, or the acrosome reaction (AR), is usually carried out by morphological evaluation under light microscopy. All the species possessing acrosome must undergo AR to fertilize the egg and AR inactivation turns sperm irreversible refractory to the oocyte. The ability the spermatozoa from any sperm sample to trigger the AR under the appropriate stimulus is another useful parameter in the evaluation of cell viability. Acrosome reaction is Ca^{2+} dependent and could be *in vitro* triggered by the divalent cation ionophore A23187 in the presence of Ca^{2+} ions in the external medium. Several studies have been developed in the sperm of several decapods as summarized by Bhavanishankar and Subramonian (protocol 35 in this Volume).

3.4 Mitochondria viability and functionality

The spermatozoa from most of the aquacultured species, from echinoderm to fish, have a small number of rounded cristate mitochondria, in accordance with the low motility duration when compared with some mammals. Exceptions are live-bearing fish, whose structure, metabolism and physiology are more similar to mammals, showing a well developed mitochondrial sheath [2]. There are not too many studies on mitochondria function and activity in fish spermatozoa, however it is obvious, extrapolating from other cell types, that it must play a key role, since they have been conserved in such a simple cell as the spermatozoon.

In mammals and due to the presence of mtDNA, it has been postulated that mitochondria could be more than a "functional engine." Perhaps mitochondria could participate in the delivery of a full set of male genetic information for a future generation [112]. However, it has been demonstrated that mitochondria and its mtDNAs are committed to degradation within the zygote [113,114], and recent studies in sea urchin revealed a mechanism similar to that described for apoptosis during the entrance of spermatozoa at fertilization (swelling, decrease in membrane potential and release of cytochrome C) [115]. Thus, mitochondria could only be present and active during live stage spermatozoa as a way of achieving fertilization, with no further role in future embryos. In this section mitochondria will only be considered as a "cell engine."

3.4.1 Methods for the analysis of mitochondrial damage

Although mitochondrial damage is frequently determined in mammalian spermatozoa, little information exists on aquatic species. Damage to mitochondria may be responsible for the decrease in the percentage of motile cells as well as for the decrease in spermatozoa ATP levels, both factors being extremely important for sperm functionallity. Different characteristics are analyzed as indicators of functional impairment: i) morphology (including, shape, size, cristae morphology, etc.), ii) mitochondrial membrane integrity [95,116-118] and iii) changes in mitochondrial membrane potential.

Morphology has been studied using electron microscopy, as was previously referred to. Nevertheless, evolution in the development of specific probes (usually fluorescent probes) has provided simpler and faster methods to evaluate not only the structure, but also the functional characteristics. Membrane-permeable lipophilic cationic fluorochromes are used as probes of membrane potential: they penetrate spermatozoa and accumulate inside mitochondria. One of these fluorochromes is the lipophilic cation compound JC-1, which has been used to assess membrane depolarization in bull, horse, human, boar, rat, deer, oyster, eel and teleost spermatozoa, (Table 1) [12,92,94,119-125]. The fluorescence of this compound depends on its concentration: JC-1 can exist in two different states, as an aggregate at high concentrations or as monomers at low concentrations of the dye. In a normal mitochondria (with polarized membrane), JC-1 penetrates the plasma membrane as a monomer and uptake by mitochondria is rapidly performed. This uptake increases concentration of the dye inside mitochondria forming aggregates which are known as J-aggregates showing red fluorescence. Nevertheless JC-1 does not accumulate in non-polarized membrane mitochondria, thus remaining in the cytoplasm as monomers, exhibiting green fluorescence. Rhodamine 123 is also a cationic fluorescent dye used as an indicator of membrane potential, providing green fluorescence to functional mitochondria. This fluorochrome has been used in the evaluation of Eastern oyster sperm [92] and obtained results showed a good correlation between mitochondrial membrane potential and sperm motility. Other similar dyes are also available, including MitoTracker Deep Red or Green, which have the same performance in accumulating inside live cell mitochondria. There are no reports on the literature on their use in fish spermatozoa, but their efficiency could be the same as the JC-1 dye. Hallap and co-workers [126] and Meseguer and co-workers [127] obtained good results on the evaluation of mitochondria activity in bull and human spermatozoa using both MitoTracker Red and Green, respectively. The evaluation of the results can thereafter be performed using flow cytometry analysis or fluorometry. Analysis by epifluorescence microscopy is difficult in several species due to the small mitochondria size, which makes identification difficult and score the events tedious.

Another approach for mitochondrial assessment is the study of mitochondrial enzyme activities, performed with commercial kits for determination and using spectophotometric methods of evaluation. Ruiz-Pesini et al. [111] studied the correlation of several mitochondrial enzyme activities (NADH dehydrogenase, succinate dehydrogenase, NADH-cytochrome C reductase, succinate-cytochrome C oxidase, citrate synthase) with the percentage of motile spermatozoa and determined that they not only correlated with motility but also with sperm viability and concentration. Some of these enzymes have also been analyzed in fish spermatozoa [128], however, due to the small number of mitochondria present in these species the accurate determination of some activities could be difficult.

More recent studies developed new reagents to evaluate mitochondrial enzymatic activity and to correlate them with motility or fertilizing capacity. As an example, the ability of spermatozoa to reduce resazurin redox dye [129,130] and methylene blue [131] was successfully used to evaluate semen quality of boar and bull. Resazurin is a non-toxic redox dye (reduction from resazurin into resorufin) that is used as an indicator of dehydrogenase activity [132].

Table 1: Species in which mitochondria have been evaluated using fluorescent dyes.

Species	Method	Objective	Reference
Rainbow trout	rhodamine 123 (Rh123)	analyse post-thaw sperm quality	[95]
Gilthead seabream	JC1	analyse fresh and post-thaw sperm quality	[89]
Striped bass	rhodamine 123 (Rh123)	damage in cryopreserved sperm	[81]
European eel	JC1	analyse fresh sperm quality	[94]
European Catfish	rhodamine 123 (Rh123)	damage in cryopreserved sperm	[117]
Eastern oyster	rhodamine 123 (Rh123)	characterize sperm quality for cryopreservation	[92]

It is well known that mitochondrial dehydrogenases are involved in many vital anabolic and catabolic processes of spermatozoa. MTT (3-(4,5-Dimethylthiazol-2-yl)-2,5-diphenyltetrazolium bromide) is a yellow water-soluble tetrazolium salt which is converted to water-insoluble purple formazan on the reductive cleavage of its tetrazolium ring by the succinate dehydrogenase system of the active mitochondria [133]. The absorbance of this colored solution can be quantified by measuring at a certain wavelength (usually between 500 and 600 nm) by a spectrophotometer. This reduction takes place only when mitochondrial reductase enzymes are active, and therefore conversion is directly related to the number of viable (living) cells. When the amount of purple formazan produced by cells treated with an agent is compared with the amount of formazan produced by untreated control cells, the effectiveness of the agent in causing the death of cells can be deduced through the production of a dose-response curve. MTT is a simple assay, a rapid and reliable method applied to the study of human, boar, equine and bovine sperm [134-136], and more recently Hamoutene et al. [137] demonstrated its application in invertebrate (sea urchin and scallops) and fish (capelins) spermatozoa.

3.5 Motility

Except for crustacea, sperm motility is likely to be the most important criteria in assessing sperm quality. Motility is by far the most widely used biomarker for sperm quality and has been applied in different situations for the analysis of sperm status. It is true that this physiological analysis does not indicate the "real" status of sperm, but can in some cases indicate the effect of a given treatment or the probability of spermatozoa reaching and fertilizing the egg. Correlations between motility and fertilizing capacity have been reported by several authors [138-141], but other researchers prevent against the relationship between motility and fertility [142,143]. This is particularly clear in species with acrosome, in which motile spermatozoa could have an impaired acrosome, being unable to fertilize.

Motility analyses have been performed in aquatic species in studies related to different subjects: from the analysis of the effect of heavy metals or other pollutants [144] to the evaluation of endocrine disrupters action [145], or the effect of cryopreservation [146]. Rurangwa et al. [60], Martinez-Pastor (personal communication) or Lahnsteiner et al. [62] also proposed the use of appropriate sperm motility assessment as a mean of selecting broodstock, considering that males producing vigorous motile spermatozoa could improve sperm quality in succeeding generations. Motility evaluation is also used to assess the effect of external factors such as stress, nutrition, husbandry conditions and environmental factors, helping to improve broodstock quality and animal welfare. Several studies have applied the analysis of sperm motility to evaluate different broodstock treatments, such as spermiation induction by hormonal therapies [147], chemical sterilization [57] or sex-reversal [48], as well as to the study of reproductive strategies such as the tactics developed by the different types of bluegill males (sneakers and parentals) [148].

It is important to know the physiology of movement in each particular species in order to adapt the method of evaluation. Species with external fertilization need dilution in an activation solution, but internal fertilizers do not. In live-bearing fish such as *Xiphophorus*, the sperm is motile upon collection before dilution, activating solutions are not necessary for motility estimates and spermatozoa sustain motility long after suspension in HBSS medium [2].

In species requiring activation, one of the key factors to proceed with an accurate evaluation is to dilute the sperm at a convenient rate. Dilution allows all the spermatozoa to be activated simultaneously and avoids high concentration in the visible field and subsequent errors in tracking. This is especially important in species with high sperm concentrations and rapid spermatozoa movement. The recommended procedure for those species requiring activation consists of a double dilution of sperm.

1. First dilution is made in a non-activation solution, to reduce sperm concentration and facilitate further activation. Non-activating solution usually mimicks seminal plasma. Attention must be paid to osmolality, pH and ion concentration, to avoid activating conditions in the solution. However, if this is not possible, in most marine species a simple NaCl or sucrose solution with osmolality similar to that of seminal plasma (around 300 mOsm/Kg), prevents sperm motility and does not affect further activation. The dilution rate depends on the species (i.e., 1:100 in rainbow trout; 1:15 in Senegalese sole) and should be the same for high or low sperm concentration samples. This will allow comparing results between samples.

2. A second dilution is made in the activating solution. In marine species, seawater is normally used and in the case of river species fresh water is the medium for spermatozoa activation. Taking into account the short duration of sperm movement in most aquatic species, activation is usually carried out adding the activating solution over a drop of diluted sperm placed on a slide, mixing them, and observing immediately under the microscope. The proportion of the activation solution in the field should also be established using the same concern as for the first dilution (i.e., 1:10 in rainbow trout; 1:5 for gilthead seabream and Senegalese sole).

3. The temperature of both sperm predilution and activation solution should be similar in all measurements and should be in accordance with the temperature of water in the broodstock tank in which fish would normally spawn. For example, for salmonids

4°C activation temperature is recommended and for some marine fish such as gilthead seabream or Senegalese sole, 19 to 22°C is suggested. Other factors affecting motility can also be important, including pH or other specific factors previously quoted.

4. When activating the sperm under the microscope "secondary movement" is produced by water movement. If this is not avoided, the automatic system used for evaluation will track these spermatozoa and integrate the results as motile spermatozoa. The use of an appropriate chamber such as a Makler chamber or dischargeable chambers will be useful in this determination. These chambers do not interfere with spermatozoa movement and set spermatozoa in one plan allowing easy visualization for tracking.

The method of choice for motility evaluation is frequently, even nowadays, the subjective visual assessment using light microscopy, depending on the ability and experience of the operator. The development of objective methods of computerised assisted evaluation (CASA), has provided a powerful tool, making motility evaluation much more accurate and giving us information about many different characteristics of individual spermatozoa movement. Moreover, CASA analysis has increased knowledge on spermatozoa physiology since several studies were conducted in the evaluation of the effect of different ions, activation solutions, plasma membrane channel inhibitors, physicochemical alterations of the external media, among others.

3.5.1 Subjective evaluation methods

The assessment of motility has essentially relied for a long time on subjective estimation of some motility characteristics [149,150] such as the percentage of motile spermatozoa [151] or the total duration of movement [152]. Some authors have used arbitrary scales of motility according to previously established criteria. Basically, these descriptive subjective scales, which score the samples from 0 to 4 or 5, classify motility in terms of percentage of moving spermatozoa in the field of view [153] or percentage of motile spermatozoa and swimming vigour [149]. Sansone et al. [154] developed a more complex system for seabass: motility was scored according to spermatozoa velocity (vigorous, rapid and forward-movement). The main problem of these scales is the low reproducibility and interpretation by different observers. Often it is difficult to establish some criteria to determine motility duration and in those species with longer duration such as grouper or herring is also time-consuming, taking into account that motility can persist for longer than one hour. Moreover, the concept of forward movement is sometimes difficult to establish and in some cases vibratory movement of spermatozoa has been considered for the analysis of total motility. The particular pattern of movement in some species can also make it difficult to apply this criterion. It is also difficult to compare scores between different samples from different days in spite of the observations being made by the same observer and under the same conditions. The estimation of the percentage of motile cells is also difficult, because the movement produced by spermatozoa can mask the number of spermatozoa not moving in the field of view, producing overestimation of motility in a given sample. Thus, depending on the species, motility assessment using this type of observation is not always reliable, and previous training of the operator is very important. However, taking into account the fact that is a rapid and cheap method of evaluation, which does not require special equipment (light microscope), subjective estimation of motility is still the method of choice in many farms to predict fertility and select males, basically in terms of motile and non-motile sperm. In some fish farms that use frozen sperm, post-thaw motility is checked to confirm the profitable use of those sperm samples to fertilize the eggs (Stolt Sea Farm, personal comment) and samples with low motility are discarded.

3.5.2 Quantitative computer-assisted assessment

To increase accuracy of sperm motility measurements, numerous attempts were made to standardize the analysis. Several systems based on computer-aided sperm analysis (CASA) were proposed (Fig. 5). These systems are the evolution of multiple photomicrography exposure and video-micrography techniques for spermatozoa tracking, using computers equipped with imaging software. A CASA system is composed of equipment used to visualize and digitalize static and dynamic sperm images and software capable of analyzing and tracking each spermatozoon. Although the development of techniques to assess spermatozoa motility have recently very much progressed, most advances have been made firstly in mammals and then applied to fish and aquatic organisms. These systems were initially developed for human male infertility and used in androgenic clinics [155-157]. The objective measurement of motility in fish sperm only arrived after some modifications in the systems to adapt them to the short motility duration and to the use of activating mediums, as well as to the high frequency of fish spermatozoa flagella beating and the type of movement, which required other specificity in contrast objectives. The first reports on objective measurement of motility were made by Cosson et al. [37] in trout sperm using stroboscopic illumination and video recording. Nowadays, more sophisticated systems make it possible to characterize several parameters in spermatozoa movement. Type of movement, trajectory and velocity can be tracked from slower to faster spermatozoa. These systems have been applied in several species such as turbot [146,158], African catfish [159], gilthead seabream [12], Sole, [13,14], red seabream, [82] within others (for more details see Table 2).

Several systems are available on the market, from basic equipment to very sophisticated systems (Table 3). However, for fish spermatozoa, a simple negative contrast microscope could be adapted with a small video camera connected to a PC (labtop or desktop). The software used can be provided by several companies (see Table 3 for some examples available on internet). Recently, free software was designed and published [160] as an alternative to high cost commercial systems and is fully capable of rapidly generating quantitative reproducible measurements for fish sperm motility. As in commercial systems a graphic user interface is employed and sperm identification and tracking are automated. As stated by the authors, since this software is an open source, independent investigation and validation of the used algorithms is possible, potentially increasing quality control for data gathering. For details on this system check Wilson-Leedy and Ingermann, [160] or http://rsb.info.nih.gov/ij/plugins/casa.html.

3.5.2.1 Parameters rendered by quantitative systems

The parameters rendered by most of the available systems provide similar information on spermatozoa characteristics. All software gives different types of parameters, from those related to the average of spermatozoa in the observed field, as percentage of motile sperm (MOT) and the percentage of progressive spermatozoa (MP), to descriptors of individual trajectory of the cells such as linearity (LIN) or of the spermatozoa velocity, such as curvilinear velocity (VCL) or straight line velocity (VSL). The type of movement and trajectory depends on the species and the activation solution, but similar aspects have been identified in many teleost. Teleost spermatozoa generally move in a straight or slightly curved trajectory, but in suboptimal conditions (end of movement, presence of toxicants, inappropriate formulation of the dilution media) movement becomes more curved or even circular [60]. The analysis of trajectory could subsequently be a good index of sperm quality or be useful for improving the composition of the dilution media used in artificial fertilization, as has been done for catfish and wolfish [52,165]. Related with velocity, VCL (actual velocity along trajectory), and VSL (distance in line from the initial to the end point divided by tracking time), are the two most widely used parameters. Both parameters are identical when movement is linear, but are different for curved or circled trajectories. Linearity (LIN) gives the relation between VSL/VCL and is a good index of spermatozoa trajectory. In some species, immediately after sperm activation both parameters are similar and tend to be different after some time and at the end of movement, indicating that spermatozoa move linearly at the beginning but tend to move in circles towards the end. VAP (velocity along a derived smooth path), is used in species with more erratic pathways such as the spermatozoa from stickleback (*G. aculeatus*) and wolfish (*A. minor*) [165,166]. In these species straightness, STR, is also used as a descriptor of the trajectory.

The analysis of the data provided by CASA systems for a sperm sample can be performed to identify spermatozoa subpopulations according to the type of movement (for example slow, rapid and ultra-rapid spermatozoa, [168]). This kind of analysis is much more precise than the evaluation of average values in the sperm sample. Any sperm dose is made from the mixture of cells whose quality and ability to fertilize is different, and the quality of the sample is probably more related to the percentage of good quality spermatozoa than to the average values of all the contained cells. The choice of parameters used for the identification of spermatozoa subpopulations shall be done with appropriate statistical procedures that must be applied to each particular system, species and laboratory, under the same conditions of observation.

The application of computer assisted sperm analysis is quite simple; however, an initial calibration of the system is required, especially in those species with high or low spermatozoa velocity. In some species such as sole, straight line velocity is so high that some of the systems available are not capable of tracking all the cells moving in the field, producing error in the analysis. To avoid those errors preliminary investigation on the physiological characteristics of fish spermatozoa activation in a given species should be done.

Table 2 - Species in which CASA analysis protocols have been established.

Species	Reference
Rainbow trout, *Oncorhynchus mykiss*	[55]
Grayling, *Thymallus thymallus*	[161]
African catfish, *Clarias gariepinus*	[52,159]
Turbot, *Scophthalmus maximus*	[158,162]
Gilthead seabream, *Sparus aurata*	[12]
Sole, *Solea senegalensis*	[13,14]
Grouper, *Epinephelus marginatus*	[233]
Red seabream, *Pagrus major*	[82]
Zebrafish, *Danio rerio*	[160]
Lake sturgeon, *Acipenser fulvescens*	[9]
Common carp, *Cyprinus carpio*	[163]
Atlantic halibut, *Hippoglossus hippoglossus*	[164]
Bluegill, *Lepomis macrochirus*	[148]
Spotted wolfish, *Anarhichas minor*	[165]
Stickleback, *Gasterosteus aculeatus*	[166,167]
European eel, *Anguilla anguilla*	[94]

3.6 Spermatozoa metabolic activity

As has been explained seminal plasma supports spermatozoa metabolism and the study of enzymatic activity and metabolic constituents both at spermatozoa and seminal plasma level could be good tools for spermatozoa status prediction. The analysis of plasma constituents has been developed in Section 3.1.1 , and we will now focus on the metabolites and enzymes involved in sperm metabolic pathways with special reference to ATP.

Table 3 - Systems available for computerized sperm motility analysis.

Type of systems	Company
Hobson Sperm Tracker	Hobson Vision Ltd. Baslow, UK
ISAS	ISAS, Valencia, Spain
medeaLAB CASA	Medical Technology Vertriebs-GMBH
Hamilton –Thornesperm analyzer	Hamilton–Thorne
Sperm class analyser- SCA 2002	Microptics, Sa, Spain
CellTrak	Motion Analysis Corporation, Santa Rosa, CA
CRISMAS	Image house medivcal
AutoSperm	2005-2007Frank Schoonjans
CEROS Sperm Analyzer	Mid Atlantic Diagnostics Inc.
Cellsoft	Cryo Resources, New York, NY
Free CASA	Free http://rsb.info.nih.gov/ij/plugins/casa.html

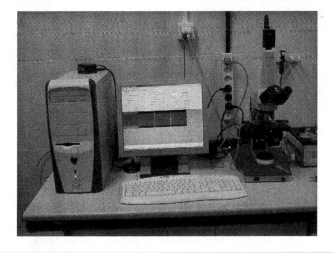

Figure 5 - Example of a computer sperm analysis system used for motility analysis in fish sperm.

As we have mentioned before some intracellular compounds are released when the cells go into the process of ageing, thus being good indicators of cell damage [54]. Therefore, a low proportion of spermatozoa to seminal plasma enzyme activity should be good predictor for high semen quality [55]. Nevertheless, cell damage can be achieved without membrane rupture and liberation of cell constituents, compromising only the internal structures of spermatozoa and affecting essential metabolic pathways, such as oxidative phosphorilation, citrate cycle, lipid metabolism, glycolysis, ATP metabolism or the respiration rate [55,62]. Some of the enzymes involved or their substrates and metabolites that have been analyzed are creatinine phosphatase, creatinine kinase, malate dehydrogenase, lactate dehydrogenase, glucose-6-P-dehydrogenase, ATP, ADP, AMP, NADH, lipids, fatty acids, glucose, lactate, piruvate, etc. [60,169]. For the evaluations whole milt can be exposed to lytic conditions such as freezing directly in liquid nitrogen, followed by thawing and homogenization to extract the components from the cells [169]. Milt can also be centrifuged and then the spermatozoa pellet resuspended in a non-activating solution and subjected to a protocol of component extraction (see Lahnsteiner et al. [55] for details).

The study of spermatozoa metabolism is closely related to the analysis of seminal plasma content and their application as a standard biomarker is not simple. Firstly, because in spite of the high number of studies on seminal plasma content, knowledge on the metabolism and metabolic pathways in spermatozoa from aquatic species is still scarce and most of the studies rely on freshwater teleosts [55,63,169]. Secondly, not all enzymatic determinations correlate with fertilization or other quality parameters despite being involved in similar processes [55], which makes some results difficult to understand. However, in the literatute there are several examples in which sperm metabolic activity has been determined in cells and plasma for a different purpose (Table 4).

Oxygen consumption is also a good indicator of spermatozoa activity, related to sperm quality [55,62].

3.6.1 ATP production and consumption in spermatozoa

ATP levels have been commonly used as a predictive parameter of sperm motility. The reason lies in the fact that the molecular mechanisms by which motility is initiated are commonly related to ATP as an energy provider [174].

Many different pathways can provide spermatozoa with energy: aerobic glycolysis, tricarboxylic acid cycle, oxidative phosphorylation, anaerobic fermentation, lipid catabolism, beta oxidation or even some authors mentioned external energy sources [128,170,173,175]. However, this last source is unlikely that could be used for spermatozoa movement in external fertilizers because there are hardly any nutrients present in their fertilization environments [175]. Oxidative phosphorylation, carboxylic acid cycle and aerobic glycolsis are considered the central energy supplying pathways for spermatozoa in some species. In salmonids ATP generation is based predominantly on oxidative phosphorylation, the glucolytic pathway being relatively minor. The endogenous ATP produced by these pathways must be accumulated, since sperm motility depends mainly on endogenous ATP stores in spermatozoa produced before activation, during the spermiation period [169].

It is known that ATP is required for the interaction between dinein and tubulin in the axonema which, in the end, will produce flagellar movement. But it should be pointed out that in many species the flagellum is separated from the mitochondrion by the cytoplasmic canal, and therefore mitochondrial ATP is not in contact with the axoneme [176] a shuttle to transport ATP close to dynein ATPase being required. Phosphocreatine could be understood as one of these shuttle molecules. The enzyme CPK catalyzes the reversible conversion of phosphocreatine (PCr) to creatine (Cr), with the associated production of ATP. In the sperm of some teleosts, the concentration of PCr decreases during the motility phase suggesting that motility depends on both ATP and PCr [177].

In some species (rainbow trout, chum salmon) flagella must be exposed to both cyclic AMP and ATP in order to become functionally motile. But the involvement of cAMP and the subsequent cascade of protein phosphorylation in the activation of fish sperm motility seem to be restricted to salmonids. The environmental factors that initiate motility vary among fishes and have been described in previous paragraphs.

Once the activation of spermatozoa motility occurs, changes in metabolites and compounds with high-energy bonds take place inside the cells (Fig.6), and a rapid depletion of ATP during movement is produced in several species: trout, carp, silurids, sea bass or turbot [176,178,179]. There are different reasons for this occurring (Fig.3). One explanation to this fact, it is the decrease in ATP/ADP ratio produced by the ATP hydrolysis which is carried out by the dinein ATPase. ATP consumption cannot be compensated by ATP production by mitochondrial oxidative phosphorilation [176]. Mitochondrial oxidative phospshorylation remains insufficient to sustain endogenous ATP stores. Another reason is that a decrease in the ATP/ADP ratio could diminish flagellar beat frequency or even cause motility arrest because ADP is a competitive inhibitor of ATP for dynein ATPase. Finally, the decrease in swimming intensity has also been correlated with the internal acidification produced by ATP hydrolysis that decreases dynein ATPase activity [176] (Fig. 3).

It is known that ATP levels can be re-established, a phenomenon called "second motility phase" or "revival" [11]. For example, in carp spermatozoa, the ATP stores restored by preventing the residual motility phase (approx. 5-10 min post-activation) in a high osmotic pressure medium (300 mOsm kg^{-1}) [67]. This high osmolality medium seems to block the axonemal machinery allowing ATP regeneration [176]. Thus the spermatozoa could have a second motility period, if external osmolality was decreased. In trout, ATP concentration decreases rapidly during the first 30 s after initiation of sperm motility in an activation medium and then slowly increases to reach its original levels 15 min after dilution in the absence of Ca^{2+} [11].

Table 4- Species in which metabolic activity has been used to assay sperm quality for different purposes.

Species	Enzymes	Objective	Reference
Rainbow trout, *Oncorhynchus mykiss*	seminal plasma and cells	characterize sperm quality for cryopreservation	[83,110,170]
Danube bleak, *Chalcalburnus chalcoides*	cells	relation between sperm metabolism, viability and motility	[170]
Bluegill, *Lepomis macrochirus*	cells	study of reproductive strategy of parental and sneaker males	[148]
Bleak *Alburnus alburnus*	seminal plasma and cells	relation between sperm metabolism and motility	[50]
African catfish, *Clarias gariepinus*	cells and seminal vesicle secretion	characterize intratesticular metabolism and seminal vesicle secretion	[44,128]
Herring, *Clupea harengus*	cells	characterize spermatozoa metabolism	[171]
Northern pike, *Esox lucius*	seminal plasma and cells	characterize sperm quality for cryopreservation	[73,172]
Turbot, *Scophthalmus maximus*	cells	relation between sperm metabolism and motility	[173]

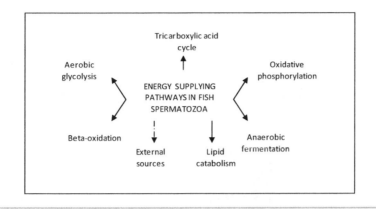

Figure 6 - Energy supplying pathways in fish spermatozoa.

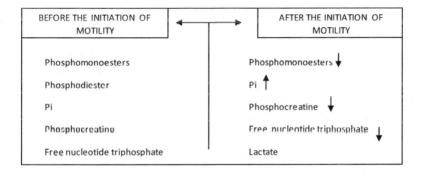

Figure 7 - Changes in metabolites and compounds with high-energy bonds prior to and after the initiation of motility.

3.6.2 Intracellular ATP and sperm quality

The correlation between ATP content and motility is generally accepted. However, this observation cannot be generalized to all cases. In the cyprinid *Alburnus alburnus*, Lahnsteiner et al. [50] failed to detect a direct relationship between ATP levels and motility. In bluegill, Burness et al. [177] found that ATP levels were also unrelated to either sperm swimming speed or percent of sperm that were motile. With these exceptions in mind, we can assume the importance of ATP as an energy source in the spermatozoa and the potential relationships between both parameters: ATP content and motility. The maintenance of high levels of ATP in sperm is considered of high importance for modulating flagellar motility [180, 181] and both parameters have been correlated in many other species [55].

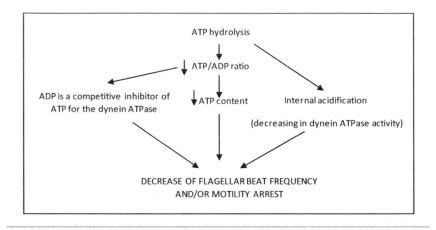

Figure 8 - ATP related facts involved in motility decrease/arrest.

Zilli et al. [181] reported a significant linear correlation between ATP concentration and fertilization rate in sea bass sperm. They even suggested the use of ATP concentration as a predictor of fertilization ability instead of sperm motility. They argued that the determination of ATP content is not as subjective as is motility determination based on microscopic observation and it is faster and less expensive in comparison with computer assisted sperm analysis system (CASA). Although the linear correlation between intracellular ATP and fertilization ability cannot be guaranteed for all samples in all species, most of the results indicate that the importance of ATP as a biomarker for semen quality is unquestionable.

3.7 DNA integrity

The final objective of spermatozoa is to deliver their genetic material to contribute to embryo development. The high degree of chromatin condensation with proteins and the presence of antioxidants in the seminal plasma protect DNA from injuries until the egg is reached. Moreover, the egg has its own molecular mechanisms to repair the possible alterations in the sperm chromatin structure to a certain degree. Nevertheless, some different factors from failures during spermatogenesis to the effect of environmental stressors to which the cell are exposed, could promote base oxidization or strand breaks in DNA.

A complete sperm evaluation should include chromatin analysis but this is rarely considered in the assessment of sperm quality of aquatic species. This kind of study is becoming more common in mammals, in which some correlation between fertility and chromatin damage has been reported [182,183]. At present, the application of these analyses in aquaculture could be related to the identification of particular problems during breeding, such as the scarce contribution of some particular males to the fertilization of females in the same tank, but is specially interesting for the evaluation of damage promoted by different factors or treatments such as the presence of toxic agents, irradiation with UV light, or the cryopreservation of cells. Some of these treatments are specifically applied with the aim of promoting DNA inactivation in motile spermatozoa, for further use in gynogenetic procedures.

Different methods are used for the evaluation of chromatin damage, most of them developed for the study of mutagenesis in other cell types and then adapted to spermatozoa. The most widely used are the TUNEL, the sperm chromatin structure assay (SCSA) and the Comet assay single cell gel electrophoresis (SCGE), and all of them detect simple or double strand breaks in the DNA double helix. In the TUNEL assay a fluorescence or radioactive labelled nucleotide is transferred to the 3' end of the DNA strand [184]. The more strand breaks, the more free 3' ends and higher fluorescence or radioactive emission is detected in the labeled cell. Nevertheless, this method is falling into disuse due to its low reproducibility and the large increase in the use of the SCSA. The sperm chromatin structure assay was developed by Evenson and colleagues [185] to evaluate male infertility in humans and is now applied worldwide, not just to men, but all kinds of mammals [183,186]. The main advantages are the simplicity of the method and the possibility of combining with flow cytometry to analyze a high number of samples and spermatozoa in a short time. SCSA is based on the characteristic methacromasy of the Acridine Orange (AO), which fluoresces green when associated to the intact double DNA helix and red when associated with denatured DNA. Cells are previously subjected to an acid denaturation treatment, in such a way that those cells with higher susceptibility to the treatment (revealing lower DNA stability) show a red nucleus, whereas the nucleus of cells with more stable chromatin fluoresces green. To our knowledge, references to the use of SCSA in aquacultured species are limited to the study of the effect of free radicals on carp sperm carried out by Zhou et al. [187]. Some unpublished attempts have been made by our group with trout sperm, but the results are not as reliable as in mammals, probably because the different chromatin structure requires some modification of the denaturation process.

The Comet assay is a versatile, sensitive and inexpensive method to evaluate strand breaks in individual cells [188]. The sperm sample is spread on a microscope slide previously covered with an agarose gel. Slides are placed in a solution to lyse the cell components leaving the DNA immobilized in the agarose. Then DNA is denatured in an alkaline solution and an electrophoresis is performed allowing DNA fragments migrate away from the main bulk of nuclear DNA. After staining with fluorescent DNA-specific stain, DNA is observed at fluorescence microscopy and cells with DNA strand breaks display a comet-like shape, with the undamaged DNA located in the head of the Comet and the fragmented DNA disperse through the comet tail (Figure 9). Comet tail length can be measured manually, but specific image analysis software provides determination of several parameters (tail moment, percentage of DNA in tail, etc.) which allow a more objective evaluation to be done. Apart from available commercial software, there are some basic free versions for comet analysis (http://www.autocomet.com/products_cometscore.php).

The Comet assay has been used to analyze DNA damage promoted by free radicals, UV irradiation, fresh storage or cryopreservation in seabass, seabream, sea lamprey and rainbow trout sperm [189-194]. Some practices (i.e., gynogenesis), require the use of viable and motile spermatozoa with an inactive nucleus. Inactivation is usually performed by irradiation, and the Comet assay can be used for monitoring the effectiveness of the method. Its use in combination with sperm motility analysis has been proposed for optimizing gynogenetic procedures in fish [191].

More sensitive methods are being used to detect other kinds of damage in the genome, such as the presence of oxidized nucleotides. The use of specific endonucleases in combination with further applications of the Comet assay [193] or the detection of damage in specific DNA sequences using qPCR [195] are promising options.

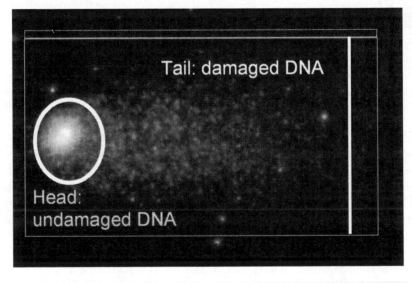

Figure 9- Nuclear DNA of a single spermatozoon after the application of the Comet assay procedure.

3.8 Fertilizing ability

Fertilization can be impaired by a relatively high number of factors, most of them unknown because so little is known about the reproductive mechanisms related to the attraction of spermatozoa to the micropyle (especially in species with external fertilization and with one or more micropyles in the egg), to the recognition and fusion of gametes and to some aspects of early embryo development.

Besides damage affecting the parameters described earlier (motility, viability, functionality of plasma membrane, DNA integrity, metabolism, etc.), which determine the principal functions of spermatozoa, some other changes in the cells can be responsible for the low capacity of spermatozoa to reach and enter the egg during the fertilization process. Thus, fertility tests have been considered the most conclusive test in the assay of sperm quality.

However, like other tests, this assay presents several disadvantages, mostly because it depends on other factors such as egg availability and quality or fertilization and egg incubation procedures. Sperm assessment is often done in farms with the objective of avoiding interference of bad quality sperm in the fertilization of eggs and to discard the possibility of having low fertility rates due to sperm quality. Thus this assay would not be the most appropriate under these conditions. However, in several research areas, fertility tests are still the principal assays performed and several points must be considered in order to validate the obtained results.

Firstly, it is very important to optimize the sperm:egg ratio. Too many spermatozoa can produce undesired effects due to competition for the micropyle or, in low quality samples, activated spermatozoa can experience problems in reaching the eggs because there are a high number of non-viable or non-motile spermatozoa present, interfering in the process of fertilization. The opposite is also undesirable; low concentrations of spermatozoa per egg can also be disadvantageous because not all cells may reach the eggs or micropyles. In some experiments performed with sperm subjected to different treatments it is difficult to define the optimal rate because when untreated sperm is used (control groups), the required spermatozoon-to-egg ratio could be different (usually lower) to the optimal one required for treated spermatozoa. As an example, with cryopreserved samples normally 100-fold, the number of spermatozoa is recommended to produce the same fertility rates as fresh sperm. However, this will depend on the quality of the used sperm, the species and volume of eggs used for fertilization. In some studies, small egg batches do not give the same results as large egg batches, probably because methods used for fertilization are not standardized.

Secondly, activation can also be crucial and influence results both in terms of its composition and used volume. The fertilization process by itself can cause damage to spermatozoa because the environment in which it takes place, both for freshwater and marine species, is undesirable. To reduce stress produced during fertilization and optimize the process, there is the possibility of design fertilization solutions that do not compromise spermatozoa viability. Basically, for marine species, the decrease in osmolality of the activation solution without compromising sperm motility could be a solution to increase fertility. In freshwater species, a similar approach has been designed but in the opposite way, increasing the osmolality of the activation medium instead of using normal freshwater. This approach produces an increase in the fertility rates of cryopreserved sperm in rainbow trout using a saline activation solution 125 mOsm/Kg [93,196-198]. The incorporation of substances that stimulate motility such as methylxanthines usually results in an increase in fertility. However, Robles et al. [48], working with rainbow trout did not obtain a significant increase in motility and fertility of sperm activated with theophyline, nor with caffeine.

Another point that must be considered in the discussion of the use of fertility tests as a way of assessing sperm quality is the parameter used to evaluate fertilization ability. Although the use of "fertility tests" is quite commonly reported in literature, some authors evaluated the fertilization rate immediately after fertilization. Assessment of hatching is more convenient, because early development could progress even upon fertilization with "abnormal" spermatozoa. In most marine species this assay could be performed until larval hatch because embryo development is quite short, but in some freshwater species such as trout, embryo development is quite long, which means that fertility tests in this case would not be the most appropriate test to be performed if the quality of sperm is to be checked in a short period of time. Both in marine and freshwater species the evaluation of fertility success is recommended, if not possible at hatch, after gastrulation, because before this stage it is difficult to detect some genetic damage provoked by bad quality sperm.

4. FACTORS AFFECTING SPERM QUALITY

4.1 Broodstock husbandry conditions, management and nutrition

The maintenance of broodstock in appropriate conditions for each species is an important requirement for the production of high quality sperm. This is related to the physico-chemical aspects of water (temperature, oxygen, salinity), photoperiod, type of tanks, density of fish among other factors. In aquacultured species it is important to establish the appropriate husbandry conditions in order to avoid influences in gamete quality.

Photoperiod and temperature are the most important factors of broodstock management affecting reproduction and gamete quality. In Senegalese sole, sperm volume decreases when broodstock water temperature is increased [13]. In rainbow trout the quality of sperm used for cryopreservation could be improved by appropriate broodstock handling. In this species, a high rearing temperature during gametogenesis followed by transfer to colder water improved sperm quality [103]. Salinity is another environmental factor which influences membrane lipid composition, especially in osmoregulatory organs [199-201]. However, in brown trout, salinity did not affect the quality of sperm since there was no effect of this factor as a direct modulator of the cholesterol/phospholipid ratio present in spermatozoa plasma membrane [104].

Several constituents of spermatozoa and seminal plasma have been associated with broodstock nutrition. Spermatozoa plasma membrane phospholipids and cholesterol levels were associated with the type of food given to male broodstock. In Senegalese sole, males fed with mussels had more cholesterol in spermatozoa plasma membrane than spermatozoa from males fed with polychaetes (Cabrita et al., unpublished results). Also, in rainbow trout, broodstook nutrition affected the spermatozoa phospholipid composition [101,202]. Other components introduced in the diet such as vitamins (E and C) can be incorporated in the composition of seminal plasma and play an important role in spermatozoa quality [203,204]. Their determinations could be used as indicators of broodstock nutritional requirements. Other substances introduced in the feeding regimes can affect negatively the quality of gametes. This is the case with gossypol, a substance present in cotton seeds used in pellet food. Gossypol impaired yellow perch spermatozoa motility and lactate dehydrogenase activity [205]. A similar effect was reported by Rinchard et al. [206] in sea lamprey, and by Lee et al. [207] in rainbow trout.

Sperm quality can also be affected by the modification of the breeding season as well as throughout the breeding period. Robles et al. [48] found that sperm from sex-reversed trout had more resistance to freezing if samples were collected during the natural breeding season than if they were collected from fish maintained in advanced photoperiod. This fact was probably related to differences in initial sperm quality from both broodstocks. Sperm production and quality also change during the breeding season. In European seabass, *Dicentrarchus labrax*, a decrease in sperm concentration was observed 2 weeks after the beginning of the reproductive season [173]. At the end of the reproductive period sperm ageing effects were also notorious in Atlantic cod [164]. Morphological and biochemical changes such as motility impairment, loss of flagella and morphological changes in the sperm head shape were the main reported changes. The same findings were recorded for sea bass, turbot, halibut and red porky [138,173,208,209]. In other species such as the Senegalese sole this effect was not so notorious and sperm quality occasionally changed during the year but not according to the reproductive season established for the species on the basis of the female spawning period [13]. However, these authors did identify two picks of main production during the year corresponding to the female reproductive period.

Stripping frequency can also affect the quality of sperm especially if a certain period of rest is not guarded. In turbot and sole, there were no significant differences in the quality of sperm stripped weekly ([209], Cabrita et al., unpublished results). In rainbow trout, sperm could be biweekly stripped without interference in sperm quality [103].

4.2 Sperm storage

Short-term storage of sperm solves problems in the management of reproduction such as asynchrony between females and males or a decrease in semen quality at the end of the reproductive season. However, several factors can affect the quality and viability of the stored sperm. Individual male variability and storage conditions are critical factors that determine the viability of sperm after short-term storage. Different factors, but especially oxidative stress, can contribute to accelerating cell ageing during short-term storage: cooling is, itself, a stressful factor that induces the generation of free radicals [210], the presence of blood cells in contaminated samples seriously increases these oxidative events [211], possible bacterial growth would also contribute in the same way and, finally, the spermatozoa metabolism itself or the impairment of mitochondrial function, increases these negative effects [193]. Oxidative stress promotes an ageing process reducing sperm motility [191] affecting the structure of lipids and the plasma membrane function [212,213] and causes serious damage to the DNA, promoting strand breaks and base oxidization [191,193]. These sperm modifications are extremely important because they could reduce the fertilization ability of the cell, but could also interfere with the normal development of future embryos.

Only noncontaminated samples can be used for short-term storage. Desiccation and bacterial growth, as well as high metabolic rates should be avoided to reduce sperm injury and proper sperm diluents and temperatures should be used to achieve this goal. Gaseous atmosphere is critical in the process [164]. However, the effect of oxygen on fish semen is contradictory. It has been reported that it has a positive effect in Salmonidae [214,215], Ictaluridae [216] and Siluridae [217] but some negative effects have been also reported in salmonid spermatozoa [218]. Antibiotics should also be employed with caution. African catfish semen for example, only tolerated gentamycin at a maximal concentration of 1 mg/ml, and the other tested antibiotics reduced sperm viability [219]. Storage temperature is also a decisive factor. African catfish semen viability was prolonged by cooling to 4°C [219]. Cooling reduces the metabolic rate and stabilizes the energy metabolism as well as motility and viability [128], but also bacterial growth decreases at low temperatures. The replenishing of the storage medium was originally presumed to be beneficial for sperm viability, however, in Atlantic halibut it resulted in earlier loss of viability of stored spermatozoa [164]. Babiak and their colleagues [164] also stated that spermatozoa sedimentation during storage does not affect spermatozoa viability for 3-4 weeks, but it is deleterious in the long run.

Cryopreservation guarantees the indefinite storage of samples. This procedure can be used to preserve gametes from genetically superior males or endangered species, aid in the transport of semen and reduce the risk of disease transmission [220]. In general most of the sperm quality indicators: sperm motility, ATP content, viability and fertility, are reduced after cryopreservation [48,89]. However, a correlation among all these parameters is not always found. The percentage of live spermatozoa in frozen/ thawed catfish sperm did not correlate with hatching rate after fertilization or with the percentage of motile spermatozoa, but negatively correlated with velocity of movement [221]. Cabrita and colleagues [89] reported that in gilthead sea bream sperm, mitochondrial status was affected by cryopreservation, since there was a decrease in sperm motility or ATP content (3.17 nmol ATP/10 spermatozoa to 1.7 nmol ATP/10 spermatozoa in 1:20 frozen samples) and an increase in the percentage of cells with mitochondrial depolarized membranes (11% for fresh and 27% for 1:20 frozen samples). However, fertility rate was similar either using fresh or frozen/thawed sperm (77 and 75% hatched larvae, respectively).

There is still another parameter to take into account after sperm cryopreservation, DNA integrity. Cabrita et al. [189] determined the degree of spermatozoal DNA damage induced by cryopreservation in two commercially cultured species, rainbow trout and gilthead sea bream. In rainbow trout there was a significant increase in the averages of fragmented DNA after cryopreservation; however, in gilthead sea bream there were no significant differences between the control samples and the cryopreserved sperm. This study demonstrates that cryopreservation can induce DNA damage at different degrees in marine and freshwater species, and corroborates that DNA integrity evaluation is an important factor that should be taken into account in the evaluation of freezing/thawing protocols, especially when sperm cryopreservation is used for gene bank purposes.

4.3 Environmental factors: water quality and presence of pollutants

The effect of pollutants can be assayed using direct gamete exposure to contaminants [222] or by *in vivo* studies, collecting sperm from animals from contaminated places. In some cases this process is more complicated due to fish migration. To avoid this problem, *in vitro* assays are often used, directly exposing sperm from normal individuals to different concentrations of contaminants. This test would have a correlation with the effects observed in animals exposed for longer time to contaminated zones.

The measurement of the activity of certain enzymes has proved to be a good biomarker for ecotoxicological evaluations since sperm cells react to contaminants in the aquatic ecosystems. Kime et al. [159] and Rurangwa et al. [148,169] have studied the effect of gamete exposure to tributyltin, a powerful toxic used for water pipes and vessels, on the motility and viability of common carp and African catfish spermatozoa, and reported that some metabolic enzymes which could be used as indicators of spermatozoa metabolism status, were affected by this toxic agent. The same approach was followed by Grzyb et al. [222] in herring, *Clupea harengus*, spermatozoa. Lactate dehydrogenase proved to be a good biomarker for sperm quality in catfish since it was released from damaged cells exposed to xenobiotics, however, the same enzyme was less affected on carp or herring spermatozoa [169,222]. The same findings were reported for marine invertebrates: sea urchin spermatozoa were more susceptible to toxics than oyster spermatozoa [223]. In the case of herring, Grzyb and co-workers [222] found creatine kinase more suitable as a biomarker for sperm cell membrane degradation by toxics. It seems that, as in other parameters of sperm quality, the choice of a good biomarker for toxicity screening based on metabolic activity is species-specific and toxicant-specific, spermatozoa from some species being more susceptible to damage than others, thus proving that preliminary tests must be performed before using certain enzymes as biomarkers.

Sperm motility is also another parameter affected by the environment that could be determined in order to check the effect of pollutants present in the aquatic environment. Motility was affected in carp, trout and catfish sperm exposed to heavy metals [144]. The exposure of fish to hormones present in water can also affect the quality of gametes and, more dramatically, reverse the sex of individuals [145].

4.4 Other aspects: stress and disease

The quality of gametes depends on the appropriate environment during gametogenesis but this may be disturbed in animals exposed to stress conditions [224]. Stress in captivity can produce several reproductive dysfunctions in male broodstock, affecting several sperm characteristics such as motility, cell viability, cell concentration, plasma composition and, at later times, even sperm production. In several flatfish species maintained in captivity, such as yellowtail flounder and turbot some of these aspects have been observed [225]. Striped and white bass males captured from the wild during the reproductive season can produce milt with non-motile sperm [226] and Atlantic halibut and plaice [227,228] released milt with such a high concentration that it does not mix with water during egg fertilization. This is probably associated with stress produced from capture or from different captivity husbandry conditions. Seminal plasma osmolality and motility activation were also affected in white bass, *Morone chrysops* due to stress associated with fish transportation [229].

Thus, since handling and transportation procedures used in aquaculture can be potentially stressful, quantitative evaluation of the effects of such procedures on sperm quality could facilitate changes in the conditions employed, so that stress could be minimized and sperm quality not affected. Another point that could be followed is to adopt a selection of breeders according to resistance to stress conditions, thus avoiding males with low quality gametes. However, few studies exist on the relationship between stress responsiveness and reproductive performance in males associated with gamete quality. Pottinger and Carrick [230] and Castranova et al. [231] found two subpopulations of rainbow trout and striped bass broodstock males, respectively, according to stress resistance (males with low stress responsiveness and males with high stress responsiveness). However, no significant differences in sperm motility and cell concentration were found between both subpopulations.

Fish disease can also affect reproduction and subsequently gamete quality, limiting male reproductive success. Sperm cells are immunologically perceived as non-self in the male reproductive tract and may therefore be attacked by the immune system. Males may consequently have to suppress their immune system in order to produce high quality sperm. This suppression may be influenced by the current level of parasite infections, suggesting that only parasite-resistant males are able to produce high quality sperm [232]. These authors found that the density of circulating granulocytes and the intensity of infection by one nematode species located outside the testes were negatively associated with sperm quality in infected male Arctic charr sampled during their spawning period. This suggests that a male extra-testicular immune environment may affect the production of high-quality sperm and that parasite infections located in the extra-testicular soma may influence sperm quality.

5. REFERENCES

1-Alfaro, J., Ulate, K., and Vargas, M., Sperm maturation and capacitation in the open thelycum shrimp *Litopenaeus* (Crustacea: Decapoda: Penaeoidea), *Aquaculture*, 270, 436, 2007.

2-Huang, C., Dong, Q., Walter, R.B., and Tiersch, T.R., Initial studies on sperm cryopreservation of a live-bearing fish, the green swordtail *Xiphophorus helleri, Theriogenology*, 62, 179, 2004.

3-Jamieson, B.G.M. and Leung, L.K.P., Introduction to fish spermatozoa and the micropyle, in: *Fish evolution and systematics: evidence from spermatozoa*, BGM Jamieson, Ed., Cambridge Univ. Press, 1991, 56.

4-Huang, C., Dong, Q., Walter, R.B., and Tiersch, T.R., Sperm cryopreservation of green swordtail *Xiphophorus helleri*, a fish with internal fertilization, *Cryobiology*, 48, 295, 2004.

5-Sarosiek, B., Ciereszko, A., Kolman, R., and Glogowski, J., Characteristics of arylsulfatase in Russian sturgeon (*Acipenser gueldenstaedti*) semen, *Comp Biochem Phys B: Biochem Mol Biol*, 139, 571, 2004.

6-Gwo, J.C., Yang, W.T., Sheu, Y.T., and Cheng, H.Y., Spermatozoan morphology of four species of bivalve (*Heterodonta, Veneridae*) from Taiwan, *Tissue and Cell*, 34, 39, 2002.

7-Lahnsteiner, F., Morphology, fine structure, biochemistry, and function of the spermatic ducts in marine fish, *Tissue and Cell*, 35, 363, 2003.

8-Darszon, A., Labarca, P., Nishigaki, T., and Espinosa, F., Ion Channels in Sperm Physiology, *Physiol Rev*, 79, 481, 1999.

9-Toth, G., Ciereszko, A., Christ, S.A., and Dabrowski, K., Objective analysis of sperm motility in the lake sturgeon, *Acipenser fulvescens*: activation and inhibition conditions, *Aquaculture*, 154, 337, 1997.

10-Morisawa, M., Suzuki, K., and Morisawa, S., Effect of potassium and osmolarity on spermatozoa motility of salmonid fishes, *J Exp Biol*, 107, 105, 1983.

11-Alavi, S.M. and Cosson, J., Sperm motility in fishes. (II) Effects of ions and osmolality: A review, *Cell Biol Int*, 30, 1, 2006.

12-Cabrita, E., Martínez-Páramo, S., Pérez-Cerezales, S., Anel, L., and Herráez, M.P., Motility of cryopreserved seabream spermatozoa: effect of channel blockers, presented at 10th International Symposium on Spermatology, Madrid, Spain, September 17-22, 2006.

13-Cabrita, E., Martínez Pastor, F., Soares, F., and Dinis, M.T., Motility activation and subpopulation analysis in *Solea senegalensis* spermatozoa, presented at 8th International Symposium on Reproductive Physiology of Fish, Saint-Malo, France, June 3-8, 2007.

14-Martinez-Pastor, F., Cabrita, E., Soares, F., and Dinis, M.T., Definition and changing patterns of *Solea senegalensis* spermatozoa after activation, presented at 10th Int. Symp. on Spermatology, Madrid, Spain, September 17-22, 2006.

15-Oda, S. and Morisawa, M., Rises of intracellular Ca^{2+} and pH mediate the initiation of motility by hyperosmolality in marine teleosts, *Cell Motil Cytoskeleton*, 25, 171, 1993.

16-Krasznai, Z., Morisawa, M., Krasznai, Z.T., Morisawa, S., Inaba, K., Bazsane, Z.K., Rubovsky, B., Bodnor, B., Borsos, A., and Marian, T., Gadolinium, a mechano-sensitive channel blocker, inhibits osmosis-initiated motility of sea-and freshwater fish sperm, but does not affect human or ascidian sperm motility, *Cell motil Cytoskel*, 55, 232, 2003.

17-Krasznai, Z., Morisawa, M., Morisawa, S., Krasznai, Z.T., Tron, L., Gaspar, R., and Marian, T., Role of ion channels and membrane potential in the initiation of carp sperm motility, *Aquat Living Resour*, 16, 445, 2003.

18-Morita, M., Takemura, A., and Okuno, M., Requirements of Ca^{2+} on activation of sperm motility in euryhaline tilapia *Oreochromis mossambicus*, *J Exp Biol*, 206, 913, 2003.

19-Krasznai, Z., Marian, T., Izumi, H., Damjanovich, S., Balkay, L., Tron, J., and Morisawa, M., Membrane hyperpolarization removes inactivation of Ca^{2+} channels, leading to Ca^{2+} influx and subsequent initiation of sperm motility in the common carp, *Proc. Natl Acad Sci USA*, 97, 2052, 2000.

20-Itoh, A., Inaba, K., Ohtake, H., Fuginuki, M., and Morisawa, M., Characterization of a cAMP-dependent protein kinase catalytic subunit from rainbow trout spermatozoa, *Biochem Biophys Res Commun*, 305, 855, 2003.

21-He, S. and Woods III, L.C., The effect of osmolarity, cryoprotectant and equilibration time on striped bass sperm motility, *J World Aquacult Soc*, 34, 255, 2003.

22-He, S., Jenkins-Keeran, K., and Woods III, C., Activation of sperm motility in striped bass via a cAMP-independent pathway, *Theriogenology,* 61, 1487, 2004.

23-Billard, R. and Cosson, J., Sperm motility in rainbow trout: Effect of pH, temperature and reproduction in fish. Basic aspects and applied aspects in endocrinology and genetics, *Les Colloques* INRA, 44, 161, 1986.

24-Kho, K.H., Satomi, T., Kazuo, I., Yoshitaka, O., and Masaaki, M., Transmembrane cell signalling for the initiation of trout sperm motility: roles of ion channels and membrane hyperpolarization for cyclic AMP synthesis, *Zoo Sci*, 18, 919, 2001.

25-Redondo-Muller, C., Cosson, M.P., Cosson, J., and Billard, R., In Vitro maturation of the potential for movement of carp spermatozoa, *Mol Reprod Dev*, 29, 259, 1991.

26-Morisawa, M., Cell signalling mechanisms for sperm activation, review, *Zool Sci*, 11, 647, 1994.

27-Butler, D.M., Allen, K.M., Garrett, F.E., Lauzon, L.L., Lotfizadeh, A., and Koch, R.A., Release of Ca^{2+} from intracellular stores and entry of extracellular Ca^{2+} are involved in sea squirt sperm activation, *Dev Biol*, 215, 453, 1999.

28-Vines, C.A., Yoshida, K., Griffin, F.J., Pillai, M.C., Morisawa ,M., Yanagimachi, R., and Cherr, G.N., Motility initiation in herring sperm is regulated by reverse sodium-calcium exchange, *PNAS,* 99, 2026, 2002.

29-Griffin, F.J., Pillai, M.C., Vines, C.A., Kaaria, J., Hibbard-Robbins, T., Yanagimachi, R., and Cherr, G.N., Effects of salinity on Sperm Motility, Fertilization and Development in the Pacific Herring, *Clupea pallasi*, *Biol Bull*, 194, 25, 1998.

30-Blaustein, M.P. and Lederer, W.J., Sodium/calcium exchange: Its physiological implications, *Physiol Rev*, 79, 763, 1999.

31-Nakajima, A., Morita, M., Takemura, A., Kamimura, S., and Okuno, M., Increase in intracellular pH induces phosphorilation of axonemal proteins for activation of flagellar motility in starfish sperm, *J Exp Biol*, 208, 4411, 2005.

32-Takai, H. and Morisawa, M., Changes in intracellular K^+ concentration caused by external osmolality changes regulates sperm motility of marine and freshwater teleosts, *J Cell Sci*, 108, 1175, 1995.

33-Boitano, S. and Omoto, C., Membrane hyperpolarization activates trout sperm without an increase in intracellular pH, *J Cell Sci*, 98, 343, 1991.

34-Hines, R. and Yashov, A., Some environmental factors influencing the activity of spermatozoa of *Mugil capito*, a grey mullet, *J Fish Biol*, 3, 123, 1971.

35-Cosson, J., The ionic and osmotic factors controlling motility of fish spermatozoa, *Aquacult Int,* 12, 69, 2004.

36-Solzin, J., Helbig, A., Van, Q., Brown, J.E., Hildebrand, E., Weyand, I., and Kaupp, U.B., Revisiting the role of H^+ in chemotactic signalling of sperm, *J Gen Physiol*, 124, 115, 2004.

37-Cosson, M.P., Billard, R., Gatti, J.R., and Christen, R., Rapid and quantitative assessment of trout spermatozoa motility using stroboscopy, *Aquaculture*, 46, 71, 1985.

38-Alavi, S.M. and Cosson, J., Sperm motility in fishes. (I) Effects of temperature and pH: A review, *Cell Biol Int*, 29, 101, 2005.

39-Billard, R. and Cosson, J., Some problems related to the assessment of sperm motility in freshwater fish, *J Exp Zool*, 261, 122, 1992.

40-Amanze, D. and Iyengar, A., The micropyle: a sperm guidance system in teleost fertilization, *Development*, 109, 495, 1990.

41-Suzuki, R., Sperm activation and aggregation during fertilization in some fish. IV the origin of sperm-stimulating factor, *Annot. Zool JPN*, 34, 24, 1961.

42-Yoshida, M., Inaba, K., Ishida, K., and Morisawa, M., Calcium and cyclic AMP mediate sperm activation, but Ca^{2+} alone contributes sperm chemotaxis in ascidian, *Ciona savignyi*, *Dev Growth Differ*, 36, 589, 1994.

43-Griffin, F.J., Vines, C.A., Pillai, M.C., Yanagimachi, R., and Cherr, G.N., The sperm motility initiation factor (SMIF) of the Pacific herring egg chorion: A minor component of major function, *Dev Growth Differ*, 38, 193, 1996.

44-Mansour, N., Lahnsteiner, F., and Patzner, R.A., Seminal vesicle secretion of African catfish, its composition, its behaviour in water and saline solutions and its influence on gamete fertilizability, *J Exp Zool*, 301, 745, 2004.

45-Zilli, L., Schiavone, R., Zonno, V., Rossano, R., Storelli, C., and Vilella, S., Effect of cryopreservation on Sea Bass Sperm proteins, *Biol Reprod*, 72, 1262, 2005.

46-Lahnsteiner, F., Characterization of seminal plasma proteins stabilizing the sperm viability in rainbow trout (*Oncorhynchus mykiss*), *Anim Reprod Sci*, 97, 151, 2007.

47-Zhao, C., Huo, R., Wang, F.-Q., Lin, M., Zhou, Z.M., and Sha, J.-H., Identification of several proteins involved in regulation of sperm motility by proteomic analysis, *Fertil Steril*, 87, 436, 2007.

48-Robles, V., Cabrita, E., Cuñado, S. and Herráez, M.P., Sperm cryopreservation of sex reversed rainbow trout (*Oncorhynchus mykiss*): parameters that affect its ability for freezing, *Aquaculture*, 224, 203, 2003.

49-Billard, R. and Zhang, T., Technique of genetic resource banking in fish, in *Cryobanking the Genetic Resource*, Watson, P.F. and Holt, W.V., Eds., Taylor and Francis, London, 2001, 144.

50-Lahnsteiner, F., Berger, B., Weismann, T., and Patzner, R.A., Motility of spermatozoa of *Alburnus alburnus* (Cyprinidae) and its relationship to seminal plasma composition and sperm metabolism, *Fish Physiol Biochem*, 15, 167, 1996.

51-Magyary, I., Urbanyl, B., and Horvath, L., Cryopreservation of common carp (*Cyprinus carpio* L.) sperm II. Optimal conditions for fertilization, *J Appl Ichthyol*, 12, 117, 1996.

52-Rurangwa, E., Volckaert, F.A.M., Huyskens, G., Kime, D.E., and Ollevier, F., Quality control of refrigerated and cryopreserved semen using computer-assisted sperm analysis (CASA), viable staining and standardized fertilization in African catfish (*Clarias gariepinus*), *Theriogenology*, 55, 751, 2001.

53-Aas, G.H., Refstie, T., and Gjerde, B., Evaluation of milt quality of Atlantic salmon, *Aquaculture*, 95, 125, 1991.

54-Ciereszko, A. and Dabrowski, K., Relationship between biochemical constituents of fish semen and fertility: the effect of short-term storage, *Fish Physiol Biochem*, 12, 357, 1994.

55-Lahnsteiner, F., Berger, B., Weismann, T., and Patzner, R.A., Determination of semen quality of the rainbow trout, *Oncorhynchus mykiss*, by sperm motility, seminal plasma parameters, and spermatozoal metabolism, *Aquaculture*, 163, 163, 1998.

56-Cabrita, E., Anel, L., and Herráez, M.P., Effect of external cryoprotectants as membrane stabilizers on cryopreserved rainbow trout sperm, *Theriogenology*, 56, 623, 2001.

57-Dabrowski, K., Glogowski, J., and Ciereszko, A., Effects of proteinase inhibitors on fertilization in sea lamprey (*Petromyzon marinus*), *Comp Biochem Physiol B: Biochem Mol Biol*, 139, 157, 2004.

58-Cabrita, E., Martínez, F., Real, M., Alvarez, R., and Herráez, M.P., The use of flow cytometry to asses membrane stability in fresh and cryopreserved trout spermatozoa, *Cryo letters*, 22, 263, 2001.

59-Ciereszko, R.E. and Dabrowski, K., Estimation of sperm concentration of rainbow trout, whitefish and yellow perch using spectrophotometric technique, *Aquaculture*, 109, 367, 1993.

60-Rurangwa, E., Kime, D.E., Ollevier, F., and Nash, J., The measurement of sperm motility and factors affecting sperm quality in cultured fish, *Aquaculture*, 234, 1, 2004.

61-Piironen, J. and Hyvärinen, H.J., Composition of the milt of some teleost fishes, *Fish Biol*, 22, 351, 1983.

62-Lahnsteiner, F., Berger, B., Weismann, T., and Patzner. R.A., Physiology and biochemical determination of rainbow trout semen quality for cryopreservation, *J Appl Aquacult*, 6, 47, 1998.

63-Billard, R., Cosson, J., Perchec, G., and Linhart, O., Biology of sperm and artificial reproduction in carp, *Aquaculture*, 129, 1, 95, 1995.

64-Loir, M., Labbé, C., Maisse, G., Pinson, A., Boulard, G., Mourot, B., and Chambeyron, F., Proteins of seminal fluid and spermatozoa in trout *Oncorhynchus mykiss*: partial characterization and variation, *Fish Physiol Biochem*, 8, 485, 1990.

65-Lahnsteiner, F., Patzner, R.A., and Weismann, T.,Testicular main ducts and spermatic ducts in some cyprinid fishes. I. Morphology, fine structure and histochemistry, *Fish Physiol Biochem*, 44, 937, 1994.

66-Ciereszko, A., Glogowski, J., and Dabrowski, K., Biochemical characterization of seminal plasma and spermatozoa of freshwater fishes, in *Cryopreservation in Aquatic Species*, Tiersch, T.R. and Mazik, P.M., Eds., World Aquaculture Society, Baton Rouge, Louisiana, 2000, 20.

67-Billard, R., Cosson, J., Crim, L.M., and Suquet, M., Sperm physiology and quality, in: *Broodstock Management and egg and larval quality*, Bromage, N. and Roberts, R., Eds., Blackwell Science, 1995, 25.

68-Lahnsteiner, F., Patzner, R.A., and Weismann, T., Energy resources of spermatozoa of the rainbow trout (*O. mykiss)* (Pisces, Teleostei), *Reprod Nutr Develop*, 33, 349, 1993.

69-Dabrowski, K. and Ciereszko, A., Ascorbic acid protects against male infertility in teleost fish, *Experientia*, 52, 97, 1996.

70-Ciereszko, A., Liu, L., and Dabrowski, K., Effects of season and dietary treatment on some biochemical characteristics of rainbow trout *Oncorhynchus mykiss* semen, *Fish Physiol Biochem*, 15, 1, 1996.

71-Glogowski, J., Babiak, I., Goryczko, K., and Dobosz, S., Activity of aspartate aminotransferase and acid phosphatase in cryopreserved trout sperm, *Reprod Fert Dev*, 8, 1179, 1996.

72-Babiak, I., Glogowski, J., Luczynski, M.J. and Luczynski, M., The effect of individual male variability on cryopreservation of northern pike, *Esox lucius* L., sperm, *Aquaculture Research*, 28, 191, 1997.

73-Glogowski, J., Babiak, I., Luczynski, M. J., and Luczynski, M., Factors affecting cryopreservation efficiency and enzyme activity in the Northern pike, *Esox lucius* sperm, *J Appl Aquacult*, 7, 53, 1997.

74-Wei, Q., Li, P., Psenicka, M., Hadi Alavi, S.M., Shen, L., Liu, J., Peknicova, J., and Linhart, O., Ultrastructure and morphology of spermatozoa in Chinese sturgeon (*Acipenser sinensis* Gray 1835) using scanning and transmission electron microscopy, *Theriogenology*, 67, 1269, 2007.

75-Marco-Jiménez, F., Pérez, L., Viudes de Castro, M.P., Garzón, D.L., Peñaranda, D.S., Vicente, J.S., Jover, M., and Asturiano, J.F., Morphometry characterisation of European eel spermatozoa with computer-assisted spermatozoa analysis and scanning electron microscopy, *Theriogenology*, 65, 1302, 2006.

76-Benetti, A.S., Santos, D.C., Negreiros-Fransozo, M.L., and Scelzo, M.A., Espermatozoal ultraestructure in three species of the genus *Uca*, Leach 1814 (Crustacea, Brachyura, Ocypodidae Micron (in press), 2007.

77-Lahnsteiner, F., Weismann, T., and Patzner, R.A., Fine structural changes in spermatozoa of the grayling, *Thymallus thymallus* (Pisces: Teleostei), during routine cryopreservation, *Aquaculture*, 103, 73, 1992.

78-Herráez, M.P., Mediavilla, M., Alvarez, R., Sánchez, A.J., Manso, A., and de Paz, P., Cellular damages caused by cryopreservation in trout semen, in *Proc. Conference Refrigeration and Aquaculture*, Bordeaux, March, 1996, 57.

79-Van Look, K.J.W. and Kime, D.E., Automated sperm morphology analysis in fishes: the effect of mercury on goldfish sperm, *J Fish Biol,* 63, 1020, 2003.

80-McAllister, B.G. and Kime, D.E., Early life exposure to environmental levels of the aromatase inhibitor tributyltin causes masculinisation and irreversible sperm damage in zebrafish (*Danio rerio*), *Aquat Toxicol,* 65, 309, 2003.

81-He, S. and Woods III, L.C., Effects of dimethyl sulfoxide and glycine on cryopreservation induced damage of plasma membranes and mitochondria to striped bass (*Morone saxatilis*) sperm, *Cryobiology,* 48, 254, 2004.

82-Liu, Q.H., Li, J., Zhang, S.C., Xiao, Z.Z., Ding, F.H., Yu, D.D., and Xu, X.Z., Flow cytometry and ultrastructure of cryopreserved red seabream (*Pagrus major*) sperm, *Theriogenology,* 67, 1168, 2007.

83-Lahnsteiner, F., Berger, B., Weismann, T., and Patzner, R.A., Changes in morphology, physiology, metabolism and fertilization capacity of rainbow trout semen following cryopreservation, *Prog Fish Culturist,* 58, 149, 1996.

84-Dong, Q., Huang, C., and Tiersch, T.R., Spermatozoal ultrastructure of diploid and tetraploid Pacific oysters, *Aquaculture,* 249, 487, 2005.

85-Tuset, V.M., Trippel, E.A., and Monserrat, J., Sperm morphometry and its influence on swimming spead in Atlantic cod, in *The 1ˢᵗ international Workshop on the biology of fish sperm,* Vodnany, August 28-30, 2007, 96.

86-Tuset, V.M., Dietrich, G.J., Wojtczak, M., Slowinska, M., Monserrat, J., and Ciereszko, A., Relationship between sperm morphology and sperm motility and fertilization ability of rainbow trout (*O. mykiss*) spermatozoa, in *The 1ˢᵗ international Workshop on the biology of fish sperm,* Vodnany, August 28-30, 2007, 98.

87-Lubzens, E., Daube, N., Pekarsky, I., Magnus, Y., Cohen, A., Yusefovich, F., and Feigin, P., Carp (*Cyprinus carpio L.*) spermatozoa cryobanks strategies in research and application, *Aquaculture,* 155, 13, 1997.

88-Garner, D.L., Dobrinsky, J.R., Welch, G.R., and Johnson, L.A., Porcine sperm viability, oocyte fertilization and embryo development after staining spermatozoa with SYBR-14, *Theriogenology,* 45, 1103, 1996.

89-Cabrita, E., Robles, V., Cuñado, S., Wallace, J.C., Sarasquete, C., and Herráez, M.P., Evaluation of gilthead seabream, *Sparus aurata,* sperm quality after cryopreservation in 5 ml macrotubes, *Cryobiology,* 50, 273, 2005.

90-Flajšhans, M., Cosson, J., Rodina, M., and Linhart, O., The application of image cytometry to viability assessment in dual fluorescence-stained fish spermatozoa, *Cell Biol Intl,* 28, 955, 2004.

91-Segovia, M., Jenkins, J.A., Paniagua-Chavez, C., and Tiersch, T.R., Flow cytometric evaluation of antibiotic effects on viability and mitochondrial function of refrigerated spermatozoa of Nile tilapia, *Theriogenology,* 53, 1489, 2000.

92-Paniagua-Chávez, C.G., Jenkins, J., Segovia, M., and Tiersch, T.R., Assessment of gamete quality for the eastern oyster (*Crassostrea virginica*) by use of fluorescent dyes, *Cryobiology,* 53, 128, 2006.

93-Cabrita, E., Alvarez, R., Anel, L., Rana, K.J., and Herraez, M.P., Sublethal damage during cryopreservation of rainbow trout sperm, *Cryobiology,* 37, 245, 1998.

94-Asturiano, J.F., Marco-Jiménez, F., Pérez, L., Balasch, S., Garzón, D.L., Peñaranda, D.S., Vicente, J.S., Viudes-de-Castro, M.P., and Jover, M., Effects of hCG as spermiation inducer on European eel semen quality, *Theriogenology,* 66, 1012, 2006.

95-Ogier de Baulny, B., Le Vern, Y., Kerboeuf, D., and Maisse, G., Flow cytometric evaluation of mitochondrial activity and membrane integrity in fresh and cryopreserved rainbow trout (*Oncorhynchus mykiss*) spermatozoa, *Cryobiology,* 34, 141, 1997.

96 Adams, S.I., Hessian, P.A., and Mladenov, P.V., Flow cytometric evaluation of mitochondrial function and membrane integrity of marine invertebrate sperm, *Invert Reprod Develop*, 44, 45, 2003.

97-Lezcano, M., Granja, C., and Salazar, M., The use of flow cytometry in the evaluation of cell viability of cryopreserved sperm of the marine shrimp (*Litopenaeus vannamei*), *Cryobiology*, 48, 349, 2004.

98-Gwo, J.C., Cryopreservation of aquatic invertebrate semen: a review, *Aquacult Research*, 31, 259, 2000.

99-Ciereszko, A. and Dabrowski, K., Spectrophotometric measurement of aspartate aminotransferase activity in mammalian and fish semen, *Anim Reprod Sci*, 38, 167, 1995.

100-Babiak, I., Glogowski, J., Goryczko, K., Dobosz, Z., Kuzminski, H., Strzezek, J., and Demianowicz, W., Effect of extender composition and equilibration time on fertilization ability and ezimatic activity of rainbow trout cryopreserved spermatozoa, *Theriogenology*, 56, 177, 2001.

101-Labbé, C., Maisse, G., Muller, K., Zachowski, S., Kaushik, S., and Loir, M., Thermal acclimation and dietary lipids alter the composition but not fluidity of trout plasma membrane, *Lipids*, 30, 23, 1995.

102-Drokin, S.I., Phospholipids and fatty acids of phospholipids of sperm from several freshwater and marine species of fish, *Comp Biochem Phys*, 104, 423, 1993.

103-Labbé, C. and Maisse, G., Influence of rainbow trout thermal acclimation on sperm cryopreservation: relation to change in the lipid composition of the plasma membrane, *Aquaculture*, 145, 281, 1996.

104-Labbe, C. and Maisse, G., Characteristics and freezing tolerance of brown trout spermatozoa according to rearing water salinity, *Aquaculture*, 201, 1, 287, 2001.

105-Cerolini, S., Zaniboni, L., Maldjian, A., and Gliozzi, T., Effect of docosahexaenoic acid and á-tocopherol enrichment in chicken sperm on semen quality, sperm lipid composition and susceptibility to peroxidation, *Theriogenology*, 66, 877, 2006.

106-Zaniboni, L., Rizzi, R., and Cerolini, S., Combined effect of DHA and á-tocopherol enrichment on sperm quality and fertility in the turkey, *Theriogenology*, 65, 1813, 2006.

107-Medina Basso, M., Eynard, A.R., and Valentich, M.A., Dietary lipids modulate fatty acid composition, gamma glutamyltranspeptidase and lipid peroxidation levels of the epididymis tissue in mice, *Anim Reprod Sci*, 92, 364, 2006.

108-Maldjian, A., Pizzi, F., Gliozzi, T., Cerolini, S., Penny, P., and Noble, R., Changes in sperm quality and lipid composition during cryopreservation of boar semen, *Theriogenology*, 63, 411, 2005.

109-Cabrita, E., Alvarez, R., Anel, L., and Herráez, M.P., The hypoosmotic swelling test performed with coulter counter: a method to assay functional integrity of sperm membrane in rainbow trout, *Anim Reprod Sci*, 55, 279, 1999.

110-Malejac, M.L., Loir, M., and Maisse, G., Qualité de la membrane des spermatozoides de la truite arc en ciel. Relation avec l'aptitude du sperme a la congelation, *Aquat Living Resour*, 3, 43, 1990.

111-Marián, T., Krasznai, Z., Balkay, L., Balázs, M., Emri, M., Bene, L., and Trón, L., Hypoosmotic shock induces an osmolality-dependent permeabilization and structural changes in the membrane of carp sperm, *J Histochem Cytochem*, 41, 291, 1993.

112-Ruiz-Pesini, E., Lapeña, A.C., Diez, C., Alvarez, E., Enriquez, A.C., and López-Pérez, M.J., Seminal quality correlates with mitochondrial functionality, *Clin Chim Acta*, 300, 97, 2007.

113-Sutovsky, P., Moreno, R.D., Ramalho-Santos, J., Dominko, T., Simerly, C., and Schatten, G., Ubiquitin tag for sperm mitochondria, *Nature*, 402, 371, 1999.

114-Nishimura, Y., Yoshinari, T., Naruse, K., Yamada, T., Sumi, K., Mitani, H., Higashiyama, T., and Kuroiwa, T., *PNAS*, 103, 1382, 2006.

115-Kazama, M., Asami, K., and Hino, A., Fertilization induced changes in sea urchin sperm: mitochondria deformation and phosphatidylserine exposure, *Mol Reprod Dev*, 73, 1303, 2006.

116-Gwo, J.C. and Arnold, C.R., Cryopreservation of Atlantic croaker spermatozoa: Evaluation of morphological changes, *J Exp Zool*, 2664, 444, 1992.

117-Ogier de Baulny, B., Labbé, C., and Maisse, G., Membrane integrity, mitochondrial activity, ATP content, and motility of the European catfish (*Silurus glanis*) testicular spermatozoa after freezing with different cryoprotectants, *Cryobiology*, 39, 177, 1999.

118-Taddei, A.R., Barbato, F., Abelli, L., Canese, S., Moretti, F., Rana, K.J., Fausto, A.M., and Mazzini, M., Is Cryopreservation a Homogeneous Process? Ultrastructure and Motility of Untreated, Prefreezing, and Postthawed Spermatozoa of *Diplodus puntazzo* (Cetti), *Cryobiology*, 42, 244, 2001.

119-Garner, D.L., Thomas, C.A., and Allen, C.H., Effect of semen dilution on bovine sperm viability as determined by dual-DNA staining and flow cytometry, *J Androl*, 18, 324, 1997.

120-Garner, D.L. and Thomas, C.A., Organelle-specific probe JC-1 identifies membrane potential differences in the mitochondrial function of bovine sperm, *Mol Reprod Dev*, 53, 222, 1999.

121-Gravance, C.G., Garner, D.L., Baumber, J., and Ball, B.A., Assessment of equine sperm mitochondrial function using JC-1, *Theriogenology*, 53, 1691, 2000.

122-Huo, L-J., Ma, X-H., and Yang, Z.M., Assessment of sperm viability, mitochondrial activity, capacitation and acrosome intactness in extender boar semen during long-term storage, *Theriogenology*, 58, 1349, 2002.

123-Martinez-Pastor, F., Johannisson, A., Gil, J., Kaabi, M., Anel, L., Paz, P., and Rodriguez-Martinez, H., Use of chromatin stability assay, mitochondrial stain JC-1, and fluorometric assessment of plasma membrane to evaluate frozen-thawed ram semen, *Anim Reprod Sci*, 84, 121, 2004.

124-Peña, F.J., Johannisson, A., Wallgren, M., and Rodriguez-Martinez, H., Antioxidant supplementation in vitro improves boar sperm motility and mitochondrial membrane potential after cryopreservation of different fractions of the ejaculate, *Anim Reprod Sci*, 78, 85, 2003.

125-Troiano, L., Granata, R.M., Cossariza, A., Kalashnikova, G., Bianchi, R., Pini, G., Tropea, F., Carani, C., and Francheschi, C., Mitochondria membrane potential and DNA stainability in human sperm cells: a flow cytometry analysis with implications for male infertility, *Exp Cell Res*, 241, 384, 1998.

126-Hallap, T., Nagy, S., Jaakma, U., Johannisson, A., and Rodriguez-Martinez, H., Mitochondrial activity of frozen-thawed spermatozoa assessed by Mitotracker Deep Red 633, *Theriogenology*, 65, 1122, 2005.

127-Meseguer, M., Garrido, N., Martinez-Conejero, J.A., Simon, C., Pellicer, A., and Remohi, J., Role of colesterol, calcium and mitochondrial activity on the susceptibility for cryodamage after a cycle of freezing and thawing, *Fertil Steril*, 81, 588, 2004.

128-Mansour, N., Lahnsteiner, F., and Berger, B., Metabolism of intratesticular spermatozoa of a tropical teleost fish (*Clarias gariepinus*), *Comp Biochem Phys B: Biochem Mol Biol*, 135, 285, 2003.

129-Foote, R.H., Resazurin reduction and other tests of semen quality and fertility of bulls, *Asian J Androl.*, 1, 109, 1999.

130-Zrimsek, P., Kunc, J., Kosec, M., and Mrkun, J., Spectrophotometric application of resazurin reduction assay to evaluate boar semen quality, *Int J Androl*, 27, 57, 2004.

131-Chandler, J.E., Harrison, C.M., and Canal, A.M., Spermatozoal methylene blue reduction: an indicator of mitochondrial function and its correlation with motility, *Theriogenology*, 54, 261, 2000.

132-Dart, M.G., Mesta, J., Crenshaw, C., and Ericsson, S.A., Modified resazurin reduction test for determining the fertility potential of bovine spermatozoa, *Arch Androl*, 33, 71, 1994.

133-Slater, T.F., Swyer, B., and Strauli, U., Studies on succinate-tetrazolium reductase systems III. Points of coupling of four different tetrazolium salts, *Biochim, Biophys Acta*, 77, 383, 1963.

134-Garzarzewicz, D., Piasecka, M., Udalu, J., Blaszczyk, B., Laszczynska, M., and Kram, A., Oxidoreductive capability of boar sperm mitochondria in fresh semen and during their preservation in BTS extender, *Reprod Biol*, 3, 161, 2003.

135-Naser-Esfahasi, M.H., Aboutorabi, R., Esfandiari, E., and Mardani, M., Sperm MTT viability assay: a new method for evaluation of human sperm viability., *J Assoc Reprod Genet*, 19, 477, 2002.

136-Aziz, D.M., Assessment of bovine sperm viability by MTT reduction assay, *Anim Reprod Sci*, 92, 1, 2006.

137-Hamoutene, D., Rahimtula, A., and Payne, J., Development of a new biochemical assay for assessing toxicity in invertebrate and fish sperm, *J Wat Res*, 34, 4049, 2000.

138-Billard, R., Cosson, J., and Crim, L.W., Motility of fresh and aged halibut sperm, *Aquat Living Resour*, 6, 67, 1993.

139-Otha, H., Shimma, H., and Hirose, K., Relationship between fertility and motility of cryopreserved spermatozoa of amago salmon *Oncorhynchus masou ishikawae*, *Fish Sci*, 61, 886, 1995.

140-Linhart, O., Rodina, M., and Cosson, J., Cryopreservation of sperm in common carp *Cyprinus carpio*: Sperm motility and hatching success of embryos, *Cryobiology*, 41, 241, 2000.

141-Lahnsteiner, F., Semen cryopreservation in the salmonidae and in the nortern pike, *Aquat Res*, 31, 245, 2000.

142-Cognie, F., Billard, R., and Chao, N., La criopreservation de la laitance de la carpe, *Cyprinus carpio*, *J Appl Ichtyol*, 5, 165, 1989.

143-Herráez, M.P., Carral, J., Alvarez., R., Sáez-Royuela, M., and de Paz, P., Efecto de distintos diluyentes y crioprotectores sobre la fertilidad del semen de la trucha común y la trucha arcoiris, in *Proc Fourth National Congress on Aquaculture*, Cerviño, A., Landin, A., de Coo, A., Guerra, A., and Torre, M., Eds., Villanova de Arosa, September 1993, 1.

144-Rurangwa, E., Roelants, I., Huyskens, G., Ebrahimi, M., Kime, D.E., and Ollevier, F., The minimum effective spermatozoa: egg ratio for artificial insemination and the effects of mercury on sperm motility and fertilisation ability in *Clarias gariepinus*, J, *Fish Biol*, 53, 402, 1998.

145-Kwon, J.Y., Haghpanah, V., Kogson-Hurtado, L.M., McAndrew, B.J., and Penman, D.J., Masculinization of genetic female nile tilapia (*Oreochromis niloticus*) by dietary administration of an aromatase inhibitor during sexual differentiation, *J Exp Zool*, 287, 46, 2000.

146-Suquet, M., Dreanno, C., Fauvel, C., Cosson, J., and Billard, R, Cryopreservation of sperm in marine fish, *Aquac Res*, 31, 231, 2000.

147-Miura, T., Yamauchi, K., Takahashi, H., and Nagahama, Y., The role of hormones in the acquisition of sperm motility on salmonid fish, *J Exp Zool*, 261, 359, 1992.

148-Burness, G., Casselman, S.J., Schulte-Hostedde, A., Moyes, C., and Montgomerie, R., Sperm swimming speed and energetics vary with sperm competition risk in bluegill (*Lepomis macrochirus*), *Behav Ecol Sociobiol*, 56, 65, 2004.

149-Guest, W.C., Avault, J.W., and Roussel, J.D., Preservation of channel catfish sperm, *Trans Am Fish, Soc*, 3, 469, 1976.

150-Billard, R., Dupont, J., and Barnabe, G., Fall in motility and survival time at low temperature of sperm of *Dicentrarchus labrax* L. (Pisces, Teleostei) during the spawning season, *Aquaculture*, 11, 363, 1977.

151-Levandusky, M.J. and Cloud, J.G., Rainbow trout (*Salmo gairdneri*) semen: effect of non-motile sperm on fertility, *Aquaculture*, 75, 171, 1988.

152-Duplinsky, P.D., Sperm motility of northern pike and chain pickerel at various pH values, *Trans Am Fish Soc*, 111, 1982.

153-McMaster, M.E., Portt, C.B., Munkittrick, K.R., and Dixon, D.G., Milt characteristics, reproductive performance, and larval survival and development of white sucker exposed to bleached kraft mill effluent, *Ecotox Environ Safe*, 23, 103, 1992.

154-Sansone, G., Fabbrocini, A., Ieropoli, S., Langellotti, A.L., Occidente, M., and Matassino, D., Effects of extender composition, cooling rate, and freezing on the motility of sea bass (*Dicentrarchus labrax, L.*) spermatozoa after thawing, *Cryobiology*, 44, 229, 2002.

155-Vantman, D., Banks, S.M., Koukoulis, G., Dennison, L., and Sherins, R.J., Assessment of sperm motion characteristics from fertile and infertile men using a fully automated computer-assisted semen analyzer, *Fertil Steril*, 51, 156, 1989.

156-Farrell, P.B., Presicce, G.A., Brockett, C.C., and Foote, R.H., Quantification of bull sperm characteristics measured by computer-assisted sperm analysis (CASA) and the relationship to fertility, *Theriogenology*, 49, 871, 1998.

157-Hirano, Y., Shibahara, H., Obara, H., Suzuki, T., Takamizawa, S., Yamaguchi, C., Tsunoda, H., and Sato, I., Relationships between sperm motility characteristics assessed by computer-aided sperm analysis (CASA) and fertilisation rates in vitro, *J Assist Reprod Genet*, 18, 213, 2001.

158-Chauvaud, L., Cosson, J., Suquet, M., and Billard, R., Sperm motility in turbot, *Scophthalmus maximus*: initiation of movement and changes with time of swimming characteristics, *Environ Biol Fis*, 43, 341, 1995.

159-Kime, D.E., Ebrahimi, M., Nysten, K., Roelants, I., Rurangwa, E., Moore, H.D.M., and Ollevier, F., Use of computer assisted sperm analysi (CASA) for monitoring the effect of pollution on sperm quality in fish; application to effects of zinc and cadmium, *Aquat Toxicol.*, 36, 223,1996.

160-Wilson-Leedy, J.G. and Ingermann, R.L., Development of a novel CASA system based on open source software for characterization of zebrafish sperm motility parameters, *Theriogenology*, 67, 661, 2007.

161-Lahnsteiner, F., Berger, B., Kletzl, M., and Weismann, T., Effect of 17-beta-estradiol on gamete quality and maturation in two salmonid species, *Aquat Toxicol*, 79, 124, 2006.

162-Suquet, M., Omnes, M.H., Normant, Y., and Fauvel, C., Assessment of sperm concentration and motility in turbot (*Scophthalmus maximus*), *Aquaculture*, 101, 177, 1992.

163-Toth, G.P., Christ, S.A., Torsella, J.A., McCarthy, H.W., and Smith, M.K., Computer-assisted sperm motion analysis of sperm from the common carp (*Cyprinus carpio*), *J Fish Biol*, 47, 986, 1995.

164-Babiak, I., Otteson, O., Rudolfsen, G., and Johnsen, S., Chilled storage of semen from Atlantic halibut, *Hippoglossus hippoglosssus* L. I: Optimizing the protocol, *Theriogenology*, 66, 2025, 2006.

165-Kime, D.E. and Tveiten, H., Unusual motility characteristics of sperm of spotted wolfish, *J Fish Biol*, 61, 1549, 2002.

166-Elofsson, H., Van look, K., Borg, B., and Mayer, I., Influence of salinity and ovarian fluid on sperm motility in the fifteen spined stickleback, *J Fish Biol*, 63, 1429, 2003.

167-Le Comber, S.C., Faulkes, C.G., Van Look, K.J.W., Holt, W.V., and Smith, C., Recovery of activity in osmotically shocked stickleback sperm: implications for pre-oviposition ejaculation, *Behaviour*, 141, 1555, 2004.

168-Martinez-Pastor, F., García Macías, V., Alvarez, M., Herráez, M.P., Anel, L., and de Paz, P., Sperm subpopulations in Iberian red deer epididymal sperm and their changes through the cryopreservation process, *Biol Reprod*, 72, 316, 2005.

169-Rurangwa, E., Biegniewska, A., Slominska, E., Skorkowski, E.F., and Ollevier, F., Effect of tributyltin on adenylate content and enzyme activities of teleost sperm: a biochemical approach to study the mechanisms of toxicant reduced spermatozoa motility, *Comp Biochem Phys*, C, 131, 335, 2002.

170-Lahnsteiner, F., Berger, B., and Weismann, T., Sperm metabolism of the teleost fishes *Chalcalburnus chalcoides* and *Oncorhynchus mykiss* and its relation to motility and viability, *J Exp Zool*, 284, 454, 1999.

171-Gronczewska, J., Zietara, M.S., Biegniewska, A , and Skorkowski, E.F., Enzyme activities in fish spermatozoa with focus on lactate dehydrogenase isoenzymes from herring *Clupea harengus*, *Comp Biochem Physiol, B: Biochem Mol Biol*, 134, 399, 2003.

172-Babiak, I., Glogowski, J., Luczynski, M.J., Kucharczyk, D., and Luczynski, M., Cryopreservation of milt of Northern pike, *J Fish Biol*, 46, 819, 1995.

173-Dreanno, C., Suquet, M., Fauvel, C., Le Coz, J.R., Dorange, G., and Quemener, L., Effect of the aging process on the quality of sea bass (*Dicentrarchus labrax*) semen, *J Appl Ichthyol*, 15, 176, 1999.

174-Tabares, C.J., Tarazona, A.M., and Ángel, M.O., Fisiología de la activación del espermatozoide en peces de agua dulce, *Rev Col Cienc Pec.*,18, 2, 2005.

175-Gwo, J.C., Ultrastructural study of osmolality effect on spermatozoa of three marine teleosts, *Tissue Cell*, 27, 491, 1995.

176-Perchec, G., Jeulin, C., Cosson, J., André, F., and Billard, R., Relationship between sperm ATP content and motility of carp spermatozoa, *J Cell Sci*, 108, 747, 1995.

177-Burness, G., Moyesb, C.D., and Montgomerieb, R., Motility, ATP levels and metabolic enzyme activity of sperm from bluegill (*Lepomis macrochirus*), *Comp Biochem Phys, A Mol Integr Physiol*, 140, 11, 2005.

178-Christen, R., Gatti, J.L., and Billard, R., Trout sperm motility. The transient movement of trout sperm is related to changes in the concentration of ATP following the activation of the flagellar movement, *Eur J Biochem*, 166, 667, 1987.

179-Dreanno, C., Suquet, M., Desbruyeres, E., Cosson, J., Le Delliou, H., and Billard, R., Effect of urine on semen quality in turbot (*Psetta maxima*), *Aquaculture*, 169, 247, 1998.

180-Zietara, M.S., Slominska, E., Rurangwa, E., Ollevier, F., Swierczynski, J., and Skorkowski, E.F, In Vitro adenina nucleotide catabolism in African catfish spermatozoa, *Comp Biochem Phys B: Biochem Mol Biol*, 138, 385, 2004.

181-Zilli, L., Schiavone, R., Zonno, V., Storelli, C., and Vilella, S., Adenosine triphosphate concentration and b-D-glucuronidase activity as indicators of sea bass semen quality, *Biol Reprod*, 70, 1679, 2004.

182-Evenson, D.P. and Wixon, R., Predictive value of the sperm chromatin assay in different populations, *Fertil Steril*, 85, 810, 2006.

183-Rybar, R., Faldikova, L., Faldyna, M., Machatkova, M., and Rubes, J., Bull and boar sperm DNA integrity evaluated by sperm chromatin structure assay in the Czech Republic, *Vet Med Czech*, 49, 1, 2004.

184-Sailer, B.L, Jost, L.K, and Evenson, D.P., Mammalian sperm DNA susceptibility to in situ denaturation associated with the presence of DNA strand breaks as measured by the terminal deoxynucleotidyl transferase assay, *J Androl*, 16, 80, 1995.

185-Evenson, D.P., Darzynkiewicz, Z., and Melamed, M. R., Comparison of human and mouse sperm chromatin structure by flow cytometry, *Chromosoma*, 78, 225, 1980.

186-Evenson, D.P., Larson, K. L., and Jost, L.K., Sperm Chromatin Structure Assay: its clinical use for detecting sperm DNA fragmentation in male infertility and comparisons with other techniques, *J Androl*, 23, 25, 2002.

187-Zhou, B., Liu, W., Siu, W.H.L., O'Toole, D., Lam, P.K.S., and Wu, R.S.S., Exposure of spermatozoa to duroquinone may impair reproduction of the common carp (*Cyprinus carpio*) through oxidative stress, *Aquat Toxicol*, 77, 136, 2006.

188-Collins, A.R., The Comet Assay for DNA Damage and Repair, *Mol Biotechnol*, 26, 249, 2004.

189-Cabrita, E., Robles, V., Rebordinos, L., Sarasquete, C., and Herraez, M. P., Evaluation of DNA damage in rainbow trout (*Oncorhynchus mykiss*) and gilthead sea bream (*Sparus aurata*) cryopreserved sperm, *Cryobiology*, 50, 144, 2005.

190-Ciereszko, A., Wolfe, T.D., and Dabrowski, K., Análisis of DNA damage in sea lamprey (*Petromyzon marinus*) spermatozoa by UV, hydrogen peroxide, and the toxicant bisazir, *Aquat Toxicol*, 73, 128, 2005.

191-Dietrich, G.J., Szpyrka, A., Wojtczak, M., Dobosz, S., Goryczko, K., Zakowski, £., and Ciereszko, A., Effects of UV irradiation and hydrogen perowide on DNA fragmentarion, motility and fertilizing ability of rainbow trout (*Oncorhynchus mykiss*) spermatozoa, *Theriogenology*, 64, 1809, 2005.

192-Labbé, C., Martoriati, A., Devaux, A., and Maisse, G., Effect of sperm cryopreservation on sperm DNA stability and progeny development in rainbow trout, *Mol Reprod Dev*, 60, 397,.2001.

193-Pérez-Cerezales, S., Martínez-Páramo, S., and Herráez, M.P., Sperm chromatin fragmentation and oxidization during short and long-term storage of sex reversed rainbow trout (*Oncorhynchus mykiss*) males, presented at 8th Int. Symp. on Reproductive Physiology of Fish, Saint Malo, France, June 3-8, 2007.

194-Zilli, L., Schiavone, R., Zonno, V., Storelli, C., and Vilella, S., Evaluation of DNA damage in *Dicentrarchus labrax* sperm following cryopreservation, *Cryobiology*, 47, 227, 2003.

195-Bennetts, L.A. and Aitken, R.J., A comparative study of oxidative DNA damage in mammalian spermatozoa, *Mol Reprod Develop*, 71, 77, 2005.

196-Dillard, R., Artificial insemination and gamete management in fish, *Mar Behav Physiol*, 14, 3, 1988.

197-Steyn, G., Van Vuren, J., and Grobler, E., A new sperm diluent for the artificial insemination of rainbow trout, *Aquaculture*, 83, 367,1989.

198-Wheeler, P.A. and Thorgaard, G.H., Cryopreservation of rainbow trout semen in large straws, *Aquaculture*, 93, 95, 1991.

199-Leray, C., Chapelle, S., Duportail, G., and Florentz, A., Changes in fluidity and 22:6n3 content in phospholipids of trout intestinal brush border membrane as related to environmental salinity, *Biochem Biophys Acta*, 778, 233, 1984.

200-Borlongan, I.G. and Benítez, L.V., Lipid and fatty acid composition of milkfish *Chanos chanos* grown in freshwater and seawater, *Aquaculture*, 104, 79, 1992.

201-Hansen, H.J.M., Olsen, A.G., and Rosenkilde, P., Formation of phosphatidylethanolamine as a putative regulator of salt transport in the gills and esophagus of the rainbow trout *Oncorhynchus mykiss*, *Comp Biochem Physiol*, 112B, 161, 1995.

202-Pustowka, C., McNiven, M.A., Richardson, G.F., and Lall, S.P., Source of dietary lipid affects sperm plasma membrane integrity and fertility in rainbow trout *Oncorhynchus mykiss* after cryopreservation, *Aquacult Res*, 31, 297, 2000.

203-Lee, K. and Dabrowski, K., Long-term effect and interactions of dietary vitamins C and E on growth and reproduction of yellow perch, *Perca flavescens*, *Aquaculture*, 230, 377, 2004.

204-Ciereszko, A. and Dabrowski, K., Sperm quality and ascorbic acid concentration in rainbow trout semen are affected by dietary vitamin C: an across-season study, *Biol Reprod*, 52, 982, 1995.

205-Ciereszko, A. and Dabrowski, K., *In vitro* effect of gossypol acetate on yellow perch (*Perca flavescens*) spermatozoa, *Aquat Toxicol*, 49, 181, 2001.

206-Rinchard, J., Ciereszko, A., Dabrowski, K., and Ottobre, J., Effects of gossypol on sperm viability and plasma sex steroid hormones in male sea lamprey, *Petromyzon marinus, Toxicology Letters*, 111, 189, 2000.

207-Lee, K.-J., Rinchard, J., Dabrowski, K., Babiak, I., Ottobre, J.S., and Christensen, J.E., Long-term effects of dietary cottonseed meal on growth and reproductive performance of rainbow trout: Three-year study, *Anim Feed Sci Tech*, 126, 93, 2006.

208-Mylonas, C.C., Papadaki, M., and Divanach, P., Seasonal changes in sperm production and quality in the red porgy *Pagrus pagrus* (L.), *Aquacult Res*, 34, 1161, 2003.

209-Suquet, M., Dreanno, C., Dorange, G., Normant, Y., Quemener, L., and Gaignon, J.L., The ageing phenomenon of turbot spermatozoa: effects on morphology, motility and concentration, intracellular ATP content, fertilization, and storage capacities, *J Fish Biol, 52*, 31, 1998.

210-Wang, A.W., Zhang, H., Ikemoto, I., Anderson, D.J., and Loughlin, K.R., Reactive oxygen species generation by seminal cells during cryopreservation, *Urology,* 49, 921, 1997.

211-Aitken, J.R. and Baker, M.A., Oxidative stress, sperm survival and fertility control, *Mol Cell Endoc*, 250, 6669, 2006.

212-Vernet, P., Aitken, R.J. and Drevet, J.R., Antioxidant strategies in the epididymis, *Mol Cell Endocrinol*, 216, 3139, 2004.

213-Herráez, M.P., Pérez-Cerezales, S., Martínez-Páramo, S., Cabrita, E., Anel, L., and Martínez-Pastor, F., Chromatin stability during short and long-term storage of sex-reversed rainbow trout sperm, (Poster), presented at World Aquaculture Society Meeting, Aquaculture 2006, Florencia, May 9-13, 2006.

214-Billard, R., Short-term preservation of sperm under oxygen atmosphere in rainbow trout *Salmo gairdneri*, *Aquaculture*, 23, 287, 1981.

215-Stoss, J. and Holtz, W., Successful storage of chilled rainbow trout (*Salmo gairdneri*) spermatozoa for up to 34 days, *Aquaculture*, 31, 269,1983.

216-Christensen, J.M. and Tiersch, T.R., Refrigerated storage of channel catfish sperm, *J World Aquac* Soc, 27, 340, 1996.

217-Sunitha, M.S. and Jayaprakas, V., Influence of pH, temperature, salinity and media on activation of motility and short term preservation of spermatozoa of an estuarine fish, *Mystus gulio* (Hamilton) (siluridae-Pisces), *Indian J Mar Scie,* 26, 361, 1997.

218-Benic, D.C., Krisfalusi, M., Cloud, J.C., and Ingermann, R.L., Short term storage of salmonid sperm in air versus oxygen, *N Am J Aquacult*, 62, 19, 2000.

219-Mansour, N., Lahnsteiner, F., and Berger, B., Characterization of the testicular semen of the African catfish, *Clarias gariepinus* (Burchell, 1822), and its short-term storage, *Aquacult Res,* 35, 232, 2004.

220-Cloud, J.C., Miller, W.H., and Levanduski, M.J., Cryopreservation of sperm as a means to store salmonid germ plasma and to transfer genes from wild fish to hatchery populations, *Prog Fish-Cult, 52*, 51, 1990.

221-Linhart, O., Rodina, M., Flajshans, M., Gela, D., and Kocour, M., Cryopreservation of European catfish *Silurus glanis* sperm: Sperm motility, viability and hatching success of embryos, *Cryobiology*, 51, 250, 2005

222-Grzyb, K., Rychlowski, M., Biegniewska, A., and Skorkowski, E.F., Quantitative deterrmination of creatine kinase release from herring (*Clupea harengus*) spermatozoa induced by tributyltin, *Comp Biochem Phys C*, 134, 207, 2003.

223-Geffard, O., Budzinski, H., Augagneur, S., Seaman, M.N.L., and His, E., Assessment of sediment contamination by spermiotoxicity and embryotoxicity bioassay with sea urchin (*Paracentrotus lividus*) and oysters (*Crassostrea gigas*), *Environ Toxicol Chem*, 20, 1605, 2001.

224-Kime, D.E. and Nash, J.P., Gamete viability as an indicator of reproductive endocrine disruption in fish, *The Science of The Total Environment,* 233, 123, 1999.

225-Zohar, Y and Mylonas, C.C., Endocrine manipulations of spawning in cultured fish: from hormones to genes, *Aquaculture*, 19, 99, 2001.

226-Berlinsky, D.L., William, K., Hodson, R.G., and Sullivan, C.V., Hormone induced spawning of summer flounder *Paralichthys dentatus*, *J World Aquac* Soc, 28, 79, 1997.

227-Vermeirssen, E.L.M., Scott, A.P., Mylonas, C.C., and Zohar, Y., Gonadotrophin-releasing hormone agonist stimulates milt fluidity and plasma concentrations of 7,20h-dihydroxylated and 5h-reduced, 3a-hydroxylated C21 steroids in male plaice (*Pleuronectes platessa*), *Gen Comp Endocrinol*, 112, 163, 1998.

228-Vermeirssen, E.L.M., Mazorra de Quero, C., Shields, R., Norberg, B., Scott, A.P., and Kime, D.E., The applications of GnRHa implants in male Atlantic halibut: effects on steroids, milt hydration, sperm motility and fertility, in *Reproductive Physiology of Fish,* Norberg, B., Kjesbu, O.S., Taranger, G.L., Andersson, E., and Stefansson, S.O., Eds., Bergen, 1999, 399.

229-Allyn, M.L., Sheehan, R.J., and Kohler, C.C., The effects of capture and transportation stress on white bass semen osmolarity and their alleviation via sodium chloride, *Trans Am Fish Soc,* 130, 706, 2001.

230-Pottinger, T.G. and Carrick, T.R., Indicators of reproductive performance in rainbow trout *Oncorhynchus mykiss* (Walbaum) selected for high and low responsiveness to stress, *Aquacult Res,* 31, 367,2000.

231-Castranova, D.A., King, V., and Woods III, L.C., The effects of stress on androgen production, spermiation response and sperm quality in high and low cortisol responsive domesticated male striped bass, *Aquaculture,* 246, 413, 2005.

232-Måsvaer, M., Liljedal, S., and Folstad, I., Are secondary sex traits, parasites and immunity related to variation in primary sex traits in the Arctic charr?, *Proc Biol Sci,* 271, 40, 2004.

233-Cabrita, E., Engrola, S., Conceição, L., Lacuisse, M., Herráez, M.P., Pousão-Ferreira, P., Marino, G., and Dinis, M.T., Preliminary attempts on the cryopreservation of dusky grouper (*Epinephelus marginatus*) sperm, presented at Aquaculture Europe 2007, Estambul, Turkey, October 24-27, 2007.

Complete affiliation:

Elsa Cabrita, Center for Marine Sciences-CCMAR, University of Algarve, Campus de Gambelas, 8000 Faro, Portugal, Phone: +351.289800900 ext. 7595, Fax: +351.289800069, e-mail: ecabrita@ualg.pt.

CHAPTER 4
EGG QUALITY DETERMINATION IN TELEOST FISH

F. Lahnsteiner, F. Soares, L. Ribeiro and M.T. Dinis

1. INTRODUCTION

Fish aquaculture production and fish recruitment can be largely affected by the quality of eggs since reproductive success depends on the quantity and/or quality of eggs produced during spawning. In recent years, efforts have been made to develop reliable parameters that enable to identify good quality eggs. According to Kjørsvik et al. [1] egg quality can be defined as the egg's potential to produce viable fry, whereas Bromage et al. [2] defined good quality eggs, for fish farming, as those exhibiting low mortalities at fertilization, eyeing, hatching and first feeding larvae.

Although eggs possess intrinsic characteristics that determine egg quality, such as maternal age, genetics, yolk composition, size, morphology, among others, there are several extrinsic factors that will influence egg quality throughout the spawning season, and consequently larvae quality, namely, ripening, manipulation, incubation conditions, water quality, among others [1,3-7].

Egg quality markers are biological parameters which are measured in the unfertilized or freshly fertilized eggs. They do not only distinguish between viable and nonviable eggs but are mainly used to estimate the percentage of eggs successfully completing the development to one of the above mentioned ontogenetic stages.

Although for important commercial fresh water fish, fry production has reached a high state of art and is a routine farming procedure, egg quality may vary considerably, mainly because of variability caused by individual female conditions and by overripening processes [1,5,8]. For cultured marine fish the fry production is still more difficult, the yield more variable and the technique restricted to a few specialized hatcheries. Although the broodstocks are able to produce large quantities of eggs, egg quality often varies greatly and in a non-controllable way (e.g., gilthead seabream, *Sparus aurata*; red seabream, *Pagrus major*; sharpsnout seabream, *Diplodus puntazzo* [9,10]; turbot, *Scophthalmus maximus* [11]; cod, *Gadus morhua* [12]; Atlantic halibut, *Hippoglossus hippoglossus* [4]; wolfish, *Anarhichas lupus* [13]. Moreover, in larvae originating from high quality egg batches viability and stress resistance may be better than in larvae from low quality egg batches [14]. Therefore, egg quality control is still more important in marine species than in fresh water species.

In earlier studies, reliable egg quality criteria was developed for several species based on empirical research, such as, egg diameter, morphology, buoyancy, fertilization rate and hatching rate, however, these parameters were insufficient and/or inaccurate for characterizing spawn quality. In recent decades, researchers developed more precise parameters to determine egg quality and parameters such as symmetry of earliest cells development, yolk composition, key enzymes of intermediary metabolism and molecular tools started to be used to characterize egg quality. Nevertheless, although these parameters can give more conclusive information on egg quality they are costly and time-consuming.

At present, the quality of the reared fish is still extremely variable, often including high and varying levels of mortality, inconsistent growth, skeletal malformations, and in flatfish, abnormal pigmentation [15,16]. In fact, the egg quality of captive fish is usually inferior when compared with wild fish, suggesting that current husbandry protocols still largely affect egg quality [1,5,17]. Moreover, relatively little is known of how egg quality varies either within or among females, or among populations during spawning season, which according to Mylonas et al. [18] might be important information to optimize egg collection and larval production in commercial hatcheries; also whether a parameter is reliable for different species and different type of eggs.

Technologies for the quality control of eggs would allow methods and conditions for broodstock maturation to improve and therefore also fry production. From the commercial point of view these technologies would represent a tool to ascertain the product quality and to keep hatcheries working efficiently [1,5,8]. Moreover, egg quality determination methods would also benefit certain scientific aspects. When performing fertilization tests under experimental conditions (e.g., for cryobiology, toxicological tests, etc.) complete test series and therefore considerable working effort may be lost due to insufficient egg quality.

Parameters for assessing egg quality (i.e., parameters which can be measured in unfertilized eggs or eggs in early developmental stages and which estimate the rate of developing embryos or larvae) are therefore necessary.

The present review summarizes methods for egg quality determination in teleost fish using morphological, morphometric, physiological, microbiological and biochemical parameters determined in ovarian fluid, unfertilized eggs, or eggs in early development stages, respectively.

2. FISH EGGS

As in other vertebrates, fish eggs are produced as a result of an encounter between eggs and sperm, which normally occurs externally, although other reproductive strategies can be found in this group [19]. In spite of the fact that both gametes contribute to egg quality, research studies have mainly focused on the maternal contribution to egg quality, since females are responsible for the synthesis of the yolk and other substances that are indispensable for a normal larval development. Nonetheless, during the last few years several studies examined the importance of males and sperm quality in egg quality [20]. Moreover, Brooks et al. [5] emphasize the importance of assessing the influence of the "male genetic factor" on egg quality, in spite of evidence of the effect of maternal genetics on egg quality.

Kamler [21] summarizes that both paternal and maternal factors contribute to total egg/larva viability, but they operate at different times and in different ways. In fact paternal effects are produced very early in ontogeny, regarding spermatozoan density and motility, whereas maternal effects reveal later, contributing to embryonic survival, which responds to egg ripeness and female age via egg matter composition.

Fish eggs are normally grouped as pelagic and demersal eggs. Freshwater species usually tend to have demersal eggs, often adhesive, e.g., Atlantic salmon (*Salmo salar*), carp (*Cyprinus carp*), etc., [22,23], although pelagic eggs might also occur, e.g., several gouramis, grass carp (*Ctenopharyngodon idella*), etc. [19,24]. Both pelagic (turbot; gilthead seabream; sea bass, *Dicentrarchus labrax*; etc.) and demersal (herring, *Clupea harengus*; several gobis; Lusitanian toadfish, *Halobatrachius didactylus*; etc.) eggs are observed [22,25] in marine the environment. In general, pelagic eggs (Fig. 1a, b) are smaller, more numerous, exhibit a thinner chorion and are more transparent (Fig. 1c) than demersal eggs [1]. Regardless of the characteristics of the egg, the most important difference between these two types of eggs is the stage of larval development at hatching, since pelagic eggs hatch into buoyant blind larvae with the mouth and anus closed, whereas demersal eggs hatch into eyed, freely swimming larvae with the ability to feed on formulated diets.

The incubation period is usually longer in demersal than pelagic eggs (strongly dependent on temperature), consequently different development stages are identified during embryonic development and external factors will affect greatly and differently larval quality in these types of fish eggs. Consequently quality parameters might vary between demersal and pelagic eggs.

For more information on fish species reproductive strategies, parental care and egg size, see the reviews by Blaxter [22], Kjørsvik et al. [1], Kamler [21].

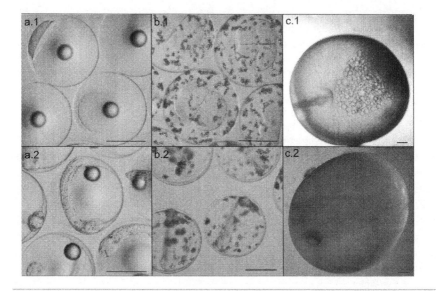

Figure 1 - Different developmental stages of pelagic (a, b) and demersal (c) eggs from marine fish species. a – *Sparus aurata*; eggs with one lipid droplet; a.1 – morula stage, a.2 – embryo stage 24 hours after morula, at 18°C incubation temperature. b- *Solea senegalensis*; eggs with several lipid droplets; b.1 – four-cells stage, b.2 – embryo stage 36h after stage b.1 at 19°C incubation temperature. c – *Halobatrachius didactylus*; c.1 – embryo stage 8 days after fertilization; c.2 – eyed stage embryo 15 days after fertilization at 23°C incubation temperature. Bar corresponds to 0.5 mm. (Photos by Laura Ribeiro, Catarina Oliveira and Wilson Pinto.)

2.1 Practical aspects of egg collections

Freshly spawned eggs are released together with ovarian fluid, the amount of which varies considerably between different species [26]. Morphological, physiological and metabolic parameters of eggs and the composition of the ovarian fluid (when sufficient amounts are available) are candidate parameters to be used as egg quality markers. The specific types of markers which can be used depend on the species and on the mode of egg collection. In many fish species, preferably in freshwater species, eggs are stripped and artificially fertilized before incubation. Therefore, individual batches of unfertilized eggs and ovarian fluid are available for egg quality determination. In other, mainly marine species, brood fish are allowed to spawn spontaneously due to their sensitivity to handling stress, and as eggs are not released all at once but in multiple small batches [27]. The eggs are collected from the tanks with egg collectors and incubated for hatching. As in the spawning tanks many individuals spawn together the collected eggs are from several individuals and not strictly in the same ontogenetic stage. This variability plays no role for hatching but complicates the egg quality determination as the investigated parameters may differ between individuals and developmental stages. Therefore, quality markers must be constant in the development stages relevant for egg quality determination or the changes during development must be at least lower than those induced by quality differences. In conclusion, egg quality assays must be developed species-specifically and for the specific mode of egg collection.

For species, where gametes can be stripped individual egg samples are collected (Fig. 2a). Ovarian fluid and eggs are separated and eggs are washed free from remnants of ovarian fluid. For species where gametes are collected by natural spawning different spawns (fertilized eggs with embryos in early ontogenetic stages) are collected from the spawning tanks via egg collectors (Fig. 2b). Then the samples are divided into sub-samples for extraction or fixation procedures and subsequent analysis. One sub-sample of each egg sample is hatched to determine embryo or larvae viability parameters. Ovarian fluid is directly analyzed. Finally, statistical correlations are determined between analyte values and viability parameters using correlation coefficients, ANOVA models, and regression models. Details on the procedures are found in the relevant original publications [9,26,28].

3. MORPHOLOGICAL PARAMETERS FOR DETERMINING EGG QUALITY

3.1 Morphology, buoyancy and appearance

The eggs of several fresh water fish species (e.g., burbot, *Lota lota*, common whitefish, *Coregonus lavaretus*; different Cyprinidae) and of most cultured marine fish species (e.g., gilthead seabream; sharpsnout seabream; common dentex, *Dentex dentex*; Atlantic cod; turbot; Senegalese sole, *Solea senegalensis*; Atlantic halibut) are small and transparent and therefore may be suitable for egg quality determination using morphological parameters. In fact, egg transparency was widely used to differentiate between good and poor quality eggs [12,29-32], since the latter are normally opaque (Fig. 3).

Egg buoyancy was also currently used to identify good quality pelagic eggs since poor quality eggs normally sank in the water column [1,31,33], and a positive correlation was obtained between buoyant eggs and hatching rate. However, in some species no correlation is observed between buoyancy and 'good' quality eggs, like Atlantic halibut [4] and other parameters must be used to identify egg quality. Despite not being the most adequate parameter to evaluate eggs quality, most fish farms hatcheries still use this methodology to separate "good eggs" from "poor eggs" since it is a very straightforward procedure.

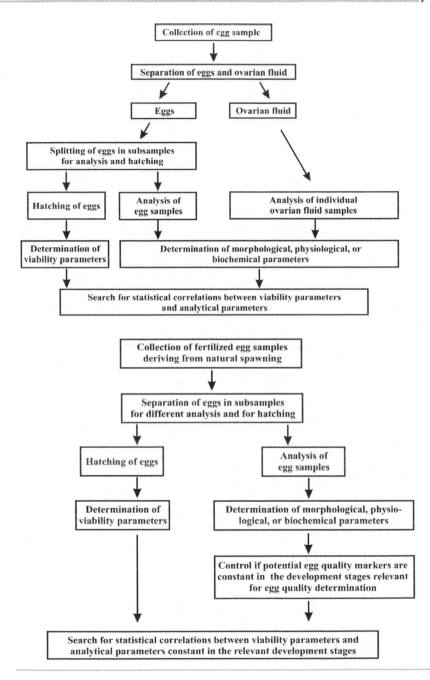

Figure 2 - Working scheme for determination of egg quality markers in fresh water species (eggs collected by stripping) (Fig. 2a) and marine species (eggs collected after natural spawning) (Fig. 2b).

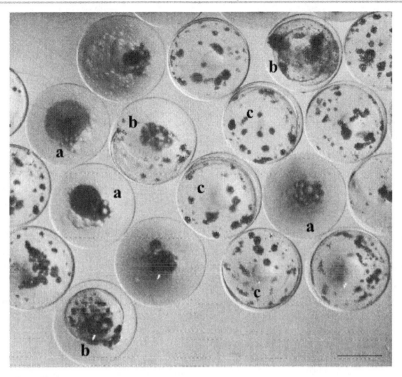

Figure 3 – Different aspects on eggs viability of *Solea senegalensis*; a – unfertilized; b – fertilized but not viable; c – fertilized and viable. Bar corresponds to 0.5 mm. (Photo by Laura Ribeiro.)

Another egg feature normally used to identify the quality of the egg was the lipid droplet position or disposition, when more than one lipid droplet was observed. Indeed, good quality eggs were reported to exhibit lipid droplets around the equatorial area [1,4,30]. However, variability was observed at hatching rate values when based on these characters.

In sharpsnout seabream and gilthead seabream, morphometric parameters were measured in floating eggs containing embryos in the first cleavage to morula stage (minimal and maximal diameter of egg, yolk, and lipid vesicle; different ratio values between these parameters describing the shape of the egg, yolk and lipid vesicle) (Fig. 4a) and it was investigated whether these parameters were correlated with the hatching rate (percentage of embryos hatching from floating eggs) [34]. From the investigated morphometric parameters the ratio value (MIDLI) of the maximal diameter of the lipid vesicle (MXDLI) to the minimal diameter of the lipid vesicle (MIDLI) [RDLV = MXDLI/MIDLI] and the lipid vesicle shape coefficient [LVSC = (MXDLV–MIDLV)/(MXDLV+MIDLV)*100] were correlated with larval survival (percentage) from viable eggs (Fig. 4b, 4c). As the shape of a lipid droplet depends on lipid type and composition, the lipid to water ratio, and the interactions with the surrounding water molecules, this parameter may be an indicator for the biochemical lipid composition. For the eggs of fresh water teleosts species no correlations have been found between morphological egg parameters and embryo or larvae viability until now.

3.2 Egg size

Egg size is frequently used as an egg quality parameter [2] because it provides an estimate of parental care investment in off-spring. Moreover, larger size eggs usually have a higher yolk content which will allow hatched larvae to survive for a longer period without food in the wild than eggs with less yolk content [22], thus increasing the chances of survival. However, studies with species maintained under favorable conditions indicated that egg size was not a determining factor for larval survival and growth rate, namely rainbow trout, *Oncorhynchus mykiss* [35], Atlantic salmon, [36], turbot, [37] and Senegalese sole [32]. Egg size is usually calculated based on egg diameter, especially if considering pelagic eggs where single diameter is commonly used. But two diameters are sometimes quoted: the longest diameter, egg length, and a second one perpendicular to it, egg breadth [38,39].

Egg size can be affected by several factors, namely nutritional status of the mother, size /age of the mother, type of spawning (stripping, natural, induced, etc), or moment of spawning (beginning, middle and end of spawning season) [8,40,41].

A decrease in egg diameter, and therefore yolk content, may occur towards the end of the spawning period for individual batch-spawning females which can be related to exhaustion of vitellogenesis reserves [1,32,42,43] and may result in a reduction in the survival percentage of the embryo and larvae [44]. In a multiple batch spawner, such as Atlantic cod, 26% of the observed variance in egg size resulted from batches-within-females [45]. Mean egg size decreased in successive batches laid by an individual female [21].

Generally, as fish size increases so does fecundity and the diameter of the eggs produced. In contrast with fish size, fish age appears to be much less important in determining fecundity, with reported increases in egg number with age probably being due to concomitant increases in fish size [46].

3.3 Fertilization rate

The fertilization rate usually indicates the percentage of fertilized eggs, expressing the successful encounter between egg and sperm. Although this parameter might be used as a quality criterion in both marine and freshwater species [40,47], inconsistent correlations with the survival rates at later stages of development were observed for some marine species [1]. According to Bromage [40], egg developmental stage used by different authors to assess fertilization rate might result in different correlations. For instance, Senegalese sole spawns naturally during the night at 18°C temperature. Consequently when eggs are sampled a developed embryo can already be identified, so the fertilization rate is calculated as the percentage of eggs with embryo within a sample of floating eggs. However, in turbot and Atlantic halibut the fertilization rate is calculated based on the number of developing eggs at the 8-cell stage expressed as a percentage of the total number of eggs, including the nonbuoyant fraction, 16 h post-fertilization [16,47]. The different methodologies use different periods of egg development so correlations between fertilization and later stages of larval development will have different results. Consequently, fertilization rate assessment needs to take into consideration the type of fertilization (induced, stripped, natural, etc.), the moment of egg collection and water temperature, among other parameters   when establishing correlations.

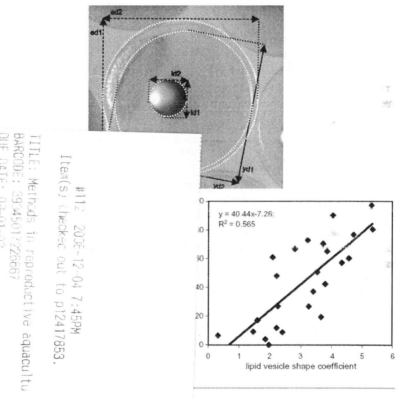

ation with egg quality. (From Lahnsteiner, F. and
h permission.) 4a. - Morphometric measurement procedures performed on *Diplodus puntazzo* eggs. Using digitized pictures the best fitting ellipse was fitted onto egg, yolk and lipid vesicle (white dotted lines) and the dimensions of the 2 major axes (= minimal and maximal diameter) were measured for each ellipse. ed1, ed2 major axis of egg; yd1, yd2 major axis of yolk; ld1, ld2 major axis of lipid vesicle;   4b - Scatter plots demonstrating the relationship between the ratio of the maximal to the minimal diameter of the lipid vesicle in viable, floating eggs in the first cleavage to gastrula stage and embryo survival to hatch in *Diplodus puntazzo*; 4c - Scatter plots demonstrating the relationship between the lipid vesicle shape coefficient in viable, floating eggs in the first cleavage to gastrula stage and embryo survival to hatch in *Diplodus puntazzo*.

Over-ripening is another important factor affecting successful fertilization, and has been the subject of several studies [1,4,31]. In fact, eggs retained in the abdominal cavity are subjected to oxygen depletion initiating tissue degeneration and other severe metabolic alterations [21]. Consequently, fertilization rates will decrease and the number of malformations usually increases [48].

The fertilization rate varies considerably during a spawn season of freshwater and marine fish species [18,47], although more consistent values of fertilization rate were observed for Atlantic halibut when temperature of broodstck was manipulated to maintain a constant value [47]. Moreover, according to Kjørsvik et al. [1] the fertility rate is usually higher when fishes spawn naturally, especially if they are batch-spawners. Nevertheless Senegalese sole fertilization rate decreased throughout the spawning season.

Several studies indicate that the fertilization rates of fishes in the wild are usually higher than captive fishes [1,21]. These observations indicate that further studies concerning broodstock husbandry conditions are still needed in order to obtain, at least, similar fertilization rates than in wild.

3.4 Blastomere morphology

Following fertilization there is a rapid series of mitotic cell divisions that partition the cytoplasm into numerous cells called blastomeres. In fish eggs the cell cleavage is restricted to the animal pole in what is referred to as the discoidal meroblastic cleavage pattern. Kjørsvik et al. [1] suggested that assessment of cell symmetry at early stages of cleavage (normal blastomeres) might be a possible general indicator of egg quality for marine fish. In blastomere morphology scoring, eggs are individually assessed at the 8-cell stage for five characteristics; symmetry, uniformity, margins, inclusion bodies and cell adhesion; the operator scores each egg using a 1 (very abnormal) to 4 (normal) qualitative scale [49]. Using blastomere morphology as quality criterion has an important economical advantage, since a few hours or days after fertilization it is possible to detect and discard bad quality spawn, thus saving energy, space and man hours. Moreover, it is also after fertilization that the majority of malformations occur, resulting in egg abortion if egg repair mechanisms do not succeed in repairing the observed malformations.

Shields et al. [49] and Kjørsvik et al. [14] reported that the blastomere morphology is a reliable parameter for egg quality determination in Atlantic cod, turbot, and Atlantic halibut. However, Vallin and Nissling [50] observed that the use of this parameter was not a consistent indicator for egg quality in Atlantic cod. Moreover, in haddock, *Melanogrammus aeglefinus*, the cleavage patterns of early embryos (8–32 blastomere stages) were investigated in correlations with the embryo survival at hatch [51]. Abnormalities in blastomere cleavage were categorized as (1) asymmetric blastomere arrangement, (2) inequality of blastomere size, (3) poor adhesion between blastomeres and (4) poor definition of blastomere margins. Low adhesion between blastomeres was related to decreased embryo viability, while asymmetry in blastomere arrangement was not [51]. Different types of cleavage abnormalities co-occurred and hierarchical multiple regressions revealed that asymmetry could be used to accurately predict hatching success, even if reduction in embryo viability was due to abnormalities other than asymmetry. Analysis of additional cleavage abnormalities suggested that complete separation between blastomeres was indicative of a very poor egg batch and resulted in little or no hatching, while cellular outcrops had no negative effect on hatching success [51].

Studies by Lahnsteiner et al. [52,53] indicated that the percentage of eggs with a normally developed morula stage (evaluation criteria: regular and symmetrical cell divisions) did not correlate with the percentage of eggs developing to the eyed embryo stage in several experiments, e.g., in different cyprinid species fertilization of eggs with cryopreserved semen resulted in high percentages of eggs reaching the morula stage but very low hatching rates as development stopped after this stage [53]. Similar results were found in cryobiological studies on burbot [52]. However, this observation might also depend on sperm alterations during freezing and thawing. In the marine fish sharpsnout seabream and gilthead seabream there was no correlation between the percentage of viable (floating) eggs in the first cleavage to morula stage and the percentage of embryos in the hatching stage or the percentage of hatched larvae [54]. The percentage of pre-hatch survival (late embryo stage before hatching) was significantly (P<0.001) correlated with the percentage of post-hatch survival (freshly hatched larvae) (*Sparus aurata*: Pearson correlation coefficient (R) = 0.887, sharpsnout seabream: R = 0.796). Similar correlations were obtained for brown trout, *Salmo trutta*, [28].

Beyond the low correlation of blastomere morphology with egg quality for some species, another factor might restrict the use of this criterion as a quality index for all fish species. Blastomere morphology is easily used when *in vitro* fertilization occurs, since a schedule can be programmed for evaluation (not readily adopted in the industry since it can consume many man hours). However, in species that spawn naturally and normally during the night, as is the case of Senegalese sole, this criterion would be difficult to follow since at egg collection an embryo is already formed. Moreover, the temperature at which spawn occurs also influences the embryogenesis development rate, making it difficult for some species to identify the different stages of development at current hatcheries practices.

3.5 Hatching rate

According to Brooks et al. [5] hatching rates and egg survival, although being the ultimate measures of egg quality tell nothing about what factors determine egg quality.

In fact, variations in egg quality leading to unpredictable hatching rates are often encountered and well demonstrated in several marine fish species, i.e., Atlantic cod [55,56], turbot [31,57], Atlantic halibut [4,58], gilthead seabream [59,60], Senegalese sole [32], wolffish [13] and in salmonids such as Atlantic salmon [61] and rainbow trout [2].

Although much effort has been made to evaluate criteria for marine egg and larval quality, the significance of poor egg quality for the results in final juvenile production has not been clarified [16].

4. PHYSIOLOGICAL PARAMETERS

When eggs of fresh water teleosts are immersed into water several changes occur which are referred to as egg water hardening and/or egg swelling. The following steps characterize this reaction [62,63]: Due to the cortical reaction (which is initiated by changes in membrane potential after contact with water) osmolytes are released from the cortical vesicles into the perivitelline space. The chorion (egg shell) is impermeable for osmolytes but permeable for molecules of lower molecular weight as water. Therefore, water is osmotically drawn into the perivitelline space. This process creates turgor pressure in the egg and the chorion is distended until a steady state between this turgor pressure and the tension force of the chorion is achieved [63]. Then the chorion hardens, which is accompanied by changes in its molecular structure [63]. The percent egg weight increase during hardening gives information about the progress of the cortical reaction, the concentration of osmotic active compounds and about the functionality of the chorion. The percent egg weight increase during hardening is significantly correlated with viability parameters of eggs in all fresh water species investigated until now: in common whitefish [64] (Fig. 5a), rainbow trout [65,66] (Fig. 5b), brown trout [28], and charr, *Salvelinus alpinus*, (Mansour, unpublished data) (investigated viability parameter: percentage of eyed stage embryos) and in different cyprinid species such as carp (Fig. 5c), silver carp, *Hypophthalmichthys molitrix*, grass carp (Fig. 5d), and Danube bleak, *Chalcalburnus chalcoides* [26]. The percent weight increase during hardening is one of the easiest egg quality assays developed until now. For determination of the egg weight increase during water hardening several unfertilized eggs which have not been in contact with water are transferred in a sieve, and remaining ovarian fluid is drained off with absorbent paper placed under the sieve. The samples are weighed to determine the weight of non-water hardened eggs. Then the eggs are incubated in glass beakers filled with water which in composition and temperature is similar to that used for hatching eggs. The eggs are re-weighed after egg hardening is terminated. This is after 120 min in Salmonidae [28,66] and after 60 min in Cyprinidae [26]. Thereby, the eggs are again placed in a sieve and the water is removed with absorbent paper and the eggs are weighed. Finally, percentile weight increase is calculated from the measured parameters. A disadvantage of the assay is that it requires long incubation times, especially in Salmonidae, as it measures only the completed reaction. Methods to shorten the assay time by measuring only the initial rate of water uptake have been described [66].

In the marine species gilthead seabream after egg activation with sea water the mean diameter of the yolk decreased for circa 8%, the egg shape changed (decrease in the ratio of the maximal:minimal diameter of the egg and in the egg shape coefficient), and the ratio value of the mean diameter of the egg to the mean diameter of the yolk increased [34]. The measured morphological changes were not related to egg quality. The decrease in mean diameter of the yolk might indicate dehydration by seawater. Also compression of yolk material due to the formation of the perivitelline space and subsequent hardening of the chorion, processes which develop a turgor pressure in the eggs [63], must be considered. These processes might also be responsible for the observed changes in egg shape from wrinkled and irregular in unfertilized eggs to unwrinkled, hard and almost spherical in fertilized eggs. On the contrary, in the Atlantic cod the egg diameter increased after activation [12,67]. In eggs of high quality cortical reaction was rapid and homogenous followed by a significant increase in osmolality and in egg diameter, but in eggs of low quality it was incomplete and the duration was approximately twice as long, and subsequently the increase in osmolality and in egg diameter was less significant [12].

5. PATHOLOGICAL ASPECTS

There are a large number of micro-organisms in the water, which increase significantly in number when eggs are incubated. During the incubation large number of bacteria approaching 500 colony forming units/mm^2 accumulate around egg surface [68]. The diverse flora that eventually develops on the egg surface reflects the bacterial composition of the water.

During hatching, various forms of interactions between bacteria and the egg surface may occur. This may result in the formation of an indigenous microflora or beginning of the infection process. Microbial growth probably occurs as a result of the increased amounts of nutrients from the metabolic by-products of the eggs, like various lipid and protein components of fish eggs [8]. There are various causes for bacterial growth in fish eggs, such as primary colonization of the dead sperm [69] or maternal infection of the eggs [70].

Several studies have reported the presence of large numbers of bacterial communities in association with the surface of fish eggs [68,70-72]. The bacteria most frequently isolated from the surface of live fish eggs are members of the genera *Cytophaga, Pseudomonas, Alteromonas* and *Flavobacterium* (Table 1).

The adherent microflora may damage developing eggs [73]. Keskin et al. [79] found bacterial loading to be more important in artificial incubation than in the wild. Poor egg quality and resulting mass mortalities have been a serious problem in larval production systems. The microflora on eggs may affect the short and long-term health of farmed fish.

Thus in aquaculture, eggs may be sterilised by various techniques to remove the adherent microflora and to prevent the transmission of bacterial pathogens during hatching and transport. Such methods (UV-irradiation, ozonisation, membrane filtration, and antibiotics) are used in intensive larviculture; however, these approaches disturb the balance of microbial communities and favour exponential growth of opportunistic bacteria. After disinfection, there is low competition for nutrients and opportunistic bacteria with high growth rates may proliferate [80].

Table 1 – Bacteria isolated from the surface of eggs from different fish species.

Fish Eggs- Species	Bacteria	Source
Halibut (*Hippoglossus hippoglossus*)	*Tenacibacter ovolyticus* *Pseudomonas* *Colwellia* *Pseudoalteromonas* *Cytophaga* *Photobacterium phosphoreum*	Hansen et al. [73] Verner-Jeffreys [74] Verner-Jeffreyset al. [75]
Cod (*Gadus morhua*)	*Alteromonas* *Pseudomonas* *Vibrio Fischeri* *Pseudoalteromonas* *Vibrio logei* *Moritella viscosa*	Hansen and Olafsen [70] Birkbeck et al. [76]
Rainbow trout (*Onchorynchus mykiss*)	*Flavobacterium psychrophilum*	Vatsos et al. [77]
Pacific threadfin (*Polydactylus sexfilis*)	*Pseudoalteromonas* *Alteromonas* *Vibrio* *Photobacterium damsela*	Verner-Jeffreys et al. [78]
Amberjack (*Seriola rivoliana*)	*Bacillus* sp. *Flavobacteriae* sp. *Sulfitobacter* sp. *Erythrobacter* *Rohodobacteria* *Colwellia*	Verner-Jeffreys et al. [78]
Turbot (*Scophthalmus maximus*)	*Acinobacter calcoacetivus* *Aeromonas hydrophila* *Acromobacter* spp. *Moraxella* spp. *Pseudomonas aeruginosa* *Pseudomonas fluorescens* *Pseudomonas* sp. *Vibrio parahaemolyticus*	Keskin et al. [79]

Use of antibiotics may result in alterations in the microflora that could be unfavourable [81]. The possibilities for using probiotics in intensive marine aquaculture eggs rearing have gained much interest in recent years, since probiotics represents a monoculture or low diversity addition of beneficial bacteria. The observed benefits of using probiotics in intensive aquaculture, indicate that eggs incubation might benefit from this procedure.

There are various interpretations of the influence of microbiological conditions on fish egg quality. Some studies show that microbial fauna does affect egg quality [79,82,83]. Some pathogenic bacteria may dissolve the chorion and zona radiata of the egg shell [73].

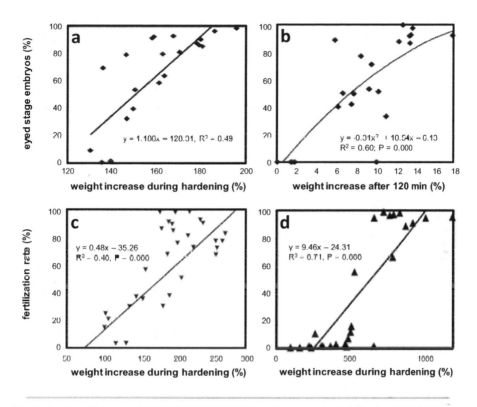

Figure 5 - Correlations established between egg quality parameters: a- Relation between the percent weight increase during hardening and the percentage of eggs developing to eyed stage embryos in *Coregonus sp.* (From Lahnsteiner, F., *Comp. Biochem. Physiol. Part B* 142:46-55, 2005. With permission.); b -. Relation between the percent weight increase during hardening and the percentage of eggs developing to eyed stage embryos in *Oncorhynchus mykiss.* (From Lahnsteiner F. and R.A. Patzner, *J. Appl. Ichthyol.* 18:24-26, 2002. With permission.); c - Relation between the percent weight increase during hardening and the fertilization rate (morula stage) in *Cyprinus carpio.* (From: Lahnsteiner, F., B. Urbanyi, A. Horvath, and T. Weismann, *Aquaculture* 195:331-352, 2001. With permission.); d - Relation between the percent weight increase during hardening and the fertilization rate (morula stage) in *Hpophthalmichthys molitrix, Ctenopharyngodon idella.*

5.1 Surface disinfection of fish eggs

The introduction of techniques for surface disinfection of marine fish eggs has been of significant importance in many hatcheries. Eggs can act as an important vehicle for the transmission of disease from parent to offspring and between hatcheries because opportunistic pathogens may be present in the surface of fish eggs [72]. Disinfection or antibiotic treatment of eggs has been widely used to reduce egg mortality and improve rearing success [84]. Agents tested included gluteraldehyde [85,86], iodophors [84], UV [87], ozone [88], hydrogen peroxide, antibiotics and sodium hypochlorite [84].

Antibiotics - The administration of antibiotics in aquaculture is not a solution, since resistance to antibiotics will develop later, [89]. Therefore, the use of antibiotics is rigorously controlled in most countries and efforts are made to minimize their use.

Glutaraldehyde - A procedure for disinfecting marine fish eggs with 400 ppm glutaraldehyde for 10 min was described by Salvesen and Vadstein [90]. Several studies have documented improved hatching, development and survival of larvae from different marine fish species after egg disinfection by this procedure [85,86,91]. Toxic effects of glutaraldehyde varied according to the egg quality and the time of the treatment [91].

Bronopol - Bronopol is used at concentrations of 200µg/ml as a treatment against a range of marine bacterial infections in farmed fertilized salmonid eggs, Atlantic salmon and rainbow trout [76].

Ozone - Improved performance of gilthead sea bream larvae after ozone disinfection of the eggs can be observed. Ozone (O_3) dissolved in seawater was evaluated, as an egg disinfectant. The results suggest that a 2 min exposure of eggs to 0.3 mg.l^{-1} O_3 would improve current protocols in marine larviculture [88].

UV irradiation - The use of UV treatment have been shown to reduce the bacterial loading of water [87] and consequently in the surface of eggs.

6. BIOCHEMICAL PARAMETERS

6.1 Ovarian fluid

In fresh water teleost fish species the ovarian fluid composition gives a good estimation of egg quality. In the lake brown trout the ovarian fluid pH, protein levels and activities of aspartate aminotransferase and ß-D-glucuronidase were significantly correlated with egg viability expressed as the number of eyed stage embryos [28]. Also in carp, silver carp, grass carp and Danube bleak, the ovarian fluid pH, protein levels and aspartate aminotransferase activity were highly correlated with the fertilization rate [26]. Examples for correlations between egg viability and ovarian fluid protein levels and pH, respectively, are shown in Fig. 6a-f. In rainbow trout in the ovarian fluid during over-ripening, levels of proteins, of esterified and non-esterified fatty acids and the activities of aspartate aminotransferase and acid phosphatase increased significantly [65]. Alterations in ovarian fluid pH and high protein levels were indicative also for over-ripe, low quality egg batches in turbot [11]. Therefore, and as these two parameters are easy to determine, they may be general markers for the egg quality of teleost fish. On the other hand, the usefulness of ovarian fluid pH as egg quality parameter could not be confirmed for Chinook salmon, *Oncorhynchus tshawytscha*, by Barnes et al. [92]. The amount of ovarian fluid is very limited in several species of fresh water fish [26] and no tool for egg quality determination exists in many marine fish as eggs are collected after natural spawning.

As seen from Fig. 6a-d the correlation between ovarian fluid proteins and viability parameters was negative in all investigated species. There are several indications that increased protein levels are due to protein leaking from the eggs into the ovarian fluid. This topic was investigated in detail in the brown trout [93]: In the ovarian fluid of this species, 12 types of proteins were determined by SDS-PAGE (Fig. 6g). Four proteins were negatively correlated with the percentage of eyed stage embryos (examples for the 68 kD and 39 kD protein see Figures 6b,c). The fact that these 4 types of ovarian fluid proteins were similar in molecular weight and staining behaviour to those occurring in the eggs is an indication that they leak out of eggs. Non-viable eggs may be degraded in the coelomic cavity and subsequently proteins may be released. Alterations of the egg plasma membrane leading to protein leakage could be an explanation, too. Also during over maturation certain types of ovarian fluid proteins increased due to protein leakage out of the eggs in rainbow trout [65]. The presented hypothesis is further supported in a study by Wojtczak et al. [94]. In this study two types of rainbow trout ova were distinguished: ova causing turbid water and having very low viability and ova that did not cause such an effect and having high viability. Protein and lipid concentrations of ovarian fluid collected from eggs causing turbid water were significantly higher than in samples collected from eggs that did not cause turbid water. The authors suggest that the turbidity of water was caused by the coagulation of egg proteins released into the ovarian fluid [94].

6.2 Biochemical composition of eggs

In teleost fish, an egg is the final product of oocyte growth and development [95]. Ovulated fish eggs only take up water and some chemicals from the water [96] and therefore all components of an egg that determine its quality, are incorporated into an egg during oogenesis. The developing oocyte synthesizes the components for DNA and protein synthesis, and mRNAs needed immediately after fertilization [97], while yolk proteins and components of the chorion are synthesized in the liver [98-100]. Deficiency or too low activity of enzymes and insufficient low metabolite levels may interrupt metabolic pathways, and lead to the accumulation or loss of distinct substrates and metabolites. Therefore, levels of metabolites or activities of enzymes might be correlated with viability parameters of embryos and larvae and subsequently be egg quality indicators.

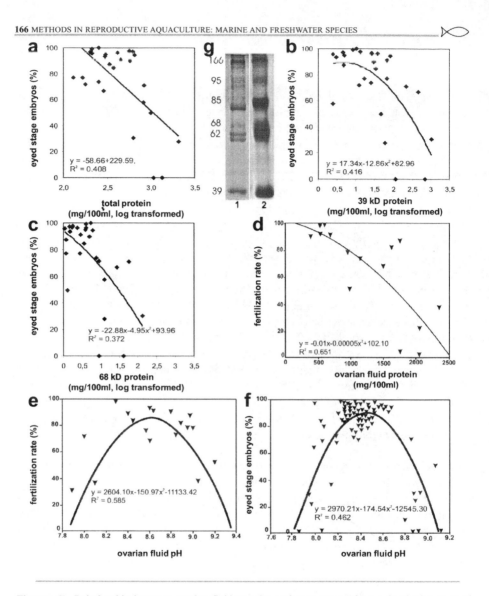

Figures 6 - Relationship between ovarian fluid proteins and percentage of eggs developing to eyed stage embryos; a - total protein, *Salmo trutta;* b - 39 kD protein, *Salmo trutta;* c - 68 kD protein, *Salmo trutta;* d - total protein, *Cyprinus carpio;* e-f - Relationship between ovarian fluid pH and percentage of eggs developing to eyed stage embryos, e - *Cyprinus carpio,* f - *Salmo trutta;* g - Qualitative protein composition of ovarian fluid from *Salmo trutta* as revealed by SDS–PAGE. Lanes 1-2: Different ovarian fluid samples stained with Coomassie blue. (Figs. 6a, b, c and g: From: Lahnsteiner, F, *Aquaculture Res.,* in press, 2007. With permission. Fig. 6f: From: Lahnsteiner, F., T. Weismann, and R.A. Patzner, Fish. Physiol. Biochem. 20:375-388, 1999. With Permission. Figs. 6d, 6e: From: Lahnsteiner, F., B. Urbanyi, A. Horvath, and T. Weismann, *Aquaculture* 195:331-352, 2001. With permission.)

Several studies have been conducted on this topic in teleost fresh water species: In lake trout, brown trout, in the freshly stripped eggs selected biochemical parameters were measured and embryo survival to the eyed stage was determined [28]. Then biochemical parameter correlations with percentage of eyed stage embryos were made to determine potential biochemical egg quality markers. Under aspects of practical application easy to measure, parameters of carbohydrate metabolism (fructose, galactose, glucose, glycogen, lactate, phosphofructokinase, pyruvate kinase, lactate dehydrogenase), lipid metabolism (cholesterol, esterified fatty acids, non-esterified fatty acids, glycerol, phospholipids, triglycerides, acetylcarnitine transferase, lipase), of protein metabolism (total protein, protease, aspartate aminotransferase), aerobic energy metabolism (malate dehydrogenase, respiration rate, NAD, NADH. ATP, acetoacetate, creatine phosphate), the biosynthetic potential (total DNA and RNA content, NADPH, and NADP-dependent isocitrate dehydrogenase), and lytic enzymes (ß-D-glucuronidase, acid and alkaline phosphatase) were measured. From the investigated metabolic parameters only activities of NADP-dependent isocitrate dehydrogenase, NAD-dependent malate dehydrogenase, respiration rate (Figure 7 a), the ratio of NADH to NAD levels, the levels of free, non-esterified fatty acids, and the ratio of non-esterified to esterified fatty acids were correlated with the number of eyed stage embryos [28].

A similar experiment was performed in Cyprinidae whereby in the carp, the silver carp, and the grass carp, the morula stage was used as viability parameter and in the Danube bleak, the eyed embryo stage. Levels of proteins, peptides, fructose, galactose, glucose, non-esterified fatty acids, esterified fatty acids, total DNA, and RNA, and activities of NADP-dependent isocitrate dehydrogenase, phosphofructokinase, pyruvate kinase, protease, lipase, NAD-dependent malate dehydrogenase, respiration rate, and aspartate aminotransferase were measured [26]. In carp, grass carp, and Danube bleak the activities of malate dehydrogenase and pyruvate kinase were correlated with the fertilization rate of the eggs, and in carp and Danube bleak with the respiration activity, too.

In a recent preliminary study using SDS-PAGE, main proteins of unfertilized eggs were qualitatively and quantitatively investigated in the brown trout, to see whether some of them were correlated with the rate of embryos reaching the eyed embryo stage [93]. In the eggs, 9 major types of proteins in the range of 95-15 kD were identified.

The 95 kD, 85 kD, 77 kD, and 39 kD protein were positively correlated with embryo survival to the eyed embryo stage. The explanatory effect of the multiple regression models was very high ($R^2 = 0.961$) indicating that distinct egg proteins are closely related with egg quality (Figure 7 b). It has to be determined in future studies if the essential proteins are yolk proteins, structural proteins or proteins involved in metabolism.

Concerning marine teleost fish species, in the gilthead seabream the metabolism of viable floating eggs was compared with the metabolism of non-viable, non-floating eggs [101]. Most importantly result of this study demonstrated that viable eggs had a well-developed carbohydrate metabolism (glycolysis, gluconeogenesis, pentose phosphate path) [101]. In non-viable eggs total monosacharides levels and the levels glucose, fructose, and galactose and the activities of pyruvate carboxylase and transaldolase were significantly decreased [101]. Therefore, carbohydrate metabolism is essential for embryogenesis. Based on these results biochemical egg quality markers were investigated in gilthead seabream [4]. Eggs obtained from natural spawning, were collected from commercial broodstocks. In the viable, floating eggs in the first cleavage to gastrula stage selected biochemical parameters (acid phosphatase, adenylate kinase, allanine aminotransferase, malate dehydrogenase, glucose-6-phosphatase, succinate dehydrogenase, transaldolase, acetyl CoA, amino acids, ATP, DNA, fructose, glucose, glucose-6-phosphate, NADH, NAD, phosphate, phospholipid, monosaccharids, sialic acid) were measured and embryo survival to hatch was determined. From the investigated parameters activities of adenylate kinase, glucose-6-phosphatase, and transaldolase, and levels of amino acids, sialic acid, glucose-6-phosphate, acid phosphatase and free monosaccharids were correlated with embryo survival to hatch (Figure 7 c,d,e). For all listed parameters the explanatory effect in simple regression models was low and precise egg quality determination was not possible with single biochemical parameters (Figure 7 c,d,e). The explanatory effect could be increased when using multiple regression models (Figure 7 f).

In sharpsnout seabream the same biochemical parameters as in gilthead seabream (acid phosphatase, adenylate kinase, glucose-6-phosphatase, and transaldolase, amino acids, monosaccharids, sialic acids) were indicators for egg quality [54]. Also for common dentex carbohydrate metabolism plays an important role during embryogenesis [102]. Finally, carbohydrate metabolism of eggs and embryos was also investigated in a fresh water species, the whitefish. In this species, metabolic pathways of glycolysis, pentose phosphate way, fructose synthesis (pylol path), and gluconeogenesis were demonstrated and several parameters of carbohydrate parameters were correlated with the percentage of eyed stage embryos developing out of the fertilized eggs, too.

In marine teleost fish the docosahexaenoic acid content, the ratio of docosahexaenoic acid:eicosapentaenoic acid and the ratio of $(n\text{-}3)$ to $(n\text{-}6)$ polyunsaturated fatty acids are commonly associated with egg quality [103,104], whereby high docosahexaenoic acid levels and high docosahexaenoic acid:eicosapentaenoic acid and $(n\text{-}3)/(n\text{-}6)$ polyunsaturated fatty acids ratios are considered desirable. Eggs from marine fish usually have a docosahexaenoic acid: eicosapentaenoic acid ratio of around 2:1 and an $(n\text{-}3)/(n\text{-}6)$ polyunsaturated fatty acids ratio between 5:1 and 10:1 [105]. Differences in lipid composition between wild and captive fish have been recorded in gilthead sea bream [106], striped bass, *Morone saxatilis* [107], turbot [108], and sea bass [109]. These differences have been attributed to broodstock diet. In the Atlantic halibut a diet with 1.8% arachidonic acid resulted in significantly higher fertilization rates, blastomere morphology scores and hatching rates than 0.4% arachidonic acid diet [110].

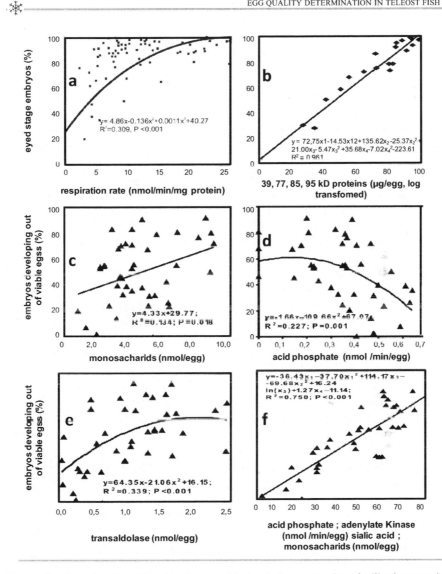

Figures 7 - Examples for relationships between biochemical parameter in unfertilized eggs and % eyed stage embryos developing out of the eggs in, a - *Salmo trutta*; b - Respiration rate, simple regression model, different egg proteins quantified by SDS-PAGE, multiple regression model; c-f - Examples for relationships between biochemical parameter in viable, floating eggs in the first cleavage to gastrula stage and the percentage of embryos developing out of viable eggs in *Sparus aurata*. c - monosacharids, d - acid phosphatase, e - transaldolase, f - multiple regression model. Note, that the explanatory effect (R^2) of all simple regression models is low and multiple regression models must be formulated to increase the explanatory effect. (Fig. 7a: Lahnsteiner, F., T. Weismann, and R.A. Patzner, Fish. Physiol. Biochem. 20:375-388, 1999. With Permission. Fig.7b: From: Lahnsteiner, F., *Aquaculture Res.*, in press, 2007. With permission. Figs. 7c-f: From: Lahnsteiner, F., and P. Patarnello, *Aquaculture* 237:443-459, 2004. With Permission.)

A ratio of docosahexaenoic acid to eicosapentaenoic acid of 2 and a ratio of eicosapentaenoic acid to arachidonic acid of 4 were associated with improved egg and larval quality [110]. In Japanese flounder (*Paralichthys olivaceus*) a diet containing 0.6% arachidonic acid improved its reproductive performance, but higher levels of arachidonic acid (1.2%) negatively affected both egg and larval quality due to a potential inhibitory effect on eicosapentaenoic acid bioconversion [111]. In the same species egg production was highest in fish fed high levels of 6.2% of *n*-3 highly unsaturated fatty acids [112]. However, egg quality parameters, such as percentage of buoyant eggs, hatching rate and percentage of normal larvae, were significantly higher in the group fed low levels (2.1%) of *n*-3 highly unsaturated fatty acids diet indicating that high levels of *n*-3 highly unsaturated fatty acids in broodstock diet negatively affect egg quality of Japanese flounder [112]. When gilthead head seabream, were fed a diet deficient in *n*-3 highly unsaturated fatty acids but rich in both oleic 18:1*n*-9 and linolenic 18:3*n*-3 acids eggs showed marked differences in lipid composition [113]. A negative correlation was detected between percentages of fertilized eggs and the levels of 18:1*n*-9, 18:3*n*-3 highly unsaturated fatty acids present in the phospholipids and between percentages of fertilized eggs and the ratio of 18:1*n*-9*n*-3 highly unsaturated fatty acids present in the phospholipids [113]. No correlation was found between the number of buoyant eggs and eicosapentaenoic 20:5*n*-3, eicosapentaenoic acid, docosahexaenoic 22:6*n*-3, docosahexaenoic acid fatty acids or total *n*-3 highly unsaturated fatty contents in eggs [113]. In cod significant correlations between hatching rate and docosahexaenoic/eicosapentaenoic acid levels were determined [114]. In the Asian sea bass, *Lates calcarifer*, the levels of total saturated fatty acids and of docosahexaenoic acid were correlated with the percentage of embryos in distinct development stages [115].

Also free amino acids play an important role in marine fish eggs whereby pelagic eggs contain up to 50% of the total amino acid pool as free amino acids [116,117]. The free amino acid pool derives from the hydrolysis of a yolk protein during final oocyte maturation [118,119]. In pelagic eggs free amino acids have important meanings as osmotic active compounds in regulating egg buoyancy [120]. During yolk resorption, the free amino acids are utilized as metabolic energy resource and reach low levels at first feeding [121-123]. They are also necessary for body protein synthesis. In gilthead seabream and sharpsnout seabream the total levels of free amino acids were correlated with embryo survival to hatch, in the Asian sea bass the levels of phosphoserin, aspartic acid, and arginine were correlated with the percentage of embryos in distinct development stages, too [115].

7. HORMONES

According to Brooks et al. [5] hormone levels might correlate with egg quality, since fish larvae are strongly affected by the action of different hormones during development. However studies on this subject are scarce and only thyroid hormones and cortisol were examined.

The importance of thyroid hormones has been widely reported during larval development. For some species it was possible to establish a correlation between thyroid hormone levels and egg quality. In fact, the pattern of variation of triiodothyronine (constant during embryogenesis, decreasing after hatching) and thyroxine (maintain constant) during embryogenesis and yolk sac resorption of chum salmon (*Oncorhynchus keta*), coincide with the beginning of endogenous production of thyroid hormones [124,125]. Moreover, higher levels of thyroid hormones in stripped sea bass egg resulted in higher larval growth rates [126,127]. Nevertheless, Tagawa and Hirano [125] working with medaka (*Oryzias latipes*) obtained identical results in hatching rate, time of hatching and survival after starvation, between eggs from normal females and eggs from females previously treated with thiourea (levels of thyroid hormones 10 times less in eggs).Thus, further studies are required to clarify the role of thyroid hormones on egg quality.

Significant amounts of cortisol were also observed in eggs and embryos, which decreased between fertilization and hatching [128-130]. In rainbow trout, this decrease seems to be related with water uptake during hardening [131]. Besides these observations no evidence exists that cortisol affects egg quality [5].

8. MOLECULAR TOOLS

Despite the importance of parental genes on egg quality, only recently some studies provided information on gene expression during embryogenesis in relation to organogenesis [5]. Moreover, despite the importance of products synthesized within the oocyte in egg quality few studies have analyzed gene expression during oogenesis. In fact, according to some authors [132-134] products synthesized in ovo, including such vital enzymes as cathepsin D, and the mechanisms controlling their expression, are likely to play a crucial role in determining egg quality.

Cathepsin D belongs to the group of aspartic proteases, proteolytic enzymes of the pepsin family [135] and some studies suggest that it is involved in the proteolysis of vitellogenin within oocytes, cleaving vitellogenin into yolk proteins and other substances essential for embryo development [136]. Considering the role of cathepsin D on vitellogenesis Brooks et al. [132] hypothesized that the activity of this enzyme might be vital for the production of a viable egg, and consequently an indicator of quality. In fact, studying rainbow trout eggs, Brooks et al. [132] observed an increase in cathepsin D mRNA expression at the beginning of vitellogenesis, which coincided with yolk protein sequester. Moreover, these authors observed that until gastrulation cathepsin D mRNA expression was not detected in either fertilized or unfertilzed eggs, indicating that there is little, if any, de novo synthesis of this message at these stages of development. From eyeing to hatching an increase was also observed, however whether this increase in cathepsin D mRNA expression is linked to yolk protein or to the formation of organs expressing cathepsin D has yet to be determined [132]. Carnevalli et al. [133] analyzing the activity of cathepsin D in gilthead seabream eggs, reinforced the ability of cathepsin to be used as an egg quality marker. In fact, higher levels of cathepsin D activity were observed in sinking eggs when compared to normal floating eggs. These data suggest that the egg sinks due to an alteration in egg osmoregulation as a result of cathepsin D proteolytic activity. Furthermore, in normal floating eggs the pattern of activity of cathepisn D matches with embryo energetic requirements, which are low during the first phase of development to increase at blastula phase [133]. Still, although this study confirms the key role of cathepsin D on vitellogenin degradation, other types of cathepsins were involved in this process [137]. In a posterior study using sea bass, Carnavelli et al. [134] reported several cathepsins in eggs. In fact, high levels of cathepsins B, D, and L were detected in the eggs, whereas no cathepsin A, C, and E activity was detected. In addition, cathepsin D was found at significantly higher levels in sinking eggs, whereas cathepsin L was more abundant in floating eggs [134]. During embryogenesis different cathepsins exhibited a specific pattern of activity suggesting that yolk degradation is tightly controlled and is assumed to be due to the action of specific proteinase inhibitors that are associated with yolk proteins [134].

Another use of molecular tools evidenced that the commonly used breeding protocols for rainbow trout can affect egg quality. Previously, Bonnet et al. [48] observed that manipulation of photoperiod, spawning induction and over-ripening resulted in a higher percentage of malformations and/or decrease in the fertility rate when compared to control treatment (without manipulation). In fact, these observations were confirmed by the analysis of egg transcriptome obtained from trout females submitted to similar treatments, where Bonnet et al. [138] observed significant changes in the egg mRNA abundance of specific genes. Moreover, a strong increase in apoliprotein C1 and tyrosine protein kinase was observed in the eggs obtained from spawning induction, although the effect of these changes on subsequent larval development is still unknown. In addition, both microarray and realtime PCR analyses showed that prohibiting egg mRNA abundance was negatively correlated with developmental success. In conclusion, Bonnet et al. [138] identified genes that might be used for a better understanding of the mechanisms associated with each type of ovulation control (i.e., hormonal, photoperiod), and in the identification of conserved mechanisms triggering the loss of egg developmental potential.

9. CONCLUSIONS

Under the assumption that egg quality markers should reveal highly significant correlations with egg viability and that they should be measurable with little effort, the percentile weight increase during hardening and the ovarian fluid pH and protein levels seem to be the most suitable quality markers for eggs of fresh water fish species (demersal), the shape of the lipid vesicle and the cleavage pattern of early embryonic cells for marine species (pelagic).

A reliable egg quality marker should fulfill the following demands: (1) It should reveal a statistically highly significant correlation with the development rate of eggs to late embryonic or early larval stages (e.g., eyed embryo stage, hatching rate, survival rate of larvae) to determine egg quality accurately, (2) it should reveal low intra-specific variations, since high variations will superpose statistical effects, and (3) it should determine objectively (4). Further, it should be measurable under routine hatchery conditions without high analytical, technical and temporal effort. The present review paper demonstrates that several morphological, physiological, and biochemical parameters of eggs and ovarian fluid, respectively, are promising candidates for egg quality determination. However, there still exist problems which restrict their routine application. Although several parameters reveal correlations with egg quality, their explanatory effect is often only low or medium in regression models allowing no exact quality determination for the full range of egg samples. Another problem is that several parameters can be determined only with considerable effort. This concerns especially the biochemical parameters of the eggs, whose determination needs sophisticated extraction and analysis procedures. Under these considerations in fresh water fish species the percentile weight increase during water hardening and the ovarian fluid pH and protein levels seem to be the most suitable egg quality markers, for marine fish eggs the shape of the lipid vesicle and the cleavage pattern of early embryos.

10. ACKNOWLEDGMENTS

Franz Lahnsteiner is grateful for financial support in egg quality studies by "Austrian Bundesministerium für Land- und Forstwirtschaft," Austrian "Fonds zur Förderung der wissenschaftlichen Forschung," and "Stiftung Aktion Österreich-Ungarn." In addition, Franz Lahnsteiner acknowledges excellent research cooperation with the fish farm Kreuzstein in St. Gilgen (Austria), Valle Ca' Zuliani hatchery in Monfalcone (Italy), Dinnyes fish farm and TEHAG fish farm in Hungary, and the Centro de Acuicultura-IRTA in Tarragona (Spain).

Laura Ribeiro was supported by a postdoctoral grant from Fundação para a Ciência e Tecnologia (SFRH/BPD/7148/2001). Florbela Soares is grateful to project MARE (22-05-01-FDR-00026) for supporting her research activity.

11. REFERENCES

1-Kjørsvik, E., Mangor Jensen, A., and Holmefjord, T., Egg quality in fishes, *Adv Mar Biol*, 26, 71, 1990.

2-Bromage, N.R., Jones, J., Randall, C., Thrush, M., Davies, B., Springate, J., Duston, J., and Barker, G., Broodstock management, fecundity, egg quality and the timing of egg production in the rainbow trout (*Oncorhiynchus mykiss*), *Aquaculture*, 100, 141, 1992.

3-Campbell, P.M., Pottinger, T.G., and Sumpter, J.P., Stress reduces the quality of gametes produced by rainbow trout, *Biol Reprod*, 47, 1140, 1992.

4-Bromage, N.R., Bruce, M., Basavaraja, N., and Rana, K., Egg quality determinants in fish: the role of overripening with special reference to the timing of stripping in the Atlantic halibut *Hippoglossus hippoglossus*, *J World Aquacult Soc*, 25, 13, 1994.

5-Brooks, S., Tyler, C.R., and Sumpter, J.P., Egg quality in fish: what makes a good egg?, *Rev Fish Biol Fisher*, 7, 387, 1997.

6-Christiansen, R. and Torrissen, O.J., Effects of dietary astaxanthin supplementation on fertilization and egg survival in Atlantic salmon (*Salmo salar L.*), *Aquaculture*, 153, 51, 1997.

7-Carrillo, M., Zanuy, S., Oyen, F., Cerda, J., Navas, J.M., and Ramos, J., Some criteria of the quality of the progeny as indicators of physiological broodstock fitness, in *Cahiers Options Méditerranéennes*, 47, Basurco, B., Ed., Mediterranean Marine Aquaculture Finfish Species Diversification, CIHEAM, Zaragoza, Spain, 2000, 61.

8-Bromage, N.R. and Roberts, R.J., *Broodstock management and egg and larval quality*, Blackwell Science, Oxford, 1994, 424.

9-Lahnsteiner, F. and Patarnello, P., Egg quality determination in the gilthead seabream, *Sparus aurata*, with biochemical parameters, *Aquaculture*, 237, 443, 2004.

10-Matsuura, S., Furuichi, M., Maruyama, K., and Matsuyama, M., Daily spawning and quality of eggs in one female red sea bream, *Suisanzoshoku*, 36, 33, 1988.

11-Fauvel, C., Omne's, M.H., Suquet, M., and Normant, Y., Reliable assessment of overripening in turbot *(Scophthalmus maximus)* by a simple pH measurement, *Aquaculture*, 117, 107, 1993.

12-Kjørsvik, E. and LÆnning, S., Effects of egg quality on normal fertilization and early development of the cod, *Gadus morhua L.*, *J Fish Biol*, 23, 1, 1983.

13-Pavlov, D.A. and Moksness, E., Production and quality of eggs obtained from wolffish *(Anarhichas lupus L.)* reared in captivity, *Aquaculture*, 122, 295, 1994.

14-Kjørsvik, E., Hoehne-Reitan, K., and Reitan, K.I., Egg and larval quality criteria as predictive measures for juvenile production in turbot *(Scophthalmus maximus L.)*, *Aquaculture*, 227, 9, 2002.

15-Gavaia, P.J., Dinis, M.T., and Cancela, M.L., Osteological development and abnormalities of the vertebral column and caudal skeleton in larval and juvenile stages of hatchery-reared Senegal sole *(Solea senegalensis)*, *Aquaculture*, 211, 305, 2002.

16-Kjørsvik, E., Hoehne-Reitan, K., and Reitan, K.I., Egg and larval quality criteria as predictive measures for juvenile production in turbot *(Scophthalmus maximus L.)*, *Aquaculture*, 227, 9, 2003.

17-Lambert, Y. and Thorsen, A., Integration of captive and wild studies to estimate egg and larval production of fish stocks, *J Northw Atl Fish Sci*, 33, 71, 2003.

18-Mylonas, C.C., Papadaki, M., Pavlidis, M., and Divanach, P., Evaluation of egg prodution and quality in the Mediterranean red porgy *(Pagrus pagrus)* during two consecutive spawning seasons, *Aquaculture*, 232, 637, 2004.

19-Bone, Q., Marshall, N.B., and Blaxter, J.H.S., Reproduction and life histories, in *Biology of Fish*, Chapman and Hall, London, 1995, 170.

20-Cabrita, E., Soares, F., and Dinis, M.T., Characterization of senegalese sole, *Solea senegalensis*, male broodstock in terms of sperm production and quality, *Aquaculture*, 261, 967, 2006.

21-Kamler, E., Parent–egg–progeny relationships in teleost fishes: an energetics perspective, *Rev Fish Biol Fish*, 15, 399, 2005.

22-Blaxter, J.H.S., Development: eggs and larvae, in *Fish Physiology*, Hoar, W.S. and Randall, D.J., Eds., Academic Press, New York, 1969, 177.

23-Jobling, M., *Environmental Biology of Fishes*, Chapman and Hall, London, 1995, 455.

24-Davis, C.C., Planktonic Fish Egg from Fresh Water, *Limnol Oceanogr*, 4, 352, 1959.

25-Russell, F.R.S., *The eggs and planktonic stages of British Marine Fishes*, Academic Press, London, New York, San Francisco, 1976, 524.

26-Lahnsteiner, F., Urbanyi, B., Horvath, A., and Weismann, T., Bio-markers for egg quality determination in cyprinid fishes, *Aquaculture*, 195, 331, 2001.

27-Chatain, B., Saroglia, M., Sweetman, J., and Lavens, P., Seabass and seabream culture: problems and prospects, Handbook of contributions and short communications, presented at the International Workshop on "Seabass and seabream culture: problems and prospects", Verona, Italy, October 16-18, 1996.

28-Lahnsteiner, F., Weismann,T., and Patzner, R.A., Physiological and biochemical parameters for egg quality determination in lake trout, *Salmo trutta lacustris*, *Fish Physiol Biochem*, 20, 375, 1999.

29-Escaffre, A.M. and Billard, R., Evolution de la fécondabilité des ovules de truite arc-en-ciel *Salmo gairdneri* laissés dans la cavité abdominale au cours de la période post-ovulatoire, *Bulletin Français de Pisciculture*, 272, 56, 1979.

30-Dinis, M.T., Methods of incubation Dover sole (*Solea solea*) eggs, in *Symposium sur l'aménagement des ressources vivantes de la Méditerranée*, Palma de Mallorca, Septembre, 1980, Relatórios de Actividades do Aquário Vasco da Gama, July 12, 1982, 8.

31-McEvoy, L.A., Ovulatory rhythms and over-ripening of eggs in cultivated turbot, *Scphthalmus maximum L.*, *J Fish Biol*, 24, 48, 1984.

32-Dinis, M.T., Ribeiro, L., Soares, F., and Sarasquete, C., A review on the cultivation potential of *Solea senegalensis* in Spain and Portugal, *Aquaculture*, 176, 27, 1999.

33 Carrillo, M., Bromage, N., Zanuy, S., Serrano, R., and Prat, F., The effect of modifications in photoperiod on spawning time, ovarian development and egg quality in the sea bass (*Dicentrarchus labrax*), *Netherlands Journal of Zoology*, 45, 204, 1989.

34-Lahnsteiner, F. and Patarnello, P., The shape of the lipid vesicle is a potential marker for egg quality determination in the gilthead seabream, *Sparus aurata* and in the sharpsnout seabream, *Diplodus puntazzo*, *Aquaculture*, 246, 423, 2005.

35 Springate, J.R.C. and Bromage, N.R., Effects of egg size on early growth and survival in rainbow trout (*Salmo gairdneri Richardson*), *Aquaculture*, 47, 163, 1985.

36 Thorpe, J.E., Miles, M.S., and Keay, D.S., Developmental rate, fecundity and egg size in Atlantic salmon, *Salmon salar L.*, *Aquaculture*, 43, 289, 1984.

37-Devauchelle, N., Alexandre, J.C., Corre, L.N., and Letty, Y., Spawning of turbot (*Scophthalmus maximus*) in captivity, *Aquaculture*, 69, 84, 1988.

38-Kato, T. and Kamler, E., Criteria for evaluation of fish egg quality, as exemplified for *Salmo gairdneri (Rich.)*, *Bull Natl Res Inst Aquacult*, 4, 61, 1983.

39-Townshend, T.J. and Wootton, R.J., Effect of food supply on the reproduction of the convict cichlid, *Cichlasoma nigrofasciatum*, *J Fish Biol*, 24, 91, 1984.

40 Bromage, N.R., Broodstock Management and seed quality – General considerations, In *Broodstock management and egg and larval quality*, Bromage, N.R. and Roberts, R.J., Eds., Blackwell Science, Oxford, 1995, 1.

41-Tamada, K. and Iwata, K., Intra-specific variations of egg size, clutch size and larval survival related to maternal size in amphidromous *Rhinogobius goby*, *Environ Biol Fish*, 73, 379, 2005.

42-Mihelakakis, A. and Kitajima, C., Spawning of the silver sea bream, *Sparus sarba*, in captivity, *Jpn J Ichthyol*, 42, 53, 1995.

43-Mihelakakis, A.,Yoshimatsu, T., and Tsolkas, C., Spawning in captivity and early life history of cultured red porgy, *Pagrus pagrus*, *Aquaculture*, 199, 333, 2001.

44-Blaxer, J.H.S., Patterns and variety in development, in: *Fish Physiology*, Hoar, W.S. and Randall, D.J., Eds., Academic Press, California, 1988, 1.

45-Chambers, R.C. and Waiwood, K.G., Maternal and seasonal differences in egg sizes and spawning characteristics of captive Atlantic cod, *Gadus morhua*, *Can J Fish Aquat Sci*, 53, 1986, 1996.

46-Bromage, N.R. and Cumaranatunga, R., Egg production in: the rainbow trout, in *Recent Advances in Aquaculture*, Muir, J.F. and Roberts, R.J., Eds., Croom Helm/Timber press, London, 1988, 63.

47-Brown, N.P., Shields, R.J., and Bromage, N.R., The influence of water temperature on spawning patterns and egg quality in the Atlantic halibut (*Hippoglossus hippoglossus L.*), *Aquaculture*, 261, 993, 2006.

48-Bonnet, E., Fostier, A., and Bobe, J., Characterization of rainbow trout egg quality: A case study using four different breeding protocols, with emphasis on the incidence of embryonic malformations, *Theriogenology*, 67, 786, 2007.

49-Shields, R.J., Brown, N.P., and Bromage, N.R., Blastomere morphology as a predictive measure of fish egg viability, *Aquaculture*, 155, 1, 1997.

50-Vallin, L. and Nissling, A., Cell morphology as an indicator of viability of cod eggs – results from an experimental study, *Fish Res*, 38, 247, 1998.

51-Rideout, R.M., Trippel, E.A., and Litvak, M.K., Predicting haddock embryo viability based on early cleavage patterns, *Aquaculture*, 230, 215, 2004.

52-Lahnsteiner, F., Mansour, N., and Weismann, T., The cryopreservation of spermatozoa of the burbot, *Lota lota* (Gadidae, Teleostei), *Cryobiology*, 45, 195, 2002.

53-Lahnsteiner, F., Berger, B., and Weismann, T., Effect of media, fertilization technique, extender, straw volume, and sperm to egg ratio on hatchtability of cyprinid embryos, using cryopreserved semen, *Theriogenology*, 60, 829, 2003.

54-Lahnsteiner, F. and Patarnello, P., Biochemical egg quality determination in the gilthead seabream, *Sparus aurata*: Reproducibility of the method and its application for sharpsnout seabream, *Puntazzo puntazzo*, *Aquaculture*, 237, 433, 2004.

55-KjÆrsvik, E., Stene, A., and Lonning, S., Morphological, physiological and genetical studies of egg quality in cod (*Gadus morhua*), in: *The Propagation of cod Gadus morhua L.*, Dahl, E., Danielssen, D. S., Mooksness, E., and Solendal, P.,Eds., Flodevigen Rapportserie, 1, 1984, 67.

56-KjÆrsvik, E., Egg quality in wild and broodstock cod *Gadus morhua L.*, *J World Aquac Soc*, 25, 22, 1994.

57-Fauvel, C., Omnes, M.H., Suquet, M., and Normant, Y., Enhancement of the production of turbot, *Scophthalmus maximus (L.)*, larvae by controlling overripening in mature females, *Aquaculture and Fisheries Management*, 23, 209, 1992.

58-Kjørsvik, E. and Holmefjord, I., Atlantic halibut (*Hippoglossus hippoglossus*) and cod (*Gadus morhua*), in *Broodstock management and egg and larval quality*, Bromage, N.R. and Roberts, R.J., Eds., Blackwell Science, Oxford, 1995, 169.

59-Tandler, A., Harel, M., Koven, W.M., and Kolkovsky, S., Broodstock and larvae nutrition in gilthead seabream *Sparus aurata* new findings on its involvement in improving growth, survival and swim bladder inflation, *Isr J Aquacult Bamidgeh*, 47, 95, 1995.

60-Fernández-Palacios, H., Izquierdo, M., Robaina, L., Valencia, A., Salhi, M., and Montero, D., The effect of dietary protein and lipid from squid and fish meals on egg quality of broodstock for gilthead seabream (*Sparus aurata*), *Aquaculture*, 148, 233, 1997.

61-Nævdal, G., Broodstock development for Norwegian salmonid aquaculture, in: *Proceedings of the special session on salmonid aquaculture*, Cook, R.H. and Pennell, W., Eds., February 16, 1989, Los Angeles, USA/World Aquaculture Society, 1831, Canadian Technical Report of Fisheries and Aquatic Sciences, 1991, 93.

62-Jaffe, L.F., The role of calcium explosions, waves, and pulses in: activating eggs, in *Biology of Fertilization*, 3, Metz, C.B. and Monroy, A., Eds., Academic Press, Orlando, Florida, 1985, 127.

63-Alderdice, D.F., Osmotic and ionic regulation in teleost eggs and larvae, in: *Fish Physiology. The Physiology of Developing Fish. Eggs and Larvae*, 11, A, Hoar, W.S. and Randall, D.J., Eds., Academic Press, London, 1988, 163.

64-Lahnsteiner, F., Carbohydrate metabolism of eggs of the whitefish, *Coregonus spp.* during embryogenesis and its relationship with egg quality, *Comp Biochem Physiol B*, 142, 46, 2005.

65-Lahnsteiner, F., Morphological, physiological and biochemical parameters characterizing the overripening of rainbow trout eggs, *Fish Physiol Biochem*, 23, 107, 2000.

66-Lahnsteiner, F. and Patzner, R.A., Rainbow trout egg quality determination by the relative weight increase during hardening: a practical standardization, *J Appl Ichthyol*, 18, 24, 2002.

67-Lønning, S. and Davenport, J., The swelling egg of the rough dab, *Hippoglossoides platessoides limandoides (Bloch)*, *J Fish Biol*, 17, 359, 1980.

68-Barker, G.A., Smith, S.N., and Bromage, N.R., The bacterial flora of rainbow trout *Salmo gairdneri Richardson*, and brown trout, *Salmo trutta L.*, eggs and its relationship to developmental success, *J Fish Dis*, 12, 281, 1989.

69-Rosenthal, H. and Odense, P., Elektronenmikroskopische Untersuchungen zur Oberflächenstruktur von Heringseiern, *Jahresbericht der Biologischen Anstalt Helgoland*, Hamburg, 44, 1986.

70-Hansen, G.H. and Olafsen, J.A., Bacterial colonization of cod (*Gadus morhua L.*) and halibut (*Hippoglossus hipoglossus*) eggs in marine aquaculture, *Appl Environ Microbliol*, 55, 1435, 1989.

71-Hansen, G.H. and Olafsen, J.A., Bacterial interactions in the early life stages of marine coldwater fish, *Microb Ecol*, 38, 1, 1999.

72-Bergh, O., Hansen, G.H., and Taxt, R.E., Experimental infection of eggs and yolk sac larvae of halibut (*Hippoglossus hippoglossus L.*), *J Fish Dis*, 15, 379, 1992.

73-Hansen, G.H., Berg, O., Michaelsen, J., and Knapskog, D., *Flexibacter ovolyticus sp. Nov.*, a pathogen of eggs and larvae of Atlantic halibut, *Hippoglossus hippoglossus L*, *Int J Syst Bacteriol*, 42, 451, 1992.

74-Verner-Jeffreys, D.W., The bacteriology of Atlantic halibut *Hippoglossus hipoglossus (L.)* larval rearing, Ph.D. thesis, University of Glasgow, 2000.

75-Verner-Jeffreys, D.W., Shields, R.J., Bricknell., I.R., and Birkbeck, T.H., Changes in the gut-associated microfloras during the development of Atlantic halibut (*Hippoglossus hipoglossus L.*) larvae in British hatcheries, *Aquaculture*, 219, 21, 2003.

76-Birkbeck, T.H., Reid, H.I., Darde, B., and Grant A.N., Activity of bronopol (Pyceze®) against bacteria cultured from eggs of halibut, *Hipoglossus hipoglossus* and cod, *Gadus morhua, Aquaculture*, 254, 125, 2006.

77-Vatsos, I.N., Thompson, K.D., and Adams, A., Adhesion of the fish pathogen *Flavobacterium psychrophilum* to unfertilized eggs of rainbow trout (*Oncorhynchus mykiss*) and n-hexadecane, *Lett Appl Microbiol*, 33, 178, 2001.

78-Verner-Jeffreys, D.W., Nakamura, I., and Shields, R.J., Egg-associated microflora of Pacific threadfin, *Polydactylus sexfilis* and amberjack, *Seriola rivoliana*, eggs. Characterisation and properties, *Aquaculture*, 253, 184, 2006.

79-Keskin, M., Keskin, M., and Rosenthal, H., Pathways of bacterial contamination during egg incubation and larval rearing of turbot (*Scophthalmus maximus*), *J Appl Ichtyol*, 10, 1, 1994.

80-Andrew, J.H. and Harris, R.F., R- and K-selection in microbial ecology, *Adv Microb Ecol*, 9, 99, 1986.

81-Skjermo, J. and Vadstein, O., Techniques for microbial control in the intensive rearing of marine larvae, *Aquaculture*, 177, 333, 1999.

82-Trust, T.J., The bacterial population in vertical flow tray hatcheries during incubation of salmonid eggs, *J Fish Res Board Can*, 29, 567, 1972.

83-Sauter, R.W., Williams, C., Meyer, E.A., Celnik, B., Banks, J.L., and Leith, D.A., A study of bacteria present within unfertilized salmon eggs at the time of spawning and their possible relation to early lifestage disease, *J Fish Diseases*, 10, 193, 1987.

84-Peck, M.A., Buckley, L., O'Bryan, L.M., Davies, E.J., and Lapolla, A.E., Efficacy of egg surface disinfectants in captive spawning Atlantic cod *Gadus morhua L.* and haddock *Melanogrammus aeglefinus L., Aquac Res*, 35, 992, 2004.

85-Salvesen, I., Oie, G., and Vadstein, O., Surface disinfection of Atlantic halibut and turbot eggs with glutaraldehyde: Evaluation of concentrations and contact times, *Aquac Int*, 5, 249, 1997.

86-Harboe, T., Huse, I., and Oie, G., Effects of egg disinfection on yolk-sac and 1st feeding stages of halibut (*Hippoglossus hippoglossus*) larvae, *Aquaculture,* 119, 157, 1994.

87-Sommer, R., Cabaj, A., Pribil, W., and Haider, T., Influence of lamp intensity and water transmittance on the UV disinfection of water, *Water Sci Technol*, 35, 113, 1997.

88-Ben-Atia, I., Lutzky, S., Barr, Y., Gamsiz, K., Shtupler, Y., Tandler, A., and Koven, W., Improved performance of gilthead sea bream, *Sparus aurata*, larvae after ozone disinfection of the eggs, *Aquac Res*, 38, 166, 2007.

89-Alderman, D.J. and Hastings, T.S., Antibiotic use in aquaculture: development of antibiotic resistance-potential for consumer health risks, *Int J Food Sci Technol,* 33, 139, 1988.

90-Salvesen, I. and Vadstein, O., Surface disinfection of eggs from marine fish: evaluation of four chemicals, *Aquaculture Int*, 3, 155, 1995.

91-Escaffre, A.M., Bazin, D., and Bergot, P., Disinfection of *Sparus aurata* eggs with glutaraldehyde, *Aquacult Int*, 9, 451, 2002.

92-Barnes, M.E., Sayler, W.A., Cordes, R.J., and Hunten, R.P., Potential indicators of egg viability in landlocked fall chinook salmon spawn with or without the presence of overripe eggs, *N Am J Aquac*, 65, 49, 2003.

93-Lahnsteiner, F., First results on a relation between ovarian fluid and egg proteins of *Salmo trutta* and egg quality, *Aquac Res*, 38, 131, 2007.

94 Wojtczak, M., Kowalski, R., Dobosz, S., Goryczko, K., KuŸminski, H., Glogowski, J., and Ciereszko, A., Assessment of water turbidity for evaluation of rainbow trout (*Oncorhynchus mykiss*) egg quality, *Aquaculture*, 242, 617, 2004.

95-Tyler, C.R. and Sumpter, J.P., Oocyte growth and development in teleosts, *Rev Fish Biol Fisher*, 6, 287, 1996.

96-Holliday, F.G.T. and Pattie Jones, M., Some effects of salinity on the developing eggs and larvae of the plaice (*Pleuronectes platessa*), *J Mar Biol Ass U.K.*, 47, 39, 1967.

97-Tata, J.R., Coordinated assembly of the developing egg, *BioEssays*, 4, 197, 1986.

98-Ng, T.B and Idler, D.R., Yolk formation and differentiation in teleost fishes, in *Fish Physiology Reproduction*, IXA, Hoar, W.S., Randall, D.J., and Donaldson, E.M., Eds., Academic Press, New York, 1983, 373.

99-Hyllner, S.J., Silversand, C., and Haux, C., Formation of the vitelline envelope precedes the active uptake of vitellogenin during oocyte development in the rainbow trout, *Oncorhynchus mykiss*, *Mol Reprod Develop*, 39, 166, 1994.

100-Oppen-Berntsen, D.O., Gram-Jensen, E., and Walther, B.T., Zona radiata proteins are synthesized by rainbow trout (*Oncorhynchus mykiss*) hepatocytes in response to estradiol-17-beta, *J Endocrin*, 135, 293, 1992.

101-Lahnsteiner, F. and Patarnello, P., Investigations on the metabolism of viable and nonviable gilthead sea bream (*Sparus aurata*) eggs, *Aquaculture*, 223, 159, 2003.

102-Giménez, G., Estévez, A., Lahnsteiner, F., Zecevic, B., Bell, J.C., Henderson, R.J., Piñera, J.A., and Sánchez-Prado, J.A., Egg quality criteria in common dentex (*Dentex dentex*), *Aquaculture*, 260, 232, 2006.

103-Izquierdo, M.S., Review article: essential fatty acid requirements of cultured marine fish larvae, *Aquacult Nutr*, 2, 183, 1996.

104-Takeuchi, T., Essential fatty acid requirements of aquatic animals with emphasis on fish larvae and fingerlings, *Rev Fish Sci*, 51, 1, 1997.

105-Sargent, J.R., Origins and functionsof egg lipids: nutritional implications, in *Broodstock Management and Egg and Larval Quality*, Bromage, N.R. and Roberts, R.J., Eds., Blackwell Science, Oxford, 1995, 353.

106-Mourente, G., Carrascosa, M.A., Velasco, C., and Odriozola, J.M., Effect of gilthead sea bream *Sparus aurata L.* broodstock diets on egg lipid composition and spawning quality, *Eur Aquac Soc Spec Pub*, 10, 179, 1989.

107-Harrel, R.M. and Woods, L.C., Comparative fatty acid composition of eggs from domesticated and wild striped bass *Morone saxatilis*, *Aquaculture*, 133, 225, 1995.

108-Silversand, C., Norberg, B., and Haux, C., Fatty-acid composition of ovulated eggs from wild and cultured turbot *Scophthalmus maximus* in relation to yolk and oil globule lipids, *Mar Biol*, 125, 269, 1996.

109-Bell, J.G., Current aspects of lipid nutrition in fish farming, in *Biology of Farmed Fish*, Black, K. and Pickering, A.D., Eds., Sheffield Academic Press, Sheffield, 1998, 114.

110-Mazorra, C., Bruce, M., Bell, J.G., Davie, A., Alorend, E., Jordan, N., Rees, J., Papanikos, N., Porter, M., and Bromage, N., Dietary lipid enhancement of broodstock reproductive performance and egg and larval quality in Atlantic halibut (*Hippoglossus hippoglossus*), *Aquaculture*, 227, 21, 2003.

111-Furuita, H., Yamamoto, T., Shima, T., Suzuki, N., and Takeuchi, T., Effect of arachidonic acid levels in broodstock diet on larval and egg quality of Japanese flounder *Paralichthys olivaceus*, *Aquaculture*, 220, 725, 2003.

112-Furuita, H., Tanaka, H., Yamamoto, T., Suzuki, N., and Takeuchi, T., Effects of high levels of n-3 HUFA in broodstock diet on egg quality and egg fatty acid composition of Japanese flounder, *Paralichthys olivaceus*, *Aquaculture*, 210, 323, 2002.

113-Almansa, E., Pérez, M.A., Cejas, J.R., Badýa, P., Villamandos, J.E., and Lorenzo, A., Influence of broodstock gilthead seabream *Sparus aurata L*. dietary fatty acids on egg quality and egg fatty acid composition throughout the spawning season, *Aquaculture*, 170, 323, 1999.

114-Pickova, J., Dutta, P.C., Larsson, P.O., and Kiessling, A., Early embryonic cleavage pattern, hatching success, and egg-lipid fatty acid composition: comparison between two cod (*Gadus morhua*) stocks, *Can J Fish Aquat Sci*, 54, 2410, 1997.

115-Nocillado, J.N., Penaflorida, V.D., and Borlongan, I.G., Measures of egg quality in induced spawns of the Asian sea bass, *Lates calcarifer (Bloch)*, *Fish Physiol Biochem*, 22, 1, 2000.

116-Rønnestad, I., Koven, W.M., Tandler, A., Harel, M., and Fyhn, H.J., Energy metabolism during development of eggs and larvae of gilthead sea bream *Sparus aurata*, *Mar Biol*, 120, 187, 1994.

117-Rønnestad, I., Koven, W.M., Tandler, A., Harel, M., and Fyhn, H.J., Utilisation of yolk fuels in developing eggs and larvae of European sea bass *Dicentrarchus labrax*, *Aquaculture*, 162, 157, 1998.

118-Matsubara, T. and Koya, Y., Course of proteolytic cleavage in three classes of yolk proteins during oocyte maturation in barfin flounder *Verasper moseri*, a marine teleost spawning pelagic eggs, *J Exp Zool*, 278, 189, 1997.

119-Matsubara, T. and Sawano, K., Proteolytic cleavage of vitellogenin and yolk proteins during vitellogenin uptake and oocyte maturation in barfin flounder *Verasper moseri*, *J Exp Zool*, 272, 34, 1995.

120-Thorsen, A. and Fyhn, H.J., Final oocyte maturation in vivo and in vitro in marine fishes with pelagic eggs, yolk protein hydrolysis and free amnio acid content, *J Fish Biol*, 48, 1195, 1996.

121-Finn, R.N., Rønnestad, I, and Fyhn, H.J., Respiration, nitrogen and energy metabolism of developing yolk-sac larvae of Atlantic halibut *Hippoglossus hippoglossus*, *Comp Biochem Physiol*, 111A, 647, 1995.

122-Finn, R.N., Fyhn, H.J., Henderson, R.J., and Evjen, M.S., The sequence of catabolic substrate oxidation and enthalpy balance of developing embryos and yolk-sac larvae of turbot *Scophthalmus maximus L*, *Comp Biochem Physiol*, 115A, 133, 1996.

123-Rønnestad, I., Thorsen, A., and Finn, R.N., Fish larval nutrition: a review of recent advances in the roles of amino acids, *Aquaculture*, 17, 201, 1999.

124-Tagawa, M. and Hirano, T., Presence of thyroxine in eggs and changes in its content during early development of chum salmon, *Oncorhynchus keta*, *General and Comparative Endocrinology*, 68, 1987.

125-Tagawa, M. and Hirano, T., Effects of thyroid-hormone deficiency in eggs on early development of the medaka, *Oryzias latipes*, *J Exp Zool*, 257, 360, 1991.

126-Brown, C.L., Doroshov, S.I., Nunez, J.M., Hadley, C., Vaneenennaam, J., Nishioka, R.S., and Bern, H.A., Maternal triiodothyronine injections cause increases in swimbladder inflation and survival rates in larval striped bass, *Morone saxatilis*, *J Exp Zool*, 248, 168, 1988.

127-Brown, C.L., Doroshov, S.I., Cochran, M.D., and Bern, H.A., Enhanced survival in striped bass fingerlings after maternal triiodothyronine treatment, *Fish Physiol Biochem*, 7, 295, 1989.

128-De Jesus, E.G., Hirano, T., and Inui, Y., Changes in cortisol and thyroid hormone concentrations during early development and metamorphosis in the Japanese flounder, *Paralichthys olivaceus*, *Gen Comp Endocr*, 82, 369, 1991.

129-Hwang, P.P., Wu, S.M., Lin, J.H., and Wu, L.S., Cortisol content of eggs and larvae of teleosts, *General and Comparative Endocrinology*, 86, 189, 1992.

130-Contreras-Sánchez, W., Schreck, C., and Fitzpatrick, M., Effect of stress on the reproductive physiology of rainbow trout, *Oncorhynchus mykiss*, in *Proceedings of the Fifth International Symposium on the Reproductive Physiology of Fish*, Goetz, F.W. and Thomas, P., Eds., Austin,Texas, USA: Fish Symposium '95, Austin, 1995, 183.

131-Brooks, S., Pottinger, T.G., Tyler, C.R., and Sumpter, J., Does cortisol influence egg quality in the rainbow trout *Oncorhynchus keta*, in: *Proceedings of the Fifth International Symposium on the Reprodutive Physiology of Fish,* Goetz, F.W. and Thomas, P., Eds., Austin, Texas, USA: Fish Symposium'95, Austin, 1995, 180.

132-Brooks, S., Tyler, C.R., Carnevali, O., Coward, K., and Sumpter, J.P., Molecular characterisation of ovarian cathepsin D in the rainbow trout, *Oncorhynchus mykiss,* Gene, 201, 45, 1997.

133-Carnevali, O., Carletta, R., Cambi, A., Vita, A., and Bromage, N.,Yolk formation and degradation during oocyte maturation in seabream *Sparus aurata*: involvement of two lysosomal proteases, *Biol Reprod*; 60, 140, 1999.

134-Carnevali, O., Mosconi, G., Cambi, A., Ridolfi, S., Zanuy, S., and Polzonetti-Magni, A.M., Changes of lysosomal enzyme activities in sea bass *Dicentrarchus labrax* eggs and developing embryos, *Aquaculture*, 202, 249, 2001.

135-Barrett, A.J., Cathepsin D and other carboxyl proteinases, in: *Proteinases in Mammalian Cells and Tissues,* Barrett, A.J., Ed., Elsevier, Amsterdam, 1977, 209.

136-Sire, M.F., Babin, P.J., and Vernier, J.M., Involvement of the lysosomal system in yolk protein deposit and degradation during vitellogenesis and embryonic-development in trout, *J Exp Zool,* 269, 69, 1994.

137-Carnevali, O., Centonze, F., Brooks, S., Marota, I., and Sumpter, J.P., Molecular cloning and expression of ovarian cathepsin D in seabream *Sparus aurata*, Biol Reprod, 66, 785, 1999.

138-Bonnet, E., Montfort, J., Esquerre, D., Hugot, K., Fostier, A., and Bobe, J., Effect of photoperiod manipulation on rainbow trout (*Oncorhynchus mykiss*) egg quality:A genomic study, *Aquaculture*, 268, 13, 2007.

Complete affiliation:

Franz Lahnsteiner, Department for Organismic Biology, University of Salzburg, Hellbrunnerstrasse 34, 5020 Salzburg, Austria, e-mail: Franz.Lahnsteiner@sbg.ac.at.

SECTION III - ARTIFICIAL FERTILIZATION IN AQUACULTURED SPECIES: FROM NORMAL PRACTICE TO CHROMOSOME MANIPULATION

CHAPTER 5
ARTIFICIAL FERTILIZATION IN AQUACULTURE SPECIES: FROM NORMAL PRACTICE TO CHROMOSOME MANIPULATION

B. Urbányi, Á. Horváth and Z. Bokor

1. INTRODUCTION

Artificial fertilization in farmed fish species throughout the world follows several routine procedures that may differ in details but follow the same general pattern. Eggs and sperm are extracted from individuals of the broodstock following induction of spawning, then gametes are mixed, activated with water or another fertilizing solution, and finally, fertilized eggs are incubated until hatching. To understand the processes that are manipulated during polyploidization, andro- and gynogenesis the physiological and genetic background of fertilization has to be discussed briefly. In this brief introduction we will concentrate on fertilization from the point of view of the egg as sperm is described in detail in several chapters of this book.

2. PHYSIOLOGICAL AND GENETIC CHANGES DURING THE PROCESS OF FERTILIZATION

2.1 Modes of reproduction

Fish exhibit a very diverse range of reproduction modes. The most common pattern of reproduction that is characteristic of most farmed fish species (and therefore is exploited in chromosome set manipulation techniques) is the fusion of male and female gametes originating from individuals of different sexes in the process of external fertilization. There are, however, several reproduction strategies that significantly differ from the described pattern. Hermaphroditism occurs among others in coral reef and deep sea fishes (including some cultured species such as the gilthead seabream, *Sparus aurata*) and can be exhibited in the form of simultaneous hermaphroditism when the fish possess ovotestes and individuals are considered both males and females or successive hermaphroditism when individuals begin their lives as males and then become females (protandrous hermaphroditism) or vice versa (protogynous hermaphroditism) [1]. Simultaneous hermaphroditism has also been observed in several sturgeon species [2,3] but it is not considered the general reproduction pattern of these species.

External fertilization is not the only method of fertilization employed by fish. Even oviparous (egg-laying) species are sub-classified into ovuliparous and zygoparous fish. Ovuliparous species release mature oocytes to be fertilized externally, whereas zygoparous fish (such as skates and some sharks) lay embryos following internal fertilization [4]. Live-bearing fish that use internal fertilization are sub-classified into either ovoviviparous fish that incubate their embryos in a modified section of the oviduct of the female, or viviparous where fertilized eggs develop in the ovary or the uterus.

Finally, some species employ an "asexual" form of reproduction when sperm of males is used only to initiate development of eggs but genetically do not contribute to the developing embryo. In several species only female individuals exist that use the sperm of related species for fertilization. A combination of viviparity and all-female "asexual" reproduction is observed in the Amazon molly (*Poecilia formosa*) [4]. A more complicated system of reproduction is characteristic of the Gibel carp (*Carassius gibelio*) where all-female triploid stocks coexist with diploid stocks containing both males and females.

2.2 Morphology of fish eggs

Most fish eggs display a similar structure: eggs are covered with an egg envelope and the egg itself is divided two distinct parts – the animal pole and the vegetal pole. The egg stores a significant amount of yolk which forces much of the cytoplasm to the outer regions of the cell. During oogenesis the fish oocyte accumulates a large number of yolk granules and lipid inclusions which in some species (e.g., *Percidae*) merge into a single oil droplet. The peripherally placed cytoplasm is thicker at a designated region which marks the animal pole. This houses the nuclear genome of the egg which is arrested in metaphase II of the second meiotic division. The peripheral cytoplasm also contains cortical alveoli which are situated beneath the internal layer of the egg envelope. Contents of the cortical alveoli are released during the cortical reaction which is part of the fertilization process.

The egg envelope (or at this moment vitelline envelope) exhibits radial striations (zona radiata) that end in microscopic pores on the egg surface. During oogenesis the egg is in contact with the surrounding follicle through the zona radiata and it also plays an important role in the formation of the perivitelline space during fertilization. The egg envelope consists of several layers, e.g., 4 in common carp [5] and in white sturgeon (*Acipenser transmontanus*) [6] or 5 in the Adriatic sturgeon (*A. naccarii*).

Most teleost eggs have a single aperture on the egg envelope, the micropyle through which the fertilizing spermatozoon can reach the egg membrane. The eggs of sturgeon species represent a notable exception as they possess several micropyles – numbers ranging from 1 to 52 were reported [7]. In several species special grooves and ridges are leading on the surface of the egg envelope towards to the micropyle [8]. A unique structure that consists of parallel ridges extending spirally from the vegetal to the animal pole and ending in the micropylar pit cover the entire surface of eggs of a *Luciocephalus* species [9]. The micropylar canal leads to a specific location on the egg membrane which is called the sperm entry site and which is characterized by several microvilli. In eggs of several fish species the first polar body can be seen on the membrane surface near the sperm entry site in the form of a round impression [10].

2.3 Process of activation of gametes

Gametes of teleosts are activated upon release into water. Motility of spermatozoa is triggered either by the decrease of osmolality or by the decrease of K^+ concentration. In some species the presence of unfertilized eggs is required for the initiation of sperm motility.

Activation of eggs is a more complicated process. Activation of eggs is on one hand a prerequisite of successful fertilization, while on the other, in most species it occurs independently of actual fertilization. Activation is a chain reaction that is associated with an increased intracellular Ca^{2+} concentration. The exact action of Ca^{2+} in egg activation and the mechanism of initiation of activation remains unsolved.

During activation yolk-free cytoplasm gathers on the animal pole and forms a blastodisc. Activation also triggers the cortical reaction which is the excretion of contents from the cortical alveoli. Cortical reaction typically starts on the animal pole and gradually expands to the entire egg surface. In certain species, however, such as the zebrafish (*Danio rerio*) cortical reaction starts randomly and occurs more or less at the same time throughout the whole of the egg surface. Excretion of cortical alveoli induces a change in osmotic concentration on two sides of the egg envelope. This triggers a water influx through the zona radiata which leads to the formation of the perivitelline space. Cortical alveoli contain among others hydrophylic polysacharides that bind to water, and thus, accelerate water influx. This influx results in a swelling of the egg envelope (which is called fertilization envelope after this process), which varies depending on the species. For example, swelling of eggs is approximately 3-6 fold in the common carp but can be as much as 50-100 fold as in case of the grass carp (*Ctenopharyngodon idella*).

Excretes of the cortical alveoli also react with the egg envelope. As a result of this process the egg envelope significantly hardens and acts as a primary protection for the developing embryo against chemical and physical reactions. It allows the influx of oxygen-rich water and the excretion of harmful metabolitic materials. Its thickness also varies depending on the species. Typically the fertilization envelope of pelagic (buoyant) eggs is thinner than that of adhesive eggs. The adhesive layer of eggs (if it exists) also forms during activation. It can spread either to the entire egg surface or only to a designated area of the egg. In the African catfish (*Clarias gariepinus*) the adhesive layer develops in the form of a ring around the animal pole and the micropyle is located at the center of this adhesive ring [9].

As it was mentioned earlier, activation of eggs occurs without sperm-egg interaction, so eggs immersed in water will eventually become activated which can be observed in the formation of the blastodisc and the perivitelline space. However, it was observed in several species that these reactions take place much later in unfertilized eggs than in those where sperm entry has occurred.

2.4 The process of fertilization in teleosts

Spermatozoa are attracted to fish eggs by chemotaxis, although, its exact function is still unclear. The presence of a chemoattractant has indirectly been demostrated in several studies. Amanze and Iyengar [8] found that spermatozoa of rosy barb (*Barbus conchonius*) are not only guided by the special grooves and ridges on the egg surface around the micropyle but the majority of sperm cells arriving in the first 30 minutes post-activation preferentially choose this guidance system. In contrast, the efficiency of this system is lost after water hardening and a much lower percentage of spermatozoa uses the guidance system 80 minutes post-fertilization.

The diameter of the micropylar canal of most species is wide enough to allow only one spermatozoon (the fertilizing one) to pass though it at a time. The micropyle of the common carp is an exception, however, as its internal aperture is wide enough to admit two cells at once [11].

As the head of the fertilizing spermatozoon fuses with the sperm entry site of the egg membrane a fertilization cone develops and extends into the micropyle. In several species the fertilization cone grows to a relatively large size and its apex is visible in the external aperture of the micropyle. At this stage the flagellum of the fertilizing spermatozoon is still visible outside the fertilization cone. Later, as the cortical reaction is initiated and the perivitelline space is forming the fertilization cone shortens considerably and finally recedes completely into the egg cytoplasm [5]. According to Wolenski and Hart [12] the formation of fertilization cone is not dependent on sperm incorporation in zebrafish.

When more than one spermatozoon merges with the sperm entry site, the phenomenon of polyspermy occurs. Polyspermy is undesirable as polyspermic embryos are not viable and die during embryogenesis. Several mechanisms are employed to prevent polyspermy in fish. As it was mentioned before the micropyle of the eggs in most species is wide enough to accomodate only one spermatozoon. Spermatozoa can only fuse with the egg membrane at the sperm entry site, and its diameter is also limited to the size of one sperm cell. Immediate formation of the fertilization cone and its extrusion into the micropyle blocks the entry of further spermatozoa and pushes supernumerary sperm cells out of the pycropylar canal. Finally the colloids excreted from the cortical alveoli can also agglutinate supernumerary spermatozoa in the perivitelline space [13].

Spermatozoa of teleost fish do not have an acrosome, sperm cells of sturgeons and paddlefishes (members of the order Acipenseriformes), however, possess a functional acrosome and an acrosome reaction is required for successful fertilization. Currently the exact mechanism of acrosome reaction is still not clear in sturgeons. Acrosome in other species contains enzymes that facilitate the penetration of the fertilizing spermatozoon through the zona pellucida of the egg. Fish eggs on the other hand, possess a micropyle that leads directly to the egg membrane, thus acrosome reaction is assumed to be unnecessary. Paradoxically, the acrosome and acrosome reaction is present in Acipenseriformes, while their eggs possess numerous micropyles. Theoretically this system is unfavorable as it facilitates the occurrence of polyspermy. Polyspermic fertilization can be induced in sturgeon, however it rarely occurs in natural reproduction. Several factors limit the occurrence of polyspermy. Sperm of Acipenseriform fish species is much less concentrated than that of other teleosts, on average it contains one billion cells per ml. Upon penetration of the fertilizing spermatozoon, cytoplasmic processes (fertilization cones) project into the micropyles from all sperm entry sites (and not only from the one that fused with the fertilizing sperm cell) [10]. Rapid exocytosis of cortical alveoli also plays an important role in prevention of polyspermy in these species [7].

2.5 Genetic changes during the process of fertilization

As it was mentioned earlier the genome of the fish egg stays arrested in metaphase II of the second meiotic division when eggs are released into water during spawning. Upon activation of the egg the division is completed and the second polar body is extruded on the animal pole. The haploid set of remaining chromosomes transforms into a female pronucleus. This process is independent of the actual fertilization and occurs in all activated eggs as it has been shown in zebrafish eggs [14].

The head of the fertilizing spermatozoon is separated from the flagellum following penetration. The nucleus then turns at an angle of 180° and at the same time starts swelling. The chromatin of the nucleus loosens and the compact nucleus of the spermatozoon head into a light vesicular male pronucleus with fine granular chromatin. A structure called the spermaster is formed around the centrosome of the sperm cell which moves ahead of the male pronucleus into the center of the animal pole where it encounters the female pronucleus. For a while both pronuclei stay in close contact and the spindle rays embrace them. Then the pronuclei slowly disintegrate and the chromosomes form their characteristic groups. Finally the spindle of the first mitotic division forms and the chromosomes become arranged along the equator and the process of mitotic divisions begins [7].

2.6 Initial development of the zygote following fertilization

Mitotic divisions occur synchronously at given time intervals until blastulation. This period is also called the period of cleavage. It has been shown that all cells divide at the same rate during cleavage and the time interval between successive divisions is also equal. This time interval is called τ_0 and it is the unit for measuring the relative duration of mitotic phases during the period of cleavage. It is also used for measuring the relative developmental periods. In poikilothermic animals it is obviously temperature-dependent but at all temperatures the interval from fertilization up to the appearance of the first cleavage equals approximately $2\ \tau_0$ in teleosts and $3\ \tau_0$ in sturgeons and paddlefishes [7].

Cleavage in teleosts is a partial division (*segmentatio partialis*) of the embryo meaning that only the animal pole of the zygote divides and forms a blastodisc and the yolk of the vegetal pole lacks cellular structure (Figure 1.). To the contrary, in Acipenseriform fish cleavage can be characterized as a complete unequal division (*segmentatio totalis inequalis*). In these fish the vegetal pole also divides, although at a slower rate and at any given point in time much larger cells are visible on the vegetal pole that on the animal one.

3. ARTIFICIAL FERTILIZATION IN CULTURED FISH SPECIES

Artificial fertilization in commercial aquaculture is part of the process of induced spawning. This activity is typically conducted with species that employ external mode of fertilization using stripped eggs and sperm.

In broader sense artificial fertilization involves several steps that are conducted in all species although with differences in the details. These steps are:

· collection of gametes
· actual fertilization of gametes
· incubation

Each of these steps has to be conducted with greatest care and accuracy in order to guarantee the highest number of surviving embryos and hatching larvae.

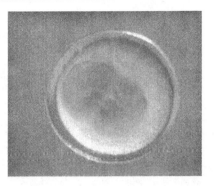

Figure 1 - A two-cell stage zygote of African catfish (*Clarias gariepinus*). Note the two blastomeres separated from the yolk of the vegetal pole and the perivitelline space around the embryo (photo by Á. Horváth).

3.1 Collection of gametes

Gametes are typically collected following induction of spawning. Spawning can be induced either using hormones such as fish pituitary extracts or synthetic LHRH or GnRH analogues or by the use of light and temperature regimes. As the details of gamete collection are described in detail in other chapters of this book, we will only summarize briefly the techniques used in commercial aquaculture.

The collection of eggs is identical for almost all fish species. Eggs are collected by stripping which means a gentle massage of the abdomen directed from the head to the tail (Fig. 2). Fish are typically anaesthetized before stripping and the belly is wiped dry. Eggs are released through the genital orifice of the females and collected into a dry bowl. Care must be taken to avoid contact of eggs with either water or urine as they can activate eggs and prevent successful fertilization.

In most fish species ovulated eggs are collected in a discrete ovary and by applying massage to the ovary the majority of eggs can be stripped out. In some species such as salmonids and Acipenseriform fish, however, eggs are released into the abdominal cavity upon ovulation. This represents a special problem in sturgeons and paddlefishes as a funnel-shaped oviduct stretches in anterior direction from the genital orifice into the body cavity and prevents the stripping of most eggs. As a solution some farmers strip the available amount of eggs and repeat this several times during the day. Several alternative methods are employed to collect eggs in these species. One of the most commonly used techniques requires the killing of females when the first released eggs appear on the bottom of the fish tank and removal of ovulated eggs from the body cavity. Others use a scalpel to make a small incision of the belly and strip the eggs trough this incision. Finally, a method called minimally invasive removal of ovulated eggs (MIST) was developed by Štich et al. [15] who introduced a small scalpel into the genital opening of the fish to cut the oviduct and thus allow the collection of eggs.

Collected eggs can vary in size and volume. In farmed species they range from 2000 eggs per kg body weight (in rainbow trout) to 100 000-200 000 eggs per kg of body weight (in common carp). Also, the number of ova in one kg of eggs can vary between 13,000-14,000 in salmonids to 1,100,000 in the silver carp (*Hypophthalmichthys molitrix*) or 1,400,000 in the pike perch (*Sander lucioperca*) (Table 1. shows reproductive data of some farmed freshwater fish species).

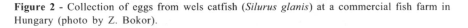

Figure 2 - Collection of eggs from wels catfish (*Silurus glanis*) at a commercial fish farm in Hungary (photo by Z. Bokor).

Collection of sperm is often more problematic and the methods are more diverse than for collecting eggs. Sperm of most fish species can be stripped similarly to eggs (by abdominal massage). However, care must be taken to avoid contamination of sperm with water or urine as similarly to eggs this would result in activation of gametes and loss in fertilizing ability.

In aquaculture practice, however, this rarely represents a real problem. Sperm of trout, for example, in commercial hatcheries is stripped directly onto the eggs. Sperm of common carp is typically stripped into a dry glass or plastic vessel and immediately used for fertilization, thus no special measures are employed to avoid contamination with urine. In cases when contact with urine should be avoided, though, different methods are used to collect sperm. In several species catheters are introduced into the sperm duct and sperm is collected directly from this duct or from the testis (Fig 3.). Catheters and methods of sperm retrieval (syringes or tubes) can be different. For sturgeons catheters with external diameter of 5 mm can be used, whereas for trout or cyprinid species silicon tubes of 1 mm in diameter are preferable. Sperm can also be collected into a saline solution which keeps the spermatozoa immotile until use for fertilization. This method of stripping is used among others for the collection of sperm from the Asian *Pangasius* catfish species in Vietnam.

Table 1 - Reproductive data of some farmed freshwater fish species (Sources:Lahnsteiner et al., 2000; Horváth and Urbány, 2000; Horváth et al., 2002)

Species	Rainbow trout	Common carp	Grass carp	Silver carp	Bighead carp	Pike perch	Catfish	Sterlet
Number of eggs per kg of female (thousands)	2	100-200	60-80	60-80	50-60	150-200	10-48	60-80
Number of ova per kg of eggs (thousands)	13-14	700-1000	800-900	900-1100	600-800	1400-2000	180-220	100-120
Eggs adhesive or not	No	Yes	No	No	No	Yes	Yes	Yes
Temperature optimum for spawning (°C)	0.3-12.8	16-22	2?-22	21-23	22-25	12-16	22-24	13-16
Volume of sperm per males (ml)	5-30	10-30	10-20	5-15	10-20	05-1.5	NA	5-20
Sperm concentration ($\times 10^9$ per ml)	4-10	80-100	50-100	50-100	50-100	15-30	NA	0.5-1.5

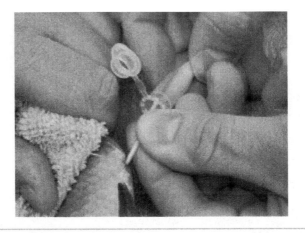

Figure 3 - Collection of sperm from the crucian carp (*Carassius carassius*) using a catheter (photo by Á. Horváth).

The collection of sperm from the sterlet (*Acipenser ruthenus*) can also be problematic. Upon removal from water males tend to contract their muscles and spray sperm out of their bodies which is then lost for spawning. Thus, extreme care must be taken to place the fish into anesthetic solution before attempting to collect sperm and the use of catheters attached to syringes is recommended.

Stripping of sperm from some catfish species is often difficult or impossible. Testis of the wels catfish (*Silurus glanis*) has a specific structure: it is folded several times within the abdominal cavity. In the African catfish (*Clarias gariepinus*) a seminal vesicle situated between the testis and the genital orifice prevents the stripping of sperm. Finally, the filamentous structure of the testis of channel catfish (*Ictalurus punctatus*) presents another difficulty for sperm collection. Wels and African catfish males are normally killed upon propagation and the testis is surgically removed from the body cavity. Afterwards it is cut to pieces and the sperm is squeezed from the testicular tissue through a mesh fabric. In case of the channel catfish sperm is extracted by crushing the testes in an immobilizing solution, e.g., in an extender for cryopreservation.

3.2 Fertilization

Upon fertilization gametes are mixed together and insemination of the eggs takes place either in the absence or the presence of an activating solution. In trout culture sperm activation and thus the fertilization of eggs occurs on contact of spermatozoa with eggs, in other species, however, an activating solution is used.

When the gametes are mixed together and the activating solution is added only later, the process is called "dry fertilization." This is the method employed in most hatcheries throughout the world. In sturgeon culture, however, "semi-dry fertilization" is recommended by some culturists in order to prevent polyspermy. In this case, sperm is first diluted in water 50-200 times and then the activated sperm is added to the eggs. As spermatozoa of Acipenseriform fish remain motile for approximately 4-6 minutes following activation, this type of fertilization does not result in a loss of motile sperm.

When eggs are mixed with an activating solution first and sperm is added only later, the procedure is called "wet fertilization." Although this method of fertilization has been reported in experimental conditions with controversial results [16,17], it is seldom used in commercial aquaculture.

Activating solution in most cases is hatchery water. Although the salt-urea solution originally developed by Woynárovich [18] was initially used for gamete activation as well as for the elimination of stickiness, currently water is used as an activating solution at most hatcheries that still use Woynárovich's technique.

Following activation pelagic eggs are allowed to swell for a short period of time and then placed into hatching facilities for incubation. In case of adhesive eggs, stickiness has to be eliminated before incubation. Stickiness reduces the supply of eggs with oxygen-rich water, facilitates the spreading of fungi such as *Saprolegnia* species and makes manipulation of eggs difficult. Different methods are employed to eliminate the adhesiveness of eggs in different species. Typically these methods use either chemical compounds or colloidal substances to bond to the sticky layer on the egg surface.

In cyprinid species Woynárovich-solution (0.4 % NaCl, 0.3 % urea) that was mentioned earlier is commonly used. Eggs are stirred in this solution for 45-60 minutes, then washed twice in a 0.5 % solution of tannic acid for 20 seconds and placed into the hatching jars. This method of fertilization has recently successfully been applied to the induced spawning of pike perch as well [19]. In Israel, cow milk is used in a dilution of 1:2 or 1:3 for the elimination of egg stickiness in the common carp [20].

Adhesiveness of catfish eggs and its effect on the development of larvae is still controversial. The eggs of African catfish have an adhesive ring around the animal pole, however, this rarely represent a problem during incubation. Thus elimination of stickiness is not necessary, Hungarian catfish farmers only use tannic acid treatment (in a concentration described above) to harden the eggs before putting them into hatching jars. Due to the very short incubation period (24-25 hours at 26°C) fungal infections do not represent a real threat. Channel catfish eggs are also allowed to stick to each other during incubation. In case of the wels catfish different methods are employed that often depend on the hatchery. According to the method used at a commercial hatchery in Szajol, Hungary eggs are fertilized using hatchery water then stirred in a solution of 0.2 % NaCl, 0.15 % urea (half of Woynárovich-solution) for 5-6 minutes, then treated with tannic acid (again, as described above) and incubated until hatching. Other farms use fine clay to cover the sticky layer on the egg surface. Exceptionally good hatch rates were observed, however, when eggs did not receive any treatment at all following activation and were allowed to stick to the inner surface of the hatching jar. It seems that wels catfish embryos are very sensitive to disturbance in the first 10-12 hours of development. Following this period eggs can be separated by gentle stirring.

Eggs of Acipenseriform species are also adhesive. Methods of the elimination of stickiness vary depending on the country or the hatchery where it is conducted and often local possibilities and conditions are adopted to hatchery use. In the United States and Western Europe the use of Fuller's earth is common. Eggs are stirred in a solution of Fuller's earth for 25-45 minutes (depending on the local practice) and then incubated in hatching jars. A special solution was developed in Hungary for the eggs of sterlet. Potato starch is dissolved in water (200-250 g of starch in one liter of water) and eggs are stirred for 90 minutes in this solution [21]. The use of other colloidal substances such as talc, fine river silt, clay or milk has also been reported. Russian scientists have recently noticed the adverse effect of elimination of egg adhesiveness, as it allows the development or preservation of defective non-sticky eggs that are carried away by the river flow in natural conditions. Therefore, in some regions of Russia experiments on semi-natural spawning in artificial riverbeds are conducted [22].

3.3 Incubation

Following fertilization fish eggs are incubated in various hatchery facilities until hatching. Hatchery design depends on local conditions and the fish species cultured at a given farm. The main units of egg incubation are various hatching vessels, jars or trays, which are almost exclusively operated on the principle of water through-flow.

Incubation period of eggs depends on the species and on the local climate. In fish, as poikilothermic animals, actual incubation time depends on the temperature of water, however, the hatching period can also be expressed in degree-days which is the temperature of incubating water multiplied by the number of days required for hatching. Typically the incubation of eggs from warm-water or tropical species requires less degree-days than that of cold-water fish. Thus, rainbow trout eggs need approximately 300-370 degree-days [23], common carp and sterlet 60-70 degree-days [18], African catfish 26 degree-days and Indian major carps (*Catla catla*, *Labeo rohita*, *Cirrhina mrigala*) only 19-20 degree-days [24]. Incubation of eggs must, however, be conducted at a temperature suitable for the species as lower temperatures may slow the development of embryos and at temperatures higher than the ideal, occurrence of malformed larvae can be more frequent.

Eggs are incubated in different types of incubators. One of the most commonly used hatching jar is called the zuger jar (Figure 4.). It resembles a bottle with its neck facing downwards and lacking a bottom. Water is pumped though the neck of the bottle at a desired flow rate and leaves through the top, thus, keeping the eggs in constant motion. Zuger jars are sold in various sizes and volumes and different materials (glass, fiberglass, metal) are employed for manufacture. These types of jars are used in countries of Europe, Asia and South America.

Figure 4 - Eggs of wels catfish in 7 liter jars. (Photo by Á. Horváth.)

Another type of hatching vessels is the McDonald-type jar which is typically made of transparent plastic and produced in volumes of approximately 7 liters. McDonald jars have cylindrical shape and a standpipe perforated at the bottom is placed into the middle of the jar. Water is pumped through the standpipe, it enters the jar through its perforations at the bottom and keeps the eggs in motion. McDonald-type hatching jars are used throughout the United States and in some Western European countries.

Trout hatcheries use different types of incubators as trout eggs tend to be very sensitive to motion in different stages of development. Some hatcheries use up-flow incubators that resemble zuger jars with a mesh in the middle of the jar to keep eggs from a distance from incoming water. Most hatcheries, however, employ variations of hatching trays where eggs are laid onto a mesh in a monolayer or only a few layers and water flows trough the mesh from the bottom and leaves on one side of the tray. Upon incubation, water should never be added to the eggs, eggs should be very gently added to water [23]. Also, trout eggs should be incubated in the dark mimicking their development in the gravel.

4. POSSIBILITIES OF CHROMOSOME SET MANIPULATION IN FISH

In vertebrates, during the regular process of fertilization the penetrating spermatozoon induces the completion of the second meiotic division, extrusion of the second polar body and the formation of haploid pronuclei. Male and female pronuclei fuse and create the first diploid nucleus of the embryo that will multiply through sequential mitotic divisions [25,26]. Thus, a genome characteristic of the given individual is formed which carries the traits determined by the same or different alleles on the loci of the two homologous chromosome sets. The diversity and variability of these traits – called genetic variability – resulted in the distribution and unparalleled adaptability of vertebrates that helped them to conquer and populate areas with different environmental conditions on Earth.

This mechanism of inheritance, the transfer of genetic information from parent to progeny is guarded by strict regulatory processes. The possibility of conducting chromosome set manipulations in animals decreases with evolutionary developemnt. The reason for this is that in animals on a higher degree of development the mechanisms protecting sexual reproduction are also more advanced. The higher the examined vertebrate class is on the evolutionary ladder, the more conservative and rigid this mechanism is. Thus, intervention into this mechanism is possible only in groups of lower vertebrates and these groups include teleosts that have great economic importance [27]. Teleosts (and in part Chondrichthyes) are excellent subjects of genetic analyses. Procedures for the manipulation of the entire genome include ploidy level alterations, gynogenesis and androgenesis [28].

In the course of chromosome set – genome – manipulations, the chromosome set of the progeny is supplied with one or several additional haploid chromosome sets through blocking a given phase of normal meiotic or mitotic division [29].

A general charactersitic of chromosome set manipulations is that the genome of gametes (either eggs or spermatozoa) is inactivated and the diploid state required for normal development is restored by applying different shocks (such as heat, cold or pressure schock), whereas, in procedures for ploidy level increase, inactivation of the gametes' genome is absent and the shock is used to multiply the chromosome set [30].

The time of the shock is largely determined by the developmental stage of the oocyte that is subjected to intervention. Genetic protocols typically intervene in two specific points in time into the development of fish eggs: the extrusion of the second polar body or the blocking of the first mitotic division.

The following procedures are typically regarded as chromosome set (genome) manipulations: induced gynogenesis and androgenesis, induced triploidization, tetraploidization or induction of higher ploidy levels, sex reversal, transfer of genetic material such as nuclei, active chromosome fragments, cloned genes and interspecific hybridization (Table 1.).

The most general problem of chromosome set manipulation procedures is the very low survival rate of gynogenic and androgenic diploid progeny which is partly due to the selection rate of recessive genes, partly to the effect of treatments [32].

The first true chromosome set manipulation was conducted by Hertwig who fertilized frog eggs following irradiation of sperm with different dosages of X-ray radiation in 1911 [33]. The results were as follows:

1. using low dosages a high number of abnormal embryos developed,

2. increasing dosages resulted in an increasing rate of abnormality and a decrease in the number of normally developing embryos,

3. at a certain dosage no progeny was observed – this was the dosage that resulted in 100% letality,

4. above this dosage embryos reappeared, most of them were deformed, however, some of them were intact.

Some of the latter ones were later grown into adulthood. Hertwig suspected that these individuals were results of gynogenesis.

Table 2 - Overview of chromosome set manipulations (according to [31])

Eggs and sperm							
Irradiation prior to fertilization	None		Sperm			Egg	
Target of treatment (shock)	Polar body	First mitotic division	Polar body	First mitotic division	First mitotic division		
Result of treatment	Triploid	Tetraploid	Heterozygous gynogenetic	Heterozygous gynogenetic	Homozygous gynogenetic	Homozygous androgenetic
Sexual characteristics	Sterile male (XXY) and sterile female (XXX)	Fertile male (XXYY) and fertile female (XXXX)	Female (XX)	Female (XX)	Female (XX)	Male (YY)

The reasons of the Hertwig-effect are known for contemporary scientists: at the 100% lethal dosage DNA is damaged to the level where it is not able to function normally, however, it disturbs ontogeny to a degree which leads to a complete stop. An irradiation with an even higher dosage leads to a DNA damage resulting in a total loss of functionality, however, nuclear proteins and cytocenter necessary for the initiation of cleavage stay intact and is thus able to regulate cell division. This experiment allowed the artificial induction of gynogenesis.

Several methods were tested for the inactivation of DNA in spermatozoa since then. Peacock [34] has listed 370 different methods including 45 physical, 93 chemical, 63 biological and 169 combined ones. Some effective methods (obviously not exclusively) are listed here:

- UV- treatment – in Rana pipiens [35]
- treatment with X-ray radiation [36]
- treatment with gamma radiation [36, 37, 38].

The production of individuals that inherit their genomes from only one parent is possible using three methods [39]:

Self-fertilization, which requires the development of induced hermaphroditism as most fish species possess separate sexes.

Induced gynogenesis which requires two parents but sperm is only needed to induce development. Thus, sperm has to be treated prior to fertilization so that paternal DNA does not contribute to the genome of the developing embryo. In order to compensate for the absence of paternal genome, however, maternal chromosome set needs to be duplicates either by the retention of the second polar body or by a treatment blocking the first mitotic division.

Induced androgenesis which also requires two parents, however, in this case only the paternal genome contributes to the formation of the embryo's chromosome set as the genome of the egg is inactivated prior to fertilization. Duplication of the haploid genome is required in this case, too, however, this is possible only by the blocking of the first mitotic division.

In the case when the genomes of both parents particpate in the formation of the progeny two types of polyploidy can be induced either by blocking the second meiotic or the first mititotic division [25]:

1. Triploids are produced by the fusion of diploid female and haploid male gametes. The retention of the second polar body in the II. meiotic division can even result in interspecific hybrids when gametes of two different species are used.

2. Tetraploids are produced when the first mitotic division of the embryo's diploid nucleus is blocked.

Another type of manipulation is targeted at the exclusion of genetic information of both gametes. In this case the nucleus of the egg is replaced with a nucleus of an embryonic cell that originates from a developing embryo. This is the method of nucleus transplantation which can be used to produce clones when several nuclei originating from the same embryo can be transplanted into recipient oocytes [40].

Finally, the method of gene transfer has to be mentioned as well, which allows the transfert of donor genes into the genome using active chromosome fragments or cloned plasmides. This way, individuals can obtain traits that are not characteristic of the given species.

All manipulations mentioned above have already been conducted in fish, moreover, results of the analyses have more or less accurately been evaluated, especially those of gynogenesis and polyploidy [41].

It is a matter of fact that the application of a cold, heat or pressure shock can result in the retention of the second polar body. Pressure shock has been described as a successful solution in all cases and apparently its optimization is also a relatively simple task. The use of a similar pressure value, 7,000-10,000 psi for a couple of minutes gives positive results even in unrelated species, such as those of mollusks, fish or amphibians. Thus, optimization concerns only the determination of the exact timing of the shock. The method can also be used for the production of triploids in fish farming.

The view that cold shock is effective primarily on warm water species and heat shock on coldwater ones proved to be a miscalculation as good results were found by using heat shock in warm water species and triploids have been hatched from salmon eggs treated with cold shock. However, optimization of the heat shock seems to be more feasible in coldwater species, and that of the cold shock in warm water fish [25,42].

The heat shock is typically effective in warm water fish above 40°C while the heat shock should be applied in salmonids below 0°C. In this case survival rate is the indicator of the effectiveness of different combinations of shock parameters. However, the number of eggs containing developing embryos is higher than that with diploids resulting from shocking. Several authors do not discriminate between haploid survivors and actual gynogenetic diploids which leads to the overestimation of success [25].

Karyological verification of results is highly recommended following optimization, especially that this is relatively simple with your larvae and gives a definite answer. It allows the detection of haploids, mosaics and unviable aneuploids.

Measurement of erythrocyte nucleus diameter is a relatively simple method for the determination of triploids, however, it can only be used for older individuals (as blood sampling is necessary) [25].

Generally, the methodology of chromosome set manipulation protocols has been finalized by the end of the 1980s. The purpose of the present chapter is to present the used methods based on earlier results, to confirm certain earlier assumptions through newer results and to give an introduction of the possibilities of potential utilization.

4.1 Polyploidization (triploidization, tetraploidization)

Triploidization is the production of individuals with three sets of chromosomes (3N). First triploid fish (plaice) were produced using a cold shock which resulted in 100% triploidy [28].

Two fundamental methods can be used to induce triploidy. In the first case, an intact egg fertilized with a similarly intact spermatozoon is shocked during the second meiotic division (triggered by the penetration of the fertilizing spermatozoon) and individuals with three chromosome sets are produced via the retention of the second polar body [43, 44]. The advantage of the method is simplicity and low cost, however, the produced stock is not 100% triploid and individuals have to be sorted according to ploidy   in case of stocks of several hundred thousand individuals this represents a serious difficulty. The other method is the crossing of tretraploid and diploid fish (4N × 2N), that results in a 100% triploid progeny stock.

Triploidization represents several great advantages in spite of the mentioned difficulties. Triploid individuals are sterile, unable to produce gametes which means that they can be stocked to natural waters without the threat of natural reproduction and disruption of ecological balance. Also, due to their faster growth which is also linked to the absence of gametes, market size can be attained within a shorter period of time.

Parameters of the shock were set by Spanish researchers [45] on turbot (*Scophthalmus maximus*). A cold shock was applied following incubation in 13-14°C water for 6.5 minutes after fertilization, with a shocking temperature of 0°C and a duration of 25 minutes. The resulting ratio of triploids was more than 90%.

A comparative study of the sperm parameters of triploid and diploid tench (*Tinca tinca*) males has revealed that sperm production of triploid males (0.05 ml in triploids vs. 0.58 ml in diploids), as well as GSI and testis weight were all significantly lower than that of diploid individuals.

Changes in thyroid hormone concentrations during the spawning season were investigated in diploid and triploid catfish (*Heteropneustes fossilis*) by Indian researchers [46]. Levels of total thyroxine (T3) and triiodothyronine (T3) were compared in the blood plasma and results confirmed the sterility of triploids. Concentrations of thyroid hormones and the activity of the motoric organs that influence the production of these hormones were several-fold higher in diploid individuals than in triploids.

Tetraploid individuals can be produced by intervention into the first mitotic division of the zygote. The formation of the membrane between the two daughter cells is blocked by a shock applied during the metaphase and thus the already duplicated genome stays in one cell [25,27,47]. Tetraploid stocks are able to reproduce, however, individuals are very sensitive and a large portion of them die during the first year of life [48]. Successful spawning of tetraploids with normal diploids results in triploid individuals. According to some studies, the increased size of the spermatozoon head in tetraploids can cause problems, as it cannot penetrate the narrow micropyle of a normal egg [25].

The other possibility of producing viable diploid or triploid individuals is to block chromosome separation in the second meiotic division, in other words the retention of the second polar body. If this process is successful, fertilization with intact sperm results in triploid larvae, whereas, fertilization with irradiated sperm yields diploid individuals.

There are three types of treatments – used originally in amphibians – that can give satisfactory results in fish: long-term cold shock, short-term heat shock and short-term hydrostatic shock.

All three types of shocks can be characterized by three parameters:
- · time elapsed between fertilization and shocking
- · temperature or pressure of the shock
- · duration of the shock

There are two possible methods for the optimization of these parameters. In the case of irradiated sperm, the effectiveness of the shock defines the number of viable gynogenic progeny. Sub-optimal shock will result in only a few or no survivors. Thus, the number of surviving larvae can be a criterion for optimization.

In case of normal fertilization diploid, a certain amount of triploids will appear. The number of triploids and thus the effectiveness of the shock has to be determined using other, more sophisticated and time-consuming procedures. These include chromosme number determination, or analyses based on the comparison of erythrocyte diameter.

The first successful experiments were carried out by Swarup [49] on three-spined stickleback (*Gasterosteus aculeatus*) and found 50% triploids among survivors of cold and heat shocks. The first experiment that resulted in 100% triploids using cold shock was reported in the plaice (*Pleuronectes platessa*) [50]. In spite of repeated efforts [51] production of high numbers of diploid gynogenic individuals in common carp was reported only relatively late [38]. In this experiment triploids were produced using heat shock at a rate of almost 100% albeit with high mortality percentage.

Cold shock is always applied at temperatures below 4°C. High percentages of triploids were produced using cold shock in catfish [52], cyprinids [53], and other species [54]. A moderate cold shock at 5-15°C was used with limited success to induce diploid gynogenesis in the paradise fish (*Macropodus opercularis*) [55] and in the rohu (*Labeo rohita*) [56] or triploidy in tilapia [57].

The use of a heat shock with temperature below 0°C gave partly successful results in the production of gynogenic rainbow trout [41]. In trout, the moderate heat shock was optimized for the first time at 26-29°C [39,41]. A high number of diploid and triploid gynogenic individuals were produced by the application of this type of shock between 0 and 30 minutes post fertilization on condition that the duration of the shock was long enough: 20 minutes at 26°C or 10 minutes at 29°C. Triploidy was induced with heat shock in Atlantic salmon, as well [58, 59].

Heat shock at temperatures higher than 40°C results in the retention of the second polar body in several warmwater fish species (zebra fish [60]; tilapia [61]; common carp [62]).

Pressure shock was used successfully for the first time in zebra fish [60], and since then it was found quite effective on a series of other fish species [58, 63, 64, 65].

The great value of fertile tetraploids is that their reproduction can be used for the production of new tetraploids and triploids as well as gynogenic and androgenic individuals as well as animals with high ploidy level [25]. Crosses of tetraploid males with diploid females in rainbow trout resulted in 40% fertilization. As the diploid spermatozoon is larger than the haploid one, it experiences difficulties in penetrating the micropyle. Thus, the use of fertile tetraploid females represents better opportunities.

Viability of sea bass (*Dicentrarchus labrax*) progeny produced with mitotic gynogenesis and tetraploidization was investigated depending on the volume of pressure shock and the timing of shock post-fertilization [66]. The results showed that a pressure lower than 70 MPa was ineffective in producing ploidy manipulated progeny while pressures between 70 and 80 MPa resulted in a low number of offspring. At optimum pressure levels (80-90 MPa) the hatch rates were $1 \pm 1\%$ for gynogenic double haploids (G2N) and $3 \pm 2\%$ for tetraploids. Following investigations on the time of shocking post fertilization and the duration of shock, authors concluded that by optimization of the protocol initial low hatch rates can be improved to reach $18 \pm 12\%$.

Tetraploid mud loach (*Misgurnus mizolepis*) was produced by Nam et al. [67] by blocking the first mitotic division. This was conducted in a combination of heat shock (40.5°C with a duration of 1-3 minutes) and cold shock (1.5°C, with a duration of 30-60 minutes). The highest number of tetraploids (14.4 %) was observed using a 2 minute heat shock and a 45 minute cold shock. The possibilities of crossbreeding tetraploid and diploid mud loach were investigated by Nam and Kim [68]. The analysis was conducted on 48 tetraploid males, of which 12 had retarded gonadal activity and only the other 36 had viable spermatozoa. Twenty-six fish produced haploid sperm which produced normal diploid progeny when used for fertilization of eggs. The sperm of 7 tetraploid males produced mosaics, aneuploid or triploid individuals and only 3 analyzed tetraploids produced diploid sperm and thus triploid progeny. The study has revealed that the utilization of tetraploid stocks is not a guarantee for creating triploid stocks.

4.2 Gynogenesis

Gynogenesis is the production of an emrbyo where the fertilizing spermatozoon stimulates the conclusion of the second meiotic division, initiates the extrusion of the second polar body and the onset of cleavage but its DNA is inactivated and the developing embryo inherits only the genotype of the female parent.

Gynogenesis is the normal method of reproduction in some fish species that occurs in natural conditions.

These fish include three species of the genus *Poeciliopsis* [69], the silver crucian carp (*Carassius gibelio*) [70] and the loach (*Misgurnus fossilis)* [71]. A general characteristic of all species is that they spawn with males of related species, whose sperm initiates embryonic development but the male genome does not take part in it.

In case of the *Poeciliopsis* meiosis is preceeded by an endomitosis. Thus, the triploid oogonium becomes hexaploid and the process of meiosis results in triploid oocytes. Triploid individuals develop from these via gynogenesis.

In case of the triploid silver crucian carp only the first polar body is extruded, thus, the oocyte and the adult gynogenic individual will become triploid.

In case of the diploid silver crucian carp meiosis takes place normally, however, it is followed by an endomitosis. Thus the gynogenic adult will become haploid.

Gynogenic populations of *Misgurnus* are also diploids. In the oogenesis of this species the second polar body is extruded in the second meiotic division but then fuses again with the oocyte. Thus the developing gynogenic individual will become diploid.

Since Hertwig's investigations several authors used sperm irradiated with X or gamma radiation for the production of gynogenic haploid fish such as the wild form of common carp [38,72], farmed common carp [36], sturgeon [51], plaice [37], medaka [73], paradse fish [55], and salmonids [37,65,74,75,76].

X and gamma irradiation are frequently used because of practical considerations as their strong penetrating capacity guarantees an even treatment of the entire sperm. The above mentioned authors typically used a dose around 100 krad. This dosage, however, resulted in a relatively low number of hatched larvae. Some authors tried to determine the dosage that results in the highest number of developing embryos using series of dosages as the number of embryos is reduced by the use of dosages that are either above or below the optimal one (Hertwig-effect).

According to the experiences of Chourrout et al. [75] on rainbow trout the use of gamma irradiation at a dose of 100 krad was clearly insufficient and the number of post-gastrula stage surviving embryos was increased at a dosage of 130 krad or higher. Another experience of the same work was that although this high dosage of irradiation completely inactivated the genome of spermatozoa, their motility decreased as well. The authors recommended the use of pooled sperm samples, as the same irradiation dosage resulted in different rates of sperm motility decrease depending on the individual.

Most authors did not verify the chromosome number of produced haploids. Males and females of different coloration were crossed by Ijiri and Egami [73] to demonstrate that some paternal traits were inherited by the progeny in spite of irradiation. Sperm of rainbow trout was irradiated using high intensity ionizing radiation by Chourrout and Quillet [39] and the chromosome sets of occurring haploid embryos were analyzed. Supernumerary chromosome fragments of paternal origin were found in cells of these embryos. The high frequency of these (two per cell) had not decreased at irradiation dosages between 100 and 135 krad. Supernumerary chromosome fragments were found at even lowered dosages [65], but not at 100 krad. According to these observations, it is not obvious that gamma irradiation is able to inactivate the genome of spermatozoa completely, thus, the presence of supernumerary chromosome fragment should be verified after each experiment. This is extremely important, because if the disturbances caused by the presence of these sequences are neutralized in the developing embryos, then on one hand, they can integrate into the genome and reduce the fertility of adult gynogenic fish, and on the other, their frequency can increase in further generations if these individuals are spawned.

4.2.1 Treatment of sperm with UV radiation

UV radiation was used for the irradiation of sperm in several species of fish such as cyprinids [54,56,62,77], medaka [73], zebra fish [60], tilapia [61], and several salmonids [47,61,78,79].

One of the greatest advantages of UV radiation is the relative ease of use. Its greatest disadvantage, however, is that only minute amounts of sperm can be irradiated because of its low penetration capacity. In order to achieve even and sufficient effect of irradiation, it is recommended to dilute sperm and expose it to radiation in a thin layer or stir during the treatment. Ordinary germicide lamps are suitable for irradiation. Sperm is typically placed at a distance of a few centimeters to ten cm below the lamp. The time duration of irradiation is several minutes.

The effect of UV radiation is different from that of ionizing radiations. Instead of the Hertwig effect the following phenomena can be observed after UV treatment:

- In case of short-term treatments mortality occurs only in later stages of development and none of the embryos display any signs of developmental defects until then. Karyological tests show that in this case treated animals include only pure diploids or pure triploids [40,78]. In case of long-term irradiation, all embryos will be haploid. This phenomenon occurs at irradiation periods betwenn 2.5 minutes to 8 minutes.
- Within this interval Chourrout [40] has found no cells containing supernumerary chromosome fragments, although, 1500 cells were analyzed for this purpose. Survival of haploids induced by UV irradiation remained at a relatively constant level, 90%, up to the last day, and some haploids even hatched.

- Ijiri and Egami [73] have not found any expression of paternal genetic markers, either, in the progeny following UV treatment of the sperm.
- The explanation of these phenomena is that in contrast to ionizing types of radiation that break the genome into fragments, UV radiation links the thymine bases to form dimers, thus, inactivating the entire genome.

The analysis of turbot gametes [80] revealed that turbot spermatozoa are of poor quality, the larvae are small and sensitive. Elimination of some risk factors in induced spawning increases the chances of success, and gynogenesis represented a possibility for solving this problem. Sperm was diluted 1:10 in Ringier-200 salt solution and UV radiation (30,000 erg/mm^2) was used to inactivate sperm genome. A cold shock was used to restore diploidy in a -1-0°C water bath for 25 minutes, at 6.5 minutes post fertilization.

Gynogenesis was conducted on shortnose sturgeon (*Acipenser brevirostrum*) in order to determine sex ratio and to obtain information on the possible formation of the sex ratio [81]. UV light was used for the irradiation of sperm (180-330 mJ/cm^2), and a pressure shock at 8500 psi for 5 minutes was applied to the eggs 20 minutes post fertilization. Ninety-five per cent of the fish reaching the age of one year were found gynogenic diploids (using flow cytometry and microsatellite DNA analysis). The sex ratio was 35 % males and 65 % females.

4.2.2 Sperm treatment with chemical mutagens

The advantage of using chemical mutagens over ionizing radiation is ease of use, whereas, over UV radiation, it is the possibility of treating large amounts of sperm at the same time.

The Hertwig effect has been observed on several occasions when chemical mutagens were used, e.g., on rainbow trout following the use of dimethyl sufate in different concentrations. [40,82]. The duration of treatment was 60-120 minutes in both studies. According to the observed results, Chourrout did not recommend the use of this chemical because, as it was expected, supernumerary chromosome fragments occurred in the genome, although with a lower frequency than in case of gamma radiation but with a reduced sperm motility. The number of developing embryos was also low and a high mortality rate was observed among them during ebryonic development. A further disadvantage is that sperm has to be centrifuged before fertilization in order to avoid contact of the chemical with eggs. Beside practical difficulties this procedure further decreases sperm motility. The use of toluidine blue that was used for treatment of medaka sperm [83] is more promising than that of dimethyl sulfate.

Several other studies were published on the use and effectiveness of chemical mutagens, however, they are now forbidden to use in most countries due to their carcinogenic nature.

4.2.3 Use of gynogenesis with genetic markers

UV radiation is very popular in the practice of induced gynogenesis. It seems to be a perfect method for neutralization of male genome. However, due to its low penetration capacity it is likely that some spermatozoa remain unaffected. Thus, "normal" individuals of both paternal and maternal origin will appear among the gynogenic ones which necessitate the verification of gynogenic individuals using genetic markers. Several studies report on the use of visible markers with mendelian inheritance. In these studies the male typically carries two dominant alleles while the female two recessive ones of the same gene [38,41,54,61,60,78].

In most species, however, only one or a few such visible genetic markers can be found and these can prove the paternal origin of only one gene. Thus, analysis should be supplemented with the use of biochemical markers [55,84-89], although, the limited number available individuals makes their utilization more difficult.

Another possibility is to use the irradiated sperm of a species whose hybrids can be distinguished from individuals with only maternally inherited genome [51,90].

The most attractive method is, however, the use of sperm from a species that cannot otherwise have viable hybrids with the targeted species. In this case survivors will in all probability be individuals of gynogenic origin [36,39,40,54,77,78,90]. Triploids, however, can occur from fertilization with any intact spermatozoon as a viable gynogenic individual is typically the result of an antimeiotic treatment, thus the viability of triploids should be analyzed, as well. For instance, the hybrid of Atlantic salmon and rainbow trout is not viable. On the other hand, viable triploids can hatch from trout eggs fertilized with salmon sperm upon induced gynogenesis [91].

There are several examples of interspecific gynogenesis. Eggs of African catfish were fertilized with irradiated sperm of rosy barb (*Barbus conchonius*) [92]. Sperm was treated with UV radiation (0.5 J/cm^2) and then heat shock was used to restore diploidy (40.5°C for 2 minutes, at 1.4 τ_0 post activation). Swim-up larvae were counted at a rate of 6.34 ± 2.35 %.

Mitotic gynogenesis was performed on sea bass by Bertotto et al. [93]. The sperm was treated with UV radiation and the diploid state of the embryos was restored by the application of a pressure shock (at 81 or 91 MPa) for the duration of 4 minutes. The percentage of hatched gynogenic double haploid (G2N) individuals was 7-18% and maternal origin of the progeny was verified using flow cytometry and microsatelite DNA markers.

Gynogenesis was conducted on Atlantic flounder (*Hippoglossus hippoglossus*) by Tvedt et al. [94]. Haploidy was verified visually according to morphological markers and using microsatellite DNA analysis. Sperm was diluted 1:80 in seminal plasma and irradiated using UV radiation at a rate of 65 mJ/cm^2. The second polar body was retained with a hydrostatic shock of 8,500 psi applied for duration of 5 minutes at an incubation temperature of 5-6°C, 15 minutes following fertilization. The technology is suitable for the production of all-female stocks in the practice.

Methods of gynogenesis were optimized in the Southern flounder (*Paralichthys lethostigma*) [95] in order to explore the background of farming monosex stocks as the faster growth and larger size of females has direct economic advantages. Sperm DNA was inactivated using UV irradiation (70 J/cm^2) for 3-4 minutes, in sea water. Diploidy was restored by applying a cold shock (0-2°C, in sea water with duration of 45-50 minutes). Successful fertilization test were carried out with Southern flounder sperm as well as with the sperm from a different species (gray mullet – *Mugil cephalus*).

4.3 Androgenesis

Androgenesis is a process all-paternal inheritance: the fertilizing spermatozoon induces cleavage, only the sperm cell's DNA controls further embryonic development and the progeny inherits only the genome of paternal origin.

Maternal genome is typically eliminated using irradiation. UV radiation is suitable for this purpose only in minute egg volumes (a few ml), whereas, gamma radiation (such as that of cobalt isotopes) is much more effective. A substantially lower radiation dosage is required for the elimination of maternal genome than for that of sperm. Following inactivation (irradiation) eggs are fertilized with untreated sperm and the resulting zygote will be haploid. This haploid genome has to be duplicated, its diploid state restored by blocking the first mitotic division (e.g., heat shock, [96]).

Thus androgenic progeny contains genes of paternal origin, exclusively. Genes located on homologous chromosomes are copies of each other as they are duplicated in a mitotic process which means that the individual is genetically homozygous [27].

This technique can also be used to produce "supermales" with YY genotype [60]. The progeny of these individuals will all be males. Authors disagree on the viability of YY individuals. Yamamoto [97] has found goldfish (*Carassius auratus*) supermales completely viable, and surviving YY fish were found in medaka (*Oryzias latipes*). Supermales can be used in the practice to produce all-male populations of fish which represents an advantage in species where males tend to grow faster than females [98].

Several methods are available in the literature for testing paternal inheritance. The simplest procedure is based on pigmentation inheritance of the progeny. This method allows the visual differentiation of offspring containing maternally or paternally inherited genes [96,99-103]. Koi males are most commonly used in common carp as the success of androgenesis can be estimated at an early stage of development (at the age of 2 weeks in some koi varieties) due to differences in pigmentation. Viability of androgenic progeny decreases with their growth [104] as the homozygosity of several genes that results from blocking of the first mitotic division leads to the expression of several lethal or sub-lethal mutations [96].

Androgenesis in chondrostean fish has been described for the first time by Grunina and Neifakh [105]. Experiments were carried out on Siberian sturgeon (*Acipenser baeri*) and a hatch percentage of 12.2% was recorded. In sturgeons, a possible marker for verification of paternal origin is albinism: it follows a simple monogenic inheritance with albinism being the recessive trait.

In teleosts, a substantial number of viable androgenic progeny was produced for the first time in salmonids. The use of Co^{60} gamma irradiation (at a rate of 36-88 krad) and pressure shock (8500-9000 psi, endomitosis) resulted in high numbers of androgenic diploid progeny in the rainbow trout and brook trout (*Salvelinus fontinalis*) [102-104,106-109].

In cyprinids, the first diploid androgenic progeny was produced for the first time in common carp by Kondoh, Sato and Tomita [99] and Grunina, Gomelsky and Neifakh [96]. UV irradiation was used by Kondoh, Sato and Tomita and 4-5 % hatch percentages were recorded. Sperm from orange colored males (b1b1b2b2 recessive homozygote) and eggs from pigmented (wild coloration) females were used. Eggs were irradiated using X-radiation at a dose of 25-30 krad. All treatments were conducted in batches of 300-500 eggs. Fertilized eggs, attached to the bottom of a Petri dish were incubated at 22.5 – 23°C. The most effective heat shock (the one that resulted in the highest number of viable progeny) was applied in a 41°C water bath for 2 minutes, 36 minutes (1.7 τ_0) post fertilization. Thus, a maximum of 10% hatch rate was observed in androgenic diploid progeny, however, some wild colored ones were found among the hatched larvae.

Similarly, sperm of b1b1b2b2 recessive homozygote "blond" males and eggs of pigmented (wild colored) females were used by Bongers et al. [110]. Unfortunately, no viable androgenic progeny has hatched from eggs of B1B1B2B2 homozygote dominant females, thus, the paternal origin of the "blond" offspring hatched from the eggs of B1b1B2b2 heterozygous dominant females is questionable. Eggs were irradiated using UV radiation at a dose of 150-300 MJ/cm^2 in an artificial ovarian fluid. All treatments were conducted in batches of 150-200 eggs and fertilized eggs were incubated at 24°C. In the experiment that gave the best results a heat shock was applied in a 40°C water bath for 2 minutes and 26, 28 or 30 minutes post fertilization. Maximum hatch rates of 7.2-18.3% were observed in androgenic diploid progeny (the hatch rate of the control treated with ovarian fluid was 86%). Among androgenic progeny 100% "blonds" hatched only when an irradiation dose of 250 MJ/cm was used and the heat shock was applied 30 minutes after fertilization (which also corresponded to approximately 1.7 τ_0), otherwise, pigmented individuals were also observed among the hatched progeny.

An inclomplete interspecific androgenesis in cyprinids has been reported for the first time by Grunina, Gomelsky and Neifakh [111] who produced hybrids of silver crucian carp (*Carassius gibelio*) and common carp as an initial step of the experiments. Interspecific hybrids were viable but only female individuals produced fertile gametes. The gametes were diploid which is a result of an "endoreduplication" during early oogenesis. Females were subjected to sex reversal and the resulting diploid sperm was used for fertilization, thus, rendering the application of heat shock unnecessary (other parameters of the method were identical to the one reported by Grunina, Gomelsky and Neifakh [98], described above). Although heat shock was not used, the hatch percentage of androgenic, diploid "nucleo-cytoplasmatic hybrids" was maximum 3%. Diploidy of the progeny was verified with chromosome number calculations (2N = 100).

A complete and successful interspecific androgenesis was reported for the first time by Bercsényi et al. [112] when irradiated eggs of common carp were fertilized with goldfish (*Carassius auratus*) sperm. Beside moprhological markers (triangle tail, red cap, bubble eyes), paternal origin of the progeny was verified using DNA analysis. The genome of common carp eggs was inactivated using ^{60}Co gamma radiation at a dose of 25 krad. At this irradiation dosage fertilization rates of 28.5 – 44 % and hatch rates of 17.5 – 28 % were observed. Heat shock was applied at 40°C for 2 minutes. The ideal timing of the heat shock was greatly influenced by the temperature of incubating water which was 21°C.

All-paternal origin of androgenic progeny was verified – beside phenotypic (such as color) markers – using genetic markers (iso- and allozymes), as well [100,107].

Combined use of sperm cryopreservation and androgenesis has been postulated by several authors [68,107] and androgenic progeny has been produced using cryopreserved sperm in rainbow trout [104] and sulver crucian carp [111]. Combination of the two procedures allows the conservation and restoration of species and varieties. This also represents a possibility to bypass the unsolved problem of egg and embryo cryopeservation in fish.

Successful androgenesis was conducted in the tiger barb (*Puntius tetrazona*) by Kirankumar and Pandian [113] and two different colorations (gray and blond) were used as markers. UV irradiation (4.2 W/m^2, for 210 seconds) was carried out to inactivate the genome of eggs from gray females. A 41°C water bath was used as a heat shock for 2 minutes to restore the diploid state of the embryos. Successful inactivation of the maternal genome was verified with expression analysis of the GFP (green fluorescent protein) gene on 16-hour-old haploid embryos, also, phenotypically (appearance of the blond coloration in the progeny and adult individuals) and with progeny tests.

4.4 Application of chromosome set manipulation protocols

Theoretically, there are no obstacles to the utilization of chromosome set manipulation protocols, however, the economic potential of these techniques is rarely exploited. The reason for this is the lack of profitability calculations on one hand, while on the other, the complicated nature of these techniques and low survival rates render them ineffective. In spite of this, numerous examples show that their introduction into aquaculture practice has an economic value and they can be used in environment protection and preservation of biodiversity. In the following, we present several practical methods that support the idea of utilization of these protocols in fish farming.

Practical utilization of chromosome set manipulation techniques is most common in salmonids [114]. Induced triploidy results in functionally sterile trout and salmon stocks which represents fundamental advantages in fish trade and farming. In case of rainbow trout – one of the most popular farmed fish species – monosex female stocks reach market size considerably faster than those containing both males and females. Individuals of these stocks do not show secondary sexual traits either, such as darker pigmentation of the skin, increased aggressiveness, decreasing meat quality and meat yield. Combination of chromosome set manipulation techniques with sex reversal offers further advantages. According to calculations the use of these protocols can result in 7-8% production increase in the common carp [115].

In Japan, several aquatic organisms are intentionally subjected to chromosome set manipulations based on profitability calculations for commercial purposes. These species include the rainbow trout and the ayu (*Plecoglossus altivelis*) in which triploidy offers the advantages of higher growth and survival rates as well as better meat quality – all resulting from the sterility of these individuals (less energy is spent on reproductive processes). In contrast to freshwater fish species where triploidy and utilization of its advantages work effectively, in marine species (such as the Japanese flounder – *Paralichthys olivaceus*, red sea bream – *Pagrus major*, Japanese parrot fish *Oplegnathus fasciatus*) commercialization is hindered by several factors. In these species, no differences were detected in the growth rates of triploid or diploid stocks. The presence of tetraploid stocks is justified by their importance of producing triploids. Such artificially fed and maintained tetraploid stocks in Japan exist only in rainbow trout [32].

American scientists [116] have started to use gynogenesis in the Southern flounder primarily for economic purposes. Females grow 2-3 times faster than males, thus farming of monosex, all-female stocks is obvious for increased profit. Also, less males of the identical species were available for induced spawning because of fishing regulations. Thus, the sperm of black sea bass (*Centropristis striata*) was used and high numbers of Southern flounder females were produced using meiotic gynogenesis.

Chromosome set manipulation techniques were used on 10 different farmed fish species in Israel, including 2 marine species. Practical utilization, however, took place only in the common carp: gynogenesis and sex-reversal of androgenic individuals (neo-males) were used in commercial spawning [117].

Triploidization represents a significant advantage in case of the cod (*Gadus morhua* L.) where the triploidy inducing effect of temperature shock was investigated [118]. Cold shock was conducted at -1.7 ± 0.1°C for 20 minutes, 2 hours post fertilization, while heat shock was applied at 20°C for 20 minutes (at the same time post fertilization). The highest percentage of triploids (66-100 %) with 10-20% survival was observed with the use of heat shock. Tetraploid individuals were observed in lower numbers when the heat shock was applied with a longer duration. Triploid individuals formed a sterile stock and their growth was faster than that of diploids.

An interesting study was conducted on meat and muscle fiber quality of 78-month old diploid and triploid rainbow trouts [119]. Triploid individuals were characterized by higher body weight and better carcass yield than diploids (6 kg vs. 4 kg, 66% vs. 52%). Meat of triploid fish had lower electric conductivity and darker, more reddish color, whereas, the fillet had a higher crude fiber and lower water content (6% vs. 3% and 68% vs. 74%). No differences were found in terms of protein content and quality. Altogether, triploid fish were found to grow faster and had a higher meat quality than diploid individuals.

Similar investigations were carried out on meiotic gynogenic and triploid sea bass [120]. Significant differences in comparison to diploid individuals were found in 35-45-month-old fish, fillet yield was higher in triploids and, within triploids it was higher in males. Triploid females tend to accumulate more abdominal fat towards the end of the growing season while the lipid content of the meat decreases. Similar observations were found in rainbow trout [74] and tilapia [121].

An improvement in carcass and meat quality of rainbow trout fillets and smoked products has also been reported by Haffray et al. [122]. They have found that fat and lipid content of triploid individuals has also decreased.

The combination of ploidy manipulation and hormonal sex reversal can represent a significant economic opportunity for fish farmers as complete all-female or all-male stock can be created. In some species where gynogenesis is applied, genetic homozygosity results in better growth parameters of up to 3 – 60 %. Meiotic gynogenic stocks of *Clarias macrocephalus* or *Paralichthys olivaceus* display a growth rate that is 18 - 35 % faster than that of diploid individuals [123].

A good example of the utilization of androgenesis was reported by David and Pandian [124] who used the sperm of dead individuals to produce viable offspring. Dead gold colored Buenos Aires tetra (*Hemigrammus caudovittatus*) individuals were stored at -20°C for 10-40 days. Oocytes of black widow tetra (*Gymnocorymbus ternetzi*) females acted as host cells and were irradiated using UV radiation (4.2 W/m²) for 3 minutes. A fertilization rate of 95% was observed in the control whereas with sperm from the stored individuals it was 19-24%. Diploidy was restored with a heat shock applied 25 minutes post fertilization at 41°C for 2 minutes. The rate of androgenic progeny was 11% when fresh sperm was used and 4% when the sperm from dead individuals was utilized.

Androgenesis is suitable for the preservation of threatened and endangered fish species as it has been reported in sturgeon [125] and the effectiveness of the method is greatly increased when cryopreserved sperm is used for fertilization. The method has been tested on stellate sturgeon (*Acipenser stellatus*), Russian sturgeon (*A. gueldenstaedtii*) and beluga (*Huso huso*). Eggs were irradiated using X-ray at a rate of 220 Gy. Incubating temperature was 19-20°C and heat shock (37°C, 150 seconds) was applied at $1.4 - 1.6 \, \tau_0$ post fertilization. When fresh sperm was used for fertilization 11% of the hatched progeny reached the age of 1 month, whereas with cryopreserved sperm this ratio was 4%.

5. REFERENCES

1-Hoar, W.S., Reproduction, in: *Fish Physiology Volume III Reproduction and Growth, Bioluminescence, Pigments and Poisons*, Hoar, W.S. and Randall, D.J., Eds., Academic Press, New York/London, 1, 1969.

2-Williot, P., Brun, R., Rouault, T., Pelard, M., Mercier, D., and Ludwig, A., Artificial spawning in cultured sterlet sturgeon, *Acipenser ruthenus* L., with special emphasis on hermaphrodites, *Aquaculture*, 246, 263, 2005.

3-Henne, J.P., Ware, K.M., Wayman, W.R., Bakal, R.S., and Horváth, Á., Synchronous hermaphroditism and self-fertilization in a captive shortnose sturgeon. *Trans. Am. Fish. Soc.*, 135, 55, 2006.

4-Coward, K., Bromage, N.R., Hibbitt, O., and Parrington, J., Gamete physiology, fertilization and egg activation in teleost fish, *Rev. Fish Biol. Fisheries,* 12, 33, 2002.

5-Linhart, O., Kudo, S., Billard, R., Slechta, V., and Mikodina, E. V., Morphology, composition and fertilization of carp eggs: a review, *Aquaculture,* 129, 75, 1995.

6-Cherr, G.N. and Clark, W.H., An egg envelope component induces the acrosome reaction in sturgeon sperm, *J. Exp. Zool.,* 234, 75, 1985.

7-Dettlaff, T.A., Ginsburg, A.S., and Schmalhausen, O.I., *Sturgeon fishes, developmental biology and aquaculture.* Springer-Verlag, Berlin, 1993.

8-Amanze, D. and Iyengar, A., The micropyle: a sperm guidance system in teleost fertilization, *Development,* 109, 495, 1990.

9-Riehl, R. and Appelbaum, S., A Unique adhesion apparatus on the eggs of the catfish *Clarias gariepinus* (Teleostei, Clariidae), *Japan. J. Ichthyol.,* 38, 191, 1991.

10-Linhart, O. and Kudo, S., Surface ultrastructure of paddlefish eggs before and after fertilization, *J. Fish Biol.,* 51, 573, 1997.

11-Kudo, S., Sperm penetration and the formation of a fertilization cone in the common carp egg. *Develop. Growth Diff.,* 22, 403, 1980.

12-Wolenski, J.S. and Hart, N.H., Sperm incorporation independent of fertilization cone formation in the danio egg. *Develop. Growth Differ.,* 30, 619, 1988.

13-Ohta, T. and Iwamatsu, T., Electron microscopic observations on sperm entry into eggs of the rose bitterling *Rhodeus ocellatus, J. Exp. Zool.,* 227, 109, 1983.

14-Wolenski, J.S. and Hart, N.H., Scanning eletron microscope studies of sperm incorporation into the zebrafish (*Brachydanio*) egg, *J. Exp. Zool.,* 243, 259, 1987.

15-Štich, L., Linhart, O., Shelton, W.L., and Mims, S.D., Minimally invasive surgical removal of ovulated eggs from paddlefish. *Aquac. Int.,* 7, 129-133, 1999.

16-Magyary, I., Urbányi, B., and Horváth, L., Cryopreservation of common carp (*Cyprinus carpio* L.) sperm II. Optimal conditions for fertilization, *J. App. Ichthyol.,* 12, 117, 1996.

17 Lahnsteiner, F., Berger, B., and Weismann, T., Effects of media, fertilization technique, extender, straw volume, and sperm to egg ratio on hatchability of cyprinid embryos, using cryopreserved semen, *Theriogenology,* 60, 829, 2003.

18-Woynárovich, E. and Horváth, L., The artificial propagation of warm-water finfishes – a manual for extension. *FAO Fish. Tech. Paper.* 1980.

19-Bokor, Z., Müller, T., Bercsényi, M., Horváth, L., Urbányi, B., and Horváth, Á., Cryopreservation of sperm of two European percid species, the pikeperch (*Sander lucioperca*) and the Volga pikeperch (*S. volgensis*), *Acta Biologica Hungarica,* 58, 199, 2007.

20-Gomelsky, B., personal communication, 2004.

21-Horváth, L., Tamás, G., and Seragrave, C., *Carp and Pond Fish Culture,* 2nd ed., Fishing News Books, Oxford, 2002.

22-Chebanov, S., personal communication, 2006.

23-Ingram, M., Farming rainbow trout in fresh water tanks and ponds, in: *Salmon and trout farming,* Laird, L. and Needham, T., Eds., Ellis Horwood, Chichester, 1988, 155.

24-Pillay, T.V.R., *Aquaculture, principles and practices,* Fishing News Books, Oxford, 1990.

25-Chourrout, D., Genetic manipulation in fish. Review of methods, in: *Selection hybridization and genetic engineering in aquaculture,* Ed. Thiews, K., Vol. II., 112, 1987.

26-Ihssen, P.E., Mckay, L.R., Mcmillan, I., and Phillips, R.B., Ploidy manipulation and gynogenesis in fishes - cytogenetic and fisheries applications, *Trans. Amer. Fish. Soc.,* 119, 698, 1990.

27-Thorgaard, G.H., Ploidy manipulation and performance, *Aquaculture,* 57, 57, 1986.

28-Purdom, C.E., Genetic engineering by the manipulation of chromosomes, *Aquaculture, 33,* 287, 1983.

29-Thorgaard, G.H., Application of genetic technologies to rainbow trout, *Aquaculture*, 100, 85, 1992.

30-Horvath, L. and Orban, L., Genome and gene manipulation in the common carp, *Aquaculture*, 129, 157, 1995.

31-Thorgaard, G.H., Biotechnological approaches to broodstock management, in *Broodstock management and egg and larval quality*, Bromage N.R., Roberts, R.J., Eds., Blackwell Science, Oxford, 76, 1995.

32-Arai, K., Genetic improvement of aquaculture finfish species by chromosome manipulation techniques in Japan, *Aquaculture*, 197, 205, 2001.

33-Hertwig, O., Die radium krankheit tierischer keim zellen, *Arch. Mikrosk. Anat.*, 77, 1, 1911.

34-Peacock, A.D., Some problems of parthenogenesis, *Advance. Sci.*, 9, 134, 1952.

35-Nace, C. W., The use of amphibians in biomedical research, in *Animal models for biomedical research III*. Proceedings of a Symposium. National Academy of Sciences, Washington, D.C., 103, 1970.

36-Stanley, J.G. and Sneed, K.Z., Artifical gynogenesis and its application in genetics and selective breeding of fish. in: *The early life history of fish*, Blaxter, J., Ed., Springer, Berlin, 1974, 527.

37-Purdom, C.E., Radiation induced gynogenesis in fish, *Heredity*, 24, 431, 1969.

38-Nagy, A., Rajki, K., Horváth, L., and Csányi, V., Investigation on carp, *Cyprinus carpio*, L., gynogenesis. *J. Fish Biol.*, 13, 215, 1978.

39-Chourrout, D. and Quillet, E., Induced gynogenesis in the rainbow trout, *Theor. Appl. Genet.*, 63, 201, 1982.

40-Chourrout, D., Techniques of chromosomes manipulation in rainbow trout: A new evaluation with caryology, *Theor. App. Genet.*, 72, 627, 1986.

41-Chourrout, D., Thermal induction of diploid gynogeneyis and triploidy in the eggs of the rainbow trout, *Reprod. Nutr. Develop.*, 22, 713, 1980.

42-Nagy, A., Genetic manipulation on warm water fish, in *Selection, Hybridization. II.*, 163, 1987.

43-Lincoln, R.F., Sexual maturation in female triploid plaice (*Pleuronectes platessa*) and plaice X flounder (*Platichthys flesus*) hybrids. *J. Fish Biol.*, 19, 415, 1981.

44-Thompson, B. Z., Wattendorf R. J., Hestand, R. S., and Underwood, J. L., Triploid grass carp production, *Prog Fish.-Cult.*, 49, 213, 1987.

45-Piferrer, F., Cal, M.R., Gomez, C., Bouza, C., and Martinez, P., Induction of triploidy in the turbot (*Scophthalmus maximus*) II. Effects of cold shock timing and induction of triploidy in a large volume of eggs *Aquaculture*, 220, 821, 2003.

46-Biswas, A., Kundu, S., Roy, S., De, J., Pramanik. M., and Ray, A.K., Thyroid hormone profile during annual reproductive cycle of diploid and triploid catfish, *Heteropneustes fossilis* (Bloch) *Gen. Comp. End.*, 147, 126, 2006.

47-Thorgaard, G.H., Allendorf, F.W., and Knudsen, K.L. Gene-centromere mapping in rainbow trout: high interference over long map distances, *Genetics*, 103, 771, 1983.

48-Cherfas, N.B., Hulata, G., and Kozinsky, O., Induced diploid gynogenesis and polyploidy in ornamental (koi) carp, *Cyprinus carpio* L. timing of heat shock during the first cleavage, *Aquaculture*, 111, 281, 1993.

49-Swarup, H., Effect of triploidy on the body size, general organization and cellular structure in *Gasterosteus aculeatus*, *J. Genet.*, 56, 143, 1959.

50-Purdom, C. E., Induced polyploidy in plaice (*Pleuronectes platessa*) and its hybrid with the flounder (*Platichthys fiesus*), *Heredity*, 29, 11, 1972.

51-Romashov, D.D., Nikolyukin, N.I., Belyaeva, V.N., and Timofeeva, N.A., Possibilites of producing diploid radiation-induced gynogenesis in sturgeons, *Radiobiologiya*, 3, 104, 1963.

52-Wolters, W. R., Libey, G. S., and Chrisman, C. L., Induction of triploidy in channel catfish, *Trans. Am. Fish. Soc.*, 110, 310, 1981.

53-Ueno, K., Sterility and secondary sexual character of triploid *Gnathopogon elongathus caerulescens*. *Fish. Genet. and Breed. Sci.*, 10, 37, 1985.

54-Suzuki, R., Oshiro, T., and Nakanishi, T., Survival, growth and fertility of gynogenetic diploid induced in the cyprinid loach, *Misgurnus anguillicaudatus*, *Aquaculture*, 48, 45, 1985.

55-Gervai, J. and Csányi, V., Artificial Gynogenesis and Mapping of Gene-centromere Distances in the Paradise Fish (*Macropodus opercularis*), *Theor. Appl. Genet.*, 68, 481, 1984.

56-John, G., Reddy, P.V.G.K., and Gupta, S.D., Artificial gynogenesis in two Indian major carps, *Labeo rohita* (Ham.) and *Catla catla* (Ham.), *Aquaculture*, 42, 161, 1984.

57-Valenti, R.J., Induced polyploidy in *Tilapia aurea* (Steindachner) by means of temperature shock treatment, *J. Fish. Biol.*, 7, 519, 1975.

58 Benfey, T.J. and Sutterlin, A.M., Growth and gonadal development in triploid landlocked Atlantic salmon (*Salmo salar*), *Can. J. Fish. Aquat. Sci.*, 41, 1387, 1984.

59-Johnstone, R., Induction of triploidy in Atlantic. salmon by heat shock, *Aquaculture*, 49, 133, 1985.

60-Streisinger, G., Walker, C., Dower, N., Krauber, D. and Singer, F., Production of clones of homozygous diploid zebra fish, *Nature*, 291, 293, 1981.

61-Chourrout, D. and Itskovitch, I., Three manipulations permitted by artifical insemination in tilapia, *Int. Symp. on Tilapia in Aquaculture*, Nazareth, Israel, 1983.

62-Hollebeq, M. G., Chourrout, D., Wohlfahrt, G., and Billard, R., Diploid gynogenesis induced by heat shocks after activation with UV-irradiated sperm in common carp, *Aquaculture*, 54, 69, 1986.

63-Chourrout, D., Pressure-induced retention of second polar body and suppression of first cleavage in rainbow trout: production of all-triploids, all-tetrapolids, and heterozygous and homozygous diploid gynogenetics, *Aquaculture*, 36, 111, 1984.

64-Lou, Y.D. and Purdom, C.E., Polyploidy induced by hydrostatic pressure in rainbow trout *Salmo gairdneri* Richardson, *J. Fish. Biol.*, 25, 345, 1984.

65-Onozato, H., The "Hertwig effect" and gynogenesis in Chum salmon *Oncorhynchus keta* eggs fertilized with Co60 gamma irradiated milt, *Bull. Jpn. Soc. Sci. Fish.*, 48, 1237, 1982.

66-Francescon, A., Libertini, A., Bertotto, D., and Barbaro, A., Shock timing in mitogynogenesis and tetraploidization of the European sea bass *Dicentrarchus labrax*, *Aquaculture*, 236, 201, 2004.

67-Nam, Y.K., Choi, G.C., and Kim, D.S., An efficient method for blocking the 1st mitotic cleavage of fish zygote using combined thermal treatment, exemplified by mud loach (*Misgurnus mizolepis*) *Theriogenology*, 61, 933, 2004.

68-Nam, Y.K. and Kim, D.S., Ploidy status of progeny from the crosses between tetraploid males and diploid females in mud loach (*Misgurnus mizolepis*) *Aquaculture*, 236, 575, 2004.

69-Schultz, R.J., Gynogenesis and Triploidy in the Viviparous Fish Poeciliopsis, *Science*, 157, 1564, 1967.

70-Lieder, U., Männchenmangel und natürliche parthenogenese bei Silberkarausche *Carassius auratus gibelio* (Vertebrata, Pisces), *Naturvissenschaften*, 42, 590, 1955.

71-Romashov, D.D. and Belyaeva, V.N., Cytology of radiation gynogenesis and androgenesis in the loach, *Dokl. Akad. Nauk SSSR*, 157, 964, 1964.

72-Romashov, D.D., Golovinskaja, K.A., Belyaeva, V.N., Bakulina, N.A., Pokrovskaja, G.L., and Cherfas, N.B., Diploid radiation gynogenesis in fish, *Biofizika*, 5, 461, 1960.

73-Ijiri, K. I. and Egami, N., Hertwig effect caused by UV-irradiation of sperm of *Oryzias latipes* (Teleostei) and its photoreactivation, *Mutation Res.*, 69, 241, 1980.

74-Lincoln, R.F. and Scott, A.P., Sexual maturation in triploid rainbow trout, *Salmo gairdneri* Richardson. *J. Fish Biol.*, 25, 385, 1984.

75-Chourrout, D., Chevassus, B., and Herioux, F., Analysis of a Hertwig effect in the rainbow trout (*Salmo gairdneri* Richardson) after fertilization with g-irradiated sperm, *Reprod. Nutr. Dévelop.*, 20, 719, 1980.

76-Refstie, T., Vassvik, V., and Gjedrem T., Induction of polyploidy in salmonids by cytochalasin B, *Aquaculture*, 10, 65, 1977.

77-Stanley, J. G., Production of hybrid, androgenic and gynogenetic grass carp and carp, *Trans. Am. Fish. Soc.*, 105, 10, 1976.

78-Chourrout, D., La gynogenése ches les vertébrés. *Reprod. Nutr. Develop.*, 22, 713, 1982.

79-Onozato, H. and Yamaha. E., Induction of gynogenesis with ultraviolet rays in four species of salmoniformes, *Bull. Jpn. Sac. Sci. Fish.*, 49, 693, 1983.

80-Piferrer, F., Cal, M.R., Gomez, C., Alvarez-Blazquez, B., and Martinez, P., Induction of gynogenesis in the turbot (*Scophthalmus maximus*): Effects of UV irradiation on sperm motility, the Hertwig effect and viability during the first 6 months of age, *Aquaculture*, 238, 403, 2004.

81-Flynn, S.R., Matsuoka, M., Reith, M., Martin-Robichaud, D.J., and Benfey, T.J., Gynogenesis and sex determination in shortnose sturgeon, *Acipenser brevirostrum* Lesuere, *Aquaculture*, 253, 721, 2006.

82-Tsoi, R.M., Action of nitrosolmethylurea and dimethylsulfate on sperm cells of the rainbow trout and the peled, *Dokl.Akad.Nauk.SSSR*, 189, 411, 1969.

83-Uwa, H., Gynogenetic haploid embryos of the medaka (*Oryzias latipes*), *Embryologia*, 9, 40, 1965.

84-Purdom, C.E., Thompson, D., and Dando, P.R., Genetic analysis of enzyme polymorphisms in plaice (*Pleuronectes platessa*), *Heredity*, 37, 193, 1976.

85-Nagy, A., Rajki, K., Bakos, J., and Csányi, V., Genetic analyses in carp (*Cyprinus carpio*) using gynogenesis, *Heredity*, 43, 35, 1979.

86-Thompson, D., Purdom, C.E., and Jones, B.W., Genetic analysis of spontaneous gynogenetic diploids in the plaice *Pleuronectes platessa*, *Heredity*, 47, 269, 1981.

87-Thompson, D., The efficiency of induced diploid gynogenesis in inbreeding, *Aquaculture*, 33, 237, 1983.

88-Thompson, D. and Scott, A.P., An analysis of recombination data in gynogenetic diploid rainbow trout, *Heredity*, 53, 441, 1984.

89-Guyomard, R., High level of residual heterozygosity in gynogenetic rainbow trout, *Salmo gairdneri*, Richardson, *Theor. Appl. Genet.*, 67, 307, 1984.

90-Stanley, J.G. and Jones, J.B., Morphology of androgenetic and. gynogenetic grass carp, *Ctenopharyngodon idella.* (Valenciennes), *J. Fish. Biol.*, 9, 523, 1976.

91-Chevassus, B., Hybridization in fish, *Genetics in Aquaculture*, 245, 1983.

92-Váradi, L., Benko, I., Varga, J., and Horváth, L., Induction of diploid gynogenesis using interspecific sperm and production of tetraploids in African catfish, *Clarias gariepinus* Burchell (1822), *Aquaculture*, 173, 401, 1999.

93-Bertotto, T.D., Cepollaro, F., Libertini, A., Barbaro, A., Francescon, A., Belvedere, P., Barbaro, I., and Colombo, L., Production of clonal founders in the European sea bass, *Dicentrarchus labrax* L., by mitotic gynogenesis, *Aquaculture*, 246, 115, 2005.

94-Tvedt, H.B., Benfey, T.J., Martin-Robichaud, D.J., McGowan, C., and Reith, M., Gynogenesis and sex determination in Atlantic halibut (*Hippoglossus hippoglossus*), *Aquaculture*, 252, 573, 2006.

95-Luckenbach, J.A., Godwin, J., Daniels, H.V., Beasley, J.M., Sullivan, C.V., and Borski, R.J., Induction of diploid gynogenesis in southern flounder (*Paralichthys lethostigma*) with homologous and heterologous sperm, *Aquaculture*, 237, 499, 2004.

96-Grunina, A.S., Gomelsky, B.I., and Neifakh, A.A., Diploid androgenesis in pond carp, *Genetika*, 26, 2037, 1990.

97-Yamamoto, T., Sex differentation, in: *Fish Physiology, Vol. 3.*, Hoar, W.A. and Randall, D.J., Eds., Academic Press, New York, 117, 1969.

98-Bongers, A.B.J., Zandieh-Doulabi, B., Richter, C.J.J., and Komen, J., Viable androgenetic YY genotypes of common carp (*Cyprinus carpio* L.), *J. Heredity*, 90, 195, 1999.

99-Kondoh, S., Sato, S., and Tomita, M., Induction of androgenetic fancy carp, *Rep. Niigata Pref. Inland Water Fish. Exp. Station*, 15, 19, 1989.
100-Gillespie, L.L. and Armstrong, J.B., Production of androgenetic diploid axolotls by supression of first cleavage, *J. Exp. Zool.*, 213, 423, 1980.
101-Gillespie, L.L. and Armstrong, J.B., Suppression of first cleavage in the Mexican axolotl (*Ambystoma mexicanum*) by heat shock or hydrostatyc pressure, *J. Exp. Zool.*, 218, 441, 1981.
102-Parsons, J.E. and Thorgaard, G.H., Induced androgenesis in rainbow trout, *J. Exp. Zool.*, 231, 407, 1984.
103-Parsons, J.E. and Thorgaard, G.H., Production of androgenetic diploid rainbow trout, *J. Hered.*, 76, 177, 1985.
104-Thorgaard, G.H., Scheerer, P.D., Herschberger, W.K., and Myers, J.M., Androgenetic rainbow trout produced using sperm from tetrapolid males show improved survival, *Aquaculture*, 85, 215, 1990.
105-Grunina, A. S. and Neifakh, A.A., Induction of diploid androgenesis in the siberian sturgeon *Acipenser baeri* Brandt, *Dev. Gen.*, 22, 20, 1991.
106-Arai, K., Onozato, H., and Yamazaki, F., Artificial androgenesis induced with gamma irradiation in masu salmon, *Oncorhyncus masou, Bull. Fac. Fish. Hokkaido Univ.*, 30, 181, 1979.
107-Scheerer, P.D., Thorgaard, G.H., Allendorf, F.W., and Knudsen, K.L., Androgenetic rainbow trout produced from inbred and outbred sperm sources show similar survival, *Aquaculture*, 57, 289, 1986.
108-Scheerer, P. D., Thorgaard, G.H., and Allendorf, F.W., Genetic Analysis of Androgenetic Rainbow trout, *J. Ex. Zool.*, 260, 382, 1991.
109-May, B., Henley, K.J, Krueger, C.C., and Gloss, S.P., Androgenesis as a mechanism for chromosome set manipulation in brook trout (*Salvelinus fontialis*), *Aquaculture*, 27, 57, 1988.
110-Bongers, A.B.J., in't Veld, E.P.C., Abo-Hashema, K., Bremmer, I.M., Eding, E.H, Komen, J., and Richter, C.J.J., Androgenesis in common carp (*Cyprinus carpio* L.) using UV irradiation in a synthetic ovarian fluid and heat shocks, *Aquaculture*, 122, 119, 1994.
111-Grunina, A. S., Gomelsky, B.I., and Neifakh, A.A., Induced diploid androgenesis in common carp and production of androgenetic hybrids between common carp and crucian carp, in: *Abstracts of 4-th International Symposium on Genetics in Aquaculture,* Wuhan, 48, 1991.
112-Bercsényi, M., Magyary, I., Urbányi, B., Orbán, L., and Horváth, L., Hatching out goldfish from common carp eggs: interspercific androgenesis between two cyprinid species, *Genome*, 41, 573, 1998.
113-Kirankumar, S. and Pandian, T.J., Production of androgenetic tiger barb, *Puntius tetrazona. Aquaculture*, 228, 37, 2003.
114-Johnstone, R., Experience with salmonid sex reversal and triploidisation technologies in the United Kingdom. *Bull. Aquac. Assoc. Can.*, 96, 9, 1996.
115-Gomelsky, B., Chromosome set manipulation and sex control in common carp: a review *Aquat. Living Resour.*, 16, 408, 2003.
116-Morgan, A.J., Murashige, R., Woolridge, C.A., Luckenbach, J.A., Watanabe, W.O., Borski, R.J., Godwin, J., and Daniels, H.V., Effective UV dose and pressure shock for induction of meiotic gynogenesis in Southern flounder (*Paralichthys lethostigma*) using black sea bass (*Centropristis striata*) sperm *Aquaculture*, 256, 290, 2006.
117-Rothbard, S., A review of ploidy manipulations in aquaculture: the Israeli experience, *Isr. J. Aquac. Bamidgeh*, 58, 266, 2006.
118-Peruzzi, S., Kettunen, A., Primicerio, R., and Kauric, G., Thermal shock induction of triploidy in Atlantic cod (*Gadus morhua* L.) *Aquac. Res.*, 38, 926, 2007.

119-Poontawee, K., Werner, C., Muller-Belecke, A., Horstgen-Schwark, G. and Wicke, M., Flesh qualities and muscle fiber characteristics in triploid and diploid rainbow trout *J. App. Ichthyol*, 23, 273, 2007.

120-Peruzzi, S., Chatain, B., Saillant, E., Haffray, P., Menu, B., and Falguiere, J.C., Production of meiotic gynogenetic and triploid sea bass, *Dicentrarchus labrax* L. 1. Performances, maturation and carcass quality *Aquaculture*, 230, 41, 2004.

121-Hussain, M.G., Rao, G.P.S., Humayun, N.M., Randall, C.F., Penmann, D.J., Kime, D., Bromage, N.R., Myers, J.M., and McAndrew, B.J., Comparative performance of growth, biochemical composition and endocrine profiles in diploid and triploid tilapia *Oreochromis niloticus* L. *Aquaculture*, 138, 87, 1995.

122-Haffray, P., Vauchez, C., Rault, P., and Reffay, M., Innovations dans les écloseries francaises depuis 1991. *La Pisciculture Francaise,* 134, 23, 1999.

123-Pandian, T.J. and Koteeswaran, R., Ploidy induction and sex control in fish, *Hydrobiologia*, 384, 167, 1998.

124-David, C.J. and Pandian, T.J., Cadaveric sperm induces intergeneric androgenesis in the fish, *Hemigrammus caudovittatus. Theriogenology*, 65, 1048, 2006.

125-Grunina, A.S., Recoubratsky, A.V., Tsvetkova, L.I., and Barmintsev, V.A., Investigation on dispermic androgenesis in sturgeon fish. The first successful production of androgenetic sturgeons with cryopreserved sperm, *Int. J. Refrigeration,* 29, 379, 2006.

Complete affiliation:

Béla Urbányi, Department of Fish Culture, Szent István University, Páter Károly u. 1., Gödöllő H-2103 Hungary e-mail: urbanyib@nt.ktg.gau.hu.

SECTION IV - SPERM AND EGG CRYOPRESERVATION

CHAPTER 6
CHILLED STORAGE OF SPERM AND EGGS

J. Bobe and C. Labbe

1. GENERAL CONSIDERATIONS

After collection, storage of gametes at sub-zero temperatures is required in several situations. Storage of gametes can eliminate the need to keep breeders available for artificial fertilization. It allows the collection chore to extend over a several day period, especially for fish caught in the wild, before the artificial fertilization can be performed in adapted facilities. The shipping of gametes from the fish farm to other breeding sites becomes possible. Some pathological diagnosis can be performed on the gametes and the results can be obtained before the gametes are released for use. In the case of sperm, the cryopreservation procedure of fish strains with many different male samples is easier to set up some time after the collection. In some species, surgical removal of the testis is the only possibility to get sperm, and storage of this testicular-extracted sperm for several days allows the subsequent handling of egg collection and fertilization. Considering the poor initial quality or the intrinsic fragility of sperm in some species, cryopreservation is difficult to achieve and chilled storage remains one mean to diffuse the genetic progress.

In almost every species bred in aquaculture, chilled storage of gametes was tempted, with more or less success depending on the physical and physiological characteristics of the cells. During gamete storage at temperatures above the freezing points, the metabolism of the cells is only slowed down, but not arrested contrarily to the -196°C storage. As a consequence, special attention has to be paid to ensure the proper functioning of cell metabolism and the maintenance of the cell function upon rewarming: the appropriate atmosphere should be provided in order to ensure that the right oxygen concentration is available; the storage medium should buffer deleterious pH variations; appropriate substrates should sustain energy metabolism of the cell; specific ionic composition should prevent gamete activation and enable cellular homeostasis with regards to water movement in and out of the cell. Specific molecules may also be added in order to protect the gametes from possible cold injury.

The way gamete quality is explored to assess storage tolerance has to be carefully considered. In sperm, most authors assess sperm motility. This parameter is very informative because different sperm subpopulations are easily detected [1]. The fertilization rate is a more integrative parameter, especially in acrosome bearing species such as oyster or sturgeon where a good motility with damaged acrosomes leads to poor fertilization rates. However, considering the large sperm/egg ratio used to fertilize eggs, often about 10^6, fertilization rate will be less sensitive to intermediate variation in sperm quality. Indeed, microscopic observation of sperm motility will deal with fractions of 200-300 spermatozoa/field whereas fertilization will deal with thousands spermatozoa per egg. It is therefore likely that a sizeable fertilization rate is achievable with only one live spermatozoa out of 100,000. One striking example is given by Rana et al. [2] in Nile tilapia (*Oreochromis niloticus*) where a 0% sperm motility grade still gave 50% fertilization rate.

In this section, egg and sperm storage will be developed separately as the requirements between the two are very different. Some general rules for gamete storage will be presented, and specific recommendations depending on the species will be outlined.

2. SPERM STORAGE

2.1 Storage temperature

Temperatures of 0°C (melting ice) to 4°C (fridge) are the most widely used as they are easy to reach and easy to regulate. Besides, such low temperature drastically reduces bacteria growth, and this may explain why temperatures below 6°C are always reported as better than higher temperatures. The best sperm storage temperature is not related to the broodstock rearing temperature. Undiluted sperm of rainbow trout (*Oncorhynchus mykiss*) will maintain its fertilizing capacity for less than one day when stored above 12°C whereas storage temperature of 0-5°C will maintain sperm fertilizing capacity for up to 8 days [3]. Sperm from warm water species such as Nile tilapia or channel catfish (*Ictalurus punctatus*) will also be efficiently stored at 0-4°C [2,4]. This tolerance of fish sperm to low temperatures differs notably from sperm of many mammalian species. Spermatozoa of aquatic species do not suffer cold shock [5,6], possibly because in these poikilothermic species, cellular membranes adapt more readily to temperature changes than in homeotherms. This is why addition of cryoprotectant to cool and store sperm at 4°C is scarcely reported. Christensen and Tiersch [4] showed that addition of methanol 5% allowed channel catfish testicular sperm (*Ictalurus punctatus*) to retain higher motility percentages for a longer storage time than did the buffer alone. However, this positive effect was attributed to a bacteriostatic effect rather than to a cold protection effect. Another indirect effect of cryoprotectant was observed by Ogier de Baulny et al. [7] in European catfish (*Silurus glanis*) where dimethyl acetamide induced a doubling of the ATP content of spermatozoa, a property which could positively improve sperm storage.

Sperm super cooling is sometimes used [8]. This method allows sample preservation at sub-zero temperatures (-2 to -5°C). The presence of cryoprotectant prevents ice crystallization in the medium, therefore avoiding one major damaging event associated with cryopreservation. In Atlantic salmon (*Salmo salar*), 70% fertilization rates were obtained after 38 days at -2°C in ethylene glycol [8]. However, this method requires a very strict control of the temperature to avoid ice crystallization if temperature decreases or cryoprotectant toxicity if temperature increases.

2.2 Duration of storage

The storage duration is mostly dependant on the storage conditions. However, the collection time in the spawning season is known to affect sperm quality in several species [9-13], and it is likely that this initial quality will affect sperm storage ability. Such an effect was observed with red porgy (*Pagrus pagrus*) where maintenance of undiluted sperm viability lasted 4 days at the beginning of the season and more than 12 days 3 months later. There is also an overall species dependence of sperm storage ability. In marine species for example, the longest reported storage ability of turbot (*Scophtalmus maximus*) and stripped bass (*Morone saxatilis*) spermatozoa were of 6-7 days [14,15] whereas some motility activation was still possible after more than 38 days of storage for cod (*Gadus morhua*) and haddock (*Melanogrammus aeglefinus*) spermatozoa [16]. The most impressive paddlefish (*Polyodon spathula*) sperm still retained some motility after 56 days storage [17]. The best results reported for salmonids vary between 4 days (*Oncorhynchus masou* [3]), 14 days (*Salmo trutta fario* [18]), and a striking 34 days for undiluted rainbow trout (*Oncorhynchus mykiss*) sperm [19].

2.3 Gaseous atmosphere and gas exchanges

It is generally recommended to allow a good gaseous exchange between sperm and the atmosphere as anoxia is always deleterious. This is why sperm layers of 2-4 mm are preferred to more fully filled tubes or bags, in order to increase the atmosphere/ sperm ratio and the atmosphere/sperm interface (see Jensen and Alderdice [20] with chum salmon –*Oncorhynchus keta*- spermatozoa). Generally, oxygen flushing in the storage container at the beginning of storage is better than or at least equivalent to air. Improvement of preservation under oxygen was demonstrated for channel catfish (*Ictalurus punctatus*) [4], rainbow trout (*Oncorhynchus mykiss*) [21], deccan mahseer (*Tor khudree*) [22], stripped bass (*Morone saxatilis*) [23], and carp (*Cyprinus carpio*) [24]. For short term storage from few hours to 2-3 days, air only is suitable in salmonids [25]. On the contrary, Chereguini et al. [14] observed that air was better than oxygen for storage of turbot (*Scophtalmus maximus*) spermatozoa. Mansour et al. [26] observed that the most immature samples of testicular African catfish (*Clarias gariepinus*) spermatozoa would be better stored under air than under oxygen whereas oxygen was favorable to storage of more mature and concentrated samples. In all, sperm of most species handles oxygen exposure without obvious oxidative stress. These spermatozoa may not bear mechanisms related to reactive oxygen species production similar to those devoted to capacitation as described in mammalian species [27,28]. Another explanation is that plasma membranes may be enriched with the antioxidants added in the diet. For example, we observed that carotenoids are present in high proportions in membrane lipids of rainbow trout (*Oncorhynchus mykiss*) spermatozoa.

2.4 Dilution and diluent composition

2.4.1 Dilution ratio

One limitation to the sperm dilution is that the medium should be adapted to prevent sperm motility activation in order to preserve the cell energy reserves. The set up of these diluents is difficult when sperm quantities are limited, when mature males are difficult to obtain, or when data on sperm physiology and seminal plasma composition are not available in a given species. This may explain why many studies on sperm storage were developed on undiluted sperm [2,29,30,31,32]. However, one advantage of sperm dilution is that it reduces sperm density, a step which is especially important when dealing with testicular sperm (Cannel catfish -*Ictalurus punctatus*- [4], African Catfish -*Clarias gariepinus*- [26], red drum -*Sciaenops ocellatus*- [33]). Besides, dilution reduces the problems associated with urine contamination [34]. Almost whenever it was tested, sperm dilution improved sperm storage duration which increased from several hours when undiluted to 21-28 day after dilution in the sturgeon *Acipencer oxyrinchus* [35], and from 1 day when undiluted to 6 days after dilution in turbot *Scophtalmus maximus* [14]. Tench (*Tinca tinca*) short term storage (few hours) was possible only on diluted sperm [36], and dilution also improved storage in striped bass (*Morone saxatilis*) [15], cod (*Gadus morhua*), haddock (*Melanogrammus aeglefinus*), and smelt (*Osmerus mordax*) [16,37].

Interestingly, sperm of some species are very sensitive to the dilution ratio, possibly because there is equilibrium between the issues which are favorable (urine dilution, sperm density reduction, pH control) and unfavorable (dilution of some important seminal plasma components) to storage. For sperm from Atlantic cod (*Gadus morhua*), haddock (*Melanogrammus aeglefinus*) and rainbow smelt (*Osmerus mordax*), a sperm:diluent ratio of 1:3 was better than 1:1, 1:2, 1:5 and 1:10 [16,37]. In brown trout (*Salmo trutta fario*) and Chinook salmon (*Oncorhynchus tschawytscha*), dilution above 1:3 decreased sperm quality after a 48h storage compared to 1:1, 1:2 and 1:3 dilutions [38]. In African catfish (*Clarias gariepinus*), 1:5 was better than 1:3 or 1:10 [39]. We believe that the optimal dilution ratio will have to be adapted to the initial sperm density and to the buffering capacity of the diluent.

2.4.2 Characteristics of the diluent

Almost every publication on sperm storage presents a different diluent composition, including within the same species. The best strategy developed by the authors was to approach the seminal fluid composition. This allows that the right ionic composition is present to prevent motility activation and the concomitant rapid exhaustion of cells energy reserves. Ions at the right concentration will also maintain osmotic equilibrium between the medium and the cells.

In salmonid sperm, motility activation is triggered by dilution of the potassium ion [40] and potassium is present in high concentration in the seminal fluid (37.3 mM). The inactivating diluent should contain at least 6mM potassium [41] to prevent motility activation. Most other fresh water and sea water species will only require that the osmolality of the storage diluent is slightly below 300 mOsm/kg for fresh water species (290 for salmonids, 295 for channel catfish -*Ictalurus punctatus*- [4,42]) and slightly above 300 for marine species (310-350 for striped bass -*Morone saxatilis*- [15,43], 320 for cod -*Gadus morhua*- and haddock -*Melanogrammus aeglefinus*- [16]). Interestingly, in oyster *Crassostrea virginica* whose body is exposed to pure sea water, the best osmolality was 833 mOsm/kg [44].

A pH 8 to 8.5 is often used, again because it is close to the seminal fluid values, although carp spermatozoa could also be stored below pH 7 [1]. In salmonids, a pH of 8-9 is absolutely required in order to maintain sperm ability to be activated [45]. Moreover, exposure of rainbow trout (*Oncorhynchus mykiss*) testicular sperm to the same pH as that in the sperm ducts which contain the most mature spermatozoa allowed the immature spermatozoa to acquire potential for motility [46]. This major property of the pH in the diluent was used by Maisse (unreleased data) to set up a commercial diluent (Storfish, IMV) which allows both storage and artificial maturation of testicular sperm in sex-reversed females.

It is suspected that Tris-HCl used as a buffering molecule will be toxic for some cells. This is why Hepes or a Hepes/Bicine combination should be preferred.

Some authors proposed to add protective molecules such as BSA, amino acids, hen yolk [26,43], glutathion [16], sucrose or glucose [16,43]. Although these molecules arc known to protect plasma membranes during cryopreservation, the improvements brought by these molecules with regards to chilled storage were not always experimentally tested. Mansour et al. [26] showed that they did not improve sperm quality after 4 days storage of African catfish (*Clarias gariepinus*) sperm. The utilization of some carbohydrates of the tricarboxylic acid cycle added to the external medium was experimentally demonstrated for several fish species [47,48]. However, their interest as energy source during chilled storage was little explored. The lactate + pyruvate combination sustained adenine nucleotide energy charge of African catfish (*Clarias gariepinus*) sperm [49] although many other tricarboxylic acids were either inefficient or toxic to the cells [26,49].

Last, the removal of seminal plasma from ejaculated sperm by centrifugation prior to dilution is a mean to obtain a constant sperm concentration: a given pellet volume will be resuspended in a given diluent volume. Depending on the sperm morphology, this centrifugation step can be very damaging. Common damage lies in flagella breakage at the basis of the head. This is especially encountered with spermatozoa whose flagella is not inserted deep within the nucleus grove as in several cyprinids [50] and some other freshwater teleosts [51,52]. Some species with a better flagella insertion such as salmonid [53] or some marine teleost fish [54] will be less sensitive to centrifugation. In salmonids, a 250g centrifugation for 20 min followed by a cautiously performed pellet resuspension in the storage diluent (no vortex) will not alter the morphology and quality of spermatozoa.

2.4.3 Antibiotics

Addition of antibiotics either to the undiluted sperm or to the storage diluent always improves storage duration, and this addition can be stated as the most important parameter for chilled storage of sperm. When Paniagua-Chavez et al. [44] tested oyster sperm storage conditions, they could almost predict the quality losses from the amount of bacteria in the samples. The combination penicillin/streptomycin is the antibiotics combination the most widely used. Addition of 50 IU/ml bipenicillin + 50 µg/ml streptomycin to undiluted carp sperm allowed motility and fertilization capacity to be maintained for more than 18 days at 4°C when the control samples lost all fertilization capacity in less than 6 days [24]. The same antibiotics concentration was used for Atlantic cod (*Gadus morhua*) and haddock (*Melanogrammus aeglefinus*) [16]. In Atlantic salmon (*Salmo salar*), higher antibiotics concentrations (125 IU/ml penicillin+125µg/ml streptomycin) were not toxic [55], and they provided the same improvement of storage duration as the one observed in carp. Paddlefish (*Polyodon spathula*) sperm storage was also improved by addition of the antibiotic combination penicillin/streptomycine [17]. In African catfish (*Clarias gariepinus*), however, addition of 25 to 50 IU/ml penicillin + 25 to 50 µg/ml streptomycine did not improve sperm quality during short term storage (4 days) and doses of 100 IU/ml + 100 µg/ml were toxic for the cells whereas addition of gentamycine sulfate at 1mg/ml did improve the motility of these stored sperms. In the same species, Christensen and Tiersch [4] improved storage duration from 3 to 8 days with an antibiotics/antimycotic cocktail containing 100 IU penicillin + 100 µg/ml streptomycine and 0.25µg/ml antimycotic amphotericin. These authors also observed some toxicity in the first few days of storage, but this slight deleterious effect was overcome by the very favorable prevention of bacterial growth.

2.5 Specifics

2.5.1 Post mortem storage

Collection of sperm on dead fish was tested, either after fishing (case of the red tilefish *Branchiostegus japonicus* [56]) or for experimental purposes in order to mimic accidental death of the males in fish farm conditions (rainbow trout -*Oncorhynchus mykiss*-, brown trout -*Salmo trutta fario*-, Chinook salmon -*Oncorhynchus tschawytscha*-, herring -*Clupea harengus*-, mackerel -*Rastrelliger neglectus*- see Billard et al. [18] for a short review). Fertilization rates dropped in the 5 to 12h after the animal death, even though the males were kept at 4°C [18]. The lack of oxygen is proposed to explain these dramatic losses compared to the fairly good preservation obtained when the sperm was collected immediately after death and stored *in vitro*. In red tilefish (*Branchiostegus japonicus*), however, sperm collected in the testis 5 to 8 hours after death could be stored *in vitro* for up to 7 days and still displayed high motility. This work confirms that motility losses induced by anoxia are reversible provided that the cells are thoroughly oxygenated. This reversibility of motility and fertilizing ability losses after oxygen deprivation was experimentally demonstrated in rainbow trout (*Oncorhynchus mykiss*) [57]. The author proposed that the losses in sperm functionality may be connected to a decrease of the seminal fluid pH induced by high levels of carbon dioxide. These data emphasize that a poor initial sperm quality can be overcome by the appropriate storage condition.

2.5.2 Storage prior to cryopreservation

In some cases, chilled storage of sperm is a mean to differ sperm collection from the cryopreservation task. When chilled storage is operated in good conditions and that sperm functionality is not altered, cryopreservation of stored spermatozoa should not impair the cryopreservation outcome. In tilapia (*Oreochromis niloticus*), fertilization rates after sperm freeze-thawing were not influenced by the storage time of undiluted spermatozoa prior to cryopreservation for up to 6 days [2]. A significant decrease was only observed after 8 days storage prior to cryopreservation. Christensen and Tiersch [42] also observed that 5 days storage of diluted channel catfish (*Ictalurus punctatus*) sperm prior to cryopreservation yielded almost the same post-thawing motility than sperm stored for one hour only. On the contrary, sperm storage for 24 and 48 h prior to freezing was not as innocuous in striped bass (*Morone saxatilis*) since post-thaw motility decreased from 50% (fresh sperm cryopreservation) to 30 % (24h pre-freezing storage) and 18 % (48 h pre-freezing storage) [43]. However, these motility losses are still compatible with a high fertilization rate which should not be very different between the groups. In rainbow trout (*Oncorhynchus mykiss*), Maisse (unreleased data) observed that post-thawing motilities of 15% still yielded 70% fertilization rates, and the results were not different between sperm cryopreserved just after collection and sperm cryopreserved after 24h in the best storage conditions (dilution 1:5 with the commercial diluent Storfish, thin layer under oxygen, 4°C). The sperm dilution ratio for a chilled storage prior to cryopreservation has to be carefully considered as the cryopreservation procedure will add another dilution step upon addition of the cryoprotectant. A too high dilution factor of the spermatozoa will require that the straw number is increased proportionally in order to get enough spermatozoa per male at thawing. Therefore, the smallest dilution ratio which will maintain sperm viability must be used for this chilled storage.

2.6 Conclusion and practical recommendations

From the data presented here, some general recommendations can be drawn to anyone attempting sperm storage at 0-4°C with a new species. It is our feeling that the most important parameter is antibiotics addition, preferably a combination of penicillin and streptomycin at concentrations not higher than 50-100 units/ml. When dilution is considered, an analysis of seminal plasma composition and physics (pH, osmolality) is an efficient strategy to set up the diluent composition. Sperm dilution at a sperm:diluent ratio 1:1 to 1:3 is to be tested, as it will generally lengthen storage duration. Higher dilution ratio (1:10) should be suitable for testicular sperm, in order to reduce cell density, and the smallest ratio should be considered when the stored cells are to be cryopreserved. In all, the best initial sperm quality is a prerequisite to an optimal chilled conservation.

3. EGG STORAGE

3.1 Storage time and temperature

In all fish species investigated, eggs released from the ovary at the time of ovulation usually undergo a decrease of their ability to be fertilized and successfully develop into a normal embryo. This phenomenon, also called "overripening" occurs more or less rapidly depending on the species. *In vivo*, this post-ovulatory ageing-induced decrease of egg quality occurs rather rapidly in most fish species. Thus, a significant decrease of egg quality is observed *in vivo* after a period of time ranging between a few hours and a few days [58-60]. In contrast to most other fish species, salmonids can hold their eggs in the body cavity for at least a week without any significant drop of egg quality [58]. It is also noteworthy that, within the same species, egg quality can be highly variable after similar post-ovulatory ageing times, as indicated by the large variability in embryonic survival observed in most studies. This suggests that the overall success of the egg holding procedure could significantly vary between different egg batches being held under similar conditions. Temperature is a determining factor of *in vivo* post-ovulatory ageing of the eggs. In cold water species, such as rainbow trout (*Oncorhynchus mykiss*), high temperature during post-ovulatory ageing have a negative impact on egg quality [58].

Several studies have shown that it is possible to store eggs *in vitro*, in ovarian or coelomic fluid. In rainbow trout (*Oncorhynchus mykiss*), comparable egg qualities were observed when eggs were held for a short period of time *in vivo* in the broodfish or *in vitro* in coelomic fluid under the same temperature [61]. However, egg quality observed after *in vitro* holding is never higher than what is observed *in vivo*. Indeed, egg fertilizability after long holding times are usually lower than what occurs *in vivo* in the broodfish [62]. Most studies have shown that the 2 key factors determining the success of *in vitro* storage in ovarian or coelomic fluid are time and temperature. A few examples of chilled storage of teleost eggs in ovarian or coelomic fluid are discussed below for several fish species.

3.1.1 Salmon

In chum salmon (*Oncorhynchus keta*), the highest fertilization rates were observed when eggs were held at 3°C [20]. A step-wise decrease of egg fertility was observed when temperature was increased to 6, 9, 12 and 15°C [20]. In Sockeye salmon (*Oncorhynchus gorbuscha*), unfertilized eggs stored at 2.9°C remained fertile much longer than eggs stored at 9.9°C [63]. Thus, for eggs stored for 22 hours, the fertilization rate observed was above 90% for both temperatures. In contrast, after 70 hours, the percentage of fertile eggs was above 90% when eggs were held at 2.9°C and below 40% when eggs were held at 9.9°C. In pink salmon, eggs stored at 8.5°C declined in fertility after 8 hours of storage while over 85% of the eggs remained fertile up to 46 hours when held at 3.2°C [63].

3.1.2 Trout

In both rainbow and brown (*Salmon trutta*) trout, unfertilized eggs can be held *in vitro* in coelomic fluid at 0°C for at least 5-7 days [64]. However a decrease of fertilization ability was observed after one day in rainbow trout but not in brown trout. In addition, egg fertilization ability remained constant for 7 days in rainbow trout while it significantly decreased in brown trout after 3 days. Thus, after 5 days at 0°C, over 60% of rainbow trout and brown trout eggs could be fertilized [64]. A similar observation was made in a second study in which unfertilized rainbow trout eggs held in coelomic fluid at 1°C for 5 days exhibited fertilization rates above 80% [65]. In a third study, the fertilization ability of rainbow trout eggs held in coelomic fluid at 1.4°C remained, in average, above 80% for 7 days. In contrast, egg fertilization ability exhibited a dramatic decrease for holding times above 11 days and after 18 days, egg fertilization ability was close to 0% [66]. In this species, it is noteworthy that embryonic survival rates recorded when eggs were held for 5-7 days in coelomic fluid at low temperatures (0-2°C) [64,66] are comparable to survival rates observed when egg were held in the body cavity at 10-12°C for the same time [58,67]. This suggests that chilled storage of unfertilized eggs in coelomic fluid is possible for several days in this species without any major negative effect on subsequent embryonic development. For longer holding times, eggs should preferably remain in the body cavity rather than held *in vitro* to maximize fertilization success.

3.1.3 Tilapia

For warm water species such as tilapia (*Sarotherodon mossambicus*) in which eggs normally develop at high temperature, egg holding at low temperatures prior to fertilization can be a major problem. In this species, unfertilized eggs can successfully be held in coelomic fluid for 1.5 hour at temperatures above 16°C [68]. In contrast, holding at 13°C or below resulted in fertilization rates below 40%. Similarly, eggs held in coelomic fluid for 19 hours exhibited fertilizations rates below 35%, the maximum being obtained at 20-22°C.

3.1.4 Carp

In common carp (*Cyprinus carpio*), short-term (5, 30 and 60 minutes) exposure of non-activated eggs to low temperatures induced a marked decrease of embryonic survival at hatching [69]. For example, after a 60-minute exposure at 0°C, a 28% hatching rate was observed while a 87% hatching rate was observed for control eggs held at 24°C. In another study carried out using koi carps (*Cyprinus carpio*) storage of pre-activated eggs at 7°C resulted in a clear decrease of embryonic survival at hatching for holding times greater that 2 hours [70].

3.1.5 Catfish

In European catfish (*Silurus glanis*), *in vitro* holding of ovulated eggs at 8, 19 and 25°C for 3.5, 8.5 or 12 hours resulted in a dramatic decrease of hatching rates. No survival was observed at 8°C while both 19 and 25°C holding temperatures resulted in low embryonic survival and an increased occurrence of malformed larvae at hatching [62].

3.2. Storage medium

Many attempts have been made to hold unfertilized eggs in artificial medium rather than in biological fluid (e.g., coelomic fluid). These attempts were driven by the need to provide a standard procedure using a controlled medium rather than rely on the availability of ovarian or coelomic fluid. Many studies were conducted in salmonids in which eggs are released into the body cavity at the time of ovulation and are bathed in a liquid called ovarian or coelomic fluid. The composition of coelomic fluid has been studied in several salmonid species [71]. Coelomic fluid pH is around 8.5 and its osmolarity ranges between 250 and 300 mOsm. In rainbow trout, unfertilized eggs can be stored in modified Cortland medium buffered with either Hepes or Tris for 24 or 48 hours at 12-13°C [72]. Embryonic survival rates recorded at hatching were similar in eggs held in artificial medium and control eggs sampled directly from the females. In contrast, in another study conducted using a very close artificial medium, a 3-day *in vitro* storage at 12°C resulted in a dramatic decrease of subsequent developmental success in comparison to eggs held *in vivo* or in coelomic fluid [61]. Together, these studies along with other data in trout Erdahl et al. [39] suggest that *in vitro* holding using very simple artificial medium is possible for salmonid eggs at least for 2 days at 10-12°C without any major effect on subsequent embryonic development. In common carp, several mineral media were successfully used to hold unfertilized eggs at ambient temperature (22-25°C) for 3 hours. Longer holding times resulted in a dramatic decrease of subsequent hatching rates and inter-female variability [73]. In contrast, when holding was carried out at 5°C, subsequent fertilization rate was extremely low (below 5%).

For both coelomic fluid and artificial media, holding procedure should limit any bacterial contamination of the eggs. It is known that trout coelomic fluid has anti-bacterial activity [74]. However, it could be beneficial to filter coelomic fluid before the beginning of holding procedure to avoid any bacterial carryover. This is especially important for long-term storage at relatively high temperature [61]. Similarly, artificial medium used for egg storage usually contains antibiotics [68,72].

Thus, while it seems that egg holding is possible in some specific mineral media (see above), exposing eggs to sperm activating or immobilizing solutions will, in contrast, result in a dramatic decrease of their ability to be fertilized within a few minutes [62]. Indeed, sperm activating solutions will most likely induce egg activation and prevent any further fertilization.

3.3 Atmosphere/Oxygenation

Oxygenation during egg holding procedure can sometimes be used to improve subsequent embryonic development. In tilapia (*Sarotherodon mossambicus*), for example, subsequent embryonic development was significantly increased from 35% to 55% if oxygenation was used at the optimal temperature of 19°C when eggs were stored for 20 hours [68]. Thus, oxygenation is probably important for warm water species in which eggs cannot be held at low temperatures. In addition, it is noteworthy that many experimental data on egg holding in coelomic fluid or mineral medium were obtained using small batches of eggs kept under good oxygenation conditions [61,72]. It is therefore probably important to ensure proper oxygenation for procedures in which large egg batches are used. It is for example possible to use sealed plastic bags containing oxygen-enriched atmosphere.

3.4 Holding and transport

One important factor that needs to be taken into account is how eggs are held during *in vitro* storage. Indeed, the type, size, shape, and volume of the container used for the holding procedure are susceptible to affect subsequent embryonic survival. While many types of containers have been used in the literature, it seems difficult to conclude about the absolute performance of each individual type of container used. However, due to the large variability of egg quality that can sometimes be observed between separate broodfish, it can be beneficial to hold egg batches originating from different broodfish in separate containers to avoid cross batches contamination induced by dead eggs.

A few data exist on post-mortem continuation of egg fertility in fish. In rainbow trout, egg collected from dead females up to 12 hours post-mortem still retained full fertilization ability (fertilization rate above 90%) when female bodies were stored at 4°C outside of the water [18]. Even though this has never been investigated in other fish species, this possibility could be kept in mind for short-time storage if eggs can not be held in ovarian fluid under satisfying conditions (e.g., in the field). It is indeed very easy to keep a dead fish in a cooler at 4°C for a few hours. Alternatively, it is possible to transport or ship the eggs. However, shipping of rainbow trout eggs under non controlled temperature conditions resulted in a marked decrease of subsequent hatching rates [66].

3.5. Original egg quality and fertilization procedure

It is obvious that the intrinsic quality of the egg will be the most determining factor of subsequent development when freshly ovulated eggs are fertilized using optimal fertilization conditions and sperm quality (e.g., motility). Thus, the original quality of the egg is susceptible to having a significant impact on the overall success of the chilled storage procedure. For example, rainbow trout [66] or brown trout [65] eggs of non optimal origin showed lower fertilization ability after chilled storage. In contrast, rainbow trout eggs exhibiting an intermediate quality at the time of collection (approx. 75% fertilization rate) showed very good storage ability at 0°C in coelomic fluid. Indeed, fertilization rates after 7 days were around 70% [64]. Together, these experiments suggest that the intrinsic quality of the egg at the time of collection is not necessarily sufficient to ensure satisfying chilled storage ability. In fact, many breeding factors can induce egg quality defects. It has also been shown that specific egg quality defects could be induced by specific husbandry practices or environmental factors [75]. It is therefore likely that storage ability could be affected by specific factors that could be different from the key factors known or suspected to trigger the quality of freshly ovulated eggs.

Finally, it is also important to use controlled fertilization procedure after holding procedure in order to maximize fertilization success. For example, very simple procedures such as rinsing the eggs before fertilization, checking spermatozoa motility, and using sufficient sperm concentration could be beneficial.

3.6 Conclusions and practical recommendations

In conclusion, it is possible to successfully store unfertilized fish eggs. The maximum storage time will depend on the species as the decrease of the egg ability to be fertilized and subsequently develop into a normal embryo occurs in a species-dependent manner during post-ovulatory ageing. For some fish species, egg storage can be carried out in coelomic or ovarian fluid or for short periods of time in artificial medium.

The use of chilled storage is possible at least in cold-water species. In that case, chilled storage will improve storage timing in comparison to ambient or "normal" water temperature (e.g., 10-12°C for salmonids).

It is also important to keep in mind that the overall success of the chilled storage procedure is sometimes highly variable depending on the original broodfish or the original egg batch; some of this variability probably being explained by intrinsic differences in egg quality or storage ability.

For warm-water species, chilled egg storage is an issue as they normally reproduce in a totally different range of temperatures. For those species, optimal storage temperature, time, and conditions (e.g., oxygenation) have to be investigated in a species-dependent manner.

Finally several practical details such as holding conditions, fertilization procedures, collection time, and transport conditions could also play a significant role in the overall success of the storage procedure.

4. REFERENCES

1-Ravinder, K., Nasaruddin, K., Majumdar, K.C., and Shivaji, S., Computerized analysis of motility, motility patterns and motility parameters of spermatozoa of carp following short-term storage of semen, *J Fish Biol*, 50, 1309, 1997.
2-Rana, K.J., Muiruri, R.M., McAndrew, B.J., and Gilmour, A., The influence of diluents, equilibration time and pre-freezing storage time on the viability of cryopreserved *Oreochromis niloticus* (L.) spermatozoa, *Aquacult Fisheries Manag*, 21, 25, 1990.
3-Scott, A.P. and Baynes, S.M., A review of the biology, handling and storage of salmonid spermatozoa, *J Fish Biol*, 17, 707, 1980.
4-Christensen, J.M. and Tiersch, T.R., Refrigerated storage of channel catfish sperm, *J World Aquacult Soc*, 27, 340, 1996.

5-Drobnis, E.Z., Crowe, L.M., Berger, T., Anchordoguy, T.J., Overstreet, J.W., and Crowe, J.H., Cold shock damage is due to lipid phase transitions in cell membranes: a demonstration using sperm as a model, *J Exp Zool*, 265, 432, 1993.

6-Labbe, C., Crowe, L.M., and Crowe, J.H., Stability of the lipid component of trout sperm plasma membrane during freeze-thawing, *Cryobiology*, 34, 176, 1997.

7-Ogier de Baulny, B., Labbe, C., and Maisse, G., Membrane integrity, mitochondrial activity, ATP content, and motility of the European catfish (*Silurus glanis*) testicular spermatozoa after freezing with different cryoprotectants, *Cryobiology*, 39, 177, 1999.

8-Truscott, B., Idler, D.R., Hoyle, R.J., and Holtz, W., Sub-zero preservation of Atlantic salmon sperm, *J Fish Res Board Can*, 25, 363, 1978.

9-Buyukhatipoglu, S. and Holtz, W., Sperm output in rainbow trout (*Salmo gairdneri*)-effect of age, timing and frequency of stripping and presence of females, *Aquaculture*, 37, 63, 1984.

10-Munkittrick, K.R. and Moccia, R.D., Seasonal changes in the quality of rainbow trout (*Salmo gairdneri*) semen: effect of a delay in stripping on spermatocrit, motility, volume and seminal plasma constituents, *Aquaculture*, 64, 156, 1987.

11-Suquet, M., Dreanno, C., Dorange, G., Normant, Y., Quemener, L., Gaignon, J.L., and Billard, R., The ageing phenomenon of turbot spermatozoa: effects on morphology, motility and concentration, intracellular ATP content, fertilization, and storage capacities, *J Fish Biol*, 52, 31, 1998.

12-Fauvel, C., Savoye, O., Dreanno, C., Cosson, J., and Suquet, M., Characteristics of sperm of captive seabass in relation to its fertilization potential, *J Fish Biol*, 54, 356, 1999.

13-Shangguan, B. and Crim, L.W., Seasonal variations in sperm production and sperm quality in male winter founder, *Pleuronectes americanus*: the effects of hypophysectomy, pituitary replacement therapy, and GnRHa treatment, *Marine Biology*, 134, 19, 1999.

14-Chereguini, O., Cal, R.M., Dreanno, C., Ogier de Baulny, B., Suquet, M., and Maisse, G., Short-term storage and cryopreservation of turbot (*Scophthalmus maximus*) sperm, *Aquat Living Resour*, 10, 251, 1997.

15-Jenkins-Keeran, K. and Woods III, L.C., An Evaluation of Extenders for the Short-Term Storage of Striped Bass Milt, *N Am J Aquacult*, 64, 248, 2002.

16-DeGraaf, J.D. and Berlinsky. D.L., Cryogenic and refrigerated storage of Atlantic cod (*Gadus morhua*) and haddock (*Melanogrammus aeglefinus*) spermatozoa, *Aquaculture*, 234, 527, 2004.

17-Brown, G.G. and Mims, S.D., Storage, transportation, and fertility of undiluted and diluted paddlefish milt, *Prog Fish Cult*, 57, 64, 1995.

18-Billard, R., Marcel, J., and Matei, D., Survie *in vitro* et *post mortem* des gametes de truite fario (*Salmo trutta fario*), *Can J Zool*, 59, 29, 1981.

19-Stoss, J. and Holtz, W., Successful storage of chilled rainbow trout (*Salmo gairdneri*) spermatozoa for up to 34 days, *Aquaculture*, 31, 269, 1983.

20-Jensen, J.O.T. and Alderdice, D.F., Effect of temperature on short-term storage of eggs and sperm of chum salmon (*Oncorhynchus keta*), *Aquaculture*, 37, 251, 1984.

21-Billard, R., Short-term preservation of sperm under oxygen atmosphere in rainbow trout (*Salmo gairdneri*), *Aquaculture*, 23, 287, 1981.

22-Basavaraja, N. and Hegde, S.N., Some characteristics and short-term preservation of spermatozoa of Deccan mahseer, Tor khudree (Sykes), *Aquacult Res*, 36, 422, 2005.

23-Jenkins-Keeran, K., Schreuders, P., Edwards, K., and Woods, L.I., The Effects of Oxygen on the Short-term Storage of Striped Bass Semen, *N Am J Aquacult*, 63, 238, 2001.

24-Saad, A., Billard, R., Theron, M.C., and Hollebecq, M.G., Short-term preservation of carp (*Cyprinus carpio*) semen, *Aquaculture*, 71, 133, 1988.

25-Bencic, D.C., Krisfalusi, M., Cloud, J.G., and Ingermann, R.L., Short-Term Storage of Salmonid Sperm in Air versus Oxygen, *N Am J Aquacult*, 62, 19, 2000.

26-Mansour, N., Lahnsteiner, F., and Berger, B., Characterization of the testicular semen of the African catfish, *Clarias gariepinus* (Burchell, 1822), and its short-term storage, *Aquacult Res*, 35, 232, 2004.

27-de Lamirande, E. and Gagnon, C., Capacitation-associated production of superoxide anion by human spermatozoa, *Free Radic Biol Med*, 18, 487, 1995.

28-Aitken, R.J., Harkiss, D., Knox, W., Paterson, M., and Irvine, D.S., A novel signal transduction cascade in capacitating human spermatozoa characterised by a redox-regulated, cAMP-mediated induction of tyrosine phosphorylation, *J Cell Sci*, 111 (Pt 5), 645, 1998.

29-Jayaprakas, V. and Bimal Lal, T.S., Factors affecting the motility and short-term storage of spermatozoa of the Indian major carps, *Labeo rohita* and *Cirrhinus mrigala*, *J Aquacult Tropics*, 11, 67, 1996.

30-Kang, K.H., Shao, M.Y., Kho, K.H., and Zhang, Z.F., Short-term storage and cryopreservation of *Urechis unicinctus* (Echiura: Urechidae) sperm, *Aquacult Res*, 35, 1195, 2004.

31-Mylonas, C.C., Papadaki, M., and Divanach, P., Seasonal changes in sperm production and quality in the red porgy *Pagrus pagrus* (L.), *Aquacult Res*, 34, 1161, 2003.

32-Rodriguez-Munoz, R. and Ojanguren, A.F., Effect of short-term preservation of sea lamprey gametes on fertilization rate and embryo survival, *J Appl Ichtyo*, 18, 127, 2002.

33-Wayman, W.R., Tiersch, T.R., and Thomas, R.G., Refrigerated storage and cryopreservation of sperm of red drum, *Sciaenops ocellatus* L, *Aquacult Res*, 29, 267, 1998.

34-Dreanno, C., Suquet, M., Desbruyeres, E., Cosson, J., Le Delliou, H., and Billard, R., Effect of urine on semen quality in turbot (*Psetta maxima*), *Aquaculture*, 169, 247, 1998.

35-Park, C. and Chapman, F.A., An Extender Solution for the Short-Term Storage of Sturgeon Semen, *N Am J Aquacult*, 67, 52, 2005.

36-Rodina, M., Cosson, J., Gela, D., and Linhart, O., Kurokura Solution as Immobilizing Medium for Spermatozoa of Tench (*Tinca tinca L.*), *Aquacult Int*, 12, 119, 2004.

37-DeGraaf, J.D. and Berlinsky, D.L., Cryogenic and Refrigerated Storage of Rainbow Smelt *Osmerus mordax* Spermatozoa, *J World Aquacult Soc*, 35, 209, 2004.

38-Erdahl, A.W., Erdahl, D.A., and Graham, E.F., Some factors affecting the preservation of salmonid spermatozoa, *Aquaculture*, 43, 341, 1984.

39-Erdahl, A.W., Cloud, J.G., and Graham, E.F., Fertility of rainbow trout (*Salmo gairdneri*) gametes: Gamete viability in artificial media, *Aquaculture*, 60, 323, 1987.

40-Morisawa, M. and Suzuki, K., Osmolality and potassium ion: Their roles in initiation of sperm motility in teleosts, *Science*, 210, 1145, 1980.

41-Morisawa, M., Okuno, M., Suzuki, K., Morisawa, S., and Ishida, K., Initiation of sperm motility in teleosts, *J Submicrosc Cytol*, 15, 61, 1983.

42-Christensen, J.M. and Tiersch, T.R., Cryopreservation of channel catfish sperm: effects of cryoprotectant exposure time, cooling rate, thawing conditions, and male-to-male variation, *Theriogenology*, 63, 2103, 2005.

43-He, S. and Woods, L.I., Effects of glycine and alanine on short-term storage and cryopreservation of striped bass (*Morone saxatilis*) spermatozoa, *Cryobiology*, 46, 17, 2003.

44-Paniagua-Chavez, C.G., Buchanan, J.T., and Tiersch, T.R., Effect of extender solutions and dilution on motility and fertilizing ability of eastern oyster sperm, *J Shellfish Res*, 17, 231, 1998.

45-Baynes, S.M., Scott, A.P., and Dawson, A.P., Rainbow trout, *Salmo gairdnerii* Richardson, spermatozoa: effects of cations and pH on motility, *J Fish Biol*, 19, 259, 1981.

46-Morisawa, S. and Morisawa, M., Induction of potential for sperm motility by bicarbonate and pH in rainbow trout and chum salmon, *J Exp Biol*, 136, 13, 1988.

47-Terner, C. and Korsh, G., The oxydative metabolism of pyruvate, acetate and glucose in isolated fish spermatozoa, *J Cell Comp Physiol*, 62, 243, 1963.

48-Mounib, M.S., Metabolism of pyruvate, acetate and glyoxylate by fish sperm, *Comp Biochem Physiol*, 20, 987, 1967.

49 Zietara, M S, Slominska, E., Swierczynski, J, Rurangwa, F, Ollevier, F., and Skorkowski, E,F, ATP Content and Adenine Nucleotide Catabolism in African Catfish Spermatozoa Stored in Various Energetic Substrates, *Fish Physiol Biochem*, 30, 119, 2004.

50-Ohta, T., Mizuno, T., and Mizutani, M.M.M., Sperm morphology and IMP distribution in membranes of spermatozoa of cyprinid fishes, *J Submicrosc Cytol Pathol*, 26, 181, 1994.

51-Ohta, T., Kato, K.H., Abe, T., and Takeuchi, T., Sperm morphology and distribution of intramembranous particles in the sperm heads of selected freshwater teleosts, *Tissue Cell*, 25, 725, 1993.

52-Gwo, J.C., Kao, Y.S., Lin, X.W., Chang, S.L., and Su, M-S., The ultrastructure of milkfish, *Chanos chanos* (Forsskal), spermatozoon (Teleostei, Gonorynchyformes, Chanidae), *J Submicrosc Cytol Pathol*, 27, 99, 1995.

53-Billard, R., Ultrastructure of trout spermatozoa: Changes after dilution and deep-freezing, *Cell Tissue Res*, 228, 205, 1983.

54-Lahnsteiner, F. and Patzner, R.A., Fine structure of spermatozoa of two marine teleost fishes, the red mullet, *Mullus barbatus* (Mullidae) and the white sea bream, *Diplodus sargus* (Sparidae), J., *Submicrosc Cytol Pathol*, 27, 259, 1995.

55-Stoss, J. and Refstie, T., Short-term storage and cryopreservation of milt from Atlantic salmon and sea trout, *Aquaculture*, 30, 229, 1983.

56-Fujinami, Y., Takeuchi, H., Tsuzaki, T., and Ohta, H., Sperm motility and short term preservation of testicular spermatozoa obtained from captured and dead red tilefish *Branchiostegus japonicas*, *B Jpn Soc Sci Fish*, 69, 162, 2003.

57-Bencic, D.C., Cloud, J.G., and Ingermann, R.L., Carbon dioxide reversibly inhibits sperm motility and fertilizing ability in steelhead (*Oncorhynchus mykiss*), *Fish Physiol Biochem*, 23, 275, 2000.

58-Aegerter, S. and Jalabert, B., Effects of post-ovulatory oocyte ageing and temperature on egg quality and on the occurrence of triploid fry in rainbow trout, *Oncorhynchus mykiss*, *Aquaculture*, 231, 59, 2004.

59-Bromage, N., Bruce, M., Basavaraja, N., Rana, K., Shields, R., Young, C., Dye, J., Smith, P., Gillespie, M., and Gamble, J., Egg quality determinants in finfish: The role of overripening with special reference to the timing of stripping in the Atlantic halibut *Hippoglossus hippoglossus*, *J World Aquacult Soc*, 25, 13, 1994.

60-Hirose, K., Machida, Y., and Donaldson, E.M., Induced ovulation of Japanes flounder (*Limanda yokohoma*) with HCG and salmon gonadotropin, with special references to changes in the quality of eggs retained in the ovarian cavity after ovulation, *Bull Jap Sc Fish*, 45, 31, 1979.

61-Bonnet, E., Jalabert, B., and Bobe, J., A 3-day *in vitro* storage of rainbow trout (Oncorhynchus mykiss) unfertilised eggs in coelomic fluid at 12 degrees C does not affect developmental success, *Cybium*, 27, 47, 2003.

62-Linhart, O. and Billard, R., Survival of ovulated oocyte of the European catfish (*Silurus glanis*) after *in vivo* and *in vitro* storage or exposure to saline solutions and urine, *Aquat Living Resour*, 8, 317, 1995.

63-Withler, F.C. and Morley, R.B., Effects of chilled storage on viability of stored ova and sperm of sockeye and pink salmon, *J Fish Res Board Can*, 25, 2695, 1968.

64-Carpentier, P. and Billard, R., Conservation à court terme des gamètes de Salmonidés à des températures voisines de 0°C, *Ann Biol Anim Biochim Biophys*, 18, 1083, 1978.

65-Billard, R. and Gillet, C., Ageing of eggs and temperature potentialization of micropolluant effects of the aquatic medium on trout gametes, *Cahier du Laboratoire de Montereau*, 12, 35, 1981.

66-Babiak, I. and Dabrowski, K., Refrigeration of rainbow trout gametes and embryos, *J Exp Zoolog A Comp Exp Biol*, 300, 140, 2003.

67-Springate, J.R.C., Bromage, N.R., Elliott, J.A.K., and Hudson, D.L., The timing of ovulation and stripping and their effects on the rates of fertilization and survival to eying, hatch and swim-up in the rainbow trout (*Salmo gairdneri* R.), *Aquaculture*, 43, 313, 1984.

68-Harvey, B. and Kelley, R.N., Short-term storage of *Sarotherodon mossambicus* ova, *Aquaculture*, 37, 391, 1984.

69-Dinnyes, A., Urbanyi, B., Baranyai, B., and Magyary, I., Chilling sensitivity of carp (*Cyprinus carpio*) embryos at different developmental stages in the presence or absence of cryoprotectants: Work in progress, *Theriogenology*, 50, 1, 1998.

70-Rothbard, S., Rubinshtein, I. and Gelman, E., Storage of common carp, *Cyprinus carpio* L., eggs for short durations, *Aquacult Res*, 27, 175, 1996.

71-Lahnsteiner, F., Weismann, T., and Patzner, R.A., Composition of the ovarian fluid in 4 salmonid species: *Oncorhynchus mykiss*, *Salmo trutta* f *lacustris*, *Salvelinus alpinus* and *Hucho huch*, *Reprod Nutr Dev*, 35, 465, 1995.

72-Goetz, F.W. and Coffman, M.A., Storage of unfertilized eggs of rainbow trout (*Oncorhynchus mykiss*) in artificial media, *Aquaculture*, 184, 267, 2000.

73-Glenn, D.W. and Tiersch, T.R., Effect of extenders and osmotic pressure on storage of eggs of ornamental common carp *Cyprinus carpio* at ambient and refrigerated temperatures, *J World Aquacult Soc*, 33, 254, 2002.

74-Coffman, M.A., Pinter, J.H., and Goetz, F.W., Trout ovulatory proteins: site of synthesis, regulation, and possible biological function, *Biol Reprod*, 62, 928, 2000.

75-Bonnet, E., Fostier, A., and Bobe, J., Characterization of rainbow trout egg quality: A case study using four different breeding protocols, with emphasis on the incidence of embryonic malformations, *Theriogenology*, 67, 786, 2007.

Complete affiliation:

Catherine Labbe, Fish Reproduction, INRA, UR 1037 SCRIBE, Campus de Beaulieu, F-35000 Rennes. e-mail: Catherine.Labbe@rennes.inra.fr

CHAPTER 7
BASIC PRINCIPLES OF FISH SPERMATOZOA CRYOPRESERVATION

J. Cloud and S. Patton

1. INTRODUCTION

Genetic conservation of existing fish stocks is an important goal in itself, and as a component of programs designed to insure a viable and sustainable fishery under changing environmental conditions. With the constant threat of losing genetic diversity in specific native fish stocks, the establishment of a program for the long-term storage of fish germ plasm would serve as a back-up or insurance for the presently ongoing conservation programs. Additionally, the preservation of specific stocks of wild fish will provide a genetic resource that may be important to future animal breeding programs of the aquaculture industry.

At present, the cryopreservation of sperm is the simplest means of storing fish germ plasm for extended periods of time. From the data of Ashwood-Smith [1] and Whittingham [2], Stoss [3] has estimated storage time for fish sperm held in liquid nitrogen to be between 200 and 32,000 years. Therefore, the time scale for the storage period is more than adequate for a germ plasm repository. The successful cryopreservation of spermatozoa from a wide variety of marine and freshwater teleost fish species [4-21] adequately demonstrate that cryopreservation of fish spermatozoa is a functional procedure and that the methodology can be utilized directly, or modified, to preserve the spermatozoa of numerous fish populations. In addition to the establishment of germ plasm repositories, this extended storage of viable fish spermatozoa can also be utilized as a component of programs producing disease-free stocks and in breeding programs in which the spawning times of males and females are not synchronized.

There are two important caveats in the development of a fish sperm bank. First, this is a genetic repository, and it will not solve any population problems of a fish stock that is decreasing, nor will it directly result in more fish in the oceans and rivers. What a sperm bank will do is guarantee that the genes, or combination of genes, that make a fish stock unique will not be lost forever and that these genes are in a form that can be easily incorporated into the germline of a species through the fertilization of eggs. Second, the cryopreservation of sperm is a potentially destructive process; therefore, the fertility of the frozen/thawed sperm will not be greater than the fertility of the starting material. Indeed, the quality of thawed sperm is usually a reflection of the quality of the sperm that was cryopreserved. Since space in liquid nitrogen storage facilities is finite and sperm cryopreservation is labor intensive, the manager of a germ plasm repository has to consider whether to save/cryopreserve poor quality milt (is the genetics of some males sufficiently unique that their milt should be stored knowing that few offspring will result ?) and needs to guard against saving a large quantity of milt from a single male because of the availability of a large volume of high quality milt.

2. CRYOPRESERVATION OF CELLS: AN OVERVIEW

Essentially there are four time periods in the cooling sequence associated with the cryopreservation of cells: (1) cooling the cells to the point of ice formation, (2) the formation of ice (with the associated heat of fusion), (3) cooling through the critical period (from about -10°C to -40°C), and (4) reduction to liquid nitrogen temperature. For some cells, the initial cooling from an ambient temperature to below zero degrees Celsius is an important consideration in the cryopreservation process. With most if not all fish sperm, the initial reduction in temperature does not appear to be a factor in the success of the cryopreservation process and will not be dealt with further; however, it must be realized cells should not be exposed to a series of temperature changes (all glassware and containers should be at the same temperature as the cells). Because the cells to be cryopreserved are in a solution containing salts and a cryoprotectant, the freezing point of the medium surrounding the cells is below zero (0°C). As the cells and their surrounding media are cooled, ice forms in the media outside the cells. With this formation of ice, the temperature of the cells and their immediate surroundings will increase because of the heat of fusion of water. The goal at this stage of the process is to have ice form near the freezing point of the extracellular solution (as opposed to supercooling the solution) and minimize the temperature change that occurs as a result of the heat of fusion.

During the critical third phase, there is a net movement of water out of the cells as the temperature is being reduced. With this formation of ice outside the cells, the amount of liquid water or solvent decreases with a resultant increase in the osmotic pressure of the extracellular fluid. Because of the difference in the osmotic pressures of the extracellular and intracellular environments, water moves out of the cell. With the net movement of water out of the cells, the intracellular salt concentration increases. The potentially detrimental effects of a higher intracellular level of salts is offset, or diminished, by the reduction in temperature. With the continued decrease in temperature, more ice is formed in the extracellular space, the osmotic pressure of the extracellular fluid increases further, and there is a continued net flow of water out of the cell. Therefore, the cooling rate during this phase needs to be slow enough to allow water to move out of the cells (if the rate of cooling is too fast, ice crystals will form inside the cell), but it must be fast enough to protect the intracellular environment from the effect of high salt concentrations.

The success of cryopreservation is dependent upon cryoprotectants, such as dimethyl sulfoxide (DMSO), glycerol, or methanol in the freezing solution. These small compounds enter the cells and are presumed to protect the cells during the dehydration process. These compounds are absolutely necessary for successful cryopreservation; however, these compounds can be toxic at concentrations required as a cryoprotectant, and may need to be added slowly at a cool temperature. The type of cryoprotective agent (DMSO, glycerol, or methanol) that is best is generally species specific; the different cryoprotectants are usually not interchangeable. However, although DMSO is the cryoprotective agent normally associated with salmonid sperm, Lahnsteiner et al. [22] and Jodun et al. [23] have demonstrated methanol to be an effective cryoprotectant for salmonids.

Other large molecules or components, such as bovine serum albumin or egg yolk, are added to the freezing solution to act as a membrane stabilizer. These components are generally benign.

The rate at which semen is cooled is a critical factor in the success of the cryopreservation process; for salmonid sperm, cooling rates of -20 to -30° C/min appear to be optimal (reviewed by Stoss, [3]). Therefore, when establishing or modifying a protocol it is helpful to monitor temperature changes within the straw. This is easily done by threading a small thermocouple into a filled (but unsealed) straw. By attaching a chart recorder to the thermocouple monitor, a cooling curve of the fluid in the straw can easily be obtained.

3. COLLECTING AND SHIPPING MILT

3.1 Collection

The following is a description of methods that we have used to handle milt from salmonids. These methods may need to be slightly modified for other species of fish.

Care should be exercised when collecting milt to ensure the highest quality possible. Males, normally anesthetized with tricaine methanosulfanate (MS-222), should be thoroughly dried and gently stripped to reduce or eliminate milt contamination from fecal matter or urine. Individual milt specimens should be placed in an appropriate container and gassed with oxygen to aid cell respiration. In our experience, plastic bags are convenient and easy to use; the bags should be placed on their sides if possible to maximize milt surface area and the exchange of gases. Oxygenated milt samples should be stored on ice (ideally maintained at 0-4° C) in an insulated ice chest for shipment to laboratory. Since ice is a solid at a temperature equal to the temperature of the freezer from which it was removed, milt should not be placed directly on ice as this may damage cells from an excessive temperature drop. A simple precaution is to use newspaper as insulating layer between the ice and milt samples.

3.2 Shipping

Sperm from many fishes will remain viable for an extended period of time after collection and as a result can be shipped easily. As described above, milt should be shipped at 0-4° C in individual containers filled with oxygen. The key to shipping milt is to remember sperm can be damaged by freezing temperatures and packaging sperm in direct contact with ice or cold pack will damage the cells. The best method is to pack the shipping container so ice is melting over the bags of milt, or provide an insulating layer between the ice/cold pack and samples to be cryopreserved.

4. ESTIMATING SPERM MOTILITY

Milt evaluation is an important component of a cryopreservation program in order to (1) cull poor quality milt samples prior to freezing and (2) to estimate the fertility of the stored sperm post-thaw. Reasons for culling poor quality milt are that the cryopreservation process is costly in terms of personnel time and materials, that space in liquid nitrogen storage tanks is finite, and that eggs to be fertilized by cryopreserved semen may be valuable. Current methods used to estimate fertility evaluation that will be described are (1) fertilization rate and (2) sperm motility.

4.1 Fertilization rate

The proportion of eggs fertilized by a given number of spermatozoa or by a standardized volume of milt has been used by numerous investigators to evaluate success of sperm cryopreservation. In most of these studies, the fertilization rate was determined by calculating the proportion of embryos that successfully developed to retinal pigmentation (eye-up) or that hatched relative to the number of eggs to which sperm was added. While these endpoints are useful for some species, the time required to reach these developmental endpoints restricts usage in other species such as the salmonids. As an alternative for those species with an extended embryology, the proportion of eggs successfully fertilized can be diagnosed after 12-24 hours of incubation by using the initial cleavage divisions as an endpoint. During this time, fertilization will result in the formation of the blastodisc and the completion of the first or second cleavage division (see image below). To more clearly visualize the cleavage furrows of the blastoderm, the embryo is treated with a solution to clear the chorion (fix the fertilized eggs with 1% acetic acid in physiological saline or Stockard's solution). If fertilization has been successful, the blastodisc will have undergone first cleavage division during this 24 hour time period. Although this procedure requires a good quality stereomicroscope and a little experience, the savings in time and incubator space is well worth the effort.

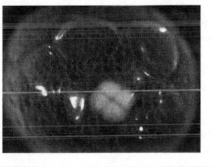

Figure 1 - 4-cell embryo (second cleavage) at approximately 10 hours post-fertilization with incubation temperature of 10°C.

4.2 Sperm motility

The successful estimation of salmonid sperm motility requires speed and experience. Sperm in seminal plasma is nonmotile and only gains motility upon activation. Although sperm can be activated by hatchery water (hypo-osmotic solution), a better choice is an iso-osmotic sperm activating solution (125 mM NaCl, 0.1 mM $CaCl_2$, 30 mM Tris-HCl, pH 9.2; [24]) at a dilution of 20 to 1. To make an estimate, place 200 µl of sperm activating solution on a microscope slide (have it centered so that it will be within the field of view when quickly moved into place; also have the microscope focused on the activation solution); add 10 µl of milt, quickly mix and view. Activated sperm should appear as a swirling mass. Start by determining if 50% or more of the cells are motile. If it appears as though greater than 50% of the sperm are motile, determine if sperm motility is greater than 75% (you may want to examine more than one sample to get an estimate of motility for which you are confident). As it turns out, sperm samples that have very high or very low motility are relatively easy to score; it is intermediate samples that are more difficult. For a more detailed description of this sperm motility assay see Terner [25].

When optimizing cryopreservation methods for different species, a more precise sperm motility assay may be necessary. While the aforementioned motility assay is simple and inexpensive, it is not a precise quantitative measure. Computer-assisted sperm analysis (CASA) is a more objective procedure. While CASA requires specialized equipment and training, it is currently the most objective and comprehensive quantification method for rapid assessment of the proportion of sperm that are motile, their velocity, and other sperm parameters that can be used to predict the fertility of semen.

As a means to reduce the cost associated with automated commercial options of CASA, Wilson-Leedy and Ingermann [26] have described an open source CASA software program. This program is available through the National Institutes of Health software Image J and is available with documentation at http://rsb.info.nih.gov/ij/plugins/casa.html. Using this system, motion of sperm can be characterized relative to time post-activation and the impact of acquisition conditions upon data analysis determined. Listed below (Table 1) are motility parameters available for examination with the open source ImageJ plugin.

See a detailed description of CASA sperm motility assays in Kime et al. [27] and Wilson-Leedy and Ingermann [26].

Sperm motility is one parameter to use in the culling process. Our experience indicates there is a precipitous drop in post-thaw motility of salmonid sperm when the motility of fresh semen is below 50%. For this reason, unless the sperm is from a valuable male where there is need to store his genetics, sperm with less than 50% motility is not banked.

5. THE FREEZING SOLUTION

In order to cryopreserve sperm, milt is normally diluted 1:3 with a freezing solution that is comprised of two components: 1) a glucose based extender with osmotic properties of seminal plasma containing a membrane stabilizer and buffering agent, and 2) a cryoprotective agent that prevents ice crystals formation during freezing.

5.1 Milt extenders

A milt extender is a physiological solution that does not induce sperm activation. The milt extenders used in freezing solutions have ranged in complexity from a solution mimicking the composition of seminal plasma to a simple glucose solution. Investigations using sperm from freshwater or marine species have demonstrated that the composition of this solution is an important factor in post-thaw fertility of sperm, and that there does not appear to be a single extender that maximizes fertility for all species. The literature shows that more complicated solutions are not necessarily the best.

Table 1: Sperm motility characteristics calculated by the ImageJ plugin.

Percent motility	Percent of tracked sperm identified as exhibiting motility during the 1 s period of analysis
Velocity curvilinear (VCL)	Total point to point distance traveled by the sperm over the time period analyzed to a per second value
Velocity average path (VAP)	Velocity over an average path, generated by a roaming average of sperm position from one-sixth of the video's frame rate, such that each point is generated by averaging the coordinates of a set number of locations on the VCL path
Velocity straight line (VSL)	Maximum distance moved on the VAP path by the sperm from the first VAP point during the video segment analyzed
Linearity (LIN)	Measure of path curvature determined by dividing VSL by VAP
Straightness (STR)	Measure of VCL side to side movement determined by dividing VAP by VCL
Duration estimate	(Percent motility at 15 s post-activation-percent motility at 55 s post-activation)/percent motility at 15 s post-activation

5.2 Cryoprotective agents

The freezing solution must contain a cryoprotective agent (see Mazur [28,29] for a detailed discussion of the need and activity of cryoprotective agents). These intracellular cryoprotectants are small compounds; because they are capable of entering sperm cells quickly and because of the small size of fish sperm, there is no need for an equilibration period. As indicated before, cryoprotectants are required for survival during the freezing and thawing phases of cryopreservation; however, these compounds can be toxic to cells and should be added to the milt at a reduced temperature. The final concentration of these compounds in the freezing solution needs to be optimized to afford the greatest protection to the cells during freezing and thawing while minimizing toxicity. To date, the specific type and concentration of cryoprotective agent that provides the maximum protection, and is least toxic, is ultimately identified by fertility testing extended milt before and after freezing.

5.3 Buffers

Buffering agents have been shown generally to greatly improve fertility of sperm post-thaw. A reduction in extracellular pH has a significant inhibitory effect on the capacity of salmonid sperm to become motile when activated [30].

5.4 Membrane stabilizers

A number of large compounds or biological mixtures have been included in the formulation of freezing solutions with the assumption that they are beneficial in protecting or stabilizing the extra-cellular surface of spermatozoa during the freezing process. Avian egg yolk has been one of these additives [10,31], and Baynes and Scott [31] provide clear evidence that this additive (5 to 20% egg yolk) provides a significant increase in post-thaw fertility. Other compounds that have been used as a membrane stabilizer for salmonid milt include bovine serum albumin and promine D [32].

5.5 Producing a simple freezing solution

This recipe will make 100 mL of a 300 mM glucose, 30 mM Tris based freezing solution; to make larger volumes increase the proportions accordingly. The glucose solution can be made ahead of time and frozen in convenient aliquots.

1- Pour 75 ml of distilled water into a 100 ml graduated cylinder, add 5.40 g glucose (dextrose) and 0.363 g Tris. Stir glucose and Tris until completely dissolved; bring the pH of the solution to 8.5.
2- Separate the egg white from the yolk of two eggs. When all the egg white has been removed, place the yolk on a large piece of filter paper (24 cm diameter). Gently rupture the membrane (the vitelline membrane and any egg white that was not removed should adhere to the filter paper) and allow the yolk to flow into a receptacle.
3- Add 13.3 ml of the egg yolk to the 75 ml glucose/Tris based freezing solution in the 100 ml graduated cylinder. Using a pipet, add 10 mL of DMSO to the glucose/Tris freezing solution and mix. Finally, add enough distilled water to make 100 mL total. Chill solution to approximately 4° C before using.

6. SEMEN STRAWS: PACKAGING AND IDENTIFICATION

Animal scientists have used polyethylene straws with a capacity of 0.25 mL (bovine mini straw) or 0.5 mL (bovine medium straw) to preserve semen of different species of domestic livestock for decades. The extended semen is drawn up into the straw and the ends are sealed. Because of the geometrical structure of the straw, the extended semen will equilibrate with the temperature of its environment very quickly; therefore, the semen in the straws will freeze and thaw readily. Because of their usage in animal husbandry, these semen straws are commercially available, can be easily labeled and can be stored efficiently. We have routinely used medium straws with a plugged end (Fig. 2).

Figure 2 - 0.5ml straws used for loading the sperm and freezing in liquid nitrogen.

We have found that it is helpful to provide a complete description of the origin of the milt in each straw. We routinely include the identification number of the male, year of collection/cryopreservation and an identification of the source of the male. Straws are printed prior to filling and are placed in the refrigerator at the same temperature as the semen and freezing solution.

6.1 Manual filling system

There are numerous automated systems available that will label, fill and seal semen straws. However, even though these instruments are very efficient and reliable, many hatcheries that may want to cryopreserve and store milt will find the investment in these automated systems beyond their budgets. Therefore, we will describe the methodology for a manual filling system.

The basic procedure used is a system originally developed for bovine semen. Ideally, this filling procedure should be conducted in a cold room at 4 ° C; however, this condition is not always available. As a minimum, all materials and containers should be stored in the refrigerator at 4° C until used, and the filling procedure conducted on a tray of ice.

Disease: The procedures described below are designed with the assumption the milt of any one male may contain a disease or organism. Therefore, nothing that was used to measure or hold milt of one male is reused for another male. Likewise, it is prudent to wash hands periodically and clean any spills during the course of the filling process.

Materials
Tubes (disposable or autoclaved) in a rack
0.5 mL semen straws
Vacuum pump or vacuum line
Filling nozzle (15 pin for medium straws)
Straw clips (1 for every 15 straws)
- Bubblers with troughs (1 bubbler and trough for each sample)

- Bubbler stand
- Sealant

Procedure:

1. Start by precooling the following:
    i) 1 ml disposable pipet for each male
    ii) 1 bubbler and bubbler trough for each male
    iii) Straws/clips (each clip holds 15 straws; each straw holds 0.5 ml semen/ freezing solution) to accommodate the milt to be cryopreserved for each male. When assembling the straws in each clip, orient the straws with the plugged ends in the same direction.
    iv) Canes and goblets for each male (should be prelabeled)
2. Connect the 15-straw filling nozzle to the vacuum pump.
3. Assemble the bubbler and trough in the bubbler stand.
4. Transfer the semen from one male to a tube. For every 1 ml of semen, add 3 ml of freezing solution (the freezing solution should be added slowly, and the tube should be agitated to facilitate mixing).
5. Thoroughly mix the semen and freezing solution.
6. Pour the semen/freezing solution mixture into the bubbler trough.
7. Place the end of the straws into the bubbler trough; attach the presealed ends of ten straws in a clip to the filling nozzle and draw the freezing solution mixture into the straws (when the mixture reaches the sealing power in the end, the liquid will stop rising in the straw, see Figure 3 below)

Figure 3 - Mechanism used for loading the sperm into straws.

8. When straws fill to the presealed end, remove the straws from the nozzle. Do not invert.
9. Place the straws onto the comb of the bubbler.

10. Remove the straws and fill the open end of the straws with sealing powder (see Figure 4 below).

11. Remove straws from the clip and wipe off excessive sealant from the straws.

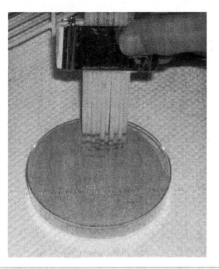

Figure 4 - Sealing powder (polivinil) used for sealing the straws.

6.2 Filling individual straws

A small number of straws can be filled using a syringe fitted with a piece of tubing (this procedure can be used to load straws without investing in the filling equipment). The tubing is attached to the plugged end of the straw, and the milt/freezing solution is drawn into the straw using the syringe, leaving an air space at the end for expansion during the freezing process. Carefully remove the straw from tubing by handling the straw by the plugged end and tamp sealing powder into the open end. Dip the unplugged end (that now has sealant in it) into a bit of water to insure the sealant has been moistened. Place these filled straws on the cooled dish until they are ready to be frozen (no more than five minutes).

7. METHODS OF FREEZING MILT

One of the more critical factors in the cryopreservation of cells is the rate at which the cells are cooled. The freeze rate must be slow enough to allow the movement of water out of the cells so that ice crystals do not form intracellularly and rapid enough so the increase in intracellular salt concentration does not damage the cellular components [28,29]. In practice, the most successful and practical method of freezing fish spermatozoa is the two step method. The first step is to reduce the temperature of the sperm from storage temperature (i.e., the normal storage temperature for salmonid spermatozoa is 4° C) to approximately – 70° C. The optimal rate of cooling for spermatozoa for this step varies among species. The second step is to plunge the frozen milt into liquid nitrogen at -196° C.

1. Straw rack/individual straws

After the straws are filled and sealed, they are placed on a straw rack; the loaded straw rack is then placed in liquid nitrogen vapor for 15 minutes. The distance above the liquid nitrogen will determine the temperature at that position and the resultant rate of cooling of the milt within the straws. Following this step, the straws are plunged into liquid nitrogen, loaded into goblets/canes, and transferred to a liquid nitrogen dewar for storage.

The optimal cooling rate is species specific. By varying the distance above the level of the liquid nitrogen, the temperature at which the straws are exposed and the cooling rate can be varied.

2. Straw rack/canes

Because handling liquid nitrogen presents some risk of injury and because fish hatcheries may not be fully equipped to handle cryopreserved materials, an alternative strategy is to freeze the straws in the goblets. After filling and sealing, straws are placed into 10 mm goblets (5/goblet). The goblets are placed on canes and transferred into liquid nitrogen vapor. Following the initial freezing, the canes are plunged into liquid nitrogen and transferred to liquid nitrogen dewars for storage. One of the big advantages of this method is that there is less handling of the straws after they have been frozen.

8. FROZEN MILT INVENTORY/RETRIEVAL

As indicated earlier, to have the greatest confidence in the system, each individual straw should be labeled to insure the identity of the genetic material inside. In addition, the goblet into which the straws will be placed should also be labeled in a similar fashion. For easier identification, each cane should contain straws from a single male. Each cane is identified using a colored tab attached to the top end; these tabs are labeled with identification and are color coded. The canes are generally filled starting with the lower goblet and then the upper goblet (the upper goblet must be placed on the cane so that the straws in the lower goblet do not float away). A cane that is partially filled can be identified by bending the tab to an upright position; as a result, straws for a male can be retrieved from the partially canes first.

A log book of the inventory should contain the identification number of the male, the date it was frozen, and a description of where the sample was obtained. It should also include the liquid nitrogen dewar and the canister number in which the canes are located.

9. REFERENCES

1-Ashwood-Smith, M.J., Low temperature preservation of cells, tissues and Organs, in: *Low Temperature Preservation in Medicine and Biology,* Ashwood-Smith, M.J. and Farrant, J., Eds., Pitman Medical Ltd., Tunbridge Wells, Kent, 1980, 19.

2-Whittingham, D.G., Principles of embryo preservation, in: *Low Temperature Preservation in Medicine and Biology,* Ashwood-Smith, M.J. and Farrant, J., Eds., Pitman Medical Lts. Tunbrdge Wells, Kent, 1980, 65.

3 Stoss, J., Fish gamete preservation and spermatozoan physiology, in: *Fish Physiol, Vol 9, Part B,* Hoar, W.S., Randell, D.J., and Donaldson, E.M., Eds., Academic Press, New York, 1983, 305.

4-Blaxter, J.H.S., Sperm storage and cross-fertilization of spring and autumn spawning herring, *Nature,* 172, 1189, 1953.

5-Graybill, J.R. and Horton, H.F., Limited fertilization of steelhead trout eggs with cryo-preserved sperm, *J. Fish. Res. Bd. Can.,* 26, 1400, 1969.

6 Ott, A.G. and Horton, H.F., Fertilization of Chinook and coho salmon eggs with cryo-preserved sperm, *J Fish Res Bd Can,* 28, 745, 1971a.

7 Ott, A.G. and Horton, H.F., Fertilization of steelhead trout (*Salmo gairdneri*) eggs with cryo-preserved sperm, *J Fish Res Bd Can,* 28, 1915, 1971b.

8-Chao, N.H., Chen, H.P., and Liao, I.C., Study on cryogenic preservation of grey mullet sperm, *Aquaculture,* 5, 389, 1975.

9-Erdahl, D.A. and Grahm, E.F., Cryopreservation of spermatozoa of the brown, brook, and rainbow trout. *Cryo-Letters* 1, 203, 1980.

10-Legendre, M. and Billard, R., Cryopreservation of rainbow trout sperm by deep-freezing, *Reprod Nutr Dev,* 20, 1859, 1980.

11 Stoss, J. and Holtz, W., Cryopreservation of rainbow trout (*Salmo gairdneri*) sperm. I. Effect of thawing solution, sperm density and interval between thawing and insemination, *Aquaculture,* 22, 97, 1981.

12-Hara, S., Canto, J.T., and Almendras, J.M.E., A comparative study of various extenders for milkfish, *Chanos chanos* (Forskal) sperm preservation, *Aquaculture,* 32, 313, 1982.

13-Harvey, B., Cryopreservation of *Sarotherodon mossambicous* spermatozoa, *Aquaculture,* 32, 313, 1983.

14-Chao, N.H., Chao, W.C., Liu, K.C., and Liao, I.C., The biological properties of black porgy (*Acanthopagrus schlegeli*) sperm and its cryopreservation, *Proc Natl Sci Counc, B. ROC,* 10, 145, 1986.

15-Bolla, S., Holmefjord, I., and Refstie, T., Cryogenic preservation of Atlantic halibut sperm, *Aquaculture,* 65, 371, 1987.

16-Leung, L., Cryopreservation of the spermatozoa of the barramundi, *Lates calcarifer* (Telostei: Centropomidae), *Aquaculture,* 64, 243, 1987.

17-Cloud, J.G., Miller, W.H., and Levenduski, M.J., Cryopreservation of sperm as a means to store salmonid germ plasm and to transfer genes from wild fish to hatchery poputlations, *The Progressive Fish-Culturist,* 52, 51, 1990.

18-Gwo, J.C., Strawn, K., Longnecker, M.T., and Arnold, C.R., Cryopreservation of Atlantic croaker spermatozoa, *Aquaculture,* 94, 355, 1991.

19-Young, J.A., Capra, M.F., and Blackshaw, A.W., Cryopreservation of summer whiting (*Sillago ciliata*) spermatozoa, *Aquaculture,* 102, 155, 1992.

20-Gwo, J.C., Cryopreservation of black grouper (*Epinephelus malabaricus*) spermatozoa, *Theriogenolgy,* 39, 1331, 1993.

21-Pillai, M., Yanagimachi, R., and Cherr, G., In vivo and in vitro initiationn of sperm motility using fresh and cryopreserved gametes from the pacific herring, *Clupea pallasi, J Exp Zool,* 269, 62, 1994.

22-Lahnsteiner, F., Weisman, T., and Patzner, R., Methanol as cryoprotectant and the suitability of 1.2 and 5 ml straws for cryopreservation of semen from salmonid fishes, *Aquac Res,* 28, 471, 1997.

23-Jodun, W.A., King, K., Farrell, P., and Wayman, W., Methanol and egg yolk as cryoprotectants for Atlantic salmon spermatozoa, *N Am J Aquacult,* 69, 36, 2006.

24-Cosson, J., Billard, R., Cibert, C., Dréanno, C., and Suquet, M., Ionic factors regulating the motility of fish sperm, in *The Male Gamete,* Gagnon, C., Ed., Cache River Press, Vienna, 1999, 161.

25-Terner, C., Evaluation of salmonid sperm motility for cryopreservation, *The Progressive Fish-Culturist,* 48, 230, 1986.

26-Wilson-Leedy, J. G. and Ingerman, R.L., Development of a novel CASA system based on open source software for characterization of zebrafish sperm motility parameters, *Theriogenology,* 67, 661, 2007.

27-Kime, D.E., Van Look, K.J.W., McCallister, B.G., Huyskens, G., Rurangwa, E., and Ollevier, F., Computer-assisted sperm analysis (CASA) as a tool for monitoring sperm quality in fish, *Comp Biochem Phys C,* 130, 425, 2001.

28-Mazur, P., Freezing of living cells: Mechanisms and implications, *Amer J Physiol,* 247C, 125, 1984.

29-Mazur, P., Principles of cryobiology, in: *Life in the Frozen State,* Fuller, B., Lane, N., and Benson, E., Eds., CRC Press, Boca Raton, 2004, 3.

30-Bencic, D.C., Cloud, J.G., and Ingerman, R.L., Carbon dioxide reversibly Inhibits sperm motility and fertilizing ability in steelhead trout (*Oncorhynchus mykiss*), *Fish Physiol Biochem,* 23, 275, 2000.

31-Baynes, S.M. and Scott, A.P., Cryopreservation of rainbow trout spermatozoa: the influence of sperm quality, egg quality and extender composition on post-thaw fertility, *Aquaculture,* 66, 53, 1987.

32-J. Stoss, J. and Holtz, W., Cryopreservation of rainbow trout (*Salmo gairdneri*) sperm. II. Effect of pH and presence of a buffer in the diluent, *Aquaculture,* 25, 217, 1981.

Complete affiliation:

Joseph G. Cloud, Ph.D., Professor of Zoology and Chair, Department of Biological Sciences, University of Idaho, Moscow, ID 83844-3051, Phone: 208-885-6388, FAX: 208-885-7905. e-mail: jcloud@uidaho.edu.

CHAPTER 8
CRYOPRESERVATION OF FISH OOCYTES

T. Zhang and E. Lubzens

1. INTRODUCTION

The ability to create cryo-banks of viable fish sperm, eggs and embryos would be one of the most powerful facilitating tools in aquaculture and species conservation. Preservation of haploid nuclear genomes of viable male and female reproductive cells, such as sperm cells or unfertilized eggs or of viable embryos, is vital if fish populations are to be maintained. While no specific problem is encountered in preservation of fish spermatozoa, the preservation of the maternal genome of fish remains a challenge. In contrast to mammalian species where methods for cryopreservation of embryos and oocytes are well established (reviewed in Ambrosini et al., [1]), successful cryopreservation methods for yolk-laden fish embryos have remained elusive. Unlike mammalian species, fish oocytes and eggs contain yolk that consists of proteins, lipids and in several species also lipid droplets. Fish oocytes are smaller than their respective eggs, constitute a single compartment and do not have permeability barriers such as the yolk syncytial layer found in embryos [2,3]. Some earlier studies showed lower membrane permeability of oocytes when compared to fully matured eggs [4-6], although more recent studies showed ovarian follicles to be more permeable to water and cryoptectants than embryos [7-9] with differences between freshwater species such as medaka and zebrafish and the marine gilthead seabream. However, fish oocyte membrane permeability may still not be high enough or their cryopreservation. The high water content of pelagic floating eggs is an additional problem for most cultured marine fish species. The term "oocyte" refers to, in most of the cases, ovarian follicles that inclue the oocyte with its attached theca and granulose cells. The term oocyte will be used here to indicate an ovarian follicle.

The main obstacles that were found in cryopreservation of fish embryos included: a) their relatively large size, resulting in a low surface to area volume ratio that could retard water and cryoprotectant efflux and influx, b) compartments such as the blastoderm and yolk, differing in water content and permeability properties, c) the plasma membrane and the yolk syncytial layer, inhibiting water and cryoprotective influx and efflux and d) chilling injury at specific developmental stages [3,10,11].

The lack of success to date in cryopreservation of fish oocytes and embryos has prompted alternative approaches to be investigated, using materials that can be successfully cryopreserved. Three procedures have been employed: (i) chimera production exploiting the ability to cryopreserve isolated blastomeres and their subsequent insertion into recipient embryos, (ii) androgenesis using frozen sperm and (iii) primordial germ cell (PGC) transplantation. Chimera production from frozen blastomeres has the advantage of banking the diploid embryonic cells comprising both male and female parental genome contributions. The procedure is a highly skilled operation involving blastomere isolation and micro-injection into selected embryos. Production of germ-line chimeras that has been reported for several fish species, depends on the transplantation of blastomeres at the blastula stage, when blastomeres still maintain pluripotency. A proportion of the transplanted cells would be expected to enter the germ line and subsequently develop into eggs. This approach was reported for chimera formation within the same species in model fish such as medaka (*Oryzias latipes*; [12]), zebrafish (*Danio rerio*; [13,14]), rainbow trout (*Oncorhynchus mykiss*; [15,16]) and loach (*Misgurnus anguillicaudatus*; [17]). Cross-species germ-cell chimera production was also reported by transplantation of triploid crucian carp (*Cyprinus carpio*) into recipient diploid goldfish (*Carassius auratus*; [18]). Successful results were mostly obtained by transplantation of the lower (or ventral) part of the donor blastomere, where primordial germ-cells were located [18]. Blastomere cryopreservation and subsequent insertion into early stage embryos do offer a route for maternal genome banking [15].

Androgenesis can provide homozygous lines for biomedical research and monostocks for commercial purposes, and improved survival rates of androgenetic individuals are being achieved [19]. However, the very nature of the procedure, the insertion of the diploidized nucleus from sperm into a denucleated oocyte, means that there is no nuclear maternal genome contribution. Several genetic factors are inherited only maternally by the oocyte such as mitochondrial DNA and mRNAs that determine the early stages of embryonic development and making vital contributions to cellular functions [20].

Animal gametes have been successfully derived from sexually undifferentiated PGCs which are the progenitors of the germ-cell lineage giving rise to either eggs or sperm [21-24]. In these studies, PGCs from donor fish were microinjected into the peritoneal cavity of recipient newly hatched fish embryos (rainbow trout), to avoid immune rejection by the recipient fish. At this stage the recipient fish were sexually undifferentiated but contained PGCs in their genital ridge. The injected PGCs migrated into the genital ridge and colonized it. The donor PGCs differentiated synchronously with the endogenous germ cells into functional sperm cells in male recipient or eggs in female recipients. The donor derived sperm cells or eggs were fully functional when the recipient fish reached sexual maturation and the donor offspring could be distinguished by specific genetic markers. Successful cryopreservation methods for PGCs were also developed, demonstrating a novel approach that facilitates the preservation of both paternal and maternal genome resources [24,25]. At present this technique has only been established for transplantation of heterologous PGCs between closely related salmonid species [26] and it is necessary to wait until the surrogate fish reaches sexual maturation for recovering the gametes of the donor fish. A slightly different approach was reported by Ciruna et al., [27], where host germ cells were completely replaced in zebrafish by donor cells, leading to efficient production of future generations of germ-line replacement offsping. In addition to the production of only donor gametes, it raises the feasibility of *in vitro* transgenesis and gene targeting technology in fish, contributing to future basic research on specific gene functions.

The problems and limitations of the current approaches to preserve maternal genetic material means there is an urgent need for developing successful protocols for the cryopreservation of fish oocytes. Cryopreservation of fish oocytes offers several advantages, such as the smaller size of the oocytes, the absence of a fully formed chorion that may reduce the rate of water and cryoprotectant movements during cryopreservation and the relatively low water content in oocytes of pelagic spawners. Early oocyte developmental stages (or ovarian follicles) in fish may therefore be most suited for cryopreservation (see Fig. 1, for examples of oocyte developmental stages of gilthead seabream and zebrafish oocytes, respectively). However, these oocytes will need to undergo *in vitro* maturation, ovulation and fertilization after cryopreservation. One of the advantages for using cryopreserved oocytes is that it leaves the option open for the spermatozoa donor, in contrast to embryos where the paternal genetic contributor has to be decided before cryopreservation.

2. STUDIES CARRIED OUT ON CRYOPRESERVATION OF FISH OOCYTES

Cryopreservation of fish gametes has been studied extensively in the last three decades and the successful cryopreservation of the spermatozoa from many species including salmonid, cyprinids, silurids and acipenseridae is well documented [28-32]. In the last 25 years, attempts to cryopreserve fish eggs and embryos have been conducted on about 20 fish species. Although eggs or embryos have been shown to survive for a short time after cooling to subzero temperatures, successful cryopreservation of fish eggs and embryos remains elusive [2,33-36]. Recent reported success for flounder (*Paralichthys olivaceous*) embryos by vitrification [37], has been disputed [38] as well as the reported success in cryopreservation of carp embryos [34]. There has been virtually no published information on the cryopreservation of ovary or unfertilised eggs of fish species until recently.

2.1 Evaluation of oocyte viability

One of the first requirements for studies on oocytes is the need to establish methods for assessing their viability in order to estimate the damage incurred by cryoprotectants, chilling, freeze-thawing procedures or any other type of manipulation. Four different tests were used to assess oocyte viability: trypan blue (TB) staining, thiazol MTT staining, fluorescein diacetate staining and *in vitro* maturation followed by observation of germinal vesicle breakdown (GVBD). TB staining is a simple method for evaluating cell membrane integrity as cells (or oocytes) take up the coloring agent if their membrane integrity is compromised. Yellow MTT (3-(4,5-Dimethylthiazol-2yl)-2,5-diphenyltetrazolium bromide, a tetrazole) is reduced to purple formazan in the mitochondria of living cells. This reduction takes place only when mictochondria reductase enzymes are active, thus indicating cell viability. Another dye uptake that can be used for assessing viability is diacetyl fluorescein. Fluorescein diacetate, a nonfluorescent derivative of fluorescein, can be transported across cell membranes and becomes fluorescent upon hydrolysis by esterases present in the cytoplasm of viable cell. Resultant fluorescein accumulates within cells and allows direct visualization by epifluorescent microscopy. Observation of the occurrence of GVBD in *in vitro* incubated oocyte as a functional test for viability is limited to stage III oocytes or later developmental stages. Selman et al. studies [39] showed that TB staining was more suitable for indicating viability of zebrafish oocytes while MTT staining was more appropriate for evaluating viability of seabream oocytes [40-43]. Genomic and proteomic tools [44,45] could also be used for verification of gene expression and protein profiles but these methods are lengthy and are not fully developed for routine use in optimising cryopreservation of oocytes.

2.2 Membrane permeability to water and cryoprotectants

Investigation into fish oocyte membrane permeability is essential for developing successful protocols for their cryopreservation. Three methods have been used in studying membrane permeability in fish oocytes: a) membrane permeability parameter estimation based on oocyte volumetric changes when exposed to cryoprotectant solutions, b) direct measurement of cryoprotectant uptake by HPLC and c) measurement of radio-labelled cryoprotectant uptake. The permeability of the zebrafish (*Danio rerio*) oocyte membrane to water and cryoprotectants has been studied and oocyte membrane permeability parameters were reported for the first time by Zhang et al [8]. This study was conducted on stage III and stage V zebrafish oocytes. Volumetric changes of stage III oocytes in different concentrations of sucrose solutions were measured after 20 min exposure at 22°C and the osmotically inactive volume of the oocytes (V_b) was determined using the Boyle Van't Hoff relationship. The osmotically inactive volume of stage III zebrafish oocytes was found to be 69.5%. The mean values ± SE of the hydraulic conductivity (L_p) were found to be 0.169 ± 0.02 and 0.196 ± 0.01 mm/min/atm in the presence of DMSO and PG, respectively at 22°C, assuming an internal isosmotic value for the oocyte of 272 mOsm/Kg. The P_s values indicative of the cryoprotectant permeability, were 0.000948 ± 0.00015 and 0.000933 ± 0.00005 cm/min for DMSO and PG, respectively. The study also showed that the membrane permeability of stage III oocytes decreased significantly with temperature. No significant changes in cell volume during methanol treatment were observed. The L_p and P_s values obtained for stage III zebrafish oocytes are generally lower than those obtained from other aquatic invertebrates and higher than those obtained with immature medaka oocytes [9] or fish embryos [46]. It was not possible to estimate membrane permeability parameters for stage V oocytes using the methods employed in this study because stage V oocytes experienced the separation of outer oolemma membrane from inner vitelline during exposure to cryoprotectants. Relatively high concentrations of DMSO (1.2- 1.8M) in unfertilized medaka (*Oryzias latipes*) eggs were reported by Routray et al. [7] using HPLC after short periods of incubation. Permeation of DMSO was significantly increased by incubation of eggs (1-3 min) in 1M trehalose followed by application of hydrostatic pressure (50 atm). Accumulation of radiolabeled methanol was also studied in zebrafish and gilthead seabream oocytes [47]. The concentrations of methanol in gilthead seabream were found to reach 1.16 M ± 0.59 M after 15 min of incubation, while zebrafish oocytes were found to contain 0.39 ± 0.04 M after 60 min of incubation. Moreover, the concentrations within gilthead seabream oocytes were found to be significantly higher than those found in fertilized eggs at various stages of embryonic development. Most of the radiolabaled methanol was recovered from the yolk compartment [48].

2.3 Sensitivity to cryoprotectant toxicity

Toxicity of cryoprotectants to zebrafish (*Danio rerio*) oocytes has been investigated by Plachinta et al. [41]. Commonly used cryoprotectants DMSO, methanol, ethylene glycol (EG), propylene glycol (PG), sucrose and glucose were studied. Stage III (vitellogenic), stage IV (maturation) and stage V (mature egg) zebrafish oocytes were incubated in Hank's medium (0.137 M NaCl, 5.4 mM KCl, 0.25 mM Na_2HPO_4, 0.44mM KH_2PO_4, 1.3 mM $CaCl_2$, 1 mM $MgSO_4$, 4.2 mM $NaHCO_3$) containing different concentrations of cryoprotectants (0.25-4M) for 30 min at room temperature. In this study, GVBD test showed that cryoprotectant toxicity to stage III zebrafish oocytes increased in the order of methanol, PG, DMSO, EG, glucose and sucrose. No Observed Effect Concentrations (NOECs) for stage III oocytes were 2 M, 1 M, 1 M, 0.5 M, <0.25 M and <0.25 M for methanol, PG, DMSO, EG, glucose and sucrose, respectively. TB test also showed that the toxicity of tested cryoprotectants increased in the same order. The sensitivity of oocytes to cryoprotectants appeared to increase with development stage with stage V oocytes being the most sensitive. Studies performed on the oocytes of the gilthead seabream showed that the tolerance of these oocytes to cryoprotectants differed from those reported for zebrafish and they tolerated 6M methanol and EG. For gilthead seabream oocytes, trypan blue staining was not suitable as an indicator of viability and MTT was found to be more reliable [40,49].

2.4 Sensitivity to chilling

Chilling sensitivity of stage III (vitellogenic) and stage V (mature) fish oocytes was studied by Isayeva et al. [50]. Oocyte viability was assessed using three different methods TB staining, MTT staining and GVBD. The results showed that zebrafish oocyte were very sensitive to chilling and their survival decreased with decreasing temperature and increasing exposure time periods at zero and subzero temperatures. Normalised survivals for stage III oocytes assessed with TB staining after exposure to 0, -5 or -10 °C for 15 or 60 min were 90.1±6.0%, 77.8±7.6%, 71.2±9.3% and 60.2±3.8%, 49.6±6.7%, 30.4±3.0%, respectively. The study found that the sensitivity of viability assessment methods increase in the order of MTT<TB<GVBD. Stage III oocytes were more susceptible to chilling than stage V oocytes, and that individual female had a significant influence on oocyte chilling sensitivity.

There was another study on chilling sensitivity of zebrafish oocyte in the literature and the work was conducted with only two types of oocytes: with or without yolk [51]. In this study Perl and Arav reported that less than 30% of large oocytes survived after 15 min exposure to 0°C while Isayeva et al. [50] reported a survival of 63.7% for stage III oocytes under similar conditions. The discrepancies of the two studies may be due to the facts that in Perl and Arav's study, cFDA (5-carboxy-fluoresceindiacetate) was used for assessment of oocyte viability (which gives ambiguous results, Isayeva et al. [50]) and mixed oocytes at different developmental stages were also used. Oocytes of cold acclimated fish carp (*Cyprinus carpio*) was also found to be chilling sensitive and their survival decreased with decreased temperature and increased exposure time [52]. Chilling sensitivity in zebrafish oocytes was linked to lipid phase transition [51].

2.5 Low temperature storage of fish gonad cells and oocytes

It has been reported that the development of fish gametes and eggs was affected by changes in temperature [53-56]. Several studies indicate that fish ova or eggs can be maintained at low temperatures in a non-frozen form for several days or weeks. Combs and Burrows [57] found that both chinook (*Oncorhynchus nerka*) and pink (*Oncorhynchus garbuscha*) salmon eggs could tolerate 1.7°C for long periods if they were initially stored at 5.6°C for a month. A further study of Combs [58] defined the stage of development at which low water temperatures could be tolerated by chinook salmon eggs as being 144 hours of incubation. Maddock [59] reported that the incubation period of brown trout (*Salmo trutta*) ova could be extended for up to 4 months at a temperature of 1.4°C if ova were initially stored at a temperature of 7.6°C for 13 days. The study of Pullin and Bailey [60] on cryopreservation of plaice (*Pleuronectes platessa*) eggs at low temperatures showed that the survival rate increased with increasing development of the eggs used. Harvey and Ashwood-Smith

[6] found that although the freezing point of isolated, unfertilised eggs of rainbow trout (*Salmo gairdneri*) is -1.7°C, these eggs supercool readily to temperatures around -20°C when ice is absent from the surrounding medium. They found that eggs cooled at 1°C/min in paraffin oil generally remain unfrozen to -18°C unless seeded. Harvey et al. [61] also reported that unfertilised salmonid ova can be stored at -1°C for up to 20 days in artificial media and in ovarian fluid. In all cases ovarian fluid preserved fertility longer than the synthetic media did. Some factors such as oxygen and bacterial infection limit the development and survival of fish eggs during storage [60,62].

2.6 Cryopreservation of fish oocytes using controlled slow cooling

In order to study cryopreservation of zebrafish (*Danio rerio*) oocytes, large numbers of oocytes need to be isolated from the ovaries for experimental use. The mechanical method used for isolating oocytes is laborious and time consuming. A new enzymatic method was developed for separation of zebrafish oocytes using hyaluronidase [63] but this method was not appropriate for the isolation of oocytes from the gilthead seabream (*Sparus aurata*) ovaries (Lubzens et al., unpublished). In zebrafish, ovaries were placed in 0.4, 0.8 or 1.6 mg/ml hyaluronidase in Hank's solution immediately after removal for 5, 10 or 20 min at 22°C. Stage I (follicle diameter < 0.14 mm), stage II (d = 0.14-0.34 mm) and stage III (d >0.5 mm) oocytes were then separated by gentle pipetting. Control oocytes were separated mechanically in Hank's solution using forceps and scissors. Stage I and II oocytes viability was assessed using trypan blue staining (0.2% for 5 min). Stage III oocytes viability was assessed using both TB staining and GVBD. For GVBD observation, oocytes were incubated in 50% L-15 medium at 25°C for 24 h. Results from these experiments showed that the optimum hyaluronidase concentration for oocytes separation was 1.6 mg/ml and the optimum treatment time was 10 min at 22°C. Under these conditions, ~80% of oocytes were separated. Survivals after enzymatic separation were $96.6 \pm 0.7\%$, $92.9 \pm 1.3\%$ and $94.6 \pm 0.9\%$ for stage I, stage II and stage III oocytes, respectively, using TB staining. These survivals were significantly higher than those obtained from mechanical separation ($83.2 \pm 2.2\%$). GVBD test showed no significant differences in stage III oocyte survival between enzymatic treatment ($35.4 \pm 2.0\%$) and mechanical separation ($37.4 \pm 1.6\%$).

Studies on cryopreservation of zebrafish (*Danio rerio*) vitellogenic oocytes using controlled slow cooling have been carried out by Plachinta et al. [42,43], Guan et al. [64]. The best results in these experiments were obtained with stage III (vitellogenic) oocytes cryopreseved in 4 M methanol + 0.2 M glucose in KCl-buffer. In these studies, $88\pm1.7\%$ oocyte viability was retained immediately after thawing as indicated by TB staining, similar to room temperature controls ($88.7\pm2.8\%$). However, after 2 h incubation at 22°C, the viability of freeze-thawed oocytes decreased to $19.4 \pm 6.6\%$. Microscopic observations showed that all oocytes became translucent after freezing. In the gilthead seabream, the best results obtained so far in slow-cooling cryopreservation procedures, were for stage III or stage IV oocytes in 3 M ethylene-glycol prepared in 75% L-15 culture medium. In these experiments, oocytes stained positively with MTT, indicating their apparent viability, immediately after thawing from liquid nitrogen, but careful microscopic observations revealed a translucent appearance at the circumference of the oocyte that increased with the incubation time. Two hours after thawing these oocytes were completely translucent and did not seem viable. On the other hand, when the same procedure was performed with smaller size stage III oocytes (less than 300 μm in diameter), the oocytes were not translucent after thawing and retained viability (as assessed by MTT staining) for at least two hours. Ice nucleation was suspected to be the main cause for the translucent appearance of oocytes as it may have damaged the oocyte internal compartments, releasing cathepsins that change the physical structure of the yolk proteins (Lubzens et al., unpublished, [48]). Cathepsin activation is associated with the translucent appearance of oocytes during normal maturation and GVBD. During GVBD, opaque oocytes turn translucent probably as a result of the proteolysis of major yolk components by activation of cathepsins and the uptake of water by oocytes of pelagic marine fish [65,66].

Studies on membrane integrity and cathepsin activities of zebrafish (*Danio rerio*) oocytes after freezing to -196°C using controlled slow cooling were carried out by Zhang et al. [67] using 2 M methanol and 2 M DMSO as cryopotectants. In these experiments, stage III oocytes were exposed to 2 M methanol or DMSO (in Hank's) for 30 min at 22°C before they were loaded into 0.5 ml plastic straws and cooled from 20ºC to -7.5ºC at 2ºC/min, manually seeded at -7.5ºC and samples held for 5 min, then cooled from -7.5ºC to -40ºC at 0.5ºC/min and -40ºC to -80ºC at 10ºC/min. Samples were then plunged into liquid nitrogen (LN). Membrane integrity and cathepsin activities of oocytes were assessed both after cryoprotectant treatments at 22°C and after freezing in LN_2. Oocytes membrane integrity was assessed using TB staining. Cathepsin B and L colorimetric analyses were performed using substrates Z-Arg-ArgNNap and Z-Phe-Arg-4MβNA-HCl, respectively. 2-Naphthylamine and 4-Methoxy-2-Naphthylamine were used as standards. Cathepsin D enzymatic activity was performed by analysing the level of hydrolytic action on haemoglobin. In this study, DMSO appeared to be a better cryoprotectant than methanol. TB staining showed that 63.0 ± 11.3% and 72.7 ± 5.2% oocytes membrane stayed intact after DMSO and methanol treatment for 30 min at 22°C while 14.9 ± 2.6% and 1.4 ± 0.8% stayed intact after freezing in DMSO or methanol to -196°C. The results also showed that the activity of cathepsin B was not affected by 2 M DMSO treatment (8.03 ± 1.02 nM/min/mg/ml of 2-Naphthylamine compared to 8.25±1 nM/min/mg/ml 2-Naphthylamine of control). However, oocyte cathepsin B activity was lowered when treated with 2 M methanol (4.13 ± 0.3 nM/min/mg/ml of 2-Naphthylamine). After freezing to -196°C, oocyte cathepsin B activity was decreased to 3.61±0.7 and 1.5±0.5 nM/min/mg/ml of 2-Naphthylamine, with 2 M DMSO and 2 M methanol, respectively. Cathepsin D activity however was decreased by both 2 M DMSO (0.48 ± 0.03 U/mg) and 2 M methanol, (0.3 ± 0.1 U/mg) treatments when compared to the control (0.7 ± 0.1 U/mg). Activities were decreased by cooling to 0.2 ± 0.03 U/mg and 0.08 ± 0.01 U/mg for 2 M DMSO and 2 M methanol, respectively. Cooling reduced activity of cathepsin L to 8.4 ± 0.8 and 3.7 ± 1 μM/min/mg/ml of 4-

Methoxy-2-Naphthylamine for DMSO and methanol, respectively, compared to $12.5 \pm$ 2.4 µM/min/mg/ml of 4-Methoxy-2-Naphthylamine for control. The results indicate that cryoprotectant and cooling treatments altered the activities of lysosomal enzymes involved in oocyte maturation and yolk mobilization.

3. FUTURE STUDIES

Unlike mammalian species, fish oocytes and eggs contain yolk that consists of proteins, lipids and in several species also lipid droplets. Fish oocytes are smaller than their embryos, constitute a single compartment and do not have permeability barriers such the yolk syncytial layer found in embryos [2,3]. Some earlier studies showed lower membrane permeability of oocytes when compared to fully matured eggs [4-6], although more recent studies showed ovarian follicles to be more permeable to water and cryoptectants than embryos [7,8,9] with differences between freshwater species such as medaka and zebrafish and marine species such as gilthead seabream. However, fish oocyte membrane permeability may need to be improved if successful cryopreservation is to be achieved. One of the approaches under intense investigation is the artificial expression of aquaporin-3, which acts as water/cryoprotectant channel, through injection of its cRNA into zebrafish embryos [68,69] or immature medaka oocytes [70]. In zebrafish embryos, membrane permeability to water and propylene glycol increased by 50% and the permeability of injected medaka oocytes was almost as high as that of controls. It remains to be shown that this procedure improves the viability of oocytes after freezing to cryogenic temperatures.

The yolk content was shown to be implicated in chilling sensitivity of zebrafish embryos [71] and possibly contributes to the chilling sensitivity of zebrafish oocytes [50]. Brief exposure to low temperatures was not found to be a problem during controlled slow cooling procedures of gilthead seabream oocytes, but damage to yolk was observed immediately after ice nucleation in relatively large vitellogenic or maturing oocytes but not in small size oocytes that just embarked on vitellognin uptake [48]. The damage to yolk may have been the consequence of injuries caused to the oocyte cytoskeleton. Further studies are needed on whether this damage could be avoided by improving permeation of cryoprotectants in larger oocytes.

Assessment of oocyte viability is a key issue in studies involving manipulation of oocytes. Vital dyes, such as TB or MTT, indicative of cell membrane integrity and active metabolic activity, respectively, are quick and simple for using in vast array of empirical experiments. The results obtained with these methods should, however, be correlated with functional viability test. GVBD in *in vitro* incubated oocytes could be used for assessing functional viability but it is limited to oocytes responding to maturation inducing hormones. Genomic and proteomic tools could also be used for optimisation of oocyte cryopreservation protocols although these testes involve relatively high costs and sample processing is lengthy and complicated.

Implantation of primordial germ cells into surrogate parents with subsequent production of xenogenic, donor-derived offspring seems to offer an exciting solution to maternal genome banking in fish. The universality of this approach requires additional studies demonstrating that it can be performed between widely divergent species and that gamete production could be carried out in surrogate parents with short generation time [26]. However, the number of produced offspring would be too small for direct cultivation in the aquaculture industry but could serve for genome banking through production of broodstock fish. The vision of large scale cryopreservation of fish oocytes or eggs, similar to that of spermatozoa cryobanking, require future progress in oocyte structural and functional studies. These should include better understanding of changes in oocyte and egg compartments and transformations of structures of lipids and yolk proteins, during chilling and cryopreservation procedures. Moreover, additional studies will be required for *in vitro* maturation and fertilization if successful cryopreservation of fish oocytes is to be achieved.

4. REFERENCES

1-Ambrosini, G., Andrisani, A., Porcu, E., Rebellato, E., Revelli, A., Caserta, A., Cosmi., E., Marci, R., and Moscarini, M., Oocyte cryopreservation: State of art, *Reprod Toxicol* 22, 250, 2006.

2-Hagedorn, M., Hsu, E.W., Pilatus, U., Wildt, D., Rall, W.F., and Blackband, S.J., Magnetic resonance microscopy and spectroscopy reveal kinetics of cryoprotectant permeation in a multicompartmental biological system, *Proc Natl Acad Sci*, 93, 7454, 1996.

3-Hagedorn, M., Kleinhanas, F.W., Artemov, D., and Pilatus, U., Characterization of a major permeability barrier in the zebrafish embryo, *Biol Reprod*, 59, 1240, 1998.

4-Prescott, D.M., Effect of activation on the water permeability of salmon eggs, *J Cell Comp Physiol*, 45, 1, 1955.

5-Loeffler, C.A. and Lovtrup, S., Water balance in the salmon egg, *J Exp Biol*, 52, 291, 1970.

6-Harvey, B. and Ashwood-Smith, M. J., Cryoprotectant penetration and suppercooling in the eggs of samonid fishes, *Cryobiology*, 19, 29, 1982.

7-Routray, P., Suzuki, T., Strussmann, C. A., and Takai, R., Factors affecting the uptake of DMSO by the eggs and embryos of medaka, *Oryzias latipes*, *Therogenology*, 58, 1483, 2002.

8-Zhang, T., Isayeva, A., Adams, S.L., and Rawson, D.M., Studies on membrane permeability of zebrafish (*Danio rerio*) oocytes in the presence of different cryoprotectants, *Cryobiology*, 50, 285, 2005.

9-Vladez, M.D. Jr., Miyamoto, A., Hara, T., Edashige, K., and Kasai, M., Sensitivity to chilling of medaka (*Orysias latipes*) embryos at various developmental stages, *Theriogenology*, 64, 112, 2005.

10-Zhang, T. and Rawson, D.M., Studies on chilling sensitivity of zebrafish (*Brachydanio rerio*) embryos, *Cryobiology*, 32, 239, 1995.

11-Zhang, T. and Rawson, D.M., Permeability of the vitelline membrane of zebrafish (*Brachydanio rerio*) embryos to methanol and propane-1,2,-diol, *Cryo-Letters*, 17, 273, 1996.

12-Wakamatsu, Y., Ozato, K., Hashimoto, H., Kinoshita, M., Sakaguchi, M., Iwamatsu, T., Hyodo-Taguchi, Y., and Tomita, H., Generation of germ-line chimeras in medaka (Oryzias latipes), *Mol Marine Biol Biotech*, 2, 325, 1993.

13-Lin, S., Long, W., Chen, J., and Hopkins, N., Production of germ-line chimeras in zebrafish by cell transplants from genetically pigmented to albino embryos, *Proc Natl Acad Sci U.S.A.*, 89, 4519, 1992.

14-Nagai, T., Otani, S., Saito, T., Maegawa, S., Inke, K., Arai, K., and Yamaha, E., Germ-line chimera produced by blastoderm transplantation in zebrafish, *Nippon Suisan Gakkaishi*, 71, 1, 2005.

15-Nilsson, E.E. and Cloud, J.G., Cryopreservation of rainbow trout (*Oncorrhynchus mykiss*) blastomeres, *Aquat Living Resour*, 6, 77, 1993.

16-Takeuchi, Y., Yoshizaki, G., and Takeuchi, T., Production of germ-line chimeras in rainbow trout by blastomere transplantation, *Mol Reprod Develop*, 59, 380, 2001.

17-Nakagawa, M., Kobayashi, T., and Ueno, K., Production of germline chimera in loach (*Misgurunus anguillicaudatus*) and proposal of new method for preservation of endangered species, *J Exp Zool*, 293, 624, 2002.

18-Yamaha, E., Kazama-Wakabayashi, M., Otani, S., Fujimoto, T., and Arai, K., Germ-line chimera by lower-part blastoderm transplantation between diploid goldfish and triploid crucian carp, *Genetica*, 111, 227, 2001.

19-Babiak, I., Dobosz, S., Goryczko, K., Kuzminski, H., Brzuan, P., and Ciesielski, S., Androgenesis in rainbow trout using cryopreserved spermatozoa: the effect of processing and biological factors, *Theriogenology*, 57, 1229, 2002.

20 Broughton, R.E., Malam, J.E., and Roe, B.A., The complete sequence of the zebrafish (Danio rerio) mitochondrial genome and evolutionary patterns in vertebrate mitochondrial DNA, *Genome Research*, 11, 1958, 2001.

21-Yoshizaki, G., Takeuchi, Y., Sakatani, S., and Takeuchi, T., Germ cell-specific expression of green flurescent protein in transgenic rainbow trout under control of the rainbow trout vasa-like gene promoter, *Int J Dev Biol*, 44, 323, 2000.

22-Yoshizaki, G., Tagi, Y., Takeuchi, Y., Sawatari, E., Kobayashi, T., and Takeuchi, T., Green fluorescent protein labeling of primordial germ cells using nontransgenic method and its application for germ cell transplantation in salmonidae, *Biol Reprod*, 73, 88, 2005.

23-Takeuchi, Y., Yoshizaki, G., Kobayashi, T., and Takeuchi, T., Mass isolation of primordial germ cells from transgenic rainbow trout carrying the green fluorescent protein gene driven by the *vasa* gene promoter, *Biol Reprod*, 67, 1087, 2002.

24-Okutsu, T., Yano, A., Nagasawa, K., Shikina, S., Kobayashi, T., Takeuchi, Y., and Yoshizaki, G., Manipulation of fish germ cell: Visualization, Cryopreservation and Transplantation, *J Rep Dev*, 52, 685, 2006.

25-Kobayashi, T., Takeuchi, Y., Yoshizaki, G., and Takeuchi, Y., Cryopreservation of trout primordial germ cells: a novel techniques for preservation of fish genetic resources, *Fish Physiol Biochem*, 28, 479, 2003.

26-Takeuchi, Y., Yoshizaki, G., and Takeuchi, T., Surrogate broodstock produces salmonids, *Nature*, 430, 629, 2004.

27-Ciruna, B., Weidinger, G., Knaut, H., Thisse, B., Thisse, C., Raz, E., and Schier, A., Production of maternal-zygote mutant zebrafish by germ-line replacement, *Proc. Natl. Acad. Sci. U.S.A.*, 99, 14919, 2002.

28-Rana, K.J. and Gilmour, A., Cryopreservation of fish spermatozoa: effect of cooling methods on the reproducibility of cooling rates and viability, in: *Refrigeration and Aquaculture Conference*, Bordeaux, March 20-22, 1996, 3.

29-Maisse, G., Cryopreservation of fish semen: a review, in: *Refrigeration and Aquaculture Conference*, Bordeaux, March 20-22, 1996, 443.

30-Tsvetkova, L.I., Cosson, J., Linhart, O., and Billard, R., Motility and fertilizing capacity of fresh and frozen-thawed spermatozoa in sturgeons *Acipenser baeri* and *A. ruthenus*, *J Applied Ichthyol*, 12, 107, 1996.

31-Magyary, I., Dinnyes, A., Varkonyi, E., Szabo, R., and Varadi, L., Cryopreservation of fish embryos and embryonic cells, *Aquaculture*, 137, 103, 1996.

32-Lahnsteiner, F., Cryopreservation protocols for sperm of salmonid fishes, in: *Cryopreservation in aquatic species*, Tiersch, T.R. and Mazik, P.M., Eds., The World Aquaculture Society, Baton Rouge, Louisiana, USA, 2000, 91.

33-Harvey, B., Cooling of embryonic cells, isolated blastoderm and intact embryos of the zebrafish *Brachydanio rerio* to -196°C, *Cryobiology*, 20, 440, 1983.

34-Zhang, X.S., Zhao, L., Hua, T.C., Chen, X.H., and Zhu, H.Y., A study on the cryopreservation of common carp (*Cyprinus carpio*) embryos, *Cryo-Letters*, 10, 271, 1989.

35-Zhang, T., Rawson, D.M., and Morris, G.J., Cryopreservation of pre-hatch embryos of zebrafish (*Brachydanio rerio*), *Aquat Living Resour*, 6, 145, 1993.

36-Robles, V., Cabrita, E., Real, M., Alvarez, R., and Herráez, M.P., Vitrification of turbot embryos: preliminary assays, *Cryobiology*, 47, 30, 2003.

37-Chen, S.L. and Tian, Y.S., Cryopreservation of flounder (*Paralichthys alivaceus*) embryos by vitrification, *Theriogenology*, 63, 1207, 2005.

38-Edashige, K., Valdez, D.M. Jr., Hara, T., Saida, N., Seki, S., and Kasai, M., Japanese Flounder (*Paralichthys alivaceus*) embryos are difficult to cryopreserve by vitrification, *Cryobiology*, 53, 96, 2006.

39-Selman, K., Wallace, R.A., Sarka, A., and Qi, X., Stage of oocyte development in the zebrafish, Brachiodanio rerio, *J Morphol*, 218, 203, 1993.

40-Lubzens, E., Pekarsky I., Blais, I., and Meiri, I., Evaluating viability of fish oocytes: a first step towards methods for cryopreservation, *Cryobiology*, 47 268, 2003.

41-Plachinta, M., Zhang, T., and Rawson, D.M., Studies on cryoprotectant toxicity to zebrafish (*Danio rerio*) oocytes, *CryoLetters*, 25, 415, 2004a.

42-Plachinta, M., Zhang, T., and Rawson, D.M., Preliminary studies on cryopreservation of zebrafish (*Danio rerio*) vitellogenic oocytes using controlled slow cooling, *Cryobiology*, 49, 347, 2004b.

43-Plachinta, M., Zhang, T., and Rawson, D.M., Studies on the effect of certain supplements in cryoprotective medium on zebrafish (*Danio rerio*) oocytes quality after controlled slow cooling, *Cryobiology*, 51, 405, 2005.

44-Knoll-Gellida A., Andre, M., Gattegno, T., Forgue, J., Admon, A., and Babin, P.J., Molecular phenotype of zebrafish ovarian follicle by serial analysis of gene expression and proteomic profiling, and comparison with the transcriptomes of other animals, *BMC Genomics*, 7, 46, 2006.

45-Lubzens, E., Gattegno, T., Pekarsky, I., Blais, I., Chapovetsky, V., and Admon., A., Proteomic analyses on the effect of cryopreservation procedures on fish oocytes, *Cryobiology*, 53, 398, 2006.

46-Zhang, T. and Rawson, D.M., Permeability of dechorionated 1-cell and 6-somite stage zebrafish (*Brachydanio rerio*) embryos to water and methanol, *Cryobiology*, 37, 13, 1998.

47-Lubzens, E., Pekarsky, I., Blais, I., Cionna, C., and Carnevali, O., Cryopreservation of oocytes from a marine fish: Achievements and obstacles, *Cryobiology*, 51, 385, 2005.

48-Zhang, T., Rawson, D.M., Pekarsky, I., Blais, I., and Lubzens, E., Low temperature preservation of fish gonad cells and oocytes, in *The Fish Oocyte: from basic studies to biotechnological applications*, Babin, P.J., Cerda, C. and Lubzens, E., Eds., Springer, 2007, 411.

49-Lubzens, E., Pekarsky, I., and Blais, I., Developing methods for cryopreservation of fish oocytes: accumulation of [14C] methanol in zebrafish and gilthead seabream oocytes, *Cryobiology*, 49, 319, 2004.

50-Isayeva, A., Zhang, T., and Rawson, D.M., Studies on chilling sensitivity of zebrafish (*Danio rerio*) oocytes, *Cryobiology*, 49 114, 2004.

51-Pearl, M. and Arav, A., Chilling sensitivity in zebrafish (*Brachiodanyo rerio*) oocyte is related to lipid phase transition, *CryoLetters*, 21, 171, 2000.

52-Dinnyes, A., Urbanyi, B., Baranyai, B., and Magyaryi, I., Chilling sensitivity of carp (*Cyprinus carpio*) embryos at different developmental stages in the presence or absence of cryoprotectants: work in progress, *Theriogenology*, 50, 1, 1998.

53-Barrett, I., Fertility of salmonoid eggs and sperm after storage, *J Fish Res Board Can*, 8, 125, 1951.

54-Marr, D.H.A., Influence of temperature on the efficiency of growth of salmonid embryos, *Nature*, 11, 957, 1966.

55-Poon, D.C. and Johnson, S., The effect of delayed fertilization on transported salmon eggs, *The Progressive Fish-Culturist*, 4, 81, 1970.

56-Jensen, J.O.T. and Alderdice, D.F., Effect of temperature on short-term storage of eggs and sperm of chum salmon (*Oncorhynchus keta*), *Aquaculture,* 37, 251, 1984.

57-Combs, B.D. and Burrows, R.E., Threshold temperatures for the normal development of chinook salmon eggs, *The Progressive Fish-Culturist*, 1, 3, 1957.

58-Combs, B.D., Effect of temperature on the development of salmon eggs, *The Progressive Fish-Culturist*, 7, 134, 1965.

59-Maddock, B. G., A technique to prolong the incubation period of brown trout ova, *The Progressive Fish-Culturist*, 36, 219, 1974.

60-Pullin, R.S.V. and Bailey, H., Progress in storing marine flatfish eggs at low temperatures, *Rapp P Reun Cons Int Explor Mer*, 178, 514, 1981.

61-Harvey, B., Kelley, R.N., and Ashwood-Smith, M.J., Permeability of intact and dechorionated zebrafish embryos to glycerol and dimethyl sulfoxide, *Cryobiology*, 20, 432, 1983.

62-Garside, E.T., Effects of oxygen in relation to temperature on the development of embryos of brook trout and rainbow trout, *J. Fish. Res. BD. Canada,* 23, 1121, 1966.

63-Guan, M., Zhang, T., and Rawson, D.M., Successful enzymatic separation of zebrafish (*Danio rerio*) oocytes using hyaluronidase, *Cryobiology*, 53, 418, 2006a.

64-Guan, M., Zhang, T., and Rawson, D.M., Effects of improved controlled slow cooling protocols on the survival of zebrafish (*Danio rerio*) oocytes, *Cryobiology*, 53, 377, 2006b.

65-Carnevali, O., Cionna, C., Tosti, L., Lubzens, E., and Maradonna, F., Role of cathepsins in ovarian follicle growth and maturation, *Gen. Comp. Endocrinol.*, 146, 195, 2006.

66-Fabra, M., Raldún, D., Dozzo, M.G., Deen, P.M.T., Lubzens, E., and Cerda, J., Yolk proteolisis and aquaporin play essential roles to regulate fish oocyte hydration during meiosis resumption, *Dev Biol*, 295, 250, 2006.

67-Zhang, T., Plachinta, M., Kopeika, J., Rawson, D.M., Cionna, C., Tosti, L., and Carnevali, O., Membrane integrity and cathepsin activities of zebrafish (*Danio rerio*) oocytes after freezing to -196°C using controlled slow cooling, *Cryobiology,* 51, 385, 2005b.

68-Hagedron, M., Lance, S.L., Fonseca, D.M., Kleinhans, F.W., Artimov, D., Fleischer, R., Hoque, A.T.M.S., Hamilton, M.B., and Pukazhenthi, B.S., Altering fish embryos with aquaporin-3: an essential step toward successful cryopreservation, *Biol Reprod*, 67, 961, 2002.

69-Lance, S.L., Peterson, A.S., and Hagedorn, M., Developmental expression of aquaporin-3 in zebrafish embryos (*Danio rerio*), *Comp. Biochem. Physiol. Part c*, 138, 251, 2004.

70-Vladez, M.D. Jr., Hara, T., Miyamoto, A., Seki, S., Jin, B., Kasai, M., and Edashige, K., Expression of aqauporin-3 improves the permeability to water and cryoprotectants of immature oocytes in the medaka (*Orysias latipes*), *Cryobiology*, 53, 160, 2006.

71-Liu, X.H., Zhang, T., and Rawson, D.M., The effect of partical removal of yolk on the chilling sensitivity of zebrafish (*Danio rerio*) embryos, *Cryobiology*, 39, 236, 1999.

Complete affiliation:

Tiantian Zhang: LIRANS Institute of Research in the Applied Natural Sciences, University of Bedfordshire, The Spires, 2 Adelaide Street, Luton, Bedfordshire LU1 5DU, UK. e-mail:tiantian.zhang@beds.ac.uk

CHAPTER 9
EMBRYO CRYOPRESERVATION: WHAT WE KNOW UNTIL NOW

V. Robles, E. Cabrita, J.P. Acker and P. Herráez

1. APPLICATIONS OF EMBRYO CRYOPRESERVATION

The development of embryo freezing technologies revolutionized breeding of domestic animal as well as human reproductive technologies in the last decades. In the seventies successful cryopreservation was achieved with mouse embryos [1] and the first report on pregnancy using frozen bovine embryos was published [2]. Then, cryopreservation procedures were developed for cattle, sheep, horse, cat, rabbit, mandrill, titi, mice, rat, monkey [3] and human embryos [4] and, more recently, in 2000, the first piglets from frozen embryos were born [5]. At present, survival rates at birth achieved with the transfer of thawed embryos are lower than that obtained with fresh ones, but the use of frozen/thawed embryos could be a routine breeding alternative in the near future for some species.

Cryopreservation represents, in the field of animal production, a tool for the preservation of maternal and paternal germplasm, for the global transport of genetic material and for the facilitation of breeding line regeneration or proliferation and provides a methodology for genetic rescue [6]. Moreover, embryo cryopreservation is an essential tool for the application of new embryo biotechnologies such as cloning and transgenics. Bovine and ovine cloned embryos produced by somatic cell nuclear transfer [7], by blastomere nuclear transfer [8] or by DNA-injection [9], have been cryopreserved for their use in embryo transfer.

As we will later explain in detail, embryos (fertilized eggs) from aquatic species, particularly from fish, present more difficulties than mammalian embryos in terms of cryopreservation. Embrios from aquatic species are much more sensitive to chilling and their structure is very complex, with several envelopes useful to face an external development without the protection of the maternal body. Successful and reproducible cryopreservation methods have never been described in teleosts and only partial success has been reported with some shellfish [10], but research in this field is encouraged by evident advantages that could be delivered in the fields of commercial aquaculture, *ex situ* conservation of biodiversity, biotechnology or ecotoxicology, which could take advantage from the preservation and permanent availability of specific and recently developed strains (including GMOs).

1.1 Commercial aquaculture

The possibility of embryo storage without a date of lapsing could be very helpful in the management of reproduction. One of the bottlenecks in aquaculture is the dependence on the breeding season for each species. Many of the commonly produced species are from template areas, showing seasonal reproduction. Environmental manipulations and hormonal treatments are common practices in order to modify the circannual rhythms of reproduction, to be independent of the natural breeding periods and to have year round production. Many advances have been done in this field and procedures for induction of spawning in non-natural breeding periods are becoming more common. Nevertheless, many resources are required to maintain and manage "out of season" breeders.

On the other hand, spawning is a critical feature that impacts the management of the work in fish farms. It is not easy to modify the time of spawning or to scale the spawning of different females in order to achieve an optimal utilization of hatchery facilities. As an example, turbot, *Scophthalmus maximus* spawn more than one million eggs and are unable to release the eggs in captivity. Eggs must be obtained by female stripping, which should occur only a few hours after egg maturation to avoid over-ripening and then must be immediately fertilized. Therefore, production of fertilized eggs is not always adjusted to the capacity of the hatchery and do not allow an optimal fry production scheduling.

The use of cryopreserved embryos for routine handling in farms, as seems to be suggested in the previous paragraphs, could require development of relatively simple methods of cryopreservation, able to provide high percentage of success in terms of embryo survival and fry production after thawing. At present, such perspective is not envisaged, but there are other applications compatible with more sophisticated methodologics or with lower rates of success, which are very interesting for commercial aquaculture. In this sense, cryopreservation is essential for the creation of Genetic Resources Banks (GRBs) that should be established for its application in genetic selection programs. Huge efforts are being made in selection programs for aquaculture: genetic characterization of the stocks and the search for favourable genetic traits are among the main areas of development. Cryostorage of embryos for an indefinite time could be a source of safe, selected stock used in disease outbreaks and catastrophes and to minimize genetic drift. In this way it is a useful tool for selection of better conditioned individuals as required but is also a new product that could be marketed, opening a new business opportunity.

1.2 Conservation of biodiversity

According to the World Conservation Union (IUCN) red list, updated September 2007 (www.iucnredlist.org), 2.514 species of teleosts, 591 of chondrosteans, 13 of lampreys and hag fish, 553 of crustacean and 218 bivalvia, are extinct or threatened in the wild. The red list includes information concerning species, but not about subspecies, varieties or geographically isolated subpopulations or stocks, which are often of particular interest as fishing resources for the local populations. The problem is particularly important in some geographical areas (more than 65% of the European fish are threatened according to Kirchofer and Hafti [11]) and affects some primary resources in certain communities. Environmental problems, such as changes in water quality and modifications in the course of rivers are the primary causes of this problem. The recovery of native populations requires a global solution. Cryopreservation of germplasm (creation of GRBs) provides a secure method for the *ex situ* preservation of endangered species or local varieties and creates the opportunity to reconstruct the original population after the environment is restored. Reproduction in captivity of some individuals for further reintroduction in the wild is not always possible and is under the risk of losses of genetic variability, as was demonstrated by Riabova and his colleagues [12] in sturgeon.

Cryopreservation of sperm is easier and much more successful than embryo freezing. Babiak et al. [13] and Grunina et al. [14] developed in trout and sturgeon, respectively, methodologies to recover a given population using cryopreserved sperm to fertilize eggs from other species in which the nucleus had been inactivated (androgenesis). This methodology, allows the recovery of the population from male nuclear DNA. The cryopreservation of embryos could provide the only possibility of preserving the male and female genomic DNA as well as the cytoplasmic components of the original population, including the important contribution of the mitochondrial DNA. Taking into account the proliferation of many aquatic species, even a rate of success as low as 1% of fry after freezing/thawing could provide individuals enough to recover stable populations.

1.3 Other fields of application

Ecotoxicology has often been postulated as one of the fields that could benefit from the cryopreservation of fish, sea urchin or shellfish embryos [15-17]. According to Billard and Zhang [18], fish eggs and embryos are of particular interest for ecotoxicological testing because of their environmental relevance and sensitivity. Oyster and clam larvae [19] or sea urchin embryos [20] have also been used in the evaluation of water pollutants. Standardization of their use as biomarkers requires the use of embryos with stable and definite characteristics, and cryopreservation could provide a source for continuous supply of this biological material to any laboratory in the world.

Other candidates for cryopreservation are the strains created by genetic modification, such as polyploids used in the aquaculture industry as well as knockout or transgenic lines developed for production in farms (i.e., transgenic salmons) or for research purposes in the fields of developmental biology, genetics or cell biology among others. The design of a cryopreservation protocol with the aim of preserving these genetically and biotechnologically valuable lines could include complex procedures in order to overcome the special difficulties that characterize fish and shellfish embryos, because it would be applied at laboratory scale by qualified technicians, in a well equipped laboratory. Moreover, for the storage of valuable strains or lines low hatching rates can be acceptable in some cases. These factors suggest than this field will probably be the first target for the application of embryo cryopreservation technologies.

2. CRYOPRESERVATION OF INVERTEBRATE EMBRYOS

In marine invertebrates, several studies have been conducted in order to cryopreserve both embryos and larvae. Recent progress in the cryopreservation of shellfish embryos has shed light on the feasibility of this technique for establishing gene banks and manipulating spawning programs as reviewed by Lin et al .[10]. Most of these studies have focused on the Pacific oyster, *Crassotrea gigas*, due to its commercial importance worldwide. First successful attempts in this species were reported by Chao et al. [21], Gwo [22] and Lin et al. [10], who analyzed the toxicity of cryoprotectants, the effect of chilling and the development of freezing protocols with the aim of developing long-term preservation techniques for use in hatchery applications. Chao et al. [21] indicated that the initial quality of embryos is an important factor for cryopreservation success. The use of embryos of good quality should improve the prospects of obtaining useful information and result in higher survival rates, whatever the freezing procedure used: stepwise freezing or vitrification. It was apparent from the results obtained by these authors that there is a wide range of optimal parameters supporting the successful freezing of late embryos. These factors include (1) the direct addition of cryoprotectants (CPAs) instead of stepwise addition, (2) the use of 2 M or 3 M DMSO, and (3) the application of 2.0, -2.5, -3.0, -4.0 and -5°C/ min as freezing rate. Regardless of the conditions the authors observed high incidence of embryos showing active rotary motion 1-2 h after thawing. Similar studies performed by the same group [10] obtained better results when a two-step freezing procedure was developed and optimized. With this procedure, samples were cooled to a temperature near to the freezing temperature (around -10°C), then were submitted to the "seeding" process, usually a slight touch with forceps to trigger the ice nucleation inside the sample, and then samples are frozen to -196°C, the temperature of liquid Nitrogen. The effects of cooling rate, choice of cryoprotectant, and seeding temperature on the survival of late-stage oyster embryos were examined and high survival rates (embryos with motion) of 78 and 83% were achieved using 2 M DMSO or glycerol, respectively, as cryoprotectant. The experimental results obtained in this study indicated that oyster embryos survive after freezing over a broad range of cooling rates ranging from 20.5 to 216°C/min. Nevertheless, survival rates have been evaluated immediately after thawing, and survival at hatch could be compromised, as has been observed in other species.

Although high rates of embryo survival after oyster embryo cryopreservation have been achieved, protocols for the cryopreservation of oyster larvae have been developed in parallel. This area of research has expanded due to potential applications in the oyster industry such as improved management and production of seedstock, increased availability and distribution of selected lines (e.g., disease-resistant lines) and development and maintenance of genetically modified stocks [17]. Moreover, this would allow the availability of high quality larvae all year round. Successful cryopreservation of Eastern oyster, *Crassostrea virginica*, larvae was firstly achieved with the work of Paniagua-Chaves and colleagues [17,23]. These authors obtained ~100% motile trochophora immediately post/thawing using 10 and 15% propylene glycol as cryoprotectant and freezing a small number of embryos (concentration of 125 larvae per macrotube). In order to transfer the method to the industry, large scale cryopreservation of larvae was attempted by these authors. However, survival of thawed larvae decreased as the number of larvae per macrotube increased, and less than 10% survival was achieved when 50,000 larvae were loaded in the same macrotube [17]. For production purposes higher survival rates should be achieved and survival at further larval stages should be analyzed. Although most of the studies have been developed in Eastern and Pacific oyster, other species have received attention, such as mangrove oyster, *Crassostrea rhizophorae* and pearl oyster, *Pinctada fucata martensii* [24,25]. Nascimento et al. [24] tested the toxic effect of cryoprotectants in mangrove oyster trochophora and obtained good results with methanol and DMSO, while Choi and Chang [25] reported survival rates after freezing/thawing of 43.1% for trochophores and 91% for D-shape larvae using sugars as cryoprotectants (0.2 M glucose or sucrose) and a freezing rate of 1°C/min.

Cryopreservation protocols have also been successfully applied to larvae of other species such as the sea urchin *Evechinus chloroticus* [26,27]. The most successful freezing regime cooled straws containing larvae in 1.5 M DMSO from 0 to -35°C at 2.5°C/min, held at -35°C for 5 min and then plunged straws into liquid nitrogen. Larval motility was high 2–4 h post-thawing but decreased markedly within 24 h. Some 4-armed pluteus larvae that survived beyond this time developed through to metamorphosis and settled.

Other species in which cryopreservation have been recently tempted for conservation purposes are the lace coral, *Pocillopora damicornis*, and mushroom coral, *Fungia scutaria* [28]. These species are facing severe environmental pressures due to changes in environment conditions such as increase of water temperature, sedimentation, nutrients, and other pollutants, causing a reduction in the overall larval recruitment of many species [29]. Preliminary studies on cryoprotectant toxicity, dry weight, water and cryoprotectant permeability using cold and radiolabeled glycerol, spontaneous ice nucleation temperatures, chilling sensitivity, and settlement of coral larvae have been performed [28]. Both species tested by these authors demonstrated a wide tolerance to cryoprotectants and a high permeability to glycerol, both data being favorable for cryopreservation suitability. However, both species were extremely sensitive to the rate of chilling and the absolute temperature, indicating that vitrification may be an alternative approach to the long-term conservation of this species and that slow freezing is likely to be unsuccessful.

Penaeid embryos and larvae have been the objective for some researchers. Gwo and Lin [30] evaluated the most suitable developmental stage for freezing penaid shrimp (*Penaeus japonicus*), analyzing the sensitivity to several cryoprotectants and the chilling effect and different freezing regimes. Methanol was found to be relatively nontoxic especially at later development stages. The survival rate of nauplii and zoea treated with 10% methanol in sea water and frozen to -15°C was 85% and some nauplii and zoea survived frozen to -25 and -196°C. However, no treatment yielded normal nauplii or zoea after freezing, demonstrating that further optimization of the protocol is required. Other penaeids are also under study and basic information about resistance to cooling or toxicity of different cryoprotectans has been evaluated in species such as *Penaeus stylirostris* and *T. byrdi* [31].

All these studies reflect that embryo and larval cryopreservation is possible in most of the analyzed species. Nevertheless, optimization of freezing protocols and standardization of evaluation procedures is required. Survival immediately after thawing does not ensure that developmental ability of embryo or larvae is not compromised. Invertebrate development is a complex process and the performance of thawed individuals should be checked and ensured before transfer of the cryopreservation protocols to higher scale.

3. PROBLEMS IN FISH EMBRYO CRYOPRESERVATION

3.1 Complex structure

In fishes, as other lower vertebrates, most of the egg cell is full of yolk and suffers partial discoidal cleavage after fertilization. The cell divisions not completely divide the egg, and only the cytoplasm of the blastodisc becomes the embryo, the yolk remaining undivided. The embryos are complex biological systems with three different compartments: a large yolk, a cellular compartment (the developing embryo or blastoderm in early development) and the periviteline space (Fig. 1). The yolk is surrounded by the yolk syncytial layer, the developing embryo is limited by its own cell plasma membranes, and the periviteline space surrounds these compartments and is, in turn, limited by the outer membrane, the chorion, which is formed from the egg viteline membrane [32,33].

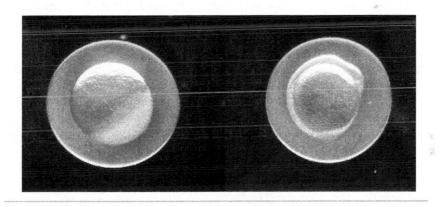

Figure 1 - Zebrafish (*D. rerio*) embryos. Right: High blastula stage; left: 5 somites stage.

Cryopreservation of any tissue or biological structures requires the use of specific cryoprotectant solutions that should permeate the cells and promote a partial sample dehydration to inhibit ice crystals growth and to protect cell components. The particular structure of fish embryos complicates embryo cryopreservation due to different factors that were also summarized by Hagedorn et al. [34]: the large overall size, and the consequent low surface/volume ratio difficult the water and cryoprotectant influx and efflux required for a successful freezing/thawing process; the large yolk compartment content, with a particular osmotic behaviour, high chilling sensitivity, and a high probability of intracellular ice formation that promotes membrane disruption. The different compartments differ in permeability properties and, finally, the membranes surrounding the embryo represent permeability barriers hindering the movement of water and cyoprotectants.

The achievement of an appropriate balance of water and cryoprotectants is a key step in the cryopreservation process and many efforts have been done in the study of embryo permeability and in the development of methods for embryo permeabilization. The chorion, the first semipermeable envelope, allows the entrance of water and other small molecules [33,35], including some cryoprotectants such as DMSO [36,37]. Some species, in particular zebrafish, *Danio rerio*, are relatively easy to dechorionize using digestion with proteases and most of the studies on fish embryo permeability have been performed with dechorionized zebrafish embryos to simplify the analysis of the results [34,38,39]. Nevertheless, most aquacultured species, such as turbot, possess envelopes resistant to dechorionization treatments [40] and the use of dechorinonized embryos for cryopreservation purposes should be discarded.

Methods applied for permeability evaluation goes from the analysis of volumetric changes undergone by the embryos submitted to different anisotonic solutions [38,41] flotation tests based on the density variations suffered by the embryos during cryoprotectant addition and water exchanges [39], the use of HPLC [36,37] or isotope-labeled cryoprotectants [42], impedance microscopy [43] or the analysis with nuclear magnetic resonance spectroscopy [44]. Most of the studies considered the embryos as a homogeneous sphere due to the difficulty in evaluating the actual characteristics in the various compartments, and reports the permeability data of the embryos without considering the differences between the blastoderm and the yolk compartment. The evaluation of the entrance of cryoprotectants separately in both compartments indicates that the water permeability is similar in the yolk sac and in the blastoderm, but the cryoprotectant permeabilty is markedly higher in the blastoderm or embryo compartment than in the yolk sac [36,44]. Other studies on the permeability of different cryoprotectants revealed that methanol, but not DMSO or propylene glycol permeated the whole embryo, including the yolk sac [44]. These results revealed that the yolk syncytial layer and the yolk sac are the main permeability barrier. Respect to the blastoderm, as reviewed by Adams et al. [39], the permeability of the egg plasma membrane decreases following activation and fertilization. Then, as the embryo develops, plasma membrane gradually increases permeability, but always remaining lower than before fertilization. This renders the blastoderm difficult to cryoprotect, especially at early developmental stages. Nevertheless, as has been pointed out, the yolk syncytial layer is much more impermeant to the cryoprotectants, indicating how difficult it is to achieve an appropriate cryoprotection inside the yolk sac and how efforts to develop permeabilization strategies suitable for both embryo compartments should be developed.

3.2 Chilling sensitivity

One of the limiting factors for cryopreservation is the chilling sensitivity. Slow freezing, usually used for cryopreservation of biological samples, requires the progressive cooling of samples from the physiological to the freezing temperature that involves the over-cooling of the sample before the start of the crystallization or freezing process. Chilling sensitivity is specific to the species and the cell type, and has been analyzed in fish embryos from zebrafish [16,34,45], rainbow trout, red sea bream, olive flounder, fathed minow, goldfish, red drum, carp or medaka [46-51]. In all of the species studied a marked sensitivity to chilling was observed in the early developmental stages that progressively decreased during embryo development. The high sensitivity to low temperatures has been related to the inhibition of metabolic and enzymatic processes, which could be especially detrimental for fast developing embryos, such as those of fish [50]. Medaka showed lower sensitivity than other analyzed species. In this model species, the survival rate of 2-4 cell embryos was affected after ony 2 min of chilling at 0°C, whereas hatching rate did not decrease when early gastrula embryos were preserved at 0 or 5°C. Other more sensitive species also showed a similar increase in cooling resistance during development. It is considered that developmental stages beyond 50% epiboly are less sensitive to chilling [16], suggesting that those later stages could be more suitable for cryopreservation. Studies on permeability, previously referred, as well as on cryoprotectant toxicity, also pointed out better results in later stages, supporting the use of advanced embryo development stages for cryopreservation.

Chilling to temperatures near 0°C can be tolerated to a certain degree and this has been applied to delay embryo development with the aim to establish protocols for embryo short-term preservation. Zhang and Rawson [45] observed that a survival of near 100% was obtained after the exposure at 0°C for 10 h of zebrafish embryos 27 to 40 h after fertilization. The addition of methanol or glycerol to the embryo medium increased embryo survival after chilling, as has been reported in carp and zebrafish [16,50]. When zebrafish embryos were exposed to 0°C for 18 h or 24 h in 1 M methanol and 0.1 M sucrose, the percentage of survival was 88% and 82%, respectively. This beneficial effect of cryoprotectant solutions could be helpful to overcome the handicap that chilling sensitivity represents for embryo freezing. Nevertheless, chilling effects accelerates rapidly at subzero temperatures and all the analyzed species died at temperatures near -30°C, making the application of slow-freezing protocols very difficult.

3.3 Embryo size

First attempts to freeze fish embryos were carried out in the late seventies and in the eighties, and were performed with salmonids (rainbow trout, brown trout, brook trout, coho salmon, or Atlantic salmon), as reviewed by Billard and Zhang [18]. These species have particularly big embryos (around 10 mm in diameter) containing a very large yolk sac. Embryo size determines the surface-to-volume ratio, which dramatically lowers with the increase of embryo diameter. Reduction of the surface-to-volume ratio results in a decrease in the exchange between the embryo and the environment, reducing the water and cryoprotectant exchange in such a way that the entrance of cryoprotectants and dehydration are very difficult processes (independently of the pemeability of the envelopes). Results obtained in these studies were discouraging and species with smaller embryos were employed in further analysis. Zebrafish are a fresh water model species easy to breed in aquaria, whose embryos are 1 mm in diameter and are likely to be the most used fish species for cryobiological studies [15,16,33,34,38,44,52-56]. Medaka, another fresh water model fish, is also being study [51,57,58]. Other marine species with relatively small embryos ($\approx$ 1 mm diameter) such as turbot [59], gilthead seabream [60], japanese flounder [61,62] or winter flounder [63] are also being used in cryobiology research.

3.4 Cryoprotectant toxicity

One of the basic information required to develop any cryopreservation procedure is the toxic effect of the cryoprotective agents (CPAs) on the sample under study. Fish embryos are very sensitive to the agents traditionally used as CPAs. Cryoprotectant toxicity increases with the time of exposure, the temperature and the concentration of the agent and changes throughout the developmental stage. It is well known that in fish, marine invertebrate and crustacean embryos, toxicity resistance to CPAs increases during embryonic development [21,50,56,64-67]. The effect of each CPA depends on their chemical properties, and is species-specific. Suzuki et al. [42] reported different results of cumulative mortality for medaka, rainbow trout, pejerrey and carp embryos exposed to concentrations of dimethyl sulfoside (DMSO) up to 5 M. The most commonly used cryoprotectants are DMSO, methanol, glycerol, propanediol, or ethylene glycol (EG), and they are typically used in the range of 1-2 M (reviewed by Billard and Zhang [18]). When compared to other CPAs, DMSO was better tolerated than EG for turbot embryos [37]; gilthead seabream embryos are more resistant to EG than to DMSO or methanol [60], and methanol and propylene glycol are less toxic to zebrafish embryos than DMSO or EG [55,56]. These data reflect the importance of specific studies to be performed for each particular species.

4. FISH EMBRYOS CRYOPRESERVATION

4.1 Slow freezing vs. vitrification

As has been explained, the process of cryopreservation is traditionally performed by a gradual decrease of temperature using linear or more complex freezing rates, from the physiological temperature to -196°C (liquid nitrogen temperature). During this process, cells have the opportunity to progressively dehydrate and incorporate cryoprotectants. The present objective for those samples that are already successfully cryopreserved is the simplification of the cryopreservation protocols and its optimization. However, basic answers are still needed for many biological samples such as teleost embryos in which some of the obstacles that slow freezing entails seem very difficult to overcome. Initial work on fish embryo freezing was focussed on the analysis of cryoprotectant toxicity and the evaluation of the slow freezing approach, establishing that the most suitable freezing rates were in the range of 0.01 and 0.75 °C/min and thawing rates between 8°C/min (in carp) and 30°C/min (zebrafish) [68]. However, all of these studies have shown that fish embryos usually cannot survive below -35°C [18] and that survivals dramatically decrease below -20°C [69], and signs of survival after freezing were never noticed. Slow freezing requires a proper dehydration of the sample, that is not achieved and Zhang et al. [70] considered that there exists an inevitable risk of intracellular ice formation.

The technique considered with highest probability of success in fish embryos is vitrification. This process implies the transformation of an aqueous solution in an amorphous solid with a vitreous aspect without ice crystal formation, due to a drastic increase in viscosity during a very fast freezing [71]. In order to consider that one solution is a vitrifying solution, it shall solidify without ice crystal formation and the recrystallization phenomena cannot happen during thawing. This is usually achieved using high CPAs concentration and high density compounds in the solution. Vitrification was first suggested by Luyet in 1937 [72] and was successfully applied with mouse embryos by Rall and Fahy in 1985 [73]. Vitrification has been applied to other multicompartmental embryos such as those from *Drosophyla* [74], or some marine invertebrates (oyster and hard clam [65]) with varying degree of success.

Recently, Gábor and Masashige [75] published an article in which they stated that they did not find any circumstance in oocyte or embryo cryopreservation where slow freezing offers considerable advantages compared with vitrification, which is in concordance with the majority of published data proving that vitrification methods are more efficient and reliable than any version of slow freezing. This study is also in concordance with the observations made by Hagedorn et al. [76] in teleost. They analyzed, in zebrafish embryos, the temperature of intracellular and extracellular ice crystals formation and conclude that intracellular ice crystals are formed at relatively high temperatures, irrespective of the presence of cryoprotectans in the freezing extender. Data suggest a very low degree of dehydration inside the embryos that promotes ice crystal growth, both in the external medium and the embryo, responsible for serious structural injuries. Consequently, they considered that vitrification, a process preventing the presence of any ice crystal, will be the best option to successfully cryopreserve teleosts embryos. From a practical point of view, vitrification will also have significant advantages for the fish farms. The absence of expensive or time-consuming technical requirements will simplify the transfer of this technique to the industry.

However, despite the fact that vitrification is considered the best option for cryopreservation, most of the rigorous studies developed in the last year, report no survival after the process. Liu et al. [56] reported 80% of zebrafish embryos that displayed intact morphology after being cryopreserved in microscope grids with 10 M methanol, but no survival was observed. Robles et al. [59] reported 49% of turbot embryos with intact morphology after vitrification in a mixture of permeable and non-permeable cryoprotectants that were progressively incorporated to the embryos in several steps and Cabrita et al. [60] obtained 28% of morphologically intact gilthead seabream embryos applying the same vitrification protocol. However, the evaluation of the morphological integrity of the embryo at light microscopy has a high subjective component that tends to be avoided. For this reason other studies performed a more objective evaluation of the vitrification protocol, Robles et al. [77] determined the activity of cytoplasmatic enzymes in control, incubated in cryoprotectants, and vitrified embryos. The aim of this study was to determine the effect of the cryoprotectants and the phenomena of freezing itself not only in the cellular metabolism, but also on the integrity of cellular membranes. Cells with seriously damaged membranes would lose part of their cytoplasmatic content, and therefore a decrease in the activity of cytoplasmatic enzymes would be recorded. The study revealed that, despite the loss of developmental ability after vitrification and thawing, embryos preserve some cellular activity. Nearly 50% of G6PDH activity was preserved in turbot embryos vitrified at 5-somite stage, indicating that total cell lysis was not produced. The decrease of enzymatic activities was more intense in zebrafish embryos, revealing a higher degree of cell lysis and, probably, a higher sensitivity to the process of vitrification. Other studies have estimated the percentage of embryonic cells that survived the process dissociating the blastomeres of the cryopreserved embryos and evaluating cell viability with fluorescent probes. Martínez-Páramo et al. (unpublished results) reported between 25 and 50% of viable cells after vitrification of zebrafish embryos in high blastula stage. This rate of cell survival indicates the appropriate cryopreservation of some of the cells but was considered incompatible with further embryo development.

The lack of success in vitrification could be due to the low entrance of cryoprotectants to the embryo compartments. Solutions used in the process are vitrifiable, but the final concentration inside one or more embryo compartments could be too low to be vitrifiable. In fact, some authors reported recrystallization inside the yolk sac during thawing [56,59], observed as embryo whitening. Recrystallization, a phenomenon that usually causes lethal effects for the cells, predominantly occurs during thawing, and should be avoided in both, slow freezing and vitrification.

Recently, low percentages of survival rates were recorded after the vitrification of embryos from a sub-arctic species, *Pleuronectes americanus* (winter flounder). Thawed embryos progressed with embryo development but none of the survivors were able to hatch [63]. This species naturally express antifreeze proteins (AFPs), necessary to survive in cold waters under 0 °C. These proteins adsorb to ice crystals and prevent ice crystal growth, but also inhibit recrystallization [78]. For this reason different studies consider the incorporation of these proteins in the cryopreservation protocols for different tissues [79,80].

4.2 What has been done until now to overcome the major problems in cryopreservation?

4.2.1 Permeability

Fish embryo cryopreservation is a controvert field due to the lack of answers to many questions, but it is generally accepted that low permeability to water and cryoprotectants is the main problem that hinders the success. What to do in order to solve this problem is a key question for cryobiologists. Studies on water and CPA permeability have shown the embryo behaviour when exposed to different cryoprotectants.

To measure water content, volumetric measures [81] and wet and dry measures have been done, however, these kind of indirect studies are more efficient when the object of study is homogeneous and behave as an ideal osmometer, and obviously it is not the case of teleost embryos. The complexity of fish embryos requires additional and more sophisticated techniques. Electron spin resonance (ESR) [44], magnetic resonance (MR) microscopy [82] and impedance microscopy [43] have been employed for this purpose. It has been stated that the reduction of the osmotically active water within the embryo, as well as the knowledge of the effective permeability of each compartment is crucial for successful cryopreservation [83]. Water permeability is different, throughout embryo development or within fish species and it has been also stated that it changes in the presence of permeating cryoprotectants [41].

As has been stated, to achieve success in cryopreservation, is not only required that the active water would be removed from the embryo, but also that cryoprotective agents would be present in all embryo compartments at enough concentration to confer an effective protection. Again, requirements vary depending on the type of freezing (slow or fast freezing), being the cryoprotectants required in a higher concentration for vitrification.

So far, all these studies have tried to find the primary barrier to cryoprotectants within the embryo, but what is that barrier? Harvey et al. [84] demonstrated using radioactively labelled (DMSO-^{14}C and glycerol-^{3}H) that the outermost barrier, the chorion, produce a delay in the free exchanges of solutes. Treatments with sodium and calcium hypochlorite and digestion with proteases, such as pronase E, have been used to increase chorion permeability, but no significant increases of DMSO influx to the embryo were observed in turbot [37]. Nevertheless, these permeabilization treatments revealed a faster efflux of DMSO when embryos were further washed for cryoprotectant removal, indicating that chorion permeabilization treatments (or dechorionization, if possible) could be necessary steps in a cryopreservation protocol [36], even when permeable cryoprotectants are used. Lubzens et al. [85] using DMSO-^{3}H observed that the CPA penetrated into the perivitelline space and some tissues but the CPA concentration in the tissues was too low to successfully protect the embryo. This fact was also inferred by Cabrita et al. [36] using HPLC. However, Hagedorn and Kleinhans [83] established that neither the blastoderm nor the yolk or the chorion, are the main barriers to the CPAs permeability, and concluded that the yolk syncytial layer is the main barrier responsible in blocking CPAs diffusion.

Knowing the problem in permeability, a solution must be found, and different alternatives to increase permeability have been suggested by different authors. Unfortunately, none of them were, in itself, the definitive solution to our problem. However, all of them have represented a step toward success. Considering that the physical barriers hinder CPA entrance, these compounds were introduced by different methods including ultrasounds, electroporation and microinjection. Bart [86] reported a several-fold increase in calcein absorption when embryos from zebrafish were treated with ultrasound without a significant increase in mortality. Electroporation has proved to have a great potential in cells, however, when it was employed with fish embryos, produced high mortality rates. Microinjection was the technique of choice for studies in CPAs incorporation in fish embryos. Janik et al. [53] and Beirão et al. [87] microinjected traditional CPAs into the yolk of zebrafish and gilthead seabream embryos, respectively, and Robles et al. [88,89] microinjected an antifreeze protein (AFPI) into the yolk sac of turbot and seabream embryos. Hagedorn et al.[82], using MR and Robles et al. [88,89], using confocal microscopy, demonstrated that, independently of the embryo species and the cryoprotectant used, the molecules that were delivered into the embryos could freely diffuse within the yolk but could not exit this compartment when embryos at late stages were treated (Fig. 2). Therefore, the microinjection of CPAs into the yolk could partially prevent ice crystal formation in this compartment but blastoderm would remain unprotected. The microinjection of AFP into the yolk sac or into the yolk sac-blastomere interface of gilthead seabream embryos significantly increased their chilling resistance [89], indicating that this approach could be included into the protocols as a complementary step to improve embryo resistance to freezing. Hagedorn et al. [52] proposed a novel solution to overcome membrane permeability problems in fish embryos. They injected zebrafish embryos with mRNA for the aquaporin 3 (AQP3) protein, a water/solute channel, and demonstrated that zebrafish embryos expressing aquaporin-3 displayed a 6-fold increase in water permeability and a 2.5-fold increase in the permeability to propylene glycol.

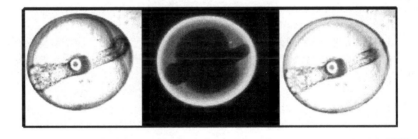

Figure 2 - Images at confocal microscopy of turbot embryos 24 h after microinjection of FITC-labelled AFPI in the periviteline space. Left: Bright light field image of the embryos. Middle: FITC fluorescence located in the chorion and the PVS. Right: Merge of both images.

Further studies of the same group [54] demonstrated that the expression of the aquaporins throughout the cell membranes within 24 h, diminishing after 96 h, without affecting embryo survival. As the aquaporin-3 protein expression diminished 4 days after transfection, this method provides a reversible mean to introduce protective molecules into cells and tissues [54]. The expression of aquaporins was later assayed in medaka oocytes [57] promoting an increase in the permeability to water and cryoprotectants, which did not affect the ability to develop to term. Thus, modified embryos are expected to dehydrate and uptake cryoprotectants more easily, being less prone to ice crystal formation, and assays on the actual benefits of this artificial expression should be evaluated.

However, all these advances were not enough to guarantee the success in cryopreservation of the embryo complex biological model. A more precise and highly sensitive technique or the combination of more than one method that allows the total control on the CPA concentration inside the different compartments would significantly simplify the situation.

4.2.2 Chilling sensitivity

As has been referred, fish embryos are characterized for having a high chilling sensitivity and, therefore, it is also generally accepted that chilling injury needs special attention when a cryopreservation protocol needs to be designed. There are two types of damage caused by low temperatures [90]: the direct damage, also called "cold shock" that are consequence of a fast chilling, and the indirect damage that depends on the freezing rates and usually are manifested after long exposures to low temperatures. Both types of damage are related to the phase transition of the lipid membranes.

The embryo sensitivity to chilling varies among species and throughout development. As has been explained, many studies state that, in most of the teleost species, embryos display better tolerance to chilling after gastrulation [34]. Zebrafish embryos display better tolerances to low temperatures 27 to 40 h after fertilization, and are particularly sensitive at 10 h.

This high sensitivity to chilling has been attributed to the high lipid content present in teleost embryos [45]. Studies on yolk removal by micro-aspiration techniques have demonstrated that a decrease in chilling sensitivity is possible by reducing the yolk content [15,91]. Considering that damage caused by chilling are associated to metabolic disorders caused by the inhibition of enzymatic process at low temperatures [50], some experiments were done in anoxia conditions to arrest development, but no improvements in chilling resistance were obtained [16].

The fact that teleost embryos display high chilling sensitivity should not be incompatible with their successful cryopreservation. Some other embryos with high chilling sensitivity, such as those from *Drosophila* and swine, have been successfully cryopreserved [74,92,93]. One way to partially overcome this problem is avoiding embryos exposure for long periods of time to the critical range of temperatures at which they are especially sensitive. The application of fast freezing rates (around 2000°C/min) used in vitrification, is the best way to reduce the cellular damage caused at these temperatures, but the use of AFPs or their expression into the embryos could also help to solve this particular obstacle.

5. BLASTOMERE, EMBRYONIC STEM CELLS AND PRIMORDIAL GERM CELL CRYOPRESERVATION: ARE THEY FEASIBLE ALTERNATIVES TO FISH EMBRYO CRYOPRESERVATION?

We have considered the whole embryo cryopreservation, however, it may be possible to cryopreserve the diploid genome of fish by using isolated blastomeres instead of intact embryos. What are the advantages of the cryopreservation of these cells, what are their survival rates, do these cells (in the case of blastomeres and stem cells) retain their pluripotency after the process? These are just some of the questions that will be considered in this section.

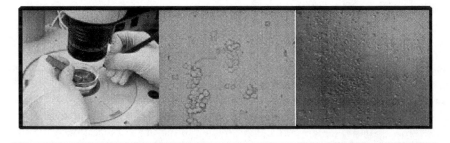

Figure 3 - Dissociation of blastomeres from zebrafish embryos. Right: manual dechorionization and yolk removal under stereoscopic microscope. Middle: dissociated blastomeres. Left: dissociated blastomeres in the culture media. (Martínez-Páramo, with permission).

The first question is easily answered. The cryopreservation of these cells benefit the conservation of the genetic patrimony, in particular in those species whose embryos cannot be successfully cryopreserved. It can contribute in the creation of gene banks for endangered species or the maintenance of commercially important farming lines. But, in addition, this technique has an important repercussion in the field of bioengineering, since we can generate a chimera introducing the blastomere into a host embryo or we can also produce a clone by nuclear transplantation into an enucleated egg. These cells can be easily transfected and were able to generate normal progenies. Moreover, Kobayashi et al. [97] described a protocol for use with rainbow trout primordial germ cells and documents the restoration of live fish from gametes derived from those cryopreserved progenitors.

Survival rates for cryopreserved blastomeres differ depending on the species and the cryopreservation protocol used (Fig. 2). Strussman [94] attempted to cryopreserve blastomeres from three species representative of marine, estuarine, and freshwater environments: whiting, *Sillago japonica,* pejerrey, *Odontesthes bonariensis,* and medaka, *Oryzias latipes,* obtaining survival rates ranging from 19.9% (whiting) and 67.4% (pejerrey), and stated that successful cryopreservation of fish blastomeres requires cooling rates around or slower than 21.0°C/min. But, are those cryopreserved blastomeres pluripotent after thawing? To examine the pluripotency of cryopreserved blastomeres, Kusuda and his colleagues [98] cryopreserved blastomeres from blastula of goldfish (*Carassius auratus*) obtaining 44 to 55% of survival after thawing. They transplanted the freezing/thawed blastomeres into a blastula of triploid crucian carp (*C. a. longsdorfii*) and obtained chimeric embryos with a survival rate of 41.6%. They observed that transplanted blastomeres were histologically identified in various organs derived from all three germ layers. A primordial germ cell differentiated from a cryopreserved blastomere was also detected suggesting with these results that blastomeres retain their pluripotency after cryopreservation and are able to differentiate into both somatic and germ cell lines.

Another important point that should be addressed in this section is the evaluation of the potential damage that can be induced by cryopreservation. In particular, when the aim of cryopreservation is gene banking for conservation programs, we should ensure that the protocol we are using for cryopreservation is not producing a significant damage on DNA. Kopeika et al. [99] analysed the mitochondrial DNA in zebrafish blastomeres after cryopreservation and observed that cryopreservation significantly increased the frequency of mutation; they state that more investigations are needed to know if such mutations interfere with overall function of the cell and to find out if similar changes occur in the nuclear DNA. A better understanding of cryopreservation effects at DNA level will be an important tool for the optimization of the protocols.

Blastomere, embryonic stem cells and PGC cryopreservation can represent a present alternative to fish embryo cryopreservation, and allows the conservation of genetically modified cells that sometimes have not been cultured yet, such as zebrafish PGCs.

6. FUTURE IN FISH EMBRYO CRYOPRESERVATION: WHAT SHOULD BE DONE NEXT

6.1 Use of new CPAs

Efforts to develop cryopreservation methods for aquatic species have focused almost exclusively on the use of penetrating or permeable cryoprotectants due to their historic use in mammalian cell and tissue cryoperservation [100,101] and their ability to confer intracellular cryoprotection. While other cryoprotectants have been proposed over the years, including nonpermeable polymers and sugars [102-104], very few have been adopted for use in clinical and/or commercial cryopreservation of mammalian cells and tissues or have been used in fish embryo preservation. As the focus of fish embryo cryopreservation is moving away from traditional cryopreservation techniques and is focusing more on newer techniques like vitrification, the limits of utilizing traditional permeable cryoprotectants have been made increasingly apparent. The search for novel cell protectants and methods for fish embryo protection are underway and early results have identified a promising alternative approach for the cryopreservation of fish embryos.

Studies have shown that in the absence of traditional cryoprotectants, comparable levels of cryoprotection can be achieved using low concentrations of intracellular protectants. Intracellular protectants were first identified by researchers studying the adaptive strategies used by natural systems that survive desiccation and freezing. Numerous plant, insect, invertebrate and vertebrate species have been shown to have the capacity to accumulate high concentrations of amphiphilic solutes (reviewed in [105,106]), proteins (reviewed in [106,107]) and/or disaccharides [108-110] in preparation of or response to severe environments. The intracellular protectant that has attracted the most attention has been the disaccharide trehalose. In 1975, Madin and Crowe reported that the nematode, *Aphelenchus avenae* accumulated large amounts of trehalose in response to desiccation [111]. This finding prompted Crowe to actively investigate the mechanism by which trehalose protected cells during freezing and desiccation. He found that trehalose could be used to preserve the structure of liposomes and proteins during freeze-drying [112,113]. However, for trehalose to be maximally effective at protecting liposomes against the damaging effects of desiccation and dry storage, it needed to be present on both sides of the membrane [112]. The protective effect of inducing intracellular trehalose accumulation intracellularly has subsequently been demonstrated by the successful freeze-drying of naturally desiccation-sensitive bacteria [114], yeast [115] and plant cells [116].

The use of intracellular sugars for the preservation of mammalian cells and tissues was first reported by Beattie and colleagues when they demonstrated the protective effect of intracellular trehalose and dimethyl sulfoxide on the cryopreservation of pancreatic islets [117]. To date, intracellular trehalose has been used to cryopreserve and/or desiccate human and porcine platelets [118-120], fibroblasts [121], keratinocytes [121], human [122] and mouse oocytes [123], kidney cells [124], hematopoietic [125] and mesenchymal stem cells [126] and red blood cells [127]. Studies on the effectiveness of intracellular trehalose in aquatic species have not been reported.

While the majority of the reports on the use of intracellular protectants in mammalian cell cryopreservation have focused on the use of intracellular sucrose and trehalose, other impermeant substances have been used as intracellular protectants. For example, artificially elevated levels of intracellular potassium have been used to protect human red blood cells during freezing and thawing [128,129]. The formation of innocuous intracellular ice in confluent cell monolayers has also been suggested as a novel "intracellular cryoprotectant" [130,131].

Finally the microinjection of naturally occurring antifreeze proteins into fish oocytes and embryos has been used to enhance the hypothermic storage and/or cryopreservation of aquatic species used in aquaculture (reviewed in [78]). These proteins, extracted from different Arctic fish, are very good candidates for being used in aquatic animals due to their lack of toxicity and their beneficial effects of freezing resistance [89]. Nevertheless, more studies are required in order to develop a simpler method of delivery which allows the AFP distribution inside all the embryo compartments as well as to determine the optimal concentration and type of AFP used in each species.

The molecules that are being developed for use as intracellular protectants have, in most cases, been used unsuccessfully in the past as impermeant cryoprotectants.

However, when present intracellularly, these normally nonpermeant molecules provide extraordinary protection to cells. We have recently presented a detailed summary of the proposed mechanisms of action of intracellular protectants and the challenges facing their use [132]. Overcoming the impermeability of fish embryos to cryoprotectants and/or developing novel means for the intracellular accumulation of normally impermeable molecules is absolutely necessary before this novel approach to fish embryo cryopreservation can be further developed.

6.2 Exploration of new techniques to increase CPA concentration into the embryos

A number of approaches have been developed over the past 10-15 years to transiently permeabilize cells and tissues (reviewed in [132-134]). Of these methods, three technologies are the most likely to emerge as the most applicable for use in fish embryo cryopreservation: genetic transfection of transmembrane pores, genetic transfection of cryoprotectant synthesis genes and/or laser-induced permeabilization.

Transmembrane pore formation: As has been previously quoted, the formation of transmembrane channels or pores in oocytes and embryos of aquatic species has been used as a technique to increase the intracellular accumulation of protective molecules. As we have explained, to improve the cryopreservation efficiency of zebrafish embryos, Hagedorn et al. demonstrated that the microinjection of the aquaporin-3 mRNA resulted in the stable expression of the aquaporin-3 channel and a significant increase in the accumulation of propylene glycol in the yolk of developing zebrafish [52]. In a similar fashion to the work with aquaporin-3, the induced expression of nonnative cryoprotectant transporters in the membrane of developing fish embryos may provide for sufficient intracellular accumulation of protective molecules to confer stable frozen storage. Recently, Kikawada et al. have reported on the successful cloning of a trehalose-specific transporter from the anhydrobiotic insect *Polypedilum vanderplanki* and the ability to induce trehalose uptake into mammalian cells by transfection of this membrane transporter into murine fibroblasts, Chinese hamster ovary cells and human hepatoma cells [135]. The use of this technology in aquatic species awaits future experimentation.

Transfection of CPA synthesis genes: The genetic engineering of mammalian cells to express foreign genes that can induce the intracellular production of protective molecules has been used to increase the tolerance of cells to the extreme environments encountered during desiccation and dry storage. Initially developed as a technique to introduce intracellular sugars into *E. coli* [136,137], the insertion of genes responsible for the coding of sucrose-6-phosphate synthase [138] and trehalose-6-phosphate synthase [124,139,140] has been conducted with mammalian cells. The gene products, when expressed in conjunction with a respective phosphatase, function to convert uridine diphosphate glucose (UDP-Glc) into sucrose and trehalose. Accumulation of millimolar concentrations of intracellular sucrose [137] and trehalose [139] have been reported using this genetic engineering approach. As advances in molecular genetics and biotechnology lead to improvements in transfection efficiencies, stability of gene expression and enhancements to overall production, the utility of genetic engineering for the cytoplasmic accumulation of protective molecules may become a valuable technique for the cryopreservation of fish embryos.

This approach could be particularly interesting for the transfection with antifreeze proteins (AFPs). AFPs natural expression seems confer a higher resistance to cryopreservation in winter flounder embryos [63], and their introduction in other species did not have negative effects on development ([89], Martínez-Paramo et al., unpublished results). Moreover, transfection of AFP genes and or promoters has been developed in several species. AFP promoters have been used for transgenesis experiences with different fishes [141], and a line of AFP transgenic salmon has been patented (United States Patent 5545808) with the aim to facilitate farming in cold waters. These experiences could be helpful in the development of methods for the artificial expression of antifreeze proteins during embryo development in other species, in order to improve its cryoresistance.

Laser-induced permeabilization: The application of high-intensity femtosecond (10^{-15} s) lasers for noninvasive reversible permeabilization of zebrafish has recently been described [142]. The focusing of femtosecond laser pulses produced from a titanium sapphire laser oscillator with a high numerical aperture microscope objective has been used to form precise sub-micron (800 nm) size optical pores in the cell membrane. Absorption of femtosecond laser pulses by nonlinear multiphoton absorption and ionization leads to multiphoton electronic excitation [143], whereby energy is transported to the liberated electrons without thermal diffusion to adjacent cellular material [144,145]. Femtosecond lasers can therefore be used to manipulate biological material with minimal damage from thermal heating [146-149]. Kohli et al. have shown that kidney epithelial cells can be selectively targeted and reversibly permeabilized with a femtosecond laser allowing for rapid cytoplasmic uptake (<0.3 s) of 0.2 M sucrose with minimal damage (>90% cell survival) [150]. Similarly, laser-induced pore formation was shown to allow for the loading of fluorescent reporter molecules and GFP-tagged DNA into chorionated and dechorionated zebrafish embryo without a significant decrease in embryo survival [142]. The contact-free and non-invasive features of laser-induced membrane permeabilization make it an attractive technology for the selective targeting of specific membranes for reversible permeabilization in the developing fish embryo.

With the development of methods for the reversible permeabilization and/or accumulation of intracellular protectants, such as sugars or AFPs, studies on the effectiveness of intracellular protectants in fish embryo cryopreservation will emerge. The practical advantages afforded by intracellular protectants are numerous. Effective at lower concentrations and with reduced chemical toxicity, intracellular protectants like trehalose, sucrose and antifreeze proteins, may not need to be removed following thawing. This would simplify processing and distribution requirements and provide a manageable method for long distance transportation and remote on-site storage of cryopreserved fish embryos. In addition, as currently identified intracellular protectants have been shown to exhibit unique physico-chemical and biophysical properties which protect cells and biomolecules during a wide range of environmental stresses,(reviewed in [132]) they may see use as "universal" protectants in the cryopreservation of gametes and reproductive tissue from a wide range of aquatic species. While intracellular protectants, such as trehalose and other disaccharides challenge our current understanding of the mechanisms of cell injury and protection, significant efforts are underway to improve our understanding of intracellular protectants and to develop these cryoprotective agents for practical use in cell and tissue cryopreservation.

6.3 Development of new methods to objectively assess cryopreservation-induced damage in fish embryos

There is still a long way to walk to succeed in fish embryo cryopreservation. There are many basic problems that have not been solved yet and they are critical for the success of the process. In this sense, it is necessary to analyze in detail the processes that take place during freezing/thawing and the specific damages that are induced in the embryos, in order to reach a gradual increase of knowledge that helps us to find a global solution. Most of the performed studies have evaluated the hatching rates or only report morphological observations, such as the chorion aspect, yolk and blastoderm opacity or the size of periviteline space, which have a high subjective component. Neither survival rate nor morphological appearance provide information about the events taking place at cellular level and the specific damage that are produced and that hinders embryo development. More attention should be paid to other parameters such as cell activity after thawing (viability, cell proliferation and survival in culture), yolk modifications during freezing that could interfere with the correct yolk uptake by cells, gene expression during early embryo development or metabolic activity after cryopreservation. This type of specific studies will contribute for sure to a better understanding of the process, and definitely demonstrate that we are working in the right direction.

6.4 Choice of the species

It is a well-known fact that sensitivity to cryopreservation is specific for the species and cell type. In the case of fish embryos we have previously described the importance of the embryo size, or the species-specific toxicity of cryoprotectants. Moreover, there are other factors that could determine differences in the suitability of each particular species for cryopreservation. Sperm from freshwater species is considered more difficult to cryopreserve than milt from marine species. This has been attributed to different mechanisms of osmorregulation according to the environment and to a different plasma membrane composition [151], and could also affect the response of embryos to cryopreservation. Nevertheless, most studies on fish embryo cryopreservation have been developed with freshwater fishes (zebrafish, medaka or carp). Robles et al. [77] working with a salt water (turbot) and a fresh water (zebrafish) species demonstrated that turbot embryos suffered a lower level of cellular damage than zebrafish embryos after vitrificacion and thawing. The best results in terms of embryo survival have been reported for two marine species: winter flounder survived but did not hatch after vitrification [63] and flounder embryos were reported to hatch after the same process [62]. Physiological temperature of the species could also be an important factor. Cooling from temperatures around 30°C to 0°C is a damaging process for cell metabolism and induces an intense oxidative stress [152,153]. It is likely that cryopreservation could be more stressful for warm water species than for cold-water ones. This could be the reason for the best results obtained with winter flounder, a sub-Arctic species [63] than with turbot (a cold water species) or the zebrafish (a warm-water fish) [77]. None of these factors have been systematically analyzed, but deserve more attention because the choice of an appropriate model is a key step for the success of the technology.

In this chapter, basic problems in fish embryo cryopreservation and recent achievements have been considered. Possible alternatives such as PGC and blastomere cryopreservation have been also suggested. However fish embryo cryopreservation is still a very desirable objective and the importance and advantages derived from it are numerous. The input from different scientific fields and the joined efforts of several multidisciplinary groups will culminate, hopefully in the near future, in the success of the technique for this complex biological model.

7. REFERENCES

1-Whittingham, D.G., Leibo, S.P., and Mazor, P., Survival of mouse embryos frozen to -196°C, *Science*, 178, 411, 1972.

2-Wilmut, I. and Rowson, I.E.A., Experiments on the low temperature preservation of cow embryos, *Vet Rec*, 92, 686, 1973.

3-Rall, W.F., Advances in the cryopreservation of embryos and prospects for the application to the conservation of salmonid fishes, in *Genetic conservation of salmonid fishes*, Cloud, J. and Thorgaard, G., Eds., Plenum Press, New York, 1992.

4-Trounsen, A. and Mohr, L.,Human pregnancy following cryopreservation, thawing and transfer of an eight-cell embryo, *Nature*, 305, 707, 1983.

5- Dobrinsky, J.R., Pursel, V.G., Long, C.R., and Johnson, L.A., Birth of piglets after transfer of embryos cryopreserved by cytoskeletal stabilisation ands vitrification, *Biol Reprod*, 62, 564, 2000.

6-Dobrinski, J.R., Advancements in cryopreservation of domestic animal embryos, *Theriogenology*, 57, 285, 2002.

7-Nguyen, B.X, Sotomaru. Y., Tani, T., Kato, Y,. and Tsunoda, Y., Efficient cryopreservation of bovine blastocysts derived from nuclear transfer with somatic cells using partial dehydration and vitrification, *Theriogenology*, 53, 1439, 2000.

8-Ushijima, H., Yamakawa, H., and Nashjima, H., Cryopreservation of bovine pre-morula stage in vitro matured/in vivo fertilized embryos after delipidation and before use in nucleus transfer, *Biol Reprod*, 60, 534, 1999.

9-Han, Y.M., Kim, S.H., Park, J.S., Park, I.Y., Kand Y.K., Lee, C.S., Koo, B.D., Lee, T.H., Yu D.Y., Kim, Y.H., Lee, K.J., and Lee, K.K., Blastocyst viability and generation of transgenic cattle following freezing of in vitro produced, DNA-inected embryos, *Anim Reprod Sci*, 63, 53, 2000.

10-Lin, T.-T., Chao, N.-H., and Tung, H.-T., Factors Affecting Survival of Cryopreserved Oyster (*Crassostrea gigas*) Embryos, *Cryobiology*, 39, 192, 1999.

11-Kirchoffer, A. and Hefti, D., in: *Conservation of endangered freshwater fish in Europe, Advances in Life Sciences*, Kirchoffer, A. and Hefti, D., Eds, Basel: Birkhäuser Verlag, 1996.

12-Riabova, G.D., Klimonov, V.O., and Rubtsova, G.A., Variation in morphometric and genetic characteristics of stellate sturgeon juveniles raised at different densities, *Genetika*, 42, 244, 2006.

13-Babiak, I., Dobosz, S., Goryczko, K., Kuzminski, H., Brzuzan, P., and Ciesielski, S., Androgenesis in rainbow trout using cryopreserved spermatozoa: the effect of processing and biological factors, *Theriogenology*, 57, 1229, 2002.

14-Grunina, A.S., Recoubratsky, A.V., Tsvetkova, L.I., and Barmintsev, V.A., Investigation on dispermic androgenesis in sturgeon fish. The first successful production of androgenetic sturgeons with cryopreserved sperm, *Int J Refrig*, 29, 3, 379, 2006.

15-Liu, X.H., Zhang, T., and Rawson, D.M., Effect of cooling rate and partial removal of yolk on the chilling injury in zebrafish (*Danio rerio*) embryos, *Theriogenology*, 55, 1719, 2001.

16-Zhang, T., Liu, X-H., and Rawson, D.M., Effects of methanol and developmental arrest on chilling injury in zebrafish (*Danio rerio*) embryos, *Theriogenology*, 59, 1545, 2003.

17-Paniagua-Chavez, C.G. and Tiersch, T.R., Laboratory Studies of Cryopreservation of Sperm and Trocophore Larvae of the Eastern Oyster, *Cryobiology*, 43, 211, 2001.

18-Billard, B. and Zhang, T., Techniques of genetic resource banking in fish, in: *Cryobanking the genetic resources: wildlife conservation for the future?*, Watson, P.F. and Holt, W.V., Eds., Taylor and Francis Books, London, 2001, 145.

19-McFadzen, I.R.D., Growth and survival of cryopreserved oyster and clam larvae along a pollution gradient in the German Bight, *Mar Ecol Prog Ser*, 91, 215, 1992.

20-Hose, J.E., Potential uses of sea urchin embryos for identifying toxic chemicals: description of a bioassay incorporating cytologic, cytogenetic and embryologic developments, *J Appl Toxicol*, 5, 245, 1985.

21-Chao, N.H., Chiang, C.P., Hsu, H.W., Tsai, C.T., and Lin, T.T.,Toxicity tolerance of oyster embryos to selected cryoprotectants, *Aquat Living Resour*, 7, 99, 1994.

22-Gwo, J.C., Strawn, K., and Arnold, C.R., Changes in mechanincal tolerance and chilling sensitivity of red drum (*Sciaenopus ocellatus*) embryos during development, *Theriogenology*, 43, 1155, 1995.

23-Paniagua-Chavez, C.G., Buchman, J.T., Supan, J. E., and Tiersch, T. R., Settlement and growth of Eastern oysters produced from cryopreserved larvae, *Cryo-Lett*, 19, 283, 1998.

24-Nascimento, A., Leite, N.L.M.B.,, Araújo, M.M.S., Sansone, G., Pereira, S.A., and Espírito Santo, M., Selection of cryoprotectants based on their toxic effects on oyster gametes and embryos, *Cryobiology*, 51, 113, 2005.

25-Choi, Y.H. and Chang, Y.J., The influence of cooling rate, developmental stage, and the addition of sugar on the cryopreservation of larvae of the pearl oyster *Pinctada fucata martensii*, *Cryobiology*, 46, 190, 2003.

26-Adams, S.L., Kleinhans, F.W., Mladenov, P.V., and Hessia, P.A., Membrane permeability characteristics and osmotic tolerance limits of sea urchin (*Evechinus chloroticus*) eggs, *Cryobiology*, 47, 1, 2003.

27-Adams, S.L., Hessian, P.A., and Mladenov, P.V., The potential for cryopreserving larvae of the sea urchin, *Evechinus chloroticus, Cryobiology*, 52, 139, 2006.

28-Hagedorn, M., Pan, R., Cox, E.F., Hollingsworth, L., Krupp, D., Lewis, T.D., Leong, J.C., Mazur, P., Rall, W.F., MacFarlane, D.R., Fahy, G., and Kleinhans, F.W., Coral larvae conservation: Physiology and reproduction, *Cryobiology*, 52, 33, 2006.

29-Fabricius, K.E., Effects of terrestrial runoff on the ecology of corals and coral reefs: review and synthesis, *Mar Pollut Bull*, 50, 125, 2005.

30-Gwo, J.-C. and Lin, C.-H., Preliminary experiments on the cryopreservation of penaeid shrimp (*Penaeus japonicus*) embryos, nauplii and zoea, *Theriogenology*, 49,1289, 1998.

31-Alfaro, J., Komen, J., and Huisman, E.A., Cooling, cryoprotectant and hypersaline sensitivity of penaeid shrimp embryos and nauplius larvae, *Aquaculture*, 195, 353, 2001.

32-Kalicharan, D., Jongebloed, W.L., Rawson, D.M., and Zhang, T., Variation in fixation techniques for field emission SEM and TEM of zebrafish (*Brachydanio rerio*) embryo inner and outer membranes, *J Electron Microsc*, 47, 645, 1998.

33-Rawson, D.M., Zhang, T., Kalicharan, D., and Jongebloed, W.L., FE-SEM and TEM studies of the chorion, plasma membrane and syncytial layers of the gastrula stage embryo of the zebrafish (*Brachydanio rerio*): a consideration of the structural and functional relationship with respect to cryoprotectant penetration, *Aquac Res*, 31, 325, 2000.

34-Hagedorn, M., Kleinhans, F.W., Wildt, D.E., and Rall, W.F., Chill sensitivity and cryoprotectant permeability of dechorionated zebrafish embryos, *Brachydanio rerio*, *Cryobiology*, 34, 251, 1997.

35-Coward, K., Bromage, N.R., Hibbitt, O., and Parrington, J., Gamete physiology, fertilization and egg activation in teleosts fish, *Rev Fish Biol Fisher*, 12, 33, 2002.

36-Cabrita, E., Robles, V., Chereguini, O., de Paz, P., Anel, L., and Herraéz., M.P., Dimethyl sulfoxide influx in turbot embryos exposed to a vitrification protocol, *Theriogenology*, 60, 463, 2003.

37-Cabrita, E., Chereguini, O., Luna, M., de Paz, P., and Herráez, M.P., Effect of different treatments on the chorin permeability to DMSO of turbot embryos (*Scophtalmus maximus*), *Aquaculture*, 221, 593, 2003.

38-Zhang, T. and Rawson, D.M., Permeability of dechorionated 1-cell and 6-somite stage zebrafish (*Brachydanio rerio*) embryos to water and methanol, *Cryobiology*, 32, 239, 1998.

39-Adams, S.L., Zhang, T., and Rawson, D.M., The effect of external medium composition on membrane water permeability of zebrafish (*Danio rerio*) embryos, *Theriogenology*, 64, 1591, 2005.

40-Robles, V., Cabrita, E., de Paz, P., and Herráez, M.P., Studies on chorion hardening inhibition and dechorionization in turbot embryos, *Aquaculture*, 262, 535, 2007.

41-Zhang, T. and Rawson, D.M., Permeability of the vitelline membrane of zebrafish (*Brachydanio rerio*) embryos to methanol and propane-1,2-diol, *Cryoletters*, 17, 273, 1996.

42-Suzuki, T., Komada, H., Takai, R., Arii, K., and Kozima, T.T., Relation between toxicity of cryoprotectant DMSO and its concentration in several fish embryos, *Fish Sci*, 61, 193, 1995.

43-Zhang, T., Wang, R.Y., Bao, Q-Y., and Rawson, D.M., Development of a new rapid measurement technique for fish embryo membrane permeability studies using impedance spectroscopy, *Theriogenology*, 66, 982, 2006.

44-Hagedorn, M., Kleinhans, F.W., Freitas, R., Liu, J., Hsu, E., Wildt, D.E., and Rall, W.F., Water distribution and permeability of zebrafish embryos, Brachydanio rerio, *J Exp Zool*, 278, 356, 1997.

45-Zhang, T. and Rawson, D.M., Studies on chilling sensitivity of zebrafish (*Brachydanio rerio*) embryos, *Cryobiology*, 32, 239, 1995.

46-Sasaki, K., Kurokura, H., and Kasahara, S., Changes in low temperature tolerance of the eggs of certain marine fish during embryonic development, *Comp Biochem Physiol*, 91A, 183, 1988.

47-Cloud, J.G., Ehrdal, A.L., and Graham, E.F., Survival and continued normal development of fish embryos after incubation at reduced temperatures, *Trans Am Fish Soc*, 117, 503, 1988.

48-Liu, K., Chou, T., and Lin, H., Cryopreservation of goldfish embryos after subzero freezing, *Aquat Living Resour.*, 6, 145, 1993.

49-Gwo, J.-C., Cryopreservation of oyster (*Crassostrea gigas*) embryos, *Theriogenology*, 43, 1163, 1995.

50-Dinnyés, A., Urbányi, B., Baranyai, B., and Magyary, I., Chilling sensitivity of carp (*Cyprinus carpio*) embryos at different developmental stages in the presence or absence of cryoprotectants: work in progress, *Theriogenology*, 50, 1, 1998.

51-Valdez, D.M., Miyamoto, A., Hara, T., Edashige, K., and Kasai, M., Sensitivity to chilling of medada (*Oryzias latipes*) embryos at various developmental stages, *Theriogenology*, 64, 112, 2005.

52-Hagedorn, M., Lance, S.L., Fonseca, D.M., Kleinhans, F.W., Artimov, D., Fleischer, R., Hoque, A.T.M.S., Hamilton, M.B., and Pukazhenthi, B.S., Altering fish embryos with aquaporin-3: An essential step toward successful cryopreservation, *Biol Reprod*, 67, 961, 2002.

53-Janik, M., Kleinhans, F. W., and Hagedorn, M., Overcoming a Permeability Barrier by Microinjecting Cryoprotectants into Zebrafish Embryos (*Brachydanio rerio*), *Cryobiology*, 41, 25, 2000.

54-Lance, S.L., Peterson, A.S., and Hagedorn, M., Developmental expression of aquaporin-3 in zebrafish embryos (*Danio rerio*), *Comp Biochem Physiol*, 138, 251, 2004.

55-Zhang, T. and Rawson, D.M., Feasibility studies on vitrification of zebrafish (*Brachydanio rerio*) embryos, *Cryobiology*, 33, 1, 1996.

56-Liu, X.H., Zhang, T., and Rawson, D.M., Feasibility of vitrification of zebrafish (*Danio rerio*) embryos using methanol, *Cryo Letters*, 19, 309, 1998.

57-Valdez, D.M., Hara, T., Miyamoto, A., Seki, S., Jin, B., Kasai, M., and Edashige, K., Expression of aquaporin-3 improves the permeability to water and cryoprotectants of immature oocytes in the medaka (*Oryzias latipes*), *Cryobiology*, 53, 160, 2006.

58-Routray, P., Suzuki, T., and Augusto, C., Strüssmann and Rikuo Takai Factors affecting the uptake of DMSO by the eggs and embryos of medaka, *Oryzias latipes*, *Theriogenology*, 58, 1483, 2002.

59-Robles, V., Cabrita, E., Real, M., Álvarez, R., and Herráez, M.P., Vitrification of turbot embryos: preliminar assays, *Cryobiology*, 47, 30, 2003.

60-Cabrita, E., Robles, V., Wallace, J.C., Sarasquete, M.C., and Herráez, M.P., Preliminary studies on the cryopreservation of gilthead seabream (*Sparus aurata*) embryos, *Aquaculture*, 251, 245, 2006.

61-Edashige, K., Valdez, D.M., Hara, T., Saida, N., Seki, S., and Kasai, M., Japanese flounder (*Paralichthys olivaceous*) embryos are difficult to cryopreserve by vitrification, *Cryobiology*, 53, 96, 2006.

62-Chen, S.L. and Tian, Y.S., Cryopreservation of flounder (*Paralichthys olivaceous*) embryos by vitrification, *Theriogenology*, 63, 1207, 2005.

63-Robles, V., Cabrita, E., Fletcher, G., Shears, M., King, M., and Herráez, M.P., Vitrification assays with a cold resistant sub-arctic fish species, *Theriogenology*, 64, 1633, 2005.

64-Simon, C., Dumont, P., Cuende, F.X., and Diter, A., Determination of suitable freezing media for cryopreservation of *Penaeus indicus* embryos, *Cryobiology*, 31, 245, 1994.

65-Chao, N.-H., Lin, T.-T., Chen, Y.-J., Hsu, H.-W., and Liao, I.-C., Cryopreservation of late embryos and early larvae in the oyster and hard clam, *Aquaculture*, 155, 31, 1997.

66-Newton, S.S. and Subramoniam, T., Cryoprotectant toxicity in penaeid prawn embryos, *Cryobiology*, 33, 172, 1996.

67-Urbanyi, B., Baranyai, B., Magyary, I., and Dinnyes, A., Toxicity of methanol, DMSO and glycerol on carp (*Cyprinus carpio*) embryos in different development stages, *Theriogenology*, 47, 408, 1997.

68-Zhang, X.S., Zhao, L., Hua, T.C., Chen, X.H., and Zhu, H.Y., A study on the cryopreservation of common carp (*Cyprinus carpio*) embryos, *Cryo-Letters*, 10, 27. 1989.

69-Gwo, J.C., Cryopreservation of eggs and embryos from aquatic organisms, in *Cryopreservation in Aquatic Species,* Tiersch, T.R. and Mazik, P.M., Eds., The World Aquaculture Society, Baton Rouge, LA, USA, 2000, 211.

70-Zhang, T., Rawson, D.M., and Morris, G.J., Cryopreservation of pre-hatch embryos of zebrafish (*Brachydanio rerio*), *Aquat Living Resour,* 6, 145, 1993.

71-Fahy, G.M., MacFarlane, D.R., Angell, C.A., and Meryman, H.T., Vitrification as approach to cryopreservation, *Cryobiology,* 21, 407, 1984.

72-Luyet, B.J., The vitrification of colloids and of protoplasm, *Biodynamica,* 1, 1, 1937.

73-Rall, W.H. and Fahy, G.M., Ice-free cryopreservation of mouse embryos at -196°C by vitrification, *Nature,* 313, 573, 1985.

74-Mazur, P., Cole, K.W., Hall, P.D., Schreuders, P.D., and Mahowald, J.W., Cryobiological preservation of *Drosophila* embryo, *Science,* 258, 1932, 1992.

75-Gabor, V. and Masashige, K., Improving cryopreservation systems, *Theriogenology,* 65, 236, 2006.

76-Hagedorn, M., Peterson, A., Mazur, P., and Kleinhans, F.W., High ice nucleation temperature of zebrafish embryos: slow-freezing is not an option, *Cryobiology,* 49, 181, 2004.

77-Robles, V., Cabrita, E., de Paz, P., Cuñado, S., Anel, L., and Heráez, M.P., Effect of a vitrification protocol on the Lactate dehydrogenase and Glucose-6-phosphate dehydrogenase and the hatching rates in zebrafish (*Danio rerio*) and turbot (*Scophthalmus maximus*) embryos, *Theriogenology,* 61, 1367, 2004.

78-Fletcher, G.L., Goddard, S.V., and Wu, Y., Antifreeze proteins and their genes: From basic research to business opportunity, *Chemtech,* 30, 17, 1999.

79-Rubinsky, B., Arav, A., and DeVries, A.L., Cryopreservation of oocytes using directional cooling and antifreeze glycoproteins, *Cryo-Lett,* 112, 93, 1991.

80-Rubinsky, B., Arav, A., and DeVries, A.L., The cryoprotective effect of antifreeze glycopeptides from Antarctic fishes, *Cryobiology,* 29, 69, 1992.

81-Leibo, S.P., Fundamental cryobiology of mouse ova and embryos, in *The freezing of mammalian embryos,* Elliot, K. and Whelan, J., Eds., Elsevier, Amsterdam, 1973. 69.

82-Hagedorn, M., Hsu, E.W., Pilatus, U. Wildt, D.E., Rall, W.F., and Blackband, S.J., Magnetic resonance microscopy and spectroscopy reveal kinetics of cryoprotectant permeation in a multicompartmental biological system, *Proc Natl Acad Sci,* USA 93, 7454, 1996.

83-Hagedorn, M. and Kleinhans, F.W., Problems and prospects in cryopreservation of fish embryos, in *Cryopreservation in Aquatic Species,* Tiersch, T.R. and Mazik, P.M., Eds., The World Aquaculture Society, Baton Rouge, LA, USA, 2000, 161.

84-Harvey, B., Kelley, R.N., and Ashwood-Smith, M.J., Permeability of intact and dechorionated zebrafish embryos to glycerol and dimethyl sulfoxide, *Cryobiology,* 20, 432, 1983.

85-Lubzens, E., Ar, A., and Magnus, Y., Steps towards cryopreservation of Japanese ornamental carp embryos: sensitivity to low temperature and permeation of ^{3}H dimethylsulfoxide (DMSO), *Aquaculture,* 1998.

86-Bart, A.N., New approaches in cryopreservation of fish embryos, in *Cryopreservation in aquatic species,* Tiersch, T.R. and Mazik, P.M., Eds., The World Aquaculture Society, Baton Rouge, LA USA, 2000, 179.

87-Beirão, J., Robles, V., Herráez, M.P., Sarasquete, C., Dinis, M.T., and Cabrita, E., Cryoprotectant microinjection toxicity and chilling sensitivity in gilthead seabream (*Sparus aurata*) embryos, *Aquaculture,* 261, 897, 2006.

88-Robles, V., Cabrita, E., Anel, L., and Herráez, M.P., Microinjection of the antifreeze protein type III (AFPIII) in turbot (*Scophthalmus maximus*) embryos: Toxicity and protein distribution, *Aquaculture,* 261, 1299, 2006.

89-Robles, V., Barbosa, V., Herráez, M.P., Matínez Páramo, S., and Cancela, L., The antifreeze protein type I (AFPI) increases seabream (*Sparus aurata)* embryos tolerance to low temperatures, *Theriogenology,* 68, 284, 2007.

90-Chen, C.P. and Walker, V.K., Cold shock and chilling tolerance in *Drosophila, J Insect Physiol,* 40, 661, 1994.

91-Liu, X.H., 2000. Studies on limiting factors relating to the cryopreservation of fish embryos, Thesis, University of Luton, UK, 318, 2000.

92-Nagashima, H., Kashiwazaki, N., Ashman, R.J., Grupen, C.G., Seamark, R.F., and Nottle, M.B., Removal of cytoplasmic lipid enhances the tolerance of porcine embryos to chilling, *Biol Reprod,* 51, 618, 1994.

93-Nagashima, H., Kashiwazaki, N., Ashman, R.J. Grupen, C.G. and Nottle, M.B., Cryopreservation of porcine embryos, *Nature,* 374, 1995.

94-Strussmann, C.A., Nakatsugawa, H., Takashima, F., Hasobe, M., Suzuki, T., and Takai, R , Cryopreservation of isolated fish blastomeres: effects of cell stage, cryoprotectant concentration and cooling rate on post thawing survival, *Cryobiology,* 39, 252, 1999.

95-Fan, L., Collodi, P., Klimanskaya, I., and Lanza, R., Zebrafish embryonic Stem Cells, *Methods in Enzymol,* 418, 64, 2006.

96-Takeuchi, Y., Yoshizaki, G., and Takeuchi, T., Generation of live fry from intraperitoneally transplanted primordia germ cells in rainbow trout, *Biol Reprod,* 69, 1142, 2003.

97-Kobayashi, T. and Takeuchi, Y., Generation of viable fish from cryopreserved primordial germ cells, *Mol Reprod Dev,* 74, 207, 2007.

98-Kusuda, S., Teranishi, T., Koide, N., Nagai, T., Arai, K., and Yamaha, E., Pluripotency of cryopreserved blatomeres of the goldfish, *J Exp Zool,* 301A, 131, 2004.

99-Kopeika, J., Zhang, T., Rawson, D.M., and Elgar, G., Effect of cryopreservation on mitochondrial DNA of zebrafish *(Danio rerio)* blastomere cells, *Mutat Res,* 570, 49, 2005.

100-Polge, C., Smith, A.U., and Parkes, A.S., Revival of spermatozoa after vitrification and dehydration at low temperatures, *Nature,* 164, 666, 1949.

101-Lovelock, J.E. and Bishop, M.W.H., Prevention of freezing damage to living cells by dimethyl sulphoxide, *Nature,* 183, 1394, 1959.

102-Sputtek, A., Singbartl, G., Langer, R., Schleinzer, W., Heinrich, H.A., and Kühnl, P , Cryopreservation of red blood cells with the non-penetrating cryoprotectant hydroxyethyl starch, *CryoLetters,* 16, 283, 1995.

103-Hubalek, Z., Protectants used in the cryopreservation of microorganisms, *Cryobiology,* 46, 205, 2003.

104-Meryman, H.T., Cryoprotective agents, *Cryobiology,* 8, 173, 1971.

105-Rice-Evans, C.A., Miller, N.J., and Paganga, G., Antioxidant properties of phenolic compounds, *Trends Plant Sci,* 2, 152, 1997.

106-Oliver, A.E., Leprince, O., Wolkers, W.F., Hincha, D.K., Heyer, A.G., and Crowe, J.H., Non-disaccharide-based mechanisms of protection during drying, *Cryobiology,* 43, 151, 2001.

107-Blackman, S.A., Obendorf, R.L., and Leopold, A.C., Desiccation tolerance in developing soybean seeds: The role of stess proteins, *Physiologia Plantarum,* 93, 630, 1995.

108-Crowe, J.H., Crowe, L.M., Tablin, F., Wolkers, W.F., Oliver, A.E., and Tsvetkova, N., Stabilization of cells during freeze-drying: The trehalose myth, in: *Life in the Frozen State,* Fuller, B.J., Lane, N., and Benson, E.E., Eds., CRC Press, New York, 2004, 581.

109-Wharton, D.A., Judge, K.F., and Worland, M.R., Cold acclimation and cryoprotectants in a freeze-tolerant antarctic nematode, *Panagrolaimus davidi, J Comp Physiol,* 170, 321, 2000.

110-Pomeroy, M.K., Siminovitch, D., and Wrightman, F., Seasonal biochemical changes in living bark and needles of red pine *(Pinus resinosa)* in relation to adaptation to freezing, *Canadian Journal of Botany,* 48, 953, 1970.

111-Madin, K.A.C. and Crowe, J., Anhydrobiosis in nematodes: Carbohydrate and lipid metabolism during drying, *J Exp Zool,* 193, 335, 1975.

112-Crowe, L.M., Crowe, J.H., Rudolph, A.S., Womersley, C., and Appel, L., Preservation of freeze-dried liposomes by trehalose, *Arch Biochem Biophys,* 242, 240, 1985.

113-Carpenter, J.F., and Crowe, J.H., Modes of stabilization of a protein by organic solutes during desiccation, *Cryobiology*, 25, 459, 1988.

114-Leslie, S.B., Israeli, E., Lighthart, B., Crowe, J.H., and Crowe,L.M., Trehalose and sucrose protect both membranes and proteins in intact bacteria during drying, *Appl Environ Microb*, 61, 3592, 1995.

115-Cerrutti, P., Segovia de Huergo, M., Galvagno, M., Schebor, C., and del Pilar Buera, M., Commercial baker's yeast stability as affected by intracellular content of trehalose, dehydration procedure and the physical properties of external matrices, *Applied Microbiology & Biotechnology*, 54, 575, 2000.

116-Leborgne, N., Teulieres, C., Rols, M.P.T.S., Teissie, J., and Boudet, A.M., Introduction of specific carbohydrates into *Eucalyptus gunnii* cells increases their freezing tolerance, *Eur J Biochem*, 229, 710, 1995.

117-Beattie, G.M., Crowe, J.H., Lopez, A.D., Cirulli, V., Ricordi, C., and Hayek, A., Trehalose: A cryoprotectant that enhances recovery and preserves function of human pancreatic islets after long-term storage, *Diabetes*, 46, 519, 1997.

118-Wolkers, W.F., Walker, N.J., Tablin, F., and Crowe,J.H., Human platelets loaded with trehalose survive freeze-drying, *Cryobiology*, 42, 79, 2001.

119-Wolkers, W.F., Walker, N.J., Tamari, Y., Tablin, F., and Crowe, J.H., Towards a clinical application of freeze-dried human platelets, *Cell Preservation Technology*, 1, 175, 2003.

120-Wolkers, W.F., Looper, S.A., Fontanilla, R.A., Tsvetkova, N.M., Tablin, F., and Crowe, J.H., Temperature dependence of fluid phase endocytosis coincides with membrane properties of pig platelets, *Biochimica et Biophysica Acta*, 1612, 154, 2003.

121-Eroglu, A., Russo, M.J., Bieganski, R., Fowler, A., Cheley, S., Bayley,H., and Toner, M., Intracellular trehalose improves the survival of cryopreserved mammalian cells, *Nat Biotechnol*, 18, 163, 2000.

122-Eroglu, A., Toner, M., and Toth, T.L., Beneficial effect of microinjected trehalose on the cryosurvival of human oocytes, *Fertil Steril*, 77, 152, 2002.

123-Eroglu, A., Lawitts, J.A., Toner, M., and Toth, T.L., Quantitative microinjection of trehalose into mouse oocytes and zygotes and its effect on development, *Cryobiology*, 46, 121, 2003.

124-Guo, N., Puhlev, I., Brown, D.R., Mansbridge, J., and Levine, F., Trehalose expression confers dessication tolerance on human cells, *Nat Biotechnol*, 18 168, 2000.

125-Buchanan, S.S., Gross, S.A., Acker, J.P., Toner, M., Carpenter, J.F., and Pyatt, D.W., Cryopreservation of stem cells using trehalose: Evaluation of the method using a human hematopoietic cell line, *Stem Cells and Development*, 13, 295, 2004.

126-Puhlev, I., Guo, N., Brown, D.R., and Levine, F., Desiccation tolerance in human cells, *Cryobiology*, 42, 207, 2001.

127-Török, Z., Satpathy, G.R., Banerjee, M., Bali, R., Little, E., Novaes, R., Ly, H.V., Dwyre, D.M., Kheirolomoom, A., Tablin, F., Crowe, J.H., and Tsvetkova, N.M., Preservation of trehalose loaded red blood cells by lyophilization, *Cell Preservation Technology*, 3, 96, 2005.

128-Meryman, H.T., Osmotic stress as a mechanism of freezing injury, *Cryobiology*, 8, 489, 1971.

129-Williams, R.J. and Shaw, S.K., The relationship between cell injury and osmotic volume reduction: II. Red cell lysis correlates with cell volume rather than intracellular salt concentration, *Cryobiology*, 17, 530, 1980.

130-Acker, J.P. and McGann, L.E., Innocuous intracellular ice improves survival of frozen cells, *Cell Transplant*, 11, 563, 2002.

131-Acker, J.P. and McGann, L.E., Protective effect of intracellular ice during freezing?, *Cryobiology*, 46, 197, 2003.

132-Acker, J.P., The use of intracellular protectants in cell biopreservation, in: *Advances in Biopreservation*, Baust, J.G., Eds., Taylor & Francis, 2006, 291.

133-Acker, J.P., Chen, T., Fowler, A., and Toner, M., Engineering desiccation tolerance in mammalian cells: Tools and techniques, in: *Life in the Frozen State,* Benson, E., Fuller, B.J., and Lane, N., Eds., CRC Press, London, 2004, 563.

134-Hapala, I., Breaking the barrier: Methods for reversible permeabilization of cellular membranes, *Crit Rev Biotechnol,* 17, 105, 1997.

135-Kikawada, T., Saito, A., Kanamori, Y., Nakahara, Y., Iwata, K., Tanaka, D., Watanabe, M., and Okuda, T., Trehalose transporter 1, a facilitated and high-capacity trehalose transporter, allows exogenous trehalose uptake into cells, *Proceedings of the National Academy of Sciences USA,* 104, 11585, 2007.

136-Kaasen, I., McDougall, J., and Strom, A.R., Analysis of the otsBA operon for osmoregulatory trehalose synthesis in *Escherichia coli* and homolgy of the OtsA and OtsB proteins to the yeast trehalose-6-phosphate synthase/phosphatase complex, *Gene,* 145, 9, 1994.

137-Billi, D., Wright, D.H., Helm, R.F., Prickett, T., Potts, M., and Crowe, J.H., Engineering desiccation tolerance in *Escherichia coli, Appl Environ Microb,* 66, 1680, 2000.

138-Potts, M., Desiccation tolerance of prokaryotes, *Microbiological Reviews,* 58, 755, 1994.

139-Garcia de Castro, A. and Tunnacliffe, A., Intracellular trehalose improves osmotolerance but not desiccation tolerance in mammalian cells, *FEBS Letters,* 487, 199, 2000.

140-Lao, G., Polayes, D., Xia, J.L., Bloom, F.R., Levine, F., and Mansbridge, J., Overexpression of trehalose synthase and accumulation of intracellular trehalose in 293H and 293HTetR:Hyg cells, *Cryobiology,* 43, 106, 2001.

141-Gong, Z., Hew, C., and Vielkino, J., Functional analysis and temporan expression of promoter regions from fish antifreeze protein genes in transgenic Japanese medaka embryos, *Mole Marine Biol Biotech,* 1, 64, 1991.

142-Kohli, V., Robles, V., Cancela, M.L., Acker, J.P., Waskiewicz, A.J., and Elezzabi, A.Y., An alternative method for delivering exogenous material into developing zebrafish embryo, *Biotechnol Bioeng* (in press), 2007.

143-Noack, J. and Vogel, A., Laser Induced Plasma Formation in Water at Nanosecond to Femtosecond Time Scales: Calculation of Thresholds, Absorption Coefficients, and Energy Density, *IEEE J Quantum Elect,* 33, 1156, 1999.

144-Chichkov, B.N., Momma, C., Nolte, S., Alvensleben, V., and Tunnermann, A., Femtosecond, picosecond and nanosecond laser ablation of solids, *Applied Physics A,* 63, 109, 1996.

145-Loesel, F.H., Fischer, J.P., Gotz, M.H., Horvath, C., Juhasz, T., Noack, F., Suhm, N., and Bille, J.F.N., Non-thermal ablation of neural tissue with femtosecond laser pulses, *Applied Physics B,* 66, 121, 1998.

146-Kohli, V., Acker, J.P., and Elezzabi, A.Y., Cell nanosurgery using ultrashort (femtosecond) laser pulses: Applications to membrane surgery and cell isolation, *Lasers Surg Med,* 37, 227, 2005.

147-König, K., Riemann, I., and Fritzsche, W., Nanodissection of human chromosomes with near-infrared femtosecond laser pulses, *Optics Letters,* 26, 819, 2001.

148-Zeira, E., Manevitch, A., Khatchatouriants, A., Pappo, O., Hyam, E., Darash-Yahana, M., Tavor, E., Honigman, A. Lewis, A., and Galun, E., Femtosecond infrared laser - An efficient and safe in vivo gene delivery system for prolonged expression, *Mol Ther,* 8, 342, 2003.

149-Heisterkamp, A., Maxwell, I.Z., Mazur, E., Underwood, J.M., Nickerson, J.A., Kumar, S., and Ingber, D.E., Pulse energy dependence of subcellular dissection by femtosecond laser pulses, *Optics Express,* 13, 3690, 2005.

150-Kohli, V., Elezzabi, A.Y., and Acker, J.P., Reversible permeabilization using high-intensity femtosecond laser pulses: An application to Biopreservation, *Biotechnol Bioeng,* 92, 7, 889, 2005.

151-Drokin, S.I., Phospholipid distribution and fatty acid composition of phosphatidylcholine and phosphatidyletalonamine in sperm of some fresh water and marine species of fish, *Aquat Living Resour,* 6, 49, 1993.

152-Wang, W.A., Zhang, H., Ikemoto, I., Anderson, D.J., and Loughlin, K.R., Reactive oxygen species generation by seminal cells during cryopreservation, *Urology*, 49, 921, 1997.

153-Voituron, Y., Sercais, S., Romestaing, C., Douki, T., and Barré, H., Oxidative DNA damage and antioxidant defenses in the European common lizard (*Lacerta vivipara*) in supercooled and frozen states, *Cryobiology*, 52, 74, 2006.

Complete affiliation:

Vanesa Robles, CMR(B)-Center of Regenerative Medicine in Barcelona, c/ Dr. Aiguader, 88, 08003 Barcelona, Spain, Phone: +34933160315, e-mail: vrobles@cmrb.eu

CHAPTER 10
CRYOBIOLOGICAL MATERIAL AND HANDLING PROCEDURES

F. Martínez-Pastor and S.L. Adams

1. INTRODUCTION

Successful germplasm-banking depends not only on lab protocols, but also on the choice of equipment to put these protocols into practice and on the good use of this equipment. Besides, handling of low-temperature material implies many risks; therefore, the staff involved in these programs must receive adequate training.

This chapter gives an overview on the choice of equipment available for sperm and embryo cryopreservation. First, we will consider the devices used for freezing and storing germplasm. Then, we will review the accessories and consumable materials available for packaging, storing and handling the germplasm samples. Next, we will approach the concept of information management systems dedicated to track the samples stored in a germplasm bank. Finally, in Section 5, we will summarise the safety procedures on liquid nitrogen use. Due to the risks that liquid nitrogen use implies, we advise reading this section in parallel to the other sections.

This chapter is mainly dedicated to those unfamiliar with the techniques and equipment related to sperm cryopreservation. Nevertheless, skilled users may find it informative.

2. FREEZING SYSTEMS AND STORAGE EQUIPMENT

2.1 Overview

Technological advances have yielded devices that allow freezing of samples using multi-step programmes and storage at very low temperatures, almost indefinitely. Simple methods can be successfully used, significantly lowering the cost of setting up a cryobank. However, there are several factors that must be considered that may restrict the type of freezing system that can be used: the material to be frozen (semen, eggs, embryos or larvae); the species; the freezing protocol (extender, handling, and thawing procedure); and the storage temperature. For instance, the need of a variable cooling rate (the velocity at which sample temperature decreases), standardization and repeatability requirements, or storage at liquid nitrogen temperatures will narrow the choice of protocols and devices that can be used.

An important fact that must be kept in mind when considering these methods is the cooling rate the protocol requires. In fact, the cooling rate depends on the difference of temperatures between the sample and the environment (Newtonian cooling). Equation (1):

$$\frac{dT(t)}{dt} = -r(T - T_{env}) \text{ (1)}$$

describes this process. It shows that the freezing rate at a given time (the left term of the equation) depends on a constant r and on the difference between the sample temperature (T) and the environment temperature (T_{env}). Simple methods that expose the sample to a fixed temperature may not be capable of keeping a constant cooling rate. Indeed, Rana and Gilmour [1] reported that the cooling rate when using the dry ice method (see Subsection 2.2) was different before and after ice nucleation, and that the slope was not linear. For achieving a linear rate, the temperature of the environment must decrease over time. This is important, as it has been shown that sperm from different fish species have different optimal cooling rates (which may or may not need to be exactly constant) [2].

Another fundamental factor in cryopreservation is the thermal mass of the samples. The thermal mass represents the capacity of the sample to absorb and hold heat. Since samples frozen in the form of pellets or in small straws have a volume of few cubic milimeters, it is very difficult to achieve the same volume in each sample. The thermal mass will influence the r in Equation 1, and it may have a large effect, specially in freezing systems that rely on passive convection. Therefore, when working with small volumes of sample, one must take into account that real freezing curves may greatly differ from theoric ones in different samples. This must be also considered when assessing the accuracy of freezing curves using thermocouples (as commented in Sections 2.2 and 2.3), since the thermal mass of the probe will be different of that of the sample.

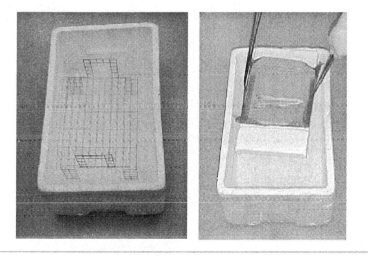

Figure.1 - LN_2 vapour freezing method: styrofoam box with home-made straw holder for placing samples above LN_2 surface (left) and ûoating device made of styrofoam and plastic net (right). *(Source:* E. Cabrita.)

2.2 Simple Freezing Methods

Simple freezing methods are those that do not require complex equipment and are therefore cheap. Control on freezing parameters is generally limited with these methods, which can be classified according to the freezing agent:
- Dry ice.
- Alcohol.
- Liquid nitrogen (LN_2).

Freezing by direct contact with dry ice is the simplest method. Dry ice is solid carbon dioxide (CO_2), which sublimates to the gas phase while maintaining a temperature of -78.5°C. Typically, droplets of the sample (extended spermatozoa or embryos) are put into small depressions on a dry ice block and allowed to freeze. Then, the frozen droplets (pellets) are quickly transferred to ultra-cold storage, usually in liquid nitrogen. The cooling rate for this method has been calculated to be around -35 C/min [3]. However, the actual cooling rate greatly depends on sample size. When using large pellets (above 100 µl) or cryovials buried in dry ice, samples will not freeze homogeneously. Rather, the core will freeze more slowly than the surface. Thus, the limitations of this method are: inability to vary the cooling rate, cooling heterogeneity due to sample size, and final temperature of less than -80° C. On the other hand, dry ice is relatively inexpensive, and the procedure is easy to carry out.

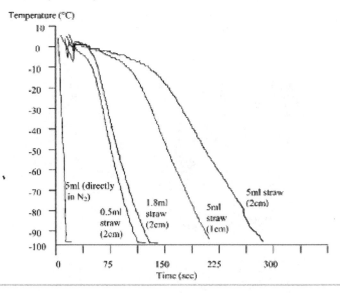

Figure 2 - Cooling curves of extended rainbow trout sperm frozen using the LN$_2$ vapour method at different heights above the LN$_2$ surface and in different packaging. *Source:* E. Cabrita et al. [12]. With permission.

A variation of the dry ice method is the cooled alcohol bath. The alcohol is poured in an insulating recipient (a Styrofoam box or a Dewar), where dry ice can be added to obtain a slurry at -78° C. Methanol, ethanol or isopropanol, at least 95% strong, are appropriate (n.b. other organic solvents, such as acetone, should not be used as these may be toxic and/or explosive). The advantages over freezing on solid dry ice are that heat transmission is faster and samples can be packaged in straws, which are then submerged into the cooled alcohol. Another advantage is that dry ice can be added slowly over time, controlling the cooling rate of the alcohol bath and, therefore, modifying the cooling rate of the samples. Nevertheless, caution must be taken when adding dry ice to alcohol, since violent bubbling may occur (see Section 5).

Liquid nitrogen (LN$_2$) vapour is a widespread freezing method, which also conveys some advantages over dry ice-based methods. Since LN$_2$ boils at -196° C, its vapour allows faster freezing rates and lower final temperatures (-160° C at 2.5 cm above LN$_2$ surface) [4]. The only material needed is an insulated container (for instance, a styrofoam box), a device to keep the samples above LN$_2$ level, and LN$_2$. The LN$_2$ is poured in the insulated container and allowed to rest, thus a temperature gradient forms from the surface of the liquid to the rim of the container (the container should be closed with a lid in order not to disturb the gradient). Sample holders are loaded with the samples (cryovials or straws) and then placed inside the container. The holders are provided with feet or floaters, which maintain the samples at a defined height above the LN_2 level. An important advantage of this method is that the samples can be subjected to different temperatures and therefore different cooling rates simply by adjusting the height of the holders above the LN_2 surface. However, it should be remembered that other factors may also inûuence the cooling rate. For example, Maxwell et al. [5], working with ram spermatozoa, reported that the optimal height above liquid nitrogen depended on the packaging type (0.5 ml straws, 0.25 ml straws, or minitubes); Figure 2 shows that the freezing curves of a given sample depend not only on the height above LN_2, but also on the volume of the sample. At the end of the freezing process, samples can be simply plunged in the liquid nitrogen below and transferred to storage tanks.

Another advantage of the LN_2 vapour method is the possibility of maintaining or increasing the freezing rate by vertically moving the sample holder during the process (placing it with strings or using clamps).

The LN_2 vapour method allows for many modifications. For instance, instead of using insulated containers, a half-full LN_2 tank may be used (see Section 2.4 for a description of liquid nitrogen tanks). In this case, the samples are mounted in a canister, which is lowered to a given height above the LN_2 level. A temperature of -190°C can be achieved in the upper part of the canisters in a half-full tank. The LN_2 tank may also be set up to permit the canister to be placed at many vertical positions, enabling different cooling rates simply by adjusting the cannister's position from the neck of the tank to the LN_2 surface. Multi-step freezing protocols may be carried out by lowering the canister into the tank over time.

LN_2 dry shippers (see Subsection 2.4) can also be used for freezing. The temperature at the bottom of a fully charged dry shipper (with no free liquid nitrogen) is usually below -150°C, and, therefore, adequate for most applications. Freezing can be carried out by simply dropping the straws to the bottom of the tank, but more complex protocols can be carried out using a canister that is lowered in a multi-step fashion [6].

These simple methods are not as suitable for standardization as are the controlled freezing methods, since their repeatability is much lower. Temperature gradients above LN_2 are easily disturbed, and positioning of sample holders or canisters above LN_2 is not always precise. Nevertheless, using thermocouple probes within control straws or cryovials [7] can be used to check system conditions beforehand, removing much of the variation and again during freezing to monitor and adjust conditions during the cooling process.

There are many papers showing the use of these methods on fish, especially regarding semen freezing on dry ice and in liquid nitrogen vapour. The efficacy of the dry ice method has been reported for semen of rainbow trout (*Oncorhynchus mykiss*) [8], sturgeon (*Acipenser fulvescens*) [9] and Northern pike (*Esox lucius*) [10]. Liquid nitrogen vapour has been tested on semen from many species including: Arctic char (*Salvelinus alpinus*) [11]; rainbow trout [12]; gilthead seabream (*Sparus aurata*) [13];

several cyprinids species [14] and African catfish (*Clarius gariepinus*) [15]. Other methods have also been tested for fish semen; for example, Aoki et al. [16] successfully froze Medaka fish (*Oryzias latipes*) semen using a dry shipper.

Figure 3 - LN$_2$-based biofreezers: Kryo 560-16® (Planer plc; upper left), EF100® (Asymptote Ltd.; upper right), DigitCool 5300® (IMV Technologies; lower left) and IceCube 15M® (SY-LAB GmbH; lower right). *Source:* Dr. F. Martínez-Pastor (Kryo 560-16®) and the respective manufacturers. With permission.

2.3 Controlled Rate Freezers

Controlled rate freezers are devices capable of lowering or increasing the temperature of biological samples at a predefined rate. These freezers usually have programming capabilities so complex freezing protocols can be carried out relatively easily. As in the case of the simple methods, there are several kinds of methods to control temperature:
- Controlled inûux of LN$_2$.
- Ultra-low freezing without LN$_2$.
- Controlled cooling of an alcohol bath.

Most freezers utilize LN$_2$ from a pressurized tank to cool the samples (Figure 3). Samples are placed in an insulated chamber, where LN$_2$ is injected using a solenoid valve, which controls LN$_2$ influx to the chamber. The system integrates a heater to increase temperature when needed, and several temperature sensors within the chamber. Depending on the program, LN$_2$ vapour is allowed into the chamber at a rate that lowers the sample temperature as desired. Temperature variations are detected by the sensors and the electronics correct deviations by increasing or suppressing LN_2 vapour influx, or by activating the heating system. Depending on the model, the final temperature reached in the chamber ranges from -100°C to 180° C. Programming is carried out either through a console integrated in the freezer, which may include a printer to display results, or alternatively through a computer.

Figure 4 - LN_2-free biofreezers: Bio-Cool® (FTS Systems; left) and EF600® (Asymptote Ltd.; center). Right photograph shows a detail of the EF600 sample plate. *Source:* From the respective manufacturers. With permission.

Recently, ultra-low freezers have been developed to carry out freezing programs and to reach temperatures below -100° C without needing LN_2 (Figure 4). The devices include a modified version of the refrigerating circuit of conventional freezers, which remove heat from the freezing chamber at the desired rate. Many models contain a cooling plate made of an alloy of very high thermal conductivity. The plate is carved with parallel grooves for supporting straws, or holes to keep cryovials. Like LN_2 vapour based freezers, temperature sensors and heating systems integrate with the refrigerating system to assure precise control of the freezing program.

Another type of controlled rate device is based on cooling an alcohol bath, where the samples are submerged. The general principle is similar to that described in Section 2.2 for alcohol-based methods, but in this case the cooling is provided by a refrigerating circuit, which removes heat from the alcohol bath at the desired rate. The bath includes a stirrer, to prevent temperature gradients forming within the liquid and final temperatures may reach -80° C. The cost of these freezers is lower than that of other devices, but they are also more basic and unable to do more complicate cooling programs.

A disadvantage of using controlled rate freezers is that most devices are large and have high power requirements so they are not practical to use in the field. However, several biofreezers have been designed with portability in mind (Fig. 5.). For example, Freeze Control® (Biogenics) consists in a small cylindrical cryochamber, where straws or cryovials are inserted. The cryochamber is put into a LN_2 cryobath, and a temperature controller regulates the freezing rate of the samples. Another option is the recently developed Planer Animal Freezer® (Planer plc). The straws are inserted in a freezing unit, which is mounted on a small LN_2 tank. The control unit can be programmed from a computer beforehand, allowing choosing among several protocols in the field without needing more equipment.

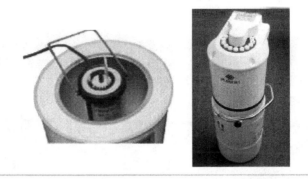

Figure 5 - Portable biofreezers: Freeze Control® (Biogenics; left) and Planer Animal Freezer® (Planer plc; right). See text for details. *Source:* From the respective manufacturers. With permission.

Many authors have tested controlled freezing methods in aquaculture (mainly LN$_2$ based), obtaining good results. Biofreezers have been used to cryopreserve prawn larvae [17], crab spermatophores [18], Pacific oyster (*Crassostrea gigas*) eggs and sea urchin sperm and larvae [19–21]. Some attempts have been carried out to cryopreserve sturgeon semen using this technology [22]. In teleostei, the use of biofreezers for sperm cryopreservation has been reported for several trout species [23-25], black porgy (*Acanthopagrus schlegeli*) [26] and catfish (*Clarias gariepinus*) [15,27,28]. Salmon blastomeres have been also frozen using a programmable freezer [29], and attempts to freeze salmon eggs have been carried out using a home-made freezer [30].

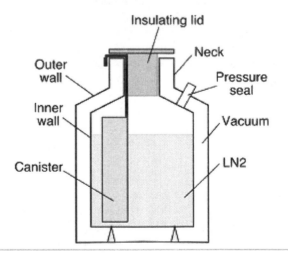

Figure 6 - Schematic diagram of a general LN$_2$ Dewar. Note the insulating structure formed by a double wall enclosing a vacuum.

2.4 Sample storage: LN$_2$ tanks and related devices

Long-term storage of cryopreserved samples can be carried out in many devices. For example, dry ice (-78.5°C) or ultra-low freezers (below -80°C) may be used for some applications. However, it is considered that "safe" storage temperatures are those below the glass transition temperature for water (-134°C, although this figure is subjected to revision). Therefore, in this chapter we will consider only those systems capable of achieving temperatures below -130°C.

The basic design of such devices is that of a Dewar flask (Fig. 6). A Dewar flask is a vessel with very good thermal insulation, capable of maintaining its contents at a constant temperature for a long period of time. The walls of the flask are a double-layer construction, with a vacuum within. Conduction and convection by the vacuum and radiation by a very reflective material coating the layers facing the vacuum prevent heat transmission. Indeed, heat transmission occurs almost solely at the neck of the Dewar, where the inner and outer walls meet.

The most widespread solution for storing cryopreserved samples consists of keeping them submerged in LN$_2$. A lot of devices are available for very different applications, thus it is worth some time choosing the most appropriate. The simplest and cheapest models are LN$_2$ tanks (Fig. 7, left). These must be checked regularly and refilled manually with LN$_2$. More sophisticated devices have a direct connection to a LN$_2$ supply system (Fig. 8, right). Electronics record LN$_2$ levels and automatically refill the device when necessary

Figure 7 - LN$_2$ refrigerators for storing samples at cryogenic temperatures (Taylor-Warton): LN$_2$ tanks (left) and automatically-refillable refrigerators (right). Canisters and other accessories are shown next to the tanks. *Source:* Taylor-Warton. With permission.

All of these devices must be provided with a system to organise the samples. In general, either a canister or rack system is used. Devices dedicated to store straws or macrotubes are provided with canisters: metal or plastic cylinders vertically positioned within the tank (Fig. 9). In the case of tanks intended to store cryovials, they are provided with a system of vertical racks with drawers, which contain cell boxes for placing the cryovials. The canister system is preferred for storing semen or embryo samples, which are generally packaged in straws. Subsection 3.4 deals more in detail with systems for organizing the samples within tanks.

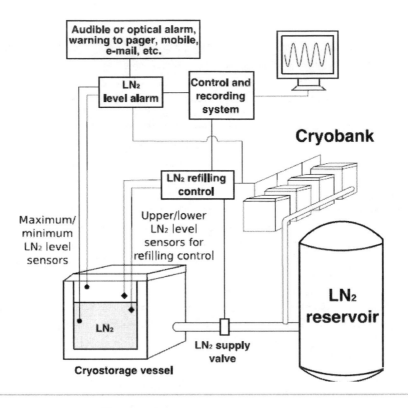

Figure 8 - Scheme of a large cryobank provided with refilling and control facilities.

Small cryobanks are usually based just on a group of LN_2 tanks (from 20 to 60 l or more). Their size assures a long duration of the LN_2 content. Smaller models of LN_2 tanks (up to 15 l) are used for transporting semen doses or embryos, but they are not appropriate for long-term storage because of faster evaporation rates. As indicated before, tanks must be checked and refilled regularly, assuring that LN_2 covers all the samples (LN_2 level checking can be carried out simply with a stick to the bottom of the tank and allowed to freeze). Even though a 40 l tank can last for more than four months before completely drying, the samples at the top of a half-dried LN_2 tank may be exposed at temperatures above -130°C (the glass transition temperature of water), thus compromising sample integrity. Tanks containing valuable samples can be provided with level probes or recording systems (see Subsection 2.5), which can prevent losses due to sudden LN_2 leakage or evaporation. The number of samples stored is highly variable and related to tank size. For instance, smaller models may contain around 50 0.5 ml straws (or around 20 2-ml cryovials), whereas larger models can hold more than 3000 0.5 ml straws (or more than 1500 2 ml cryovials).

LN_2 tanks can be quite easily moved. Thus, a cryobank based in this kind of tank does not require as much space or room as when larger and more expensive devices are used. The tanks can fit in a number of places, and may be moved to wherever is most convenient. Furthermore, the location of a whole cryobank can be changed relatively easily (between buildings and, even taking adequate safety measures, distant locations). Nonetheless, it must be taken into account that careless handling (sudden jarring, bumping, etc.) can deteriorate the tank walls, resulting in vacuum loss. Increased LN_2 evaporation, as a result of vacuum loss, causes frost to appear, especially around the neck of the tank. If this occurs, the tanks need to be replaced immediately.

Figure 8 shows how a large cryobank may be organized. In this case, cryostorage vessels (we will use this term to refer to vessels much larger than LN_2 tanks and usually provided with electronics and refilling systems) are connected to a LN_2 distribution system. The valves controlling the LN_2 supply are managed by an electronic system, which is provided with information from LN_2 sensors in the tanks. A low-level sensor triggers the opening of the LN_2 valve if the LN_2 level drops below the position of the samples, and a second sensor sends a signal to close the valve when the LN_2 level reaches a safe height. Cryostorage vessels may be provided with a backup system using sensors that trigger an alarm if the LN_2 reaches a maximum or minimum level because of valve failure. For centralised and fully automated systems, the sensors may be connected to a computer or communication device (see Subsection 2.5). This kind of large and automated cryobank is considerably more expensive than those based in LN_2 tanks. However, they have a number of important advantages: significantly more samples can be stored per container; staff no longer needs to perform continuous surveillance; risk of sample loss due to LN_2 evaporation is decreased; and samples are permanently covered by LN_2. The sample capacity is much higher than LN_2 tanks, with capacity for 10,000 to more than 80,000 vials, or a comparable number of straws.

LN_2 storage has the advantage of keeping samples at temperatures that are always below -180°C. However, there is potential for biological cross contamination between samples (see Subsection 5). If this is of concern, then storage must be carried out in LN_2 vapour. The general structure of the tank or container is similar to those dedicated to LN_2 storage, but, in this case, LN_2 is limited to the lower portion of the chamber and is not in contact with the samples. LN_2 evaporates and its vapour maintains the samples at temperatures well below -100°C. The main disadvantage of this system is temperature heterogeneity. Temperature gradients can develop from the bottom to the top of the chamber, and a deficient insulation or even opening of the lid can cause marked temperature differences between locations within the chamber. As a result, some samples may be exposed to temperatures above -130°C. Even though big containers include sensors such as the ones for LN_2 storage, temperature heterogeneity may be difficult to control and record. To minimise this, many manufacturers provide canisters and racks made of alloys with high heat transfer ratios, helping to homogenise the temperature within the chamber [31]. Of course, if a tank is full of liquid nitrogen the temperature will be -196°C. If vapour storage is used, then conductive bars (e.g., a Spine Chiller, Planer plc) can be used to cool the top of the tank by conduction, giving a typical temperature there of -150°C. Low loss dewars (e.g. , Air Liquide RGB range) are designed to achieve

-190°C at the top, which is only 6°C differential from the base.

There are ultra low freezers that are available which can achieve temperatures below -130°C without using liquid nitrogen, offering a viable alternative to LN$_2$-based cryobanks. According to manufacturers, these mechanical freezers have all the advantages of storage in LN$_2$ vapour (low contamination hazard), but have better temperature homogeneity within their chamber, therefore assuring that all the samples are held at a similar temperature.

Figure 9 - A canister full of straws being lifted to the mouth of a LN$_2$ tank. The straws are organised within hexagonal visotubes, which are located within a goblet. Note the handles of the other canisters, secured into grooves around the mouth of the tank. *Source:* Dr. F. Martínez-Pastor. With permission.

2.5 Other equipment

There is a wide choice of accessories for cryobanking (consumable materials will be treated in Section 3):
- LN$_2$ tanks for keeping a LN$_2$ stock can be used to carry out cryopreservation procedures or to refill the tanks storing the samples. These types of tanks are of simpler design than tanks for sample storage, being without canisters or racks. Some models may have a LN$_2$ withdraw device which enables easier dispensing. The device allows a small amount of pressure to build up in the dewar which then forces LN$_2$ through a dispensing hose. Safety valves prevent overpressurisation. Some LN$_2$ -based biofreezers use a similar pressurising system to obtain the LN$_2$ flux necessary to run.
- A large pressurised LN$_2$ tank (5 m³ and above) is of course necessary when a cryobank is based on storage devices provided with automatic refilling. Similarly,

this type of tank is also handy for refilling storage tanks in small cryobanks. However, the cost of such a tank may be prohibitive unless the cryobank is well funded, or belongs to a university or large organization where the cost may be shared among many departments or users. In the case of automatic refill systems, the cost of tubing and maintenance must also be taken into consideration. A medium-sized pressurised LN_2 tank may be a more viable alternative where budget constraints prevent the purchase of a large LN_2 tank but when a LN_2 stock is still needed.

▪ LN_2 level probes are useful add-ons for storage systems. The simplest probes can be adapted to many models of liquid nitrogen tank, triggering an alarm if the LN_2 level drops below that of the probe (Fig. 10). More sophisticated devices can record the LN_2 level and send messages to mobile devices (pager, mobile phone, or PDA). In tanks storing samples in LN_2 vapour, temperature sensors can be very effective, not only warning when the chamber warms above a critical temperature, but also keeping a record of temperature variation. As mentioned in Subsection 2.4, control and recording devices are often built-in in most advanced LN_2 tanks. Many devices can be networked to a computer, a whole cryobank can be surveyed, controlled, and all the events duly recorded without human intervention (e.g., Assure24seven® system, Planer plc).

▪ Oxygen monitors are safety devices that monitor oxygen deficiency, which can occur as a result of nitrogen gas accumulation (see Section 5). These devices are highly advisable in laboratories and rooms where LN_2 is being used or stored, especially if an area is not well ventilated or a large number of LN_2 tanks are present. Detectors can be worn by staff or be fixed strategically on walls or ceilings. Like other monitoring devices, sophisticated models can communicate to networks or mobile devices

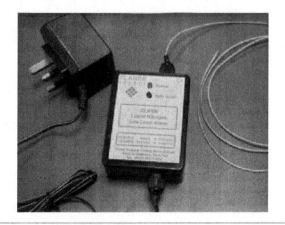

Figure 10 - A low-level sensor and monitor for generic LN_2 tanks. A built-in alarm system is triggered if the LN_2 level goes below the sensor position. *Source:* Planer plc. With permission.

3. CRYOPRESERVATION SUPPLIES AND CONSUMABLES

3.1 Overview

Apart from the equipment used to freeze and store the samples, it is also important to carefully consider the type of material used to package and identify them. Not all applications will require state of the art material, but inadequate choice can undermine the efficiency of a cryobank or eventually hamper its development. Here, we will only consider material related to packaging, organization and identification of the samples. We will not review chemicals, media or other types of consumables used for cryopreservation.

3.2 Straws and cryovials

Straws and cryovials are used to package extended semen, eggs, embryos or larvae, allowing individual identification, better organization and dosage than if using pellets, for instance. Straws are plastic cylinders with very short diameters relative to their length (Fig. 11). The large surface to volume ratio means heat transfer rates during freezing and thawing is high. Smaller straws can usually hold volumes of 0.25 or 0.5 ml. Macrotubes are used when freezing larger volumes, and can hold up to 5 ml. Since large volumes are usually necessary for aquaculture applications, macrotubes are often preferred. Although, heat transfer rates must also be taken into consideration.

Another difference between macrotubes and smaller straws is the sealing system. Macrotubes have both ends free, and are generally sealed using steel or plastic balls. Straws have a small amount of sealing powder enclosed between two cotton stoppers at one end and filling is carried out applying a negative pressure at that end. When the medium moistens the polyvinyl, it polymerises to plug the end. The other end is then sealed using sealing rods, balls, sealing powder, heat or ultrasound. Sealing rods, balls and sealing powder are available in different colours, which can be used for sample identification. Heat and ultrasound are also used for sealing. In fact, heat sealing is used in applications with strict sterility requirements.

Straws and macrotubes can be identified either by hand-writing or printing. The first option is feasible when low numbers of straws are being frozen, for example, in experiments to distinguish between treatments. When larger numbers of straws are being used, or when standards must be adhered to (quality, sterility, etc.), or complex information must be included (e.g., barcodes), then automatic printing must be used. Straw printers are discussed in Subsection 3.3, whereas sample identification is treated in Subsection 4.2.

Cryovials are usually made of plastic and are cylindrical with a hemispherical base and a lid screwed to the body. They are available in several volumes, up to 4.5 ml. The junction of the lid and the vial is usually sealed by a rubber o-ring which prevents LN_2 entering the vial. If cryovials are stored in contact with LN_2, then manufacturers recommend further sealing of the cryovials inside plastic sleeves (e.g., cryoflex, Nalg Nunc International). Entrapment of LN_2 inside the vials may cause pressure build-up during thawing resulting in explosions and/or sample loss. Cryovials usually have a designated area that can be written on to identify samples. Another option is to use special labels (resistant to ultra-low temperatures), which can be hand-written or printed on.

Management of straws and cryovials must be carried out with care once frozen. Tweezers should be cooled with LN_2 prior to handling samples to prevent thawing or recrystallization in the area where the tweezers touch the sample. Transferring frozen material between different locations should be performed quickly, keeping the cryovials or straws immersed in LN_2, where possible. Canisters should not be lifted higher than the mouth of a LN_2 tank, as this risks exposing samples to temperatures higher than -100° C and can also cause thawing or recrystallization in the straws or vials.

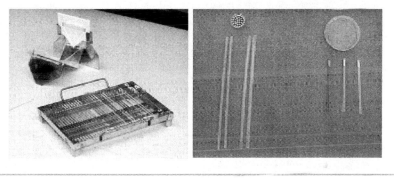

Figure 11 - Left: 0.25-mL straws on a metallic straw holder ready to be cryopreserved; next to the holder there is a boat with a plastic comb, used to remove excess sample from the straws which enables sealing of the open end. Right: comparison between macrotubes (2 mL) and straws (0.25 mL); two sealing materials are shown: metallic balls for the macrotubes and polyvinyl powder for the straws. *Source:* Dr. F. Martínez-Pastor and Dr. E. Cabrita. With permission.

3.3 Automatic fillers and printers

Straw printers and automatic straw fillers can save large amounts of time, assist with cryobank automation and improve sample tracking. Straw printers use inkjet or thermal transfer technologies to mark the straws with several lines of text, including serial numbers, special characters and barcodes. Cryobank management can benefit from straw printing, since each sample is accompanied by a complete description (e.g., date of collection and freezing, protocol used, source, etc.) and printing can be carried out quickly; some printers are able to print tens to hundreds of straws per minute. Newer printers can be managed from a computer, and the printing software linked to cryobank management systems. Samples can then be easily tracked by giving each sample a unique barcode. On thawing, samples can be immediately identified simply by using a barcode reader that is linked up to a computer managing the cryobank.

Automatic and semi automatic straw fillers are especially useful when handling large numbers of straws. A negative pressure system is used to fill the straws, and many can also seal the straw ends, either using ultrasound or balls. Some models also have an integrated straw printer.

3.4 Organization: canisters, racks, goblets and visotubes

Cryostorage vessels contain either canisters or racks in which to store and organise samples. As mentioned previously, canisters are typically used to store straws, and racks to store cryovials. Generally, only cryovials up to 2 ml may be stored using the rack system.

Canisters are cylindrical containers, made of either metal or plastic, and vertically positioned in storage vessels. They can be lowered or raised with the aid of handles, which, in the case of LN_2 tanks, are inserted in radial grooves on the mouth of the tank (Fig. 9). They contain the rest of the organizational elements which make up the storage system. The samples, generally in straws or macrotubes, are kept in cylindrical plastic containers called goblets which are of slightly smaller than the diameter of the canister. The canister can be vertically subdivided by simply piling one goblet on top of another. The height, size and number of canisters depend on the size of the storage vessel. Small LN_2 tanks generally contain one to six canisters that are capable of holding only one goblet, whereas large tanks can hold two goblets. Large refrigerators used for maintaining larger cryobanks may contain more than 100 canisters, each capable of holding up to three goblets.

Macrotubes are usually stored within the goblets, without further subdivisions. Cryovials may be also stored in goblets, but they need to be mounted onto aluminium cryocanes to prevent them from ûoating in LN_2 and to facilitate sample identification. Straws are usually stored within plastic tubes termed visotubes. Visotubes are available in different colors and shapes (usually round, hexagonal or triangular). They maybe also written on to further help identify them. In this way, they subdivide goblets and allow the easy identification of samples or groups of samples. Whereas goblets may be perforated to allow liquid or vapour nitrogen to enter, visotubes are not, thus they are able to keep LN_2 around the samples during manipulation. Figure 9 shows many straws in a canister, which are within several hexagonal visotubes. Recently, other systems to store straws, such as cassettes, have been developed.

The other organizational system, intended mainly to store cryovials, is based on racks. Racks are vertical metallic structures that have high heat conductivity and contain many drawers. Boxes with subdivisions containing cryovials may be stored within the drawers. The racks are provided with handles on the top to lift and lower them into the LN_2 tank or cryostorage vessel. Both the canister and rack systems are arranged in such a way that they enable samples within a cryobank to be readily located. In the canister system, a sample can be located by its Vessel-Canister-Goblet-Visotube identification, whereas in the rack system, it is the Vessel-Rack-Drawer-Box identification, and column and row within the box, which would enable the sample to be located. These coordinates must be precisely noted whenever a sample enters the cryobank, or is moved within the bank.

Both canisters and racks define a coordinate system that allows samples to be located precisely at any time.

3.5 Gloves, goggles, tweezers and other material

Gloves and goggles or face masks are compulsory whenever frozen material or LN_2 are being used. Gloves should be well insulated for work involving LN_2 manipulation and also loose fitting so that they can be removed quickly if necessary in an emergency. Goggles and face masks protect the eyes from LN_2 splattering and other elements projected by the expansion of boiling LN_2 when taking samples out of cryostorage vessels.

Tweezers are used to safely manipulate straws, cryovials and visotubes. As mentioned in Subsection 3.2, they must be cooled prior to use, at least to the same temperature as the samples. It is important that they provide a tight grip and are long enough to perform manipulations without having to lift the samples out of the cryostorage vessel.

Many other items are available for cryopreservation work. For instance, Figure 11 (left) shows a comb for removing the excess sample from manually filled straws, in order to leave an empty space for sealing. Handling of LN_2 may also require other items such as funnels, hoses and other instruments whose description is outside the scope of this text.

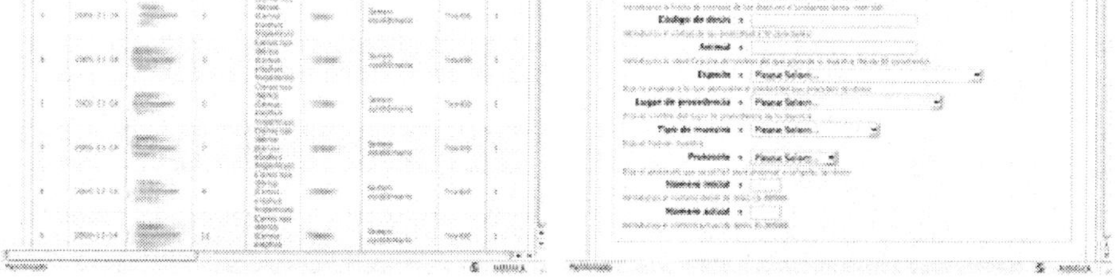

Figure 12 - Screenshots of the cryobank database of the Reproductive Biology Group (National Wildlife Research Institute, Spain). This is a personalized system based on LAMP and the Dataface framework (http://fas.sfu.ca/dataface). The configuration enables the database to be managed by one computer but enabling access from any other computer connected to the same network. The access is controlled by user permissions and passwords. All the components are open source software (http://www.fsfeurope.org).

4. STORAGE MANAGEMENT AND RECORD KEEPING

4.1 Overview

It is important to maintain accurate records and inventory control for cryopreserved samples. The information that should be recorded may include details such as:
- the sample collection date,
- the preservation date,
- the method of preservation,
- the location and identification of the material,
- the number of straws or cryovials cryopreserved for each sample and the number remaining and, any information about the post-thaw viability of a particular sample.

4.2 Spreadsheets

Spreadsheets are an ideal way to keep track of the location of straws and cryovials in small cryobanks such as an experimental situation where only a small number of samples are usually stored at any one time. For instance, a spreadsheet may be set up containing one datasheet per LN_2 tank. Each datasheet may then contain one registry per sample, which would show sample identification, location (canister, goblet, visotube or rack, drawer, box coordinates), number of straws, etc.

A system should be in place to ensure that samples cannot be banked or thawed without records being updated, and that only authorised staff can modify the spreadsheet.

4.3 Specialized database systems

In situations where: a large numbers of samples are stored; where quality standards must be achieved; or when storing sensitive material, specialized database systems must be used. These systems have a specific user interface, such as list, edit and search forms and reports. Most of them have been developed following the server-client paradigm, which means that data can be viewed and modified from a computer (client) that may be different from the one containing the programs and database (server). This enables the database to be used from another laboratory or from the field if necessary.

These systems include built-in access control so only authorised users can utilise the system. Database access can be logged, which enables error tracking and fixing to be carried out easily.

There are many commercial systems available. For instance, KryoTrak Inventory Management Software (www.kryotrak.com; Planer plc), Cryo Bio System™ storage management software (IMV Technologies) or IDA (Minitüb GmbH; specific for bull semen). Nevertheless, personalised software may be easily developed combining a webserver and a database engine, and using web programming languages such as PHP or ASP. For small cryobanks, a system based on open-source software can be delivered in a very short time using available database management frameworks to build the interface at relatively low cost (Fig. 12).

5. LIQUID NITROGEN HANDLING AND SAFETY PROCEDURES

5.1 Overview

Safe handling and use of LN_2 in cryopreservation involves understanding its physical properties and the potential hazards that are associated with them. When working with liquid nitrogen, it is imperative that safety precautions be followed at all times. The safety precautions are common-sense procedures to avoid possible injury or damage. Anyone using liquid nitrogen should be fully familiar with all aspects of safe handling practice before undertaking any cryopreservation work. This section is essentially a summary compiled from MSDS, safety standards and safety precaution material and is intended to outline best practice for working with liquid nitrogen.

5.2 Physical properties

There are two important properties of liquid nitrogen that present potential hazards. Firstly, liquid nitrogen is extremely cold. At atmospheric pressure, the boiling point of liquid nitrogen is 196°C. Secondly, a small volume of liquid nitrogen produces a large volume of nitrogen gas; the expansion ratio is 696.5:1 at 21.1 C. Some other important physical properties include:
- Appearance: Colourless liquid
- Odour: Odourless
- FreezingPoint: -210°C
- Flammability: Nonflammable

5.3 Associated hazards

There are six principal areas of hazard associated with the use of liquid nitrogen:
- Asphyxiation
- Cold burns/frostbite
- Cold embrittlement
- Pressure buildup and explosions
- Oxygen enrichment
- Boiling and splashing

Asphyxiation. The air we breathe is made up of 78% nitrogen, 21% oxygen and 1% other gases. Because liquid nitrogen expands significantly when it converts to nitrogen gas, it can displace oxygen in confined or poorly ventilated areas. If the oxygen concentration is sufficiently reduced, there is a risk of asphyxiation (suffocation). Furthermore there may be little or no warning. A person can become unconscious and then die as a result of nitrogen asphyxiation, without having sensed any warning symptoms such as dizziness.

Cold burns/frostbite: Contact with liquid nitrogen or cold vapour can cause serious injury to the skin and/or eyes. Brief contact may cause burns similar to heatburns and prolonged or severe exposure can lead to frostbite. Eyes are particularly vulnerable to injury. Contact with liquid nitrogen or cold vapour can cause the eye fluids to freeze, resulting in permanent damage. Uninsulated containers or equipment cooled by liquid nitrogen can stick fast to skin when touched and tear flesh when trying to remove.

Cold embrittlement: At extremely low temperatures, many materials that are normally ductile at room temperature (e.g., rubber, plastic, carbon steel) become brittle and will fracture readily under stress.

Pressure build up and explosions: Because the expansion ratio of liquid nitrogen to nitrogen gas is large (1:696.5), inadequate venting of liquid nitrogen containing vessels can cause pressure to build up relatively quickly. Over-pressurisation can result in an explosion. This may occur for example, if moisture from the air is allowed to condense and freeze around the opening of a storage vessel, effectively sealing it off from the atmosphere.

Oxygen enrichment: Liquid nitrogen boils at a lower temperature than liquid oxygen (-196°C versus -183°C). It can therefore condense oxygen from the atmosphere to produce an oxygen-enriched condensate. This may occur, for example, when liquid nitrogen is transferred through uninsulated metal pipes. The higher oxygen content increases the flammability of many materials creating potentially explosive conditions.

Boiling and splashing: Liquid nitrogen boils and can splash when added to a warm container or when warm objects are placed into it. Hollow rods or tubes should not be used as dipsticks. When a warm tube is inserted into liquid nitrogen, gasification and rapid expansion of liquid inside the tube will cause liquid to shoot out and spurt from the top of the tube and may cause injury.

5.4 Safety precautions

Personal protection: Personal protective equipment should be worn to prevent accidental contact with liquid nitrogen or equipment in contact with liquid nitrogen. The appropriate level of protective equipment is dependent on what type of handling is being carried out and the quantity of liquid nitrogen involved.

Eyes: A protective face shield, safety googles or safety glasses with side shields should be worn when handling liquid nitrogen. When filling dewars or transferring liquid nitrogen between containers, a protective face shield should be worn.

Hands: Ensure that jewellery such as rings, bracelets, and watches are removed. Gloves should be nonabsorbant, insulated (e.g., nylon cryo-gloves) and should be loose fitting so that they can be easily removed in the event of a spill. In some instances, gloves may be restrictive and tongs or long forceps alone may be adequate.

Clothing: High-topped impervious closed-in foot wear should be worn. Trousers should be cufless and worn outside the shoes. Long sleeve shirts or sweaters are recommended for arm protection. When handling large quantities of liquid nitrogen, an impervious long apron is also recommended.

Storage: Liquid nitrogen should be stored in purpose-built insulated containers such as dewars that have either a loose fitting cap or pressure relief valve to allow nitrogen gas to escape. Ice plugs must not be allowed to accumulate in the neck or on the lid of the dewar as this may cause overpressurisation.

Liquid nitrogen containers should always be stored in well ventilated areas to prevent the build up of evaporated nitrogen gas. The volume of the room, room ventilation and volume of cryogenic liquid being stored should be considered collectively. A risk assessment should be carried out calculating the oxygen concentration of the room in the worst case scenario (i.e., the entire contents of the liquid nitrogen container(s) is lost). This could occur for example if a dewar loses its vacuum. The oxygen concentration may be calculated using the formula

$$\%O_2 = \frac{100 \times V_0}{V_r} \quad (2)$$

where V_0 is given by the equation

$$V_0 = 0.21(V_r - V_g) \quad (3)$$

V_r is the volume of room in cubic metres and V_g is the volume of nitrogen gas released in cubic metres. Vg is given by the equation

$$V_g = VLN \times 696.5/1000 \quad (4)$$

and *VLN* is the liquid volume capacity of the storage container in litres. If the oxygen level could fall below 18%, then suitable control measures should be implemented; for example, finding a better ventilated room, installing extra oxygen deficiency alarms (an alarm should always be used), or using smaller storage dewars.

Transport: Liquid nitrogen comes under Dangerous Goods Class 2.2. The requirements for safe transport of liquid nitrogen on land are set out in relevant national, state or territory regulations and must be complied with when liquid nitrogen is transferred between facilities. Liquid nitrogen should not be transported in closed vehicles; an open vehicle, trailer or a vehicle compartment that is vented to the outside and separated from the drivers compartment (e.g., a canopied utility) should be used. The liquid nitrogen containers should be properly packaged and identified and separated from other incompatible dangerous goods when transporting. The containers should be kept upright at all times and carried securely (e.g., using brackets, strops or tie downs). Care should be taken when handling to protect the vacuum insulation system. Emergency response information (e.g., a MSDS sheet) should be carried during transport and safe handling practice followed. A dry shipper may be a safer option that a LN_2 refrigerator for transporting samples.

Liquid nitrogen containers should not be transported in passenger lifts accompanied by personnel.

5.5 Emergency procedures and first aid

Spillage: In the event of a spill, ventilate the area and evacuate all personnel. Wear appropriate personal protective equipment and monitor oxygen levels in confined spaces. If necessary, a self contained breathing apparatus should be worn.

Unconscious person: If a person working with liquid nitrogen starts to become dizzy or loses consciousness, remove the person to a well ventilated area. Ensure your own safety first and wear a self contained breathing apparatus if necessary. Keep the victim warm and seek medical attention immediately. Apply artificial respiration if breathing has stopped. Give oxygen if available.

Cold burns/frostbite: Warm the affected area as rapidly as possible by immersing in water that is close to body temperature (37°C) for 15 min. Remove clothing carefully if necessary. Do not use water that is hotter than 44°C and do not rub the affected area before or after rewarming. Protect any damaged tissue by covering loosely with a dressing and seek medical assistance. If eyes are affected, flush with warm water for 15 min and seek medical assistance.

5.6 Additional information and references

- Australia/New Zealand Standard AS/NZS 2243.2, 1997. Safety in laboratories - Part 2: Chemical aspects.
- BOCgases, LiquidNitrogen Material Safety Data Sheet.
- BOCgases, Care with cryogenics: the safe useoflow temperature liquefied gases.
- Land Transport New Zealand, Factsheet 68: Dangerous goods transported as tools-of-trade.

- Northeastern University: http://www.ehs.neu.edu/cryo/cryogen.htm
- Taylor-Wharton, TW-10 Handle with Care booklet.
- University of London: http://www.ucl.ac.uk/efd/safety services www/guidance/ liquid nitrogen/index.htm

6. COMPANY DIRECTORY

In this section we have compiled a directory with the web pages of many manufacturers and distributors of cryogenic equipment, from which we have obtained many data and graphics for this chapter.
- Air Liquide: http://www.airliquide.com
- Asymptote Ltd: http://www.asymptote.co.uk
- Biogenics: http://www.biogenics.com
- BOC Gases: http://www.boc.com
- FTS technologies:http://www.ftssystems.com
- IMV Technologies: http://www.imv-technologies.com
- MINITÜ GmbH:http://www.minitube.de Planerplc:http://www.planer.co.uk
- Revco: http://www.revco-sci.com/
- SY-LAB GmbH: http://www.sylab.com
- Taylor-Wharton: http://www.taylor-wharton.com

7. ACKNOWLEDGMENTS

The authors thank the respective companies for allowing use of graphic material to illustrate this chapter. Special thanks to Geoffrey Planer (Planer plc) for his support providing bibliographic and graphic material, and to Geoffrey Planer, Paul Lakra and Steve Butler for reviewing the draft.

8. REFERENCES

1-Rana, K. and Gilmour, A., Cryopreservation of fish spermatozoa: effect of cooling methods on the reproducibility of cooling rates and viability, in *Proceedings of the Refrigeration and Aquaculture Conference*, 3, 1996.
2-Suquet, M., Dreanno, C., Fauvel, C., Cosson, J., and Billard, R., Cryopreservation of sperm in marine fish, *Aquac Res*, 31, 231, 2000.
3-Stoss, J. and Donaldson, E., Preservation of fish gametes, in: *Proceedings of the International Symposium on Reproductive Physiology of Fish, PUDOC*, 114, Wageningen, Germany, 1982.
4-Mottershead, J., Frozen semen preparation and use, *Canadian Morgan*, Nov./Dec, 2000.
5-Maxwell, W.M., Landers, A.J., and Evans, G., Survival and fertility of ram spermatozoa frozen in pellets, straws and minitubes, *Theriogenology*, 43, 1201, 1995.
6-Roth, T.L., , Bush, L.M., Wildt, D.E., and Weisset, R.B., Scimitar-horned oryx (Oryx dammah) spermatozoa are functionally competent in a heterologous bovine in vitro fertilization system after cryopreservation on dry ice, in a dry shipper, or over liquid nitrogen vapor, *Biol Reprod*, 60, 493, 1999.

7-Cabrita, E., Robles, V., Alvarez, R., and Herráez, M.P., Cryopreservation of rainbow trout sperm in large volume straws: application to large scale fertilization, *Aquaculture*, 201, 301, 2001.

8-Legendre, M. and Billard, R., Cryopreservation of rainbow trout sperm by deep-freezing, *Reprod Nutr Dev*, 20, 1859, 1980.

9-Ciereszko, A., Dabrowski, K., and Ochkur, S.I., Characterization of acrosin-like activity of lake sturgeon (*Acipenser fulvescens*) spermatozoa, *Mol Reprod Dev*, 45, 72, 1996.

10-Babiak, I., Glogowski, J., and Luczynski, M.J., The effect of egg yolk, low density lipoproteins, methylxanthines and fertilization diluent on cryopreservation efficiency of northern pike (*Esox lucius*) spermatozoa, *Theriogenology*, 52, 473, 1999.

11-Mansour, N., Richardson, G.F., and McNiven, M.A., Effect of extender composition and freezing rate on post-thaw motility and fertility of Arctic char, *Salvelinus alpinus*(l.), spermatozoa, *Aquac Res*, 37, 862, 2006.

12-Cabrita, E., Martínez, F., Real, M., Alvarez, R., and Herráez, M.P., The use of flow cytometry to assess membrane stability in fresh and cryopreserved trout spermatozoa, *Cryo Lett*, 22, 263, 2001.

13-Cabrita, E., Robles, V., Cuñado, S.,Wallace, J.C.,Sarasquete, C., and Herraez, M.P., Evaluation of gilthead sea bream, *Sparus aurata*, sperm quality after cryopreservation in 5 ml macrotubes, *Cryobiology*, 50, 273, 2005.

14-Lahnsteiner, F., Berger, B., Horvath, A., and Urbanyi, B.,Cryopreservation of spermatozoa in cyprinid fishes, *Theriogenology*, 54, 1477, 2000.

15-Viveiros, A.T., So, N., and Komen, J., Sperm cryopreservation of African catfish, *Clarias gariepinus*: cryoprotectants, freezing rates and sperm:egg dilution ratio, *Theriogenology*, 54, 1395, 2000.

16-Aoki, K., Okamoto, M., Tatsumi, K., and Ishikawa, Y., Cryopreservation of Medaka spermatozoa, *Zool Sci*, 14, 641, 1997.

17-Arun, R. and Subramoniam, T., Effect of freezing rates on the survival of penaeid prawn larvae: a parameter analysis, *Cryo Lett*, 18, 359, 1997.

18-Jeyalectumie, C. and Subramoniam, T., Cryopreservation of spermatophores and seminal plasma of the edible crab *Scylla serrata*, *Biol Bull*, 177, 247, 1989.

19-Adams, S.L., Hessian, P.A., and Mladenov, P.V., The potential for cryopreserving larvae of the sea urchin, *Evechinus chloroticus*, *Cryobiology*, 52, 139, 2006.

20-Adams, S.L., Hessian, P.A., and Mladenov, P.V., Cryopreservation of sea urchin (*Evechinus chloroticus*) sperm, *Cryo Letters*, 25, 287, 2004.

21-Tervit, H.R., Adams, S.L., Roberts, R.D., McGowan, L.T., Pugh, P.A., Smith, J.F., and Janke, A.R., Successful cryopreservation of Pacific oyster (*Crassostrea gigas*) oocytes, *Cryobiology*, 51, 142, 2005.

22-Billard, R., Cosson, J., Noveiri, B.S., and Pourkazemi, M., Cryopreservation and short-term storage of sturgeon sperm, a review, *Aquaculture*, 236, 1, 2004.

23-Erdahl, D. and Graham, E, Cryo-preservation of spermatozoa of the brown, brook and rainbow trout, *Cryo Lett*, 1, 203, 1980.

24-Cabrita, E., Alvarez, R., Anel, L., Rana, K.J., and Herraez, M.P., Sublethal damage during cryopreservation of rainbow trout sperm, *Cryobiology*, 37, 245, 1998.

25-Salte, R., Galli, A., Falaschi, U., Fjalestad, K.T., and Aleandri, R., A protocol for the on-site use of frozen milt from rainbow trout (*Oncorhynchus mykiss* Walbaum) applied to the production of progeny groups: comparing males from different populations, *Aquaculture*, 231, 337, 2004.

26-Chao, N., Fish sperm cryopreservation in Taiwan: technology advancement and extension efforts, *Bull Inst Zool Acad Sin Monogr*, 16, 263, 1991.

27-Rurangwa, E., Volckaert, F., Huyskens, G.; Kime, D., and Ollevier, F., Quality control of refrigerated and cryopreserved semen using computer-assisted sperm analysis (CASA), viable staining and standardized fertilization in African catfish (*Clarias gariepinus*), *Theriogenology*, 55, 751, 2001.

28-Viveiros, A.T.M., Lock, E.J., Woelders, H., and Komen, J., Influence of cooling rates and plunging temperatures in an interrupted slow-freezing procedure for semen of the African catfish, *Clarias gariepinus*, *Cryobiology*, 43, 276, 2001.

29-Kusuda, S., Teranishi, T., and Koide, N., Cryopreservation of chum salmon blastomeres by the straw method, *Cryobiology*, 45, 60, 2002.
30-Harvey, B. and Ashwood-Smith, M.J., Cryoprotectant penetration and supercooling in the eggs of salmonid fishes, *Cryobiology*, 19, 29, 1982.
31-Rowley, S.D. and Byrne, D.V., Low-temperature storage of bone marrow in nitrogen vapor-phase refrigerators: decreased temperature gradients with an aluminum racking system, *Transfusion*, 32, 750, 1992.

Complete affiliation:

Felipe Martínez-Pastor, National Wildlife Research Institute (IREC) (CSIC-UCLM-JCCM), and Game Research Institute (IDR), 02071, Albacete, Spain. e-mail: Felipe.Martinez@uclm.es.

SECTION V – PROTOCOLS FOR SPERM CRYOPRESERVATION

CRYOPRESERVATION OF SPERM OF LOACH (*Misgurnus fossilis*)

J. Kopeika, E. Kopeika, T. Zhang and D. Rawson

1. INTRODUCTION

Loach, weather fish (*Misgurnus fossilis*, Linnaeus 1758) is a member of the Cobitidae family, order Cypriniformes, class Actinopterygii. Its natural habitats are still waters of lakes or river backwaters with muddy or sandy-mud bottoms in which it buries itself, and it is distributed across Central and Eastern Europe from France to Russia. Loach feed on insect larvae and small molluscs. It spawns reddish-brown eggs, which are laid on water plants, from April to June, and hatching takes place after eight to ten days, at a temperature of 14° - 21°C. The weatherfish is largely nocturnal in habit, and lies buried during the day, and in the stagnant, marshy conditions in which it lives, there is often an oxygen deficiency. Male fish develop small tubercles on their body that distinguish them from females. It has little commercial value, but it is a valuable object of study and is widely used for scientific research in Eastern Europe.

The short embryological period, high transparency of embryos, easy conditions for laboratory maintenance, and large quantity of gametes and embryos produced by individual fish make this species valuable for research on reproduction, embryology as well as cytology and genetics.

2. PROTOCOL FOR FREEZING AND THAWING

2.1 Maintenance of fish under laboratory conditions

Collecting of fish for laboratory maintenance is possible from late October, when its natural habitats are covered with the first ice. In this study the loach (*Misgurnus fossilis*) were collected from their natural habitat in the river Severniy Donez in the Ukraine, and maintained according to Neifakh [1]. Briefly, males and females are kept separately under hibernation conditions (4°C water in the dark). No feeding is needed under these conditions. Aged tap water is used and changed once a week.

2.2 Fish handling and semen collection

- Three days before an experiment, males and females are moved to separate fish tanks at 18°C under natural light conditions. No food is provided during this period.
- The fish are then injected with chorionic gonadotropin (Profasi) (300 U per female and 100 U per male) [1]. 1,500 U of dry hormone is dissolved in 200 ml of deionised water.

- The fish (Fig. 1) are taken from the water and dried with tissue paper.
- The injection is made intramuscularly in the cranial part of the body. The injection is performed slowly and gentle rubbing is applied afterwards to the injection site to avoid the injected hormone leaking due to muscle contraction.
- After injection the fish are put back into the water.
- Ovulation normally occurs 38-43 h after injection. Although the male fish were injected with a relatively small amount of hormone in this study, they could even be used without injecting since the male testes contain mature sperm from late autumn [1].

Figure 1 - Weather fish loach (*Misgurnus fossilis*).

2.3 Freezing and thawing procedures

The protocol of loach sperm cryopreservation described below was created on the basis of a previously designed one for carp sperm [2]. The present protocol was created for field conditions where there is no easy access to specialised equipment such as a programmable freezer.

- To prepare the sperm suspension the male fish are sacrificed and the testes removed. Care must be taken to avoid milt contact with water.
- Milts (Fig. 2) were cut into small pieces avoiding contamination with urine, faeces and blood.
- The homogenised milt from one male is suspended in 3.6 ml of extender (39 mM tris-base, 29 mM NaCl, 11 mM $CaCl_2$, 2 mM NaHCO3, 2 mM KCl, 139 mM mannitol, 24 mM sucrose, pH 8.1). 20% Me_2SO made with extender described above is used as cryoprotectant media. The equivalent volume of 20% Me_2SO is added very slowly to 3.6 ml of sperm suspension (making a final concentration of 10% Me_2SO).
- In order to avoid osmotic shock the cryoprotectant is poured gradually on the wall of the dish with the sperm suspension while slowly vortexing [3].
- Samples are left at 4°C for 30 min. Cryoprotectant treated sperm are placed into eppendorf tubes and frozen in one of the two possible ways described below:

- Freezing in liquid nitrogen vapour from 4°C to –15°C at a rate of 1-5°C/min, and from –15°C to –70°C at 15-20°C/min followed by direct immersion into liquid nitrogen. The freezing rate is monitored with a thermocouple and recorded by means of H-307 plotting. The rate of cooling is regulated by changing the distance between the samples and liquid nitrogen surface.
- Freezing in a cooling bath: place the eppendorf tubes into the spirit bath at a temperature of –20°C for 3 minutes, place in the bath at –40°C for 3 minutes and then plunge into liquid nitrogen.
- Thawing in both cases is carried out by placing the tubes in a water bath at 40°C for 30 seconds and then in air at 21°C.

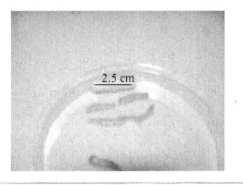

Figure 2 - Freshly extracted milts of weather loach (*Misgurnus fossilis*).

2.4 Postthaw sperm evaluation

- Quality control of the cryopreservation procedure is mainly assessed on the basis of fertilisation success.
- The motility rate of sperm is screened under a microscope immediately after thawing.
- The sperm drop is placed on the glass slide and mixed with an equal drop of activating medium (42 mM $NaHCO_3$). Motility rates were in the range of 40-60% depending on initial quality of sperm (80-90%).
- For a more accurate assessment of sperm quality after cryopreservation, the treated sperm was used for artificial fertilisation described below and the survival of embryos was monitored until hatching.

2.5 Artificial fertilization

- All the eggs were collected just before fertilisation. For this, the females were taken out of the water. All water was carefully wiped from the fish body.

- The eggs were stripped into clean, dry Petri dishes with the help of abdominal massage distributing pressure from the cranial end of abdomen towards the caudal end (Fig. 3). The diameter of a mature egg is 1.17-1.30 mm [1]. One female can produce up to ten thousand eggs.
- Egg quality was assessed visually and only those of good quality (transparent, yellowish in colour and without blood) were used in the experiments.
- The eggs were fertilised by mixing approximately 1,000 eggs with 1.0 ml fresh or 1.4 ml cryopreserved sperm suspension and 5 ml of activating medium (42 mM $NaHCO_3$) in Petri dishes. After mixing the reproductive cells the Petri dish was slowly vortexed for 2-3 minutes. Higher volumes of cryopreserved sperm suspension were used to retain motile sperm to egg ratio.
- The eggs were then left at 21°C to develop.
- Embryo survival and development was assessed by visual inspection at different stages of development until hatching every 2-3 hours. Embryo water was changed after each inspection. Unfertilised eggs and dead embryos became opaque due to necrosis and were removed during each inspection.

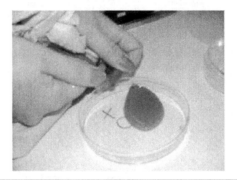

Figure 3 - Collection of loach eggs.

3. GENERAL CONSIDERATIONS

The extensive study of embryo survival derived from cryopreserved sperm has been undertaken and described elsewhere [4,5]. The survival of embryos derived from fresh sperm may vary from 68% to 92% whereas the survival of those derived from the cryopreserved sperm may be in a range from 53 to 90% (Table 1). The survival of embryos was strongly affected not only by the cryopreservation procedure, but also by the treatment of sperm with different cryoprotectants, and by female and male individuals [5]. Strong interindividual variation of cryoresistance was established in the present experiments.

Table 1- General sperm characteristics in the loach

	Fresh sperm	Cryopreserved sperm
Sperm density	3×10^6 sperm/ml	
Motility of sperm	80-90%	40-60%
pH	7.52	
Survival of embryos at hatching	68-92%	53-90%

4. REFERENCES

1-Neifakh, A.A., Weather loach *Misgurnus fossilis,* in *Objects of biology of development. Problems of biology of development*, Astaurov, B.L., Ed., Nauka, Moscow, 1975, 308, (in Russian).

2-Kopeika, E.F., Instruction on low temperature preservation of carp sperm, Moscow, VNRO in aquaculture, 11, 1986 (in Russian).

3-Kopeika, E.F. and Novikov, A.N., Cryopreservation of fish sperm, in *Cryopreservation of cell suspensions,* Zuzaeva, A.A., Ed., Kiev, "Naukova dumka", 1983, 204, (in Russian).

4-Kopeika Yu., E., Grischenko, V.I., Kopeika, E.F., Linnik, T.P., Bibenko, O.V., and Zintchenko, A.A., Effect of DMSO and its decay products on spermatozoa and survival of loach (*Misgurnus fossilis* L.) embryos, *Problems of Cryobiology,* 4, 45, 2002.

5-Kopeika, J., Kopeika, E., Zhang, T., and Rawson, D.M., Studies on the genotoxicity of dimethyl sulfoxide, ethylene glycol, methanol and glycerol to loach (*Misgurnus fossilis*) sperm and the effect, *Cryoletters,* 24(6), 365, 2003.

Complete affiliation:

Julia Kopeika, Luton Institute for Research in the Applied Natural Sciences, University of Luton, The Spires, 2 Adelaide street, Luton, Bedfordshire, LU1 5DU, UK
Julia_kop@yahoo.com.

ZEBRAFISH SPERM CRYOPRESERVATION WITH N,N-DIMETHYLACETAMIDE

J.P. Morris IV, A. Hagen and J.P. Kanki

1. INTRODUCTION

Zebrafish provide exceptionally large clutches of rapidly developing, externally fertilized embryos, permitting easy experimental access and manipulation. The optical clarity of the embryos allows tissue development and internal organogenesis to be directly observed. These qualities make the model particularly amenable to large scale forward and reverse genetic screens and have established it as a powerful tool for investigating vertebrate development and even modeling human disease [1]. Despite the relatively low cost of maintaining zebrafish lines, experimental scope is often limited by facility size. Sperm cryopreservation helps to alleviate this problem by removing males from a system while maintaining their reproductive capacity. Reliable and efficient cryopreservation techniques provide "genetic insurance" for stock and mutant lines, help maintain genetic diversity, and provide a convenient form for distributing lines via shipping to other laboratories [2-5]. The technique also facilitates reverse genetic strategies, such as Targeting Induced Lesions in Genomes (TILING) that can be limited If restricted to using only living libraries [6].

Successful sperm cryopreservation has been reported with methods that use methanol and powdered milk as cryoprotectants and microcapillaries as cryovessels [2-5]. However, our laboratory and others have observed inconsistent results with these methods [7,8]. Furthermore, the reported techniques can be inconvenient because they require complex manipulations of small volumes and are not compatible with large-scale protocols. By performing a broad evaluation of cryoprotectants and sperm diluents and investigating the effect of increasing the volume of freezing medium on fertilization capacity, we developed a method that provides efficient recovery and an increased number of archived samples. We found that cryopreserving homogenized, whole testes in 10% N,N-dimethylacetamide (DMA) diluted in Buffered Sperm Motility Inhibiting Solution (BSMIS) permits efficient embryo recovery from four, 50 µl frozen samples [7,8].

2. PROTOCOL FOR FREEZING AND THAWING

Despite the relative simplicity of this technique, sperm freezing is still a time-consuming protocol requiring a high level of preparation, organization and some technical proficiency. Therefore, it is advisable to practice this protocol with dispensable fish before attempting to cryopreserve critical lines. To increase the success of cryopreserving a line, males should be generally healthy and between the ages of 4-18 months. We also suggest that the males mate successfully several times prior to the procedure. Solutions, labels for cryotubes and adequate space in liquid nitrogen freezers should be prepared in advance. Required materials, equipment and chemicals are shown in Table 1, along with vendors, (catalogue numbers) we typically use. Figure 1 is an example of how we set up our equipment for the procedure. It is critical to use a setup that allows one to: 1) keep equipment dry and 2) have reagents and materials quickly accessible, as these procedures work best by minimizing working time.

2.1 Materials and Chemicals

Table 1 - Equipment used for sperm cryopreservation.

Materials	Qty	Detail	Vendor
Freezing			
Dry Ice		2" Pellets	
Styrofoam Box	1	10"x10"x10"	
Centrifuge Tube	1	15 ml, Conical	Falcon
Centrifuge Tubes	6	50 ml, Conical	Falcon
Cryotubes	4/fish	1.5 ml, Conical-Bottom	(#3471)
Microcentrifuge Tubes	1/fish	1.5 ml, Sterile	Scientific (#1615-5500)
Homogenization pestles	1/fish	Compatible with 1.5 ml tubes	Fisher (#K7495211590)
Liquid Nitrogen			
Liquid Nitrogen Container	1	4 l Portable Dewar Flask	Nalge Nunc. (#4150)
Cryocanes	1/fish	Aluminum, 5-Place	Nalge Nunc (#5015-0001)
Pipet Tips		200 µl and 1000 µl	
Ice Bucket	1		
Ice			
Pipetmen		200 µl and 1000 µl	
Dissection			
Microscope	1	Stereo Dissection	
Paper Towels	box	Kimwipes	Kimb.-Clark
Forceps	2	Stainless, #3C	Roboz (#RS-5043
Scissors	1	3.75", Angled	Roboz (#RS-5928)
Scalpel	1/fish		
Fish Handling			
Fish Net	1		
Fish Container	1	2.5 l Mating Cage	
Glassware	1	325 ml Crystalization Dish	Pyrex (#3410-100)
Soup Spoon	1	Plastic	
Spatula	1	Metal	
Chemicals			
N, (DMA)		Molecular Biology Grade	Sigma (#D5511)
3-Aminobenzoic Acid Ethyl Ester		Tricaine	Sigma (#A5040)
BSMIS		See "Solutions"	
E3 Egg Water		See "Solutions"	

2.2 Solutions

- BSMIS (buffered sperm motility inhibiting solution) [9]:
  75 mM NaCl, 70 mM KCl, 2 mM $CaCl_2$, 1 mM $MgSO_4$, and 20 mM Tris, ph 8 in sterile, autoclaved water. For 500 ml: 2.12 g NaCl, 2.61 g KCl, 0.11 g $CaCl_2$ (for $CaCl_2$-$2H_2O$, add 0.15 g), 0,12 g $MgSO_4$ x $7H_2O$, 1.21 g Tris, bring pH to 8. BSMIS can be stored for several months at 4°C.
- Egg Water [3]: E3 egg water: 5 mM NaCl, 0.17 mM KCl, 0.33 mM $CaCl_2$ x $2H_2O$, 0.33 mM $MgSO_4$ x $7H_2O$. For 60X stock (1 l): 17.53 g NaCl, 0.76 g KCl, 2.91 g $CaCl_2$ x $2H_2O$, 4.88 g $MgSO_4$ x $7H_2O$. Add salts one at a time, can heat water to help dissolve into solution.
- Tricaine Stock Solution: 4 mg/ml 3-Aminobenzoic Acid Ethyl Ester. For 100 ml: Add 400 mg 3-Aminobenzoic Acid Ethyl Ester to 97.9 ml ddH_2O. Add 2.1 ml 9.1M tris solution to bring pH to ~7.3. Tricaine solution can be stored for several months at -20°C.
- Tricaine Working Solution: 0.16 mg/ml Tricaine, pH 7. For 100 ml: Add 4 ml of Tricaine Stock to 96 ml water.
- Freezing Medium: 11.2% DMA in BSMIS. For 901 µl: Add 101 µl DMA to 800 µl BSMIS solution. Vortex vigorously for 2-3 minutes, assuring that DMA is completely in solution. Each set of samples from a male requires 180 µl freezing medium; for example, prepare 2 ml of stock for samples from 10 fish. Freezing medium should be made fresh each time the protocol is executed. Keep Freezing medium on ice in 15 ml tubes throughout the procedure.

2.3 Setup

- At least 30 minutes before freezing, fill the Styrofoam box with dry ice pellets. For each male to be processed, insert two 50 ml centrifuge tubes until only the cap is visible. Loosen the caps so that they can be easily removed. Set 200 µl pipetman to 50 µl, and set 1000 µl pipetman to 180 µl. It is convenient to use different size pipetmen so that they are easily distinguishable during the procedure.
- For each male, aliquot 20 µl of BSMIS into a 1.5 ml microcentrifuge tube and store on ice. Prepare fresh freezing medium and store on ice in a 15 ml conical tube. Organize four 1.5 ml cryotubes for each male with the appropriate fish ID and freezer storage position designated on a cryosafe label.
- Prepare 100 ml of Tricaine Working Solution in a 325 ml crystallization dish. Fill Dewar flask with liquid nitrogen. It is important to designate a 'wet' area and 'dry' working area. Sperm motility appears to be required for fertilization and sperm have a finite motile capacity after exposure to water. Therefore, sperm samples must not come into contact with water during preparation.
- Arrange equipment to optimize working quickly and efficiently (Fig. 1).

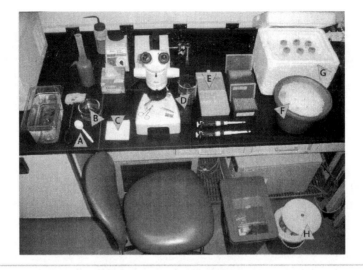

Figure 1 - Cryopreservation setup. Arrange reagents and equipment for efficient use during the protocol. Keep fish tank, net, tricaine on 'wet' side of the bench, while keeping the microscope and right side of the bench dry. A) Plastic spoon; B) Tricaine; C) Kim wipes to dry fish; D) Microfuge tube homogenizers; E) Labeled cryotubes; F) Chilled microfuge tubes with BSMIS; G) Chilled 50ml tubes on dry ice; H) Liquid nitrogen-containing Dewar.

2.4 Freezing procedure

- Remove twist-off caps from 4 labeled cryotubes. Ensure that forceps, scalpel and any equipment coming in contact with sperm are dry.
- Anesthetize the male fish in the crystallization dish and use the plastic spoon to scoop out the male. Tip-off excess water from spoon and place fish on several dry Kimwipes. Gently pat the fish dry, moving the male to different locations on the Kimwipe until water is no longer absorbed. Transfer the fish to a fresh Kimwipe on the dissection scope. Do not squeeze the fish forcefully as this could cause the sperm to be milked.
- Decapitate and remove the tail with a scalpel. Open the body cavity with a pair of scissors and splay the stomach open using the forceps. Gently remove the gut. The two lobes of the testes should be visible along the ventral sides of the air-bladders (Fig. 2). Carefully remove the testes. We have observed best recovery from testes that are large and white. As expected, necrotic, gray, stringy and/or cancerous testes yield low recovery rates. It is sometimes easier to remove the testes after removing the air bladder.
- Immediately place the testes into one of the 20 µl aliquots of chilled BSMIS. Gently homogenize with a pestle by turning the pestle 4-5 times clockwise and 4-5 times counterclockwise. Hold the sperm-BSMIS solution up to a light to

visually ensure that no large tissue clumps remain. This step should take no longer than 15 seconds; take care not to over-homogenize.

- Immediately add 180 μl freezing medium to the sperm solution and mix by pipeting up and down until the solution is uniformly cloudy. Pipet 50 μl of sperm/freezing medium solution into the four properly labeled opened 1.5 ml cryotubes. This step should take no longer than 15 seconds.

- Close the cryotubes tightly to ensure a good seal, and immediately place them in pairs in the 50 ml centrifuge tubes on dry ice. Ensure that both cryotubes are resting side-by-side in the bottom of the 50 ml centrifuge tube. Incubate on dry ice for thirty minutes. Extending or decreasing incubation on dry ice for five minutes does not appear to affect sperm viability, although we aim for thirty minutes to maximize consistency between samples.

- After thirty minutes, immediately immerse the samples in liquid nitrogen. We use cryocanes to organize samples in the portable containers. It also may be

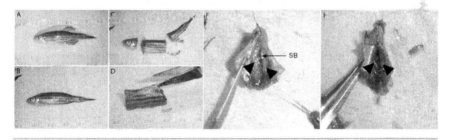

advisable to store samples in separate liquid nitrogen freezers to minimize the risk of losing all archived samples to equipment failure.

Figure 2 - Testes dissection. A) Plastic spoon to remove fish from anesthesia; B) Pat dry on kimwipes; C) Decapitate and remove tail; D) Cut along ventral midline; E) Splay and remove gut, revealing clear swim bladder (SB, which can also be removed) and bilateral testes (arrowheads) that are elongated and white; F) Example of bad testes (arrowheads) that are grey and necrotic.

2.5 *In vitro* Fertilization from Frozen Samples

We recommend keeping a designated stock of healthy, mating fish in the laboratory for *in vitro* fertilization (IVF). Females should be mated regularly and successfully produce viable clutches prior to their use for IVF. It may increase yields by allowing females to rest for approximately two weeks prior to squeezing and two to three weeks after squeezing. Clutch size and quality are critical to the success of the IVF. Do not IVF with precious sperm samples using fewer than 400-500 healthy squeezed eggs.

- Additional Equipment and Solutions: 37°C Water Bath, 28°C Incubator, Egg Water, Spatula.

- The afternoon before IVF, set up males and females in mating chambers equipped with dividers so the fish cannot mate. Set up at least three times the number of females than samples to be used for fertilization. Generally, we plan on going through seven females per line we would like to reconstitute.
- At least 30 minutes prior to IVF, incubate 3 ml of BSMIS per sample to be reconstituted in a 37°C water bath.
- Remove the samples to be used for fertilization from the liquid nitrogen freezer and place them immediately into a portable liquid nitrogen container.
- Separate females from males and anesthetize females in tricaine until their gills slow and swimming subsides. Dry females carefully and place in a Petri dish under a stereo-dissecting microscope. Apply gentle pressure to the abdomen, front to back. Eggs should emerge from the anal pore. Collect clutches until at least 4-500 eggs can be pooled, before proceeding. Work quickly to keep eggs from drying out and pile egg clutches on top of each other gently with a spatula. Once the first clutch is obtained, we aim to harvest the remaining necessary eggs in less than 5-6 minutes.
- Viable eggs will be yellow and will stick together. Up to four females can be anesthetized simultaneously to help speed up this process. Females recovering from squeezing may benefit from adding "Stress Coat" (Aquarium Pharmaceutical Inc.) to the water in the recovery tank.
- Remove a cryotube from liquid nitrogen and remove the cap. Immediately add 1ml of 37°C BSMIS to sample. Mix the sample with a pipet to thaw and apply the entire solution to the egg mass. This process should take no more than 10-15 seconds. Immediately add 1 ml of egg water to the eggs to activate the sperm. Rinse the cryotube with 1ml and add to the eggs. After 5 minutes, or after the chorions have separated from the egg mass, fill the Petri dish half way with egg water. Incubate the dish at 28°C. After 2-3 hours, remove debris and separate the fertilized clutch into groups of 100-150 eggs in clean Petri dishes. Calculate fertilization percentage after 18 hours.

3. GENERAL CONSIDERATIONS

Our protocol provides an efficient technique for archiving sperm samples from large numbers of males. The following general considerations are guidelines for troubleshooting and further optimization of the technique.

3.1 Sperm Quality Evaluation

To develop our protocol, we evaluated the ability of a number of cryoprotectants with differing permeability coefficients to maintain sperm percent motility after freezing and thawing [7]. Sperm percent motility was used as a metric of sperm quality because our preliminary experiments and other work have shown a positive correlation between sperm percent motility and fertilization capacity. However, we did not extensively investigate the correlation between post-thaw sperm percent motility and fertilization capacity. Percent motility was decreased due to freezing with all cryoprotectants investigated (Table 2), indicating an increase in the number of dead sperm in thawed samples with respect to fresh samples. Sperm must interact with and enter the micropyle chorion passage for fertilization to occur [9], a process that has a significant kinetic element as it occurs in free solution space. Dead sperm and cellular debris may affect sperm movement or block the micropyle. Therefore, further optimization of the protocol may focus on post-thaw sperm quality evaluation to develop relationships between motility, or other metrics, and fertilization capacity. Methods that do not rely on direct microscopic counting of sperm, such as computer aided sperm analysis [8] and flow-cytometric analysis of membrane integrity using fluorescent compounds excluded from intact cells [10] might be useful for post-thaw sperm evaluation. Better post-thaw quality evaluation and its relationship to pre-freeze quality and fertilization may provide a method to best determine what samples to freeze and what samples may require larger egg clutches to achieve appropriate fertilization.

Table 2 - Cryopreserved Sample Post-Thaw Motility

Cryoprotectant	Permeability Coefficient (P x 10^{-5} cm/s)	Post-Thaw Percent Motility (n=3)				Average Post-Thaw Percent Motility (n=12)
		5%	10%	15%	20%	
DMA	14.7 ± 0.37	8.8 ± 1.2	11.6 ± 5.8	12.1 ± 4.9	7.2 ± 0.6	9.9 ± 3.9
Ethylene Glycol	3.4 ± 0.1	7.7 ± 1.3	9.0 ± 6.8	7.4 ± 2.6	6.8 ± 0.4	7.6 ± 3.3
Methanol	11.4 ± 0.4	3.1 ± 2.4	8.9 ± 4.5	9.9 ± 2.2	9.3 ± 2.2	7.5 ± 3.7
DMSO	1.30 ± 0.1	8.0 ± 2.1	6.0 ± 1.0	7.6 ± 2.2	6.8 ± 0.8	7.1 ± 1.6
Glycerol	0.58 ± 0.04	5.7 ± 0.6	7.4 ± 0.2	8.4 ± 1.8	5.3 ± 0.4	6.7 ± 1.5

3.2 Freezing Medium and Diluent

We found that 10% and 15% DMA provided a trend towards high post-thaw percent motility (Table 2), and that sperm frozen in 10% and 15% DMA provided significantly increased fertilization capacity over sperm frozen with methanol and powdered milk (Table 3). Samples frozen in 10% DMA displayed significantly lower variability than samples frozen in 15% DMA; therefore, we suggest using 10% DMA. However, sperm frozen in 10% DMA did not display significantly higher fertilization than sperm frozen in 15% DMA indicating that DMA concentrations between 10% and 15% may approach the equilibrium between DMA toxicity and its maximal freezing protection capacity. Thus, titrating the DMA concentration may further optimize the protocol.

DMA has been used as a cryoprotectant for sperm from other fish species, and has been implicated to increase post-thaw intracellular ATP concentration in European

Catfish sperm [11]. Decreased ATP concentration can reduce sperm motility and postthaw fertilization capacity. Other cryoprotectants and additives that may help maintain or enhance ATP may be good candidates for further optimizing our technique.

Zebrafish sperm motility is activated by changes in intracellular potassium concentration and can be affected by pH [12]. We have had success with BSMIS at both pH 8 and pH 10 (unpublished data), but have not performed titration experiments to identify an optimal pH within this range.

Table 3 - *In Vitro* Fertilization with Cryopreserved Samples

Protocol (n=8)	Freezing Medium	Clutch Size	Recovered Embryos	Percent Fertilization
Capillary/ 20µL	10% Methanol & 15% Powdered Milk/GRS	442.1 ± 172.8	1.0 ± 1.2	0.2 ± 0.2
Capillary/ 20µL	10% DMA/HS	169.6 ± 124.6	27.9 ± 32.3	14.0 ± 10.1
Capillary/ 20µL	● 15% DMA/HS	312.1 ± 142.7	26.0 ± 15.9	10.7 ± 12.2
Protocol 1/ 200µL	10% DMA/BSMIS	478.4 ± 192.2	38.6 ± 10.6	10.2 ± 4.1
Protocol 1/ 200µL	15% DMA/BSMIS	436.6 ± 147.0	47.6 ± 42.4	9.3 ± 5.5

3.3 Sample Volume and Pooling

Our protocol was developed to archive samples from single males. Total sperm from a single male frozen in 20 µl freezing medium yielded no significant difference in fertilization rate when compared to 50 µl aliquots of sperm frozen in 200 µl of freezing solution (Table 1). Therefore, our protocol has not exceeded a dilution that limits fertilization capacity and the number of samples yielded from a single male may be increased.

Our lab has also successfully fertilized from sperm frozen after pooling from multiple males while maintaining DMA concentration and freezing medium: sperm dilution. This technique is useful for archiving mixed samples from stock lines and promoting genetic diversity in reconstituted lines.

3.4 Clutch Size

Our experiments emphasized large clutch sizes to maximize embryo recovery. Although clutch size was positively correlated with fertilization rate in some of our trials, this relationship was not observed in all our fertilization experiments. This indicates that large clutches do not necessarily enhance fertilization, but best exploit frozen sperm fertilization capacity. If recovery will be performed frequently, we suggest keeping a stock of healthy females dedicated exclusively for *in vitro* fertilization and allowing 2-3 weeks recovery between egg collections.

3.5 Sample Longevity and Storage

We have achieved successful fertilization from samples in liquid nitrogen storage for up to three years (unpublished data) without apparent loss in fertilization capacity. These findings suggest that samples frozen with our technique remain viable as long as their storage conditions remain stable. We have also observed no apparent difference in fertilization between samples stored in the vapor or the liquid phases of our freezers, although most cryotube vendors suggest that samples be stored in the vapor phase since tubes can explode upon thawing.

4. REFERENCES

1-Berghmans, S., Jette, C., Langenau, D., Hsu, K., Stewart, R., Look, T., and Kanki, J.P., Making waves in cancer research: new models in the zebrafish, *Biotechniques*, 39, 227, 2005.
2-Harvey, D., Norman Kelley, R., and Ashwood-Smith, M.J., Cryopreservation of zebrafish spermatozoa using methanol, *Can. J. Zool.*, 60, 1867, 1982.
3-Westerfield, M., *The Zebrafish Book: A guide for the laboratory use of zebrafish (Danio rerio)*, University of Oregon Press, Eugene, OR., 2000.
4-Ransom, D.G. and Zon, L.I., Collection, storage, and use of zebrafish sperm, *Methods Cell. Biol.*, 60, 365, 1999.
5-Brand, M., Granato, M., and Nusslein-Volhard, C., *Zebrafish: A practical approach*, Oxford University Press, New York, NY, 2002.
6-Wienholds, E., Schulte-Merker, S., Walderich, B., and Plasterk, R.H.A., Target-selected inactivation of the zebrafish rag1 gene, *Science*, 297, 99, 2002.
7-Morris IV, J.P., Berghmans, S., Zahrieh, D., Neuberg, D.S., and Kanki, J.P., Zebrafish sperm cryopreservation with N,N-dimethylacetamide, *Biotechniques*, 35, 956, 2003.
8-Berghmans, S., Murphey, R.D., Wienholds, E., Neuberg, D., Kutok, J.L., and Fletcher, C.D., Zebrafish sperm cryopreservation, *Methods Cell. Biol.*, 77, 645, 2004.
9-Lahnsteiner, F., Berger, B., Horvarth, A., Urbanyi, B., and Weismann, T., Cryopreservation of spermatozoa in cyprinid fishes, *Theriogenology*, 54, 1477, 2000.
10-Segovia, M., Jenkins, J.A., Paniagua-Chávez, T., and Tiersch, T.R., Flow cytometric evaluation of antibiotic effects on viability and mitochondrial function of refrigerated spermatozoa of Nile tilapia, *Theriogenology*, 15, 1489-99, 2000.
11-Baulny, B.O., Labbe, C., and Maisse, G., Membrane integrity, mitochondrial activity, ATP content, and motility of the European catfish (*Silurus glanis*) testicular spermatozoa after freezing with different cryoprotectants, *Cryobiology*, 39, 177, 1999.
12-Takai, H. and Morisawa, M., Change in intracellular K^+ concentration caused by external osmolality change regulates sperm motility of marine and freshwater teleosts, *J Cell Sci.*, 108, 1175, 1995.

Complete affiliation:
John P. Kanki, Dana-Farber Cancer Institute, Department of Pediatric Oncology, Harvard Medical School, Boston, MA 02115, USA. e-mail: john_kanki@dfci.harvard.edu.

SPERM CRYOPRESERVATION FOR LIVE–BEARING FISHES OF THE GENUS *Xiphophorus*

Q. Dong, C. Huang, L. Hazlewood, R.B. Walter and T.R. Tiersch

1. INTRODUCTION

Swordtails and platyfish of the genus *Xiphophorus* are viviparous teleost of the family Poecilliidae. They are valuable models for biomedical research, especially for cancer genetics. This animal model was one of the first to prove that certain cancers are inherited diseases. In addition, they are also valued as ornamental fish for the aquarium trade because of vibrant body coloration and a long sword–like tail (in males). Species of this genus usually attain sexual maturity at 10-12 weeks of age [1]. The sperm of *Xiphophorus* fishes are different in structure (e.g., head shape) and physiology (e.g., energy metabolism) from the sperm of oviparous fishes. Sperm of internally fertilizing species possess atypical features such as well-developed mitochondrial sheaths in the midpiece of spermatozoa and glycolytic activity comparable to that of mammalian sperm, which may be adaptations for movement or long-term survival in the female reproductive tract.

The increased research and commercial value, and the continuous decline of diversity in the wild of these fish, expand the need to preserve their genetic resources. Despite study of sperm cryopreservation in some 200 species of freshwater and marine fishes, sperm cryopreservation has just begun in live bearers including *Xiphophorus* [2]. Cryopreservation protocols have been developed to address their small body size and limited sperm volume (e.g., 5-10 µl per fish), and post-thaw motility as high as > 70% has been observed with *X. helleri* [3] and *X. couchianus* [4]. Recently, live young were produced with cryopreserved sperm from *X. helleri* (our unpublished data).

2. PROTOCOL FOR FREEZING AND THAWING

2.1 Equipment

- Sperm collection: 1 liter beakers, Petri dish, Kimwipes, stereomicroscope, tweezers, straight microscissors, styrofoam box with crushed ice, permanent marker, resealable, plastic bags, plastic goblet, 1.5 ml centrifuge tubes and top-loading balance.
- Motility estimation: micropipette and tips (10 µl), glass microscope slide, microscope with darkfield and 20 x objective.
- Straw filling: 0.25 ml straws, goblets, cane holders, PVC powder, paper towels, and filling tool (Fig. 1).

- Freezing: a controlled-rate freezer with accompanying low-pressure liquid nitrogen cylinder, liquid nitrogen storage dewar, and cryogloves.
- Sample shipping: shipping Dewar (CP35, Taylor-Wharton, Theodore, Alabama), plastic cable tie wraps, and tape.
- Thawing: Safety glasses, water bath with temperature controls (Model 1141, VWR Scientific, Niles, Illinois), long tweezers, scissors, paper towels, and 1.5 ml microcentrifuge tubes.
- Artificial insemination: 1 liter beakers, Petri dish, cotton batting, micropipette tips, scalpel, artificial insemination tool (Fig. 2), glassware, and 10 liter aquaria.

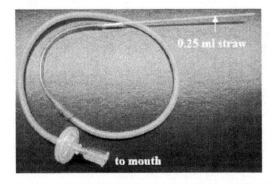

Figure 1 - Device used to fill straws.

2.2 Reagents

- Anaesthetic: 0.01% tricaine-methane sulfonate (Western Chemical Inc., Ferndale, WA)
- Hanks' balanced salt solution (HBSS) at 300 mOsm/kg (8.0 g NaCl, 0.4 g KCl, 0.16 g $CaCl_2 \cdot x\, 2H_2O$, 0.2 g $MgSO_4 \cdot x\, 7\, H_2O$, 0.06 g Na_2HPO_4, 0.06 g KH_2PO_4 0.35 g $NaHCO_3$ and 1.0 g $C_6H_{12}O_6$ in sufficient distilled water to yield a final volume of 1,000 ml)
- Cryoprotectant: glycerol

2.3 Semen collection

- Transfer the fish from breeding tank to 1 liter beaker and anesthetize the fish in 0.01% tricaine-methane sulfonate for 2 min.
- Dry the fish with Kimwipes and place them in a Petri dish.
- Carefully open the abdomen with the straight microscissors under a stereomicroscope and use tweezers to remove adherent tissue around the testis.

Mature testes usually have a milky appearance and are located above the gonopodial area.

- Use the tweezers to grasp the posterior tubular portion of the testes and remove them from the body cavity while taking care to avoid sperm release.
- Place the testes in resealable plastic bags, weigh, add sufficient HBSS based on a 1:10 ratio of testis weight to HBSS volume, seal the bag, and crush the testis with the plastic goblet to release sperm.
- Transfer the sperm solution from the plastic bag to 1.5 ml centrifuge tubes, and place it on crushed ice prior to use.

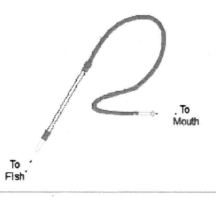

To
Mouth

To
Fish

Figure 2 - Artificial insemination tool (*pthoto:* Steven Kazianis, Texas State University- San Marcos, with permission).

2.4 Sperm motility estimation

- Take 10 ml HBSS from a stock solution stored at –20°C and allow it warm to room temperature.* Evaluate the HBSS with microscopy for presence of bacterial contamination.
- Dilute samples to attain a concentration of 10^7 cells/ml before motility estimation. Generally a dilution ratio of sperm to HBSS (fresh samples) or HBSS-cryoprotectant (thawed samples) of 1:100 will yield the desired concentration.
- Equilibrate sperm suspensions in the centrifuge tubes at room temperature for 5 min before motility estimation for samples after thawing or refrigeration.
- Gently mix the sperm suspension with HBSS with the pipette tip and evaluate motility by estimating at 200 x magnification, using darkfield microscopy, the percent of sperm that are actively moving in a forward direction.
- Estimate the percent motility in increments of 5%. Samples with motility below 5% that have motile sperm are recorded as 1%.

*It is recommended to use the same activation solution throughout a single working season. HBSS can be aliquotted into small volumes (such as 10 ml) and stored at –20°C.

2.5 Freezing and thawing procedures

- Mix sperm suspensions with an equal volume of freshly prepared 28% glycerol (to yield a final concentration of 14%).
- Load the sperm suspension (80 µl) into 0.25 ml French straws with the filling tool (Figure 1). Place eight 0.25 ml straws into a 10 mm plastic goblet, and attach two goblets to a 10 mm aluminium cane. After equilibrating at 4 °C for 10 min, transfer the canes to the freezing chamber.
- Cool the samples from 5°C to –80°C at 20-25°C/min, and hold at –80°C for 5 min.
- Remove samples swiftly from the freezing chamber and immediately plunge them into liquid nitrogen in a storage dewar.
- Thaw samples after a minimum of 12 h of storage in liquid nitrogen.
- Remove individual straws from the 10 mm goblets with the long tweezers.
- Thaw in a 40°C water bath for 7 s. Wipe straws with a paper towel and cut the PVC powder sealed ends. Empty the straws into 1.5 ml microcentrifuge tubes by cutting the cotton plug end to release the contents into the tubes.
- Dilute the sperm suspension by adding an equal volume of HBSS immediately after thawing.
- Centrifuge the sperm suspension at 1000 g for 10 min at 4°C, remove the supernatànt, and resuspend the sperm pellets with HBSS at the desired concentration.

2.6 Artificial insemination (AI):*

- Prepare the micropipette tips by cutting off most part of the fine end (leave ~ 2 m) with the scalpel at a 90° angle, and ensure that there are no rough edges.
- Anesthetize the virgin females in 0.01% tricaine-methane sulfonate for 2 min.
- Place the sedated female on the wet cotton pad, and position her so the urogenital opening (anterior to the anal fin) faces the operator.
- Pipette the thawed sperm suspension** into the micropipette tip and hold the solution in place (avoid air bubbles in the tip).
- Hold the female in place with one hand, and hold the artificial inseminator (Figure 2) with the other hand. Gently insert the plastic micropipette tip into the urogenital opening at a 45° angle.
- After encountering some resistance, push the micropipette tip a little bit further, and gently eject the sperm solution into the female.
- Acclimate the females in a glass beaker with equal volumes of water from the original tank and the new tank for 1 h before transfer into the new 10 liter aquarium.
- For fresh sperm controls, obtain sperm either by dissecting the testis as described above or by stripping and repeat the AI procedure.

*For detailed instructions see the website www.xiphophorus.org/ai.htm.
**Use sperm that is as concentrated as possible (10^9 cells/ml) to increase the success of AI.

3. GENERAL CONSIDERATIONS

In contrast to sperm of fishes that have external fertilization, sperm from *X. helleri* and *X. couchianus* can remain continuously motile for as long as two weeks when stored at 4°C.

Limitations of sperm volume can be overcome by pooling the milt from several fish, and by increasing the ratio of sperm-to-extender to 1:100 without significant loss of sperm motility. The use of an 80 ml loading volume in 0.25 ml straws would further help to maximize the sperm volume for evaluation of multiple parameters during protocol optimization process. Due to the need for specialized devices and technical skill to perform artificial insemination, and the fact that embryonic development in the female is not easily monitored, artificial insemination in live-bearers poses great challenges for post-thaw sperm quality assessment. One quick alternative method to evaluate the outcome of artificial insemination is to check embryonic development by dissecting females within 26-30 days after insemination. In general, thawed sperm is used for crosses with females of a different species (within the genus *Xiphophorus*), which enables verification that the offspring (hybrids) resulted from cryopreserved sperm rather than from sperm stored within the females from males (e.g., siblings) of the same species. Overall, the protocol presented here will yield high post-thaw motility for *X. helleri* and *X. couchianus*. To increase the success of artificial insemination, the use of concentrated sperm is generally recommended for fresh and thawed sperm samples. For thawed samples, sperm can be concentrated by centrifugation, which allows removal of the cryoprotectant solution and supernatant, and replacement with fresh HBSS at smaller volume.

4. ACKNOWLEDGMENTS

This work was supported by USPHS grants, RR-17072 from the National Center for Research Resources and CA-75137 from the National Cancer Institute, with additional support provided by the Roy F. and Joanne Cole Mitte Foundation and the U.S. Department of Agriculture. This manuscript has been approved for publication by the Director of the Louisiana Agricultural Experiment Station as number 05-11-2007.

5. REFERENCES

1-Dawes, J.A., *Livebearing Fishes. A Guide to Their Aquarium Care, Biology and Classification*, Blandford, London, England, 1991, 240.

2-Huang, C., Dong, Q., Walter, R., and Tiersch, T., Initial studies on sperm cryopreservation of a live-bearing fish, the green swordtail *Xiphophorus helleri, Theriogenology*, 62, 179, 2004.

3-Huang, C., Dong, Q., and Tiersch, T., Sperm cryopreservation of green swordtail *Xiphophorus helleri*, a fish with internal fertilization, *Cryobiology*, 48, 295, 2004.

4-Huang, C., Dong, Q., Walter, R., and Tiersch, T., Sperm cryopreservation of a live-bearing fish, the platyfish *Xiphophorus couchianus, Theriogenology*, 62, 971, 2004.

Complete affiliation:

Terrence R. Tiersch, Aquaculture Research Station, Louisiana Agricultural Experiment Station, Louisiana State University Agricultural Center, Baton Rouge, Louisiana 70820 USA. ttiersch@agcenter.lsu.edu.

USE OF POST-THAW SILVER CARP (*Hypophtalmichthys molitrix*) SPERMATOZOA TO INCREASE HATCHERY PRODUCTIONS

B. Alvarez, A. Arenal, R. Fuentes, R. Pimentel, Z. Abad and E. Pimentel

1. INTRODUCTION

Silver carp (*H. molitrix*; Valenciennes, 1844) is a native species of China and Eastern Siberia, introduced around the world for aquaculture and the control of algal blooms, and feeds on phytoplankton and zooplankton [1]. It is among 3 or 4 species of cyprinids whose world production in aquaculture exceeds 1 million tons per year. Silver carp production has increased almost five-fold in two years (1998-2000) becoming the top cultured species in 2002 for the first time [2].

Silver carp farming is the largest food fish aquaculture industry in Cuba. Egg fertilization is totally artificial, and males only produce spermatozoa between March and October. Cryopreservation of silver carp spermatozoa reduces the number of males needed, minimizes handling stress through less frequent stripping, facilitates artificial propagation when eggs are available and promotes genetic and breeding studies.

At the "Hidráulica-Cubana" fish hatchery, where we developed the cryopreservation of silver carp spermatozoa, an additional reproductive cycle has been implemented since this technology was first used in November 1999. Fry production has increased by 20% because fertilization can be performed even when fresh sperm is not available; and the dispersal of genetically improved germplasm of silver carp in Cuba has been made possible. Our study is the first to report a reliable protocol for the fry production from post-thaw silver carp spermatozoa.

2. PROTOCOL FOR FREEZING AND THAWING

2.1 Semen collection and motility estimation

- Silver carp spermiation is induced by an intramuscular injection with carp hypophyseal extract (2 mg/kg body weight) and the semen is obtained 6 h later. Semen is collected by abdominal massage of each individual (i.e., about 20 ml for each male) and put it into a 50 ml Corning tube. The semen is placed on crushed ice immediately after collection.
- Motility is estimated by placing a 1 µl sample of semen onto a glass microscopeslide, activating it with 100-200 µl of fishpond water, and viewing right after at x 100 magnification with dark-field microscopy. Observations begin as soon as the semen and solution are mixed thoroughly at room temperature. Motility is determined subjectively, and can be expressed by 5 arbitrary scores, with 0 representing no motile spermatozoa and scores 1 to 4, representing > 0 to

25%, > 25 to 50%, > 50 to 75% and > 75%, respectively, of spermatozoa with progressive movement. Only samples with motility equal to or greater than 75% are recommend for cryopreservation experiments.

- Visual sperm motility is not a reliable indicator of the fertilizing ability of post-thaw fish spermatozoa [3]; however, sperm motility is necessary for fertilization in fish [4].
- Differential staining eosin 0.67% [weigh/volume] and nigrosin 10% [weight/volume] [5] can be used to characterize silver carp spermatozoa. The sperm concentration can be expressed as the number of spermatozoa per semen milliliter.
- Generally, the mean silver carp sperm concentration is $1.2 \pm 0.5 \times 10^{10}$ spermatozoa per ml. Sperm is diluted 1:3, to the final concentration per straw estimated as 4×10^9 per ml of the solution. The sperm/egg ratio, producing the optimal hatching rates of *H. molitrix* fry under hatchery conditions, is 10^5 spermatozoa per egg [6].

2.2 Freezing and thawing procedures

- Before freezing, sperm is diluted in the solution of NaCl 68.38 mmol/l, sodium citrate 27.20 mmol/l, and dextrose 11.01 mmol/l, and in dimethyl sulphoxide (DMSO) as cryoprotectant at a final concentration of 10% [volume/volume]. The solution and cryoprotectant are added to the sperm slowly and mixed carefully. The semen dilution proportion (volume of solution + volume cryoprotectant: volume of semen) is 2 : 1.
- The mixture is divided into several 1.8 ml straws adding 1.5 ml of the mixture per straw; and equilibrated for 5 minutes at 25°C. The straws are held 4 cm over the liquid nitrogen surface for 25 minutes; and then slowly immersed in it.
- Thawing is performed in a 30°C water bath for a few seconds. Motility and hatching rates are checked and determined. Different temperatures have been used to thaw cyprinid semen, 40°C (*C. carpio* L.), 30°C (*C. carpio*), and 25°C (*C. chalcoides*) [7-9] The temperature used in our protocol has a higher thawing rate, from –196°C to 4°C (state of complete liquefaction) than 20°C and similar to 25°C [9]. Thawing from -196°C to 4°C is generally considered a critical phase, because of potential re-crystallization [6].

2.3 Fertilization

- Female silver carp are injected twice with carp hypophyseal extract (2 mg/kg body weight) and stripped 12 h later. The eggs are kept at room temperature while being used for no longer than 45 minutes.
- After thawing, one straw of cryopreserved spermatozoa can fertilize 60,000 eggs. Giving an egg to a spermatozoa ratio of 1:100,000 (this is the ratio used to produce the optimal hatching rates of *H. molitrix* fry at Hidráulica-Cubana fish hatchery for fresh spermatozoa).

- The sperm and eggs are mixed with a turkey feather; the sperm is activated by diluting slowly up to 1:1 with fishpond water. The spawns are put in an Amur incubator (about 100 l). The hatching rates are calculated as the percentage of fry obtained from eggs from frozen-thawed.

3. GENERAL CONSIDERATIONS

The results from our study corroborate previous reports on species-specific requirements for cryoprotectors and extenders in fish. It was found that DMSO was a successful cryoprotectant, whereas glycerol and methanol failed to cryopreserve *H. molitrix* spermatozoa.

No significant differences in the hatching rate of fry were observed when eggs were fertilized with either fresh or frozen-thawed spermatozoa tested simultaneously under hatchery conditions, using our protocol (Table 1). However, the hatching rates of fry from both fresh spermatozoa and frozen spermatozoa were variable among the batches. Since no significant differences in the motility of spermatozoa from frozen thawed sperm aliquots were observed, we infer that the egg quality is the cause of the differences between batches.

Since a percentage of spermatozoa die during freezing and thawing, the effective insemination ratio for frozen spermatozoa should be higher. However, using our protocol with 1:100,000 of egg:spermatozoa ratio, the same hatching efficiency was obtained regardless of whether frozen or fresh semen was used.

More than 200 fish species with external fertilization have been tested for sperm cryopreservation [10] with a very high percent (75-100) of motility in the thawed samples. However, for several species, fertility studies have not succeeded or have not yet been made. The ultimate criterion for sperm quality evaluation is the ability of spermatozoa to fertilize eggs, leading to normal embryonic development and healthy fry.

In our studies we have developed a reliable protocol for silver carp semen cryopreservation under hatchery conditions. We have increased hatchery productions and germplasm has improved. Moreover, conservation without further development of the breed or without expected future use is not a desirable strategy.

Table 1 Motility and hatching rates using post thaw *H. molitrix* spermatozoa after a year in liquid Nitrogen storage

Experiments[1]	Spermatozoa [2,3]	Time of storage in liquid nitrogen	Motility score[4]	Hatching rates (%) [5,6]
1	Frozen	365 days	4	50.6 ± 2.1 [a]
	Control	-	4	51.2 ± 3.1 [a]
2	Frozen	270 days	4	51.2 ± 2.3 [a]
	Control	-	4	53.0 ± 2.5 [a]
3	Frozen	270 days	4	54.0 ± 2.1 [a]
	Control	-	4	55.2 ± 2.3 [a]
4	Frozen	270 days	4	49.4 ± 3.1 [a]
	Control	-	4	48.6 ± 3.3 [a]
5	Frozen	270 days	4	43.5 ± 2.1 [b]
	Control	-	4	45.1 ± 1.3 [b]
6	Frozen	270 days	4	44.6 ± 1.6 [b]
	Control	-	4	45.2 ± 2.3 [b]
7	Frozen	270 days	4	47.8 ± 4.1 [a,b]
	Control	-	4	49.2 ± 3.5 [a,b]

[1] Each experiment was conducted separately.
[2] Ten percent of DMSO was used as cryoprotectant for spermatozoa.
[3] Equilibration time: 5 minutes at room temperature.
[4] Motility scores: 0 represents no motile spermatozoa 1 to 4, representing > 0 to 25%, > 25 to 50%, > 50 to 75%, and > 75%, respectively, of spermatozoa with progressive movement (n = 5).
[5] Hatching rates are expressed as mean ± standard deviation (n = 5).
[6] Values with the same superscript are not significantly different (P $>$ 0.01). Relative quantities were transformed by angular arcsin [square-root (P)] transformation, and metric data were tested for normality. Data were assessed by one-way analysis of variance (ANOVA), followed by Tukey HSD multiple range test.

4. REFERENCES

1-Billard, R., Les poissons d'eau douce des rivières de France. Identification, inventaire et répartition des 83 espèces, Delachaux & Niestlé, Lausanne, 1997,192.

2-Sugiyama, S., Staples, D., and Funge-Smith, S., Status and potential of fisheries and aquaculture in Asia and the Pacific, in *Asia-Pacific Fishery Commission*, FAO, Rap Publication, 2004, 25.

3-Gwo, J.C., Strawn, K., Longnecker, M.T., and Arnold, C.R., Cryopreservation of Atlantic croaker spermatozoa, *Aquaculture*, 94, 355, 1991.

4-Gwo, J.C., Cryopreservation of black grouper *(Epinephelus Malabaricus)* spermatozoa, *Theriogenology*, 39, 6, 1331, 1993.

5-Björndahl, L., Söderlund, I., and Kvist, U., Evaluation of the one-step eosin-nigrosin staining technique for human sperm vitality assessment, *Hum Reprod*, 18, 4, 813, 2003.

6-Alvarez, B., Fuentes, R., Pimentel, R., Abad, Z., Cabrera, E., Pimentel, E., and Arenal, A., High Fry Production Rates Using Post-Thaw Silver Carp (*Hypophtalmichthys molitrix*) Spermatozoa Under Farming Conditions, *Aquaculture*, 220, 195, 2003.

7-Babiak, I., Glogowsky, J., Bruzka, E., Szumiec, J., and Adamek, J., Cryopreservation of sperm of common carp, *Cyprinus carpio*, *Aquac Res*, 28, 567, 1995.

8-Magyary, I., Urbányi, B., and Horváth, L., Cryopreservation of common carp (*C. carpio L.*) sperm; II. Optimal conditions for fertilization, *J. Appl. Ichthyol.*, 12, 117, 1996.

9-Lahnsteiner, F., Berger, B., Horvath, A., Urbanyi, B., and Weismann, T., Cryopreservation of spermatozoa in cyprinid fish, *Theriogenology*, 54, 1477, 2000.

10-Blesbois, E. and Labbé., C., Main improvements in semen and embryo cryopreservation for fish and fowl, in *Workshop on Cryopreservation of Animal Genetic Resources in Europe*, D. Planchenault ed., ISBN 2-908447-25-8, Paris, 2003, 55.

Complete affiliation:

Amilcar Arenal, Center for Genetic Engineering and Biotechnology. P.O. Box 387, C.P. 70 100, Camagüey, CUBA. e-mail: amilcar_arenal@yahoo.com, Phone: 53 32 26 10 14, Fax: 53 32 26 15 87

CRYOPRESERVATION OF COMMON CARP SPERM

B.Urbányi and Á. Horváth

1. INTRODUCTION

Common carp (*Cyprinus carpio*) is by far the most important European farmed freshwater fish species with an annual yield of approximately 220 thousand tons [1]. Several varieties and breeds are cultured throughout the world as a food fish and as an ornamental species (koi carp). In Europe it is consumed as a traditional Christmas and Easter food chiefly in the countries of Central Europe where it is farmed in polyculture in earthen ponds and reaches market size (1-1.5 kg) in 2-3 years.

The typical reproductive season of common carp starts in the spring when water temperature reaches 17-20°C. Carp is either spawned in designated small earthen ponds and the fry is collected for further growing later or more commonly spawning is induced by hormonal stimulation in hatchery conditions.

Several methods for the cryopreservation of common carp sperm have been developed [2-7] and are currently utilized in cryopreserved gene banks in several countries such as Israel [8] and the Czech Republic [1]. A cryopreserved gene bank is currently being developed under our supervision in the Research Institute for Fisheries, Aquaculture and Irrigation (HAKI) in Szarvas, Hungary, using the cryopreservation method described below.

2. PROTOCOL FOR FREEZING AND THAWING

2.1 Semen collection and motility estimation

- Sperm can be collected from common carp males in the spawning season without hormonal stimulation, however, the volume and quality of sperm can be enhanced with the injection of carp pituitary extract (2 mg per kg of body weight) or different GnRH/LHRH analogues (such as 1 pelled per kg of body weight of Ovopel GnRH analogue) 20-24 hours prior to the planned stripping.
- Sperm is collected into dry vessels (beakers, glasses, centrifuge tubes) by a gentle pressure on two sides of the belly. Special attention should be paid to avoid contamination with urine and feces in samples designated for cryopreservation as these can lead to the activation of spermatozoa. Contamination with blood also occurs frequently; however, its negative impact on sperm quality has not been demonstrated yet.
- Motility of sperm can be estimated following dilution at a 1:99 ratio in an immobilizing medium of 200 mM KCl, 30 mM Tris (pH 8.0) and activation of this

mixture on a microscope slide in a ratio of 1:19 with an activating medium of 45 mM NaCl, 5 mM KCl, 30 mM Tris, pH 8.0 [9] at 200 × magnification.

2.2 Freezing and thawing procedures

- Various extenders are employed for the fertilization of common carp sperm such as extenders 1 and 2 by Kurokura et al. [3] and modified Kurokura's extender 2 by Magyary et al. [10] (for composition see Table 1). We prefer to use an extender composed of 350 M glucose and 30 mM Tris, pH is set at 8.0 with concentrated hydrochloric acid. The choice of tested cryoprotectants includes DMSO, dimethyl acetamide (DMA) and methanol in concentrations between 5-25%. We use methanol in a final concentration of 10% (volume to volume). Sperm is diluted with the freezing diluent in a ratio of 1:9, thus 1000 μl of diluted sperm contains 800 μl of extender, 100 μl of cryoprotectant and 100 μl of sperm.
- Diluted sperm is loaded into 0.5 ml French straws (IMV, France). Freezing is carried out in the vapor of liquid nitrogen, in a styrofoam box. Nitrogen is poured into the box and a styrofoam frame (length: 250 mm, width: 175 mm, height: 30 mm) is placed onto the surface of nitrogen. Freezing time is 3 minutes. Following freezing the straws are plunged into liquid nitrogen.
- Following storage the straws are thawed in a 40°C water bath. The thawing time for 0.5 ml straws is 13 seconds.
- Alternatively, as in case of sturgeons, other straw types such as the 1.2 ml (IMV, France) and 4 ml straws (Minitüb, Germany) can be used for the cryopreservation of common carp sperm. Freezing in these straws is carried out similarly to regular 0.5 ml French straws with the exception that the freezing time for 4 ml straws is 5 minutes.
- Thawing of these straws is also carried out in a 40°C water bath for 20 seconds in case of 1.2 ml straws and 35 seconds in case of 4 ml straws.

2.3 Fertilization

- Dry fertilization method is used for the fertilization of common carp eggs. One 0.5 ml straw can be used to fertilize 10 g of common carp eggs. The thawed sperm is mixed with the eggs and activated by the addition of water (1 ml of water is added to 10 g of eggs).
- Approximately two minutes following activation Woynárovich-solution (0.4 % NaCl, 0.3 % urea) can be added to the eggs. The remaining procedures of egg manipulation and incubation are conducted according to regular hatchery practices.

3. GENERAL CONSIDERATIONS

The method described above is a result of a general simplification of methods used earlier in our group for common carp sperm cryopreservation [10]. In this process of simplification several details have been modified to fit work at fish farms and to improve the quality of frozen-thawed sperm. Thus, we have found that the simple glucose extender yielded higher fertilization and hatch rates than the multi-component modified Kurokura's extender [6] and we also replaced the computer controlled freezer with the styrofoam box.

Table 1 - Extenders used for common carp sperm cryopreservation

	Concentration (mg/100 ml)							
	NaCl	KCl	CaCl$_2$	MgCl$_2$	NaHCO$_3$	Glucose	Tris	Reference
Kurokura's extender 1	750	20	20		20			[3]
Kurokura's extender 2	440	620	22	8	20			[3]
Modified Kurokura's extender 2	360	1000	22	8	20			[10]
Glucose extender						6300	363	[6]

Application of sperm cryopreservation to hatchery practice is especially problematic in case of common carp. The hatchery routine requires the fertilization of large volumes of eggs (1-1.5 kg) at one time. The sperm volume needed for fertilization of 1 kg of eggs is 5-10 ml which means that 100-200 regular 0.5 ml straws would have to be thawed and manipulated at the same time to fertilize this amount of eggs which is practically impossible. This is the reason why we are looking for alternative solutions for common carp sperm such as the use of larger (1.2 and 4 ml) straws. We tested both straw types fertilizing 0.5, 1 and 1.5 times the amount of eggs that were found suitable for 0.5 ml French straws, thus, 10, 20 and 30 g of eggs were fertilized with one 1.2 straw and 40, 80 and 120 g of eggs with one 4 ml straw. According to our results both types were found suitable for the cryopreservation of common carp sperm and yielded hatch rates that were not significantly different from the control (Fig. 1.) although the ideal volume of eggs that can be fertilized with one straw varied with the straw type.

When using 4 ml straws attention must be paid to the appropriate sealing of straws as they are sold without the cotton and polymer plug used in regular 0.25 ml, 0.5 ml or 1.2 ml straws. Manufacturers offer metal and plastic balls that can be inserted into one end of the straws however these balls can shoot out of the straws during handling or thawing of straws causing serious injuries. Heat sealing of straws is another possibility however we found that the straws tend to break during thawing at the sealing causing loss of sperm due to leakage into the water bath. The best solution we found so far is a home-made plug of sealing powder between two layers of cotton tissue similar to the one in conventional straws (Fig. 2.).

Jelly-like agglutination of sperm is a typical phenomenon observed when sperm of cyprinid fish species is frozen in sugar-based extenders. When analyzed under the microscope individual motile spermatozoa are released from the agglutinated mass upon activation. The phenomenon, however, does not affect the fertilizing ability of frozen sperm, we have actually found that sperm frozen in sugar-based extenders yielded significantly higher fertilization and hatch rates than ionic extenders where this agglutination of sperm was not observed [6].

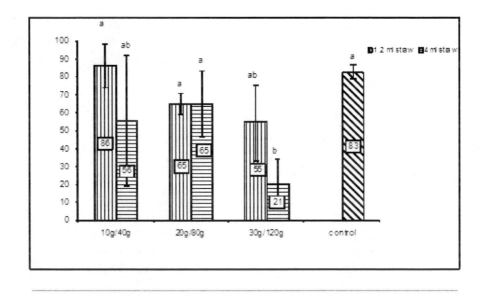

Figure 1 - Hatching rates of common carp eggs fertilized with cryopreserved sperm frozen in 1.2 ml and 4 ml straws. Egg volumes of 10, 20 and 30 g were fertilized with one 1.2 ml straw and egg volumes of 40, 80 and 120 g were fertilized with one 4 ml straw. In the control 80 g of eggs were fertilized with 400μl of fresh sperm. Columns marked with the same letter are not significantly different (P>0.05, n = 4).

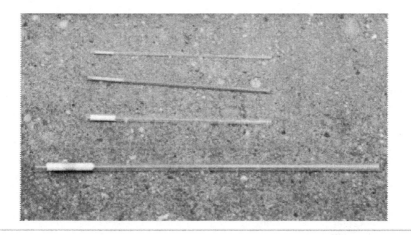

Figure 2 - Different straw types used for the cryopreservation of fish sperm (from top to bottom): 0.25 ml straw, 0.5 ml straw, 1.2 ml straw and 4 ml straw with a home-made cotton plug.

4. ACKNOWLEDGMENTS

This publication was supported by the Bolyai János Research Scholarship of the Hungarian Academy of Sciences and the post doctoral fellowship of the Hungarian Ministry of Education number 28345/2003.

5. REFERENCES

1-Linhart, O., Rodina, M., and Cosson, J., Cryopreservation of sperm in common carp *Cyprinus carpio*: sperm motility and hatching success of Embryos, *Cryobiology*, 41, 241, 2000.

2-Moczarski, M., Deep freezing of carp (*Cyprinus carpio* L.) sperm, *Bull. Acad. Pol. Sci. Ser. Sci. Biol*, 24, 187, 1987.

3-Kurokura, H., Hirano, R., Tomita, M., and Iwahashi, M., Cryopreservation of carp sperm, *Aquaculture*, 37, 267, 1984.

4-Cognie, F., Billard, R., and Chao, N.H., La cryoconservation de la laitance de la carpe, *Cyprinus carpio, J Appl Ichthyol*, 5, 165, 1989.

5-Babiak, I., Glogowski, J., Brzuska, E., Szumiec, J., and Adamek, J., Cryopreservation of sperm of common carp, *Cyprinus carpio*,. *Aquac Res*, 28, 567, 1997.

6-Horváth, Á., Miskolczi, E., and Urbányi, B., Cryopreservation of common carp sperm. *Aquat Living Resour*, 16, 457, 2003.

7-Warnecke, D. and Pluta, H., Motility and fertilizing capacity of frozen/thawed common carp (*Cyprinus carpio* L.) sperm using dimethyl-acetamide as the main cryoprotectant, *Aquaculture*, 215, 167, 2003.

8-Lubzens, E., Daube, N., Pekarsky, I., Magnus, Y., Cohen, A., Yusefovich, F., and Feigin, P., Carp (*Cyprinus carpio* L.) spermatozoa cryobanks. Strategies in research and application, *Aquaculture*, 155, 13, 1997.

9-Saad, A. and Billard, R., Spermatozoa production and volume of semen collected after hormonal stimulation in the carp, *Cyprinus carpio, Aquaculture*, 65, 67, 1987.

10 Magyary, I., Urbányi, B., and Horváth, L., Cryopreservation of common carp (*Cyprinus carpio* L.) sperm II. Optimal conditions for fertilization, *J Appl Ichthyol,* 12, 117, 1996.

Complete affiliation:
Béla Urbányi Ph.D, Department of Fish Culture, Szent István University, Páter Károly u. 1. Gödöllő H-2103 Hungary; Phone: +36-28-522-000 ext. 1659; Fax: +36-28-410-804. e-mail: Urbanyi.Bela@mkk.szie.hu

SPERM CRYOPRESERVATION FROM SEA PERCH *(Lateolabrax japonicus)*

S-L. Chen, X.S. Ji and Y-S. Tian

1. INTRODUCTION

Sea perch (*Lateolabrax japonicus*), which naturally inhabits in the coastal areas of China, Japan and the Korea peninsula, has been cultivated widely in China in recent years and a great deal of research has also been conducted on the control of reproduction [1]. The long-term preservation of sea perch sperm will simplify the production of selected strains and facilitate genetic selection of beneficial traits for intensive commercial farming [2]. In addition, the wild stock of sea perch is declining because of excessive exploitation and environmental pollution. Constructing a cryobank of sea perch semen is becoming more and more urgent.

Sea perch sperm freezing protocol has been developed using 2 ml cryovial with the aim of application in large scale in hatchery. The experimental results demonstrated that the protocol resulted in high thawing motility and fertility and could be applied on a commercial scale in sea perch hatcheries.

2. PROTOCOL FOR FREEZING AND THAWING

2.1 Equipment

- Glass bottles (25 ml)
- 15 liter liquid nitrogen jar
- Cryo biological storage system
- micropipettes (1 µl, 200 µl)
- Microscope
- pH meter
- Papers for cleaning mucus
- Towels

2.2 Reagents

- Liquid nitrogen
- NaCl
- NaH_2PO_4
- $NaHCO_3$
- KCl
- $CaCl_2 \times 2H_2O$
- $CaCl_2 \times 2H_2O$

- MgCl$_2$ x 6H$_2$O
- D-Glucose
- DMSO (dimethylsulfoxide)
- Distilled water

2.3 Semen and eggs collection

- Semen is collected by syringe after applying gentle abdominal pressure and then transferred into 25 ml bottles. The semen concentration is determined using the equation SC = (0.806 OD-0.032) x 10^8, where SC and OD are respectively semen concentration (spz/ml) and optical density at a 260 nm wavelength [3].
- Eggs are collected by abdominal pressure of the females and placed in a 500 ml beaker. About 5 ml eggs were added to a 250 ml beaker containing 200 ml fresh seawater to test egg quality. If the majority of the eggs were floatable, this batch of eggs can be used in the experiment.

2.4 Pre-freezing sperm assessment

- Sperm were activated by adding a drop of seawater on the sperm spread on a glass slide. After activation, sperm motility was immediately determined under light microscope (100 x). The motility was expressed by values from 0 to 5, with 0 representing no motile spermatozoa, and 5 from 80% to 100% sperm showing progressive movement. To validate the motility measurement, preliminary tests were carried out using three samples.

2.5 Freezing/Thawing procedure

- Semen from each male was separately cryopreserved using the "three step" method developed for Chinese carps [4-6]. In brief, within 10 min of harvesting, semen was mixed with precooled (4°C) extender MPRS (NaCl 60.35 mM, NaH$_2$PO$_4$ 1.8 mM, NaHCO$_3$ 3mM, KCl 5.23mM, CaCl$_2$ x 2H$_2$O 1.13mM, MgCl$_2$ x 6H$_2$O 1.13 mM, D-Glucose 55.55 mM) containing cryoprotectant DMSO (10%) at a ratio of 1:1 (extender:semen) and then was equilibrated at 4°C for 30 min. The equilibrated semen was transferred into 2 ml cryovials (Nalgen Company, USA). Then the cryovials filled with diluted sperm were transferred into gauze pocket with a dimension of 6 cm in width and 9 cm in length and was equilibrated for 10 min in liquid nitrogen vapour 6 cm above the liquid nitrogen surface, and then equilibrated for 5 min on the surface of liquid nitrogen, finally immersed in liquid nitrogen.
- For thawing, the gauze pocket containing cryovials was removed from liquid nitrogen and equilibrated for 5 min in liquid nitrogen vapors and then removed from the liquid nitrogen container. Semen was thawed in a water bath at 37°C

and thawing time for 2 ml volume of diluted semen was 2.5 min. The motility score of frozen-thawed sperm was determined as described above.

2.6 Fertilization

- Collect female eggs into a glass beaker. Precaution should be taken to avoid eggs contacting with water. In general, use pooled eggs from several females.
- Add sperm (150 μl for fresh sperm and 300 μl for frozen sperm) to the eggs (40 ml, approximately 40,000 eggs) and gently mix before activation with sea water of about 80 ml.
- After 5-10 min, add 400 ml sea water to the sperm-eggs mixture.
- Incubate the fertilized eggs in small incubators (10 liter) until hatching. When the eggs developed to gastrula stage, the fertilization rate (fertilization rate = number of gastrula stage eggs/number of eggs) can be determined. The hatching rate is expressed as the percentage of hatched larvae in the fertilized eggs.

3. GENERAL CONSIDERATIONS

3.1 Determination of freezing rate

When the 2 ml cryovials containing 1 ml diluted semen were equilibrated at 6 cm above the surface of liquid nitrogen for 10 min, a linear decrease of temperature from 16°C to −15°C at a cooling rate of 31°C/min was obtained. Freezing point occurred after 65 s at −15°C and the temperature surged to −12°C. The freezing rate from −12°C to −180°C was determined to be −18.6°C/min. Similar temperature profiles were obtained when the cryovials containing 1 ml diluted semen were equilibrated at 2 cm or 13 cm above the surface of liquid nitrogen. However, their freezing points and temperature surges were different. When the cryovials were equilibrated at 2 cm above the surface of liquid nitrogen, the freezing point occurred after 55 s at −25°C and the temperature surges were up to 9°C (from −25°C to −16°C), which was equal to those equilibrating at 13 cm above the surface of liquid nitrogen except their freezing time occurred after 70 s. Equilibrating at 6 cm above the surface of the liquid nitrogen for 10 min resulted in 73.3% ± 5.7 motility score, which was higher than that equilibrating at 2 cm (41.7% ± 10.6) and 13 cm (48.3% ± 2.9) (p<0.05) .

3.2 Effect of semen volume in cryovials

The post-thaw motility was not reduced significantly when the volume of diluted semen in the cryovials was increased from 0.5 ml to 1.0 ml (p>0.05). However, the post-thaw motility of 1.8 ml diluted semen in the cryovials was significantly lower than that of 0.5 ml after 30 s (p<0.05). In addition, the increases of volume of diluted semen in the cryovials also prolonged the thawing time. The thawing time was 50 s, 1.5 min and 2.5 min for 0.5 ml, 1 ml and 1.8 ml diluted semen volume, respectively.

3.3 Fertilization trials

Our cryopreservation protocol was applied in sea perch hatchery and resulted in 84.8% fertilization rate and 70.1% hatching rate, which resembled the rates obtained with fresh sperm ($81.0\% \pm 4.6$ and $87.2\% \pm 3.2$) ($p>0.05$) [7]. Under various sperm/egg ratios, the fertilization and hatching rates of frozen semen cryopreserved for three days in liquid nitrogen were not significantly different from that of fresh sperm ($p>0.05$). After being cryopreserved for one year in liquid nitrogen, the fertilization rates of frozen-thawed sperm was still not significantly different from that of fresh sperm ($p>0.05$). In fertilization trials of 440 ml eggs with 3.5 ml frozen semen cryopreserved for one year in liquid nitrogen, a 83.5% fertilization rate and 90.0% hatching rate were obtained, which was similar to the control ($96.8\% \pm 2.3$ and $87.2\% \pm 3.1$) ($p>0.05$). As a result, about 300,000 fry hatched out.

4. ACKNOWLEDGMENTS

This work was supported by a grant from the State 863 High-Technology R&D Project of China (2001AA621100) and supported in part by E-Institute of Shanghai Municipal Education Commission.

5. REFERENCES

1-Sun, G.Y., Zhu, Y.Y., Zhou, Z.L., and Chen, J.G., The reproductive biology of *Lateolabrax japonicus* in the Yangze river estuary and Zhejiang offshore waters, *J Fish China*, 18, 18 [in Chinese], 1994.
2-Chen, S.L., Progress and prospect on cryopreservation of fish gametes and embryos, *J. Fish. China*, 26(2), 161, 2002.
3-Fauvel, C., Suquet, M., Dreanno, C., Zonno, V., and Menu, B., 1. Cryopreservation of sea bass (*Dicentrarchus labrax*) spermatozoa in experimental and production simulating conditions, *Aquat Living Resour*, 11, 387, 1998.
4-Chen,S.L., Liu, X.T., Lu, D.C., Zhang, L.Z., Fu, C.J., and Fang, J.P., Cryopreservation of spermatozoa of silver carp, common carp, blunt snout bream and grass carp, *Acta Zool Sinica*, 38, 413, Chinese with English abstract, 1992.
5-Chen, S.L., Liu, X.T., Lu, D.C., Fu, C.J., Zhang, L.Z., and Fang J.P., Activation and insemination methods of cryopreserved sperm from several Chinese carps, *J Fish China* ,16, 337, Chinese with English abstract, 1992.
6-Chen, S.L., Ji, X.S., Yu, G.C., Tian, Y.S., and Sha, Z.X., Cryopreservation of sperm from turbot (*Scophthalmus maximus*) and application to large-scale fertilization, *Aquaculture*, 236, 547, 2004.
7-Ji, X.S, Chen, S.L, Tian, Y.S, Yu, G.C, Sha, Z.X, Xu, M.Y, and Zhang, S.C., Cryopreservation of sea perch (*Lateolabrax japonicus*) spermatozoa and feasibility for production-scale fertilization, *Aquaculture,* 241, 517, 2004.

Complete affiliation:
Prof. Dr. Songlin Chen, Yellow Sea Fisheries Research Institute, Chinese Academy of Fisheries Sciences, Nanjing Road 106, 266071 Qingdao, China.Email: chensl@ysfri.ac.cn

SEMEN CRYOPRESERVATION OF PIRACANJUBA (*Brycon orbignyanus*), AN ENDANGERED BRAZILIAN SPECIES

A.T.M. Viveiros and A.N. Maria

1. INTRODUCTION

Freshwater fish fauna in Brazil is particularly diverse and many species are not found naturally outside South America. Many of these species rely on seasonal flooding for reproductive migration (the *piracema*) and for access to seasonal lagoons for rearing of larvae and juveniles. These fish species are sensitive to the detrimental effects of hydroelectric dams, urbanization, agriculture and introduced species [1]. Many populations have already declined precipitously, and others are likely to become threatened or endangered in the near future. The piracanjuba *Brycon orbignyanus* is one of these endangered species. Furthermore, piracanjuba has a great potential for aquaculture as its meat has a salmon-like pink color, very tasty and high priced and its aggressive behavior is appreciated for recreational fishing. Piracanjuba spawn once a year during the spawning season (October to February). Males and females can be easily stripped of semen and eggs after hormone treatments.

Gene banks are an appropriate component of conservation efforts and are of particular importance for ensuring genetic diversity in captive breeding programs, which are carried on while other conservation measures are developed. A few studies concerning piracanjuba semen cryopreservation have been carried out. To the best of our knowledge, the first study was conducted at the Energetic Company of Minas Gerais (CEMIG) in Conceição das Alagoas, Brazil, as a MSc project [2]. Semen cryopreserved in glucose-egg yolk-DMSO produced a post-thaw motility of 40%.

The protocol described here was developed during the past three spawning seasons [3,4] and will be used to set up a gene bank for piracanjuba.

2. PROTOCOL FOR FREEZING AND THAWING

2.1. Materials to induce spermiation and to collect semen

- Carp pituitary extract (cPE) to facilitate stripping of semen
- 1 ml syringes with needles to inject cPE
- Tissue to dry fish skin
- 15 ml beakers to collect semen
- Polystyrene box to keep semen in ice bath during handling
- Crushed ice
- Haemocytometer chamber (Neubauer, Bürker) to estimate sperm concentration

- Microscope and microscope slides to check sperm motility
- NaCl 50 mM (pH = 7.6; 92 mOsm kg^{-1}) as activating solution of sperm motility

2.2. Materials to freeze/thaw semen

- Physiological saline NaCl 0.9% as semen extender, pH 7.6, 285 mOsm kg^{-1}
- pH meter to adjust semen extender pH
- Osmometer to check semen extender osmolality
- Hen egg yolk
- Methylglycol (ethylene glycol monomethyl ether; CH$_3$O(CH$_2$)$_2$OH) as cryoprotectant agent
- Test tubes to prepare freezing solutions
- Micropipettes: 1 µl, 200 µl and 1000 µl
- Basic laboratory glass material
- 0.5 ml straws
- Liquid nitrogen
- Vapor nitrogen vessel, dry-shipper (inner temperature: -170°C)
- Liquid nitrogen vessel
- Water bath
- Thermometer
- Tissue to dry straws

2.3. Materials to collect and fertilize eggs

- Carp pituitary extract to induce ovulation and facilitate stripping of eggs
- 1 ml syringes with needles to inject cPE
- Tissue to dry fish skin
- A basket to collect eggs
- Scale to weigh eggs
- Incubation system with 26-27°C water

2.4 Fish handling and semen collection

- Inject piracanjuba male with a single dose of cPE 4 mg per kg body weight (BW) to induce spermiation and facilitate stripping of semen. At 26-27°C, hand stripping of semen is possible after 5 h.
- Semen should be collected in a dry beaker (avoid any contact with water). As semen volume is larger (more than 10 ml), collect one ejaculate into three or four beakers, as one may become contaminated with urine or faeces and should be discarded.

2.5 Prefreezing sperm evaluation

- Evaluate sperm motility under microscopy 200 x magnification. First, observe 5 µl of semen on a microscope slide. Sperm cells should be immotile, as fish spermatozoa are immotile in seminal plasma. Any sperm movement suggests water or blood contamination and the sample should be discarded. Activate the immotile samples by adding 15 µl of NaCl 50 mM and estimate the percentage of moving cells. Only samples with at least 80% of moving cells should be used for cryopreservation.
- Determine sperm concentration in a haemocytometer chamber. The number of sperm cells per mL will be used to calculate the minimum sperm:egg ratio during artificial fertilization.

2.6 Freezing and thawing procedures

- Prepare semen extender (physiological saline) at least 24 h in advance. Then adjust pH to 7.6 and check osmolality which should be around 285 mOsm kg^{-1}.
- In a dry and sterile beaker, prepare semen solution by adding first physiological saline (75%), egg yolk (5%), methylglycol (10%) and then semen (10%). The final ratio is 1 semen:7.5 extender:0.5 yolk:1 methylglycol. Mix gently. Equilibration time is not needed.
- Aspirate diluted semen into 0.5 ml straws, close each straw with a metal globe, load each rack with 10 straws and place them inside the vapor nitrogen vessel.
- After 20-24 h, transfer all racks into the liquid nitrogen vessel.
- Thaw a maximum of 5 straws simultaneously. Remove the straws from the liquid nitrogen, and plunge them into a 60°C-water bath for 8 seconds, and then dry with tissue.
- Cut the straw globe end with scissors and collect post-thaw semen into a small beaker. Add another 0.5 ml of physiological saline for each thawed straw (final volume will be 5 ml when 5 straws are thawed) immediately so that the cryoprotectant becomes less toxic to sperm cells. This volume is enough to fertilize 5 g of piracanjuba eggs (*ca* 5,000 eggs), if initial sperm concentration varies from 2.8 x 10^9 to 10.8 x 10^9 spermatozoa per ml.

2.7 Post-thaw sperm evaluation

- To assess post-thaw sperm quality, motility can be estimated using the same subjective procedure described for fresh semen. To quantify sperm motility, the use of a computer-assisted sperm analysis (CASA) system provides objective methods of assessment, using at least a dozen computer-calculated motility characteristics. For further details, refer to Rurangwa et al. [5].

- To assess sperm membrane integrity, dilute 5 μl of post-thaw sperm in 10 μl of eosin-nigrosin (5% B eosin, 10% nigrosin, pH = 6.9) on a microscope slide. Then count at least a total of 150 cells (stained and unstained). Stained cell displays disrupted membrane (and thus can be considered as a dead cell), while unstained cell displays membrane integrity. Calculate the percentage of live unstained spermatozoa.

2.8 Fertilization

- Even when the most standardized protocol is used, sperm cells lose quality after freezing and thawing processes. Sperm quality continues to decrease after thawing, so post-thaw sperm should be used for fertilization as soon as possible.
- Inject female piracanjuba with two doses of cPE 0.5 and 5 mg per kg BW, in a 12 h interval, to induce ovulation. At 26-27°C, hand stripping of eggs is possible after 5 h. Eggs should be collected in a dry plastic beaker (avoid any contact with water).
- Spread post-thaw semen of 5 straws (already rediluted in 2.5 ml of physiological saline) over 5 g of eggs and mix for 5-10 seconds. Activate sperm motility with 10-20 ml of tank water, and gently mix for 1 minute. Then, to complete egg hydration, add 20 ml of tank water for 2 minutes. Finally, transfer fertilized eggs to a funnel incubation system at 26-27°C. Larvae will hatch in about 16 h, but should remain inside the incubators until first feeding (3 days after hatching).
- The artificial fertilization procedure should not last longer than one hour as egg quality decreases fast after stripping.

3. GENERAL CONSIDERATIONS

The volume of piracanjuba semen is large, varying from 10 to 20 ml or more. This amount is enough for immediate fertilization of eggs from 2 or 3 females. The left over semen should be cryopreserved. This would guarantee the genetic variation of piracanjuba in a gene bank and permit the exchange of cryopreserved semen among laboratory units and fish cultures.

The hatching rates produced with piracanjuba semen cryopreserved according to the protocol described here are high enough to guarantee the use of this technique as a conservation method for endangered piracanjuba populations. The use of cryopreserved semen on a commercial scale, however, still needs further investigation. Semen of different individuals could be pooled before freezing to avoid interference of individual semen quality. The use of larger semen containers such as 4 ml macrotubes reduces the number of tubes to store and allows fertilization of a larger volume of eggs with semen cryopreserved in only one straw. However, a higher sperm:egg ratio should be used on a commercial scale, as fertilization carried out under laboratory conditions allows the use of fewer sperm cells to fertilize one egg.

4. ACKNOWLEDGMENTS

We thank the graduate students A.V. Oliveira, L.H. Orfão, Z.A. Isaú and R.V. Araujo for assistance on experiments.

5. REFERENCES

1-Carolsfeld, J., Godinho, H G., Zaniboni Filho, E., and Harvey, D.J., Cryopreservation of sperm in Brazilian migratory fish conservation, *J Fish Biol*, 63, 472, 2003.

2-Bedore, A.G., Characteristics and cryopreservation of pacu-caranha (*Piaractus mesopotamicus*) and piracanjuba (*Brycon orbignyanus*) semen, MSc thesis, Federal University of Minas Gerais, Belo Horizonte, Brazil. 1999, 53.

3-Maria, A.N, Extenders and cryoprotectant for refrigeration and cryopreservation of piracanjuba (*Brycon orbignyanus*) semen, MSc thesis, Federal University of Lavras, Lavras, Brazil, 2005, 71.

4-Maria, A.N., Viveiros, A.T.M., Freitas, R.T., and Oliveira, A.V., Extenders and cryoprotectants for cooling and freezing of piracanjuba (*Brycon orbignyanus*) semen, an endangered Brazilian teleost fish, *Aquaculture*, 260, 1-4, 298, 2006.

5-Rurangwa, E., Volckaert, F.A.M., Huyskens, G., Kime, D.E., and Ollevier, F., Quality control of refrigerated and cryopreserved semen using computer-assisted sperm analysis (CASA), viable staining and standardized fertilization in African catfish (*Clarias gariepinus*), *Theriogenology*, 55, 751, 2001.

Complete affiliation·

Ana T M Viveiros, Animal Sciences Department, Federal University of Lavras, DZO - UFLA, Caixa postal 3037, 37200-000, Lavras, MG, Brazil. Phone: +55 35 38291223 Fax: +55 35 38291231, e-mail: ana.viveiros@ufla.br.

CRYOPRESERVATION OF ENDANGERED FORMOSAN LANDLOCKED SALMON (*Oncorhynchus masou formosanus*) SEMEN

J-C. Gwo, H.Ohta, K. Okuzawa, L-Y. Liao and Y-F. Lin

1. INTRODUCTION

Formosan landlocked salmon is the Pacific salmon showing the most southern distribution of the Pacific salmon and is only found in the upper streams of the Tachia River, Taiwan. The population is at highest risk of extinction due to a combination of low effective population size, overfishing and altered native habitats [1]. The existing population is estimated at less than 1,000 fish. Sustained maintenance of this population may depend entirely on stocked fish originating from aquaculture systems in the near future. The use of frozen sperm provides a practical means of increasing the genetically effective population size, especially when new broodstock is established for cultivation, and it can also help maintain the genetic diversity and integrity of the populations. The combination of sperm cryopreservation with androgenesis can offer a way of regenerating a stock using only sperm as a source of nuclear genomic material.

2. PROTOCOL FOR FREEZING AND THAWING

2.1 Fish handling and semen collection

- Mature wild populations of Formosan landlocked salmon (*O. masou formosanus*) and cultivated stocks of Amago salmon (*O. masou ishikawae*) were obtained from the Chichawan Stream, Shei-Pa National Park, Taiwan, and from captive brood fish at the Inland Station, National Research Institute of Aquaculture, Japan, respectively.
- Semen was stripped from at least 5 males and pooled in equal amounts during the breeding season (late October). The semen was stored on crushed ice, and sperm motility and concentration were determined within 3-4 h of sampling.

2.2 Freezing and thawing procedures

- Three cryoprotectants at 10% concentration were compared: DMSO (dimethyl sulfoxide), DMA (dimethyl acetamide), and methanol. All chemicals were reagent grade (Sigma, St. Louis, MO, USA), and water was deionized and then glass-distilled.
- Semen dilution ratio (volume extender + volume cryoprotectant: volume semen) was 6 for all cryoprotectants, 1:5. Only semen showing more than 90% progressively motile sperm and vigorous forward motility when diluted with

ISER (Isotonic Solution for Egg Rinsing; NaCl 154.7 mM, KCl 3.2 mM, CaCl$_2$ 2.3 mM; [2] were used in this study. All semen samples were processed within 10 min of sampling.

- The general freezing protocol follows the methods of Ohta et al., [3]. Briefly, part semen was diluted with 5 parts extender, consisting of 90% 300 mM glucose with 10% cryoprotectant. Extended semen was immediately pelleted (0.1 ml) on dry ice. The elapsed time, from dilution of milt to dropping on dry ice, was less than 30 s. After freezing (about 5 min), the frozen samples were immersed in liquid nitrogen.
- One pellet (0.1 ml) was rapidly thawed with 5 ml of thawing solution (120 mM NaHCO$_3$) at room temperature (25°C). Thawing of the pellet was accelerated by mixing the thawing pellets with a vortex mixer.
- A drop of sperm suspension was put on a glass slide placed on the stage of a microscope at a magnification of 200 x. The movement of sperm in the thawing solution was recorded by videomicroscopy from 5 to 10 s right after thawing of the pellet, and percent motility was measured by tracing the location of the heads of over 50 randomly chosen sperm on a TV monitor [4].

2.3 Fertilization

- Eggs were stripped from 2 mature Formosan landlocked salmon females.
- Two pellets were thawed and immediately mixed with 5 g of eggs (about 50 eggs) by gentle stirring. The fertilization capacity of the sperm was determined by inseminating eggs at a sperm-to-egg ratio of 4.2 x 10^6:1 (calculated from the average sperm concentration of 6.4 x 10^9/ml for studied population of Formosan landlocked salmon).
- Other pellets of Formosan landlocked salmon were used to fertilize fresh Amago salmon ova. Twenty captive 2-year-old Amago salmon brood fish at Japan Inland Station of National Research Institute of Aquaculture were used as egg donors. Ovulated Amago salmon eggs were obtained from 5 mature females, and then pooling and rinsing them with ISER.
- The following fertilization tests were evaluated: Amago salmon egg x Amago salmon semen; Amago salmon egg x cryopreserved Amago salmon semen; and Amago salmon egg x cryopreserved Formosan landlocked salmon semen.
- Amago semen was frozen and thawed as described above, and immediately after thawing, two pellets were mixed with 5 g of eggs (about 50 eggs) by gentile stirring. We use the same sperm-to-egg ratio as Formosan landlocked salmon for the fertilization trials.
- Fertilization rate was determined by percent eyed embryos. Some juveniles obtained by intersubspecies hybridization (Amago salmon eggs x cryopreserved Formosan landlocked salmon semen) were reared for 4 months.

3. GENERAL CONSIDERATIONS

Percent motilities of Formosan landlocked salmon were 14.5%, 9.9%, and 4.6% for semen frozen in 10% cryoprotectants DMSO, DMA, and methanol, respectively (Table 1). The lowest post-thaw percent motility for semen of Formosan landlocked salmon and of Amago salmon were observed with 10% methanol as the cryoprotectant. However, methanol gave a significantly better post-thaw percent motility in Amago salmon than in Formosan landlocked salmon. The highest post thaw fertilization rate for semen of Formosan landlocked salmon was obtained with 10% DMSO, as the cryoprotectant. Methanol and DMA were less effective. The fertilization capacity of frozen-thawed Formosan landlocked salmon is comparable to that of fresh Formosan landlocked salmon semen (94.5% to 96.2%). In Amago salmon all 3 cryoprotectants showed no significant difference in post-thaw fertilization rates. The intersubspecies fertility of the cryopreserved sperm was 71%-84% (Amago salmon egg x cryopreserved Formosan landlocked salmon semen) which was comparable to those obtained with Amago salmon egg x cryopreserved Amago salmon semen (83%-88%). Morphologically, Amago salmon have red spots on their body sides, whereas Formosan landlocked salmon do not. The intersubspecies hybrid, produced with frozen sperm, appeared to be normal and looked superficially like the paternal Formosan landlocked salmon. No red spots were observed on their bodies.

The freezing protocol developed for Masu salmon semen [3,5] appears applicable to Formosan landlocked salmon. In addition, use of DMA and methanol as cryoprotectants with 300 mM glucose resulted in low post-thaw sperm percent motility but high fertilization rates in the present study. One possible reason for the high fertilization rate with the low percent motility cryopreserved sperm in the present study may be that few eggs were inseminated. Fertilization capacity of cryopreserved sperm is clearly lower than that of fresh sperm, and increasing the number of freeze-thawed sperm appears to compensate for their decrease in fertilizing ability [6-9]. Lahnsteiner at al. [8,9] recommended a 10^6:1 sperm-to-egg ratio for various salmonid species. We used 4.2×10^6:1 sperm-to-egg ratio in the present study.

Although freezing and thawing rates are among the most critical variables affecting the success of cryopreservation, they are the least standardized variables in fish sperm cryopreservation studies, perhaps because with pelleted semen they are difficult to measure or control. Scott and Baynes [10] calculated the optimum freezing rate for salmonid sperm to be between -30 and -160°C/min. The pellet method is most widely used in salmon hatcheries because of its favorable post-thaw fertility, convenience, and simplicity. Pellet size and the freezing temperature (temperature reached before immersion in liquid nitrogen) influence the post-thaw fertility in the pellet method [3,11]. Otha et al. [3] estimated the optimum freezing rate to be between -29.9 and -92.6°C/min, and that reaching a temperature of -70°C was a prerequisite for both Masu and Amago salmon sperm. In general, the freezing rate of 50 μl extended semen dropped on a dry ice block (-79°C) is about 30°C/min [10,12], which may explain the successful results achieved by the pellet technique in most salmonids. Recently the straw method also has become popular for freezing salmonid sperm [8,9,13]. The extenders developed for the pellet method may, however, need to be adjusted for the straw method. The possible cause of the discrepancies between the 2 freezing techniques (pellet and straw) requires further study.

In Formosan landlocked salmon as well as in Amago salmon, the lowest percent post-thaw sperm motility was obtained with 10% methanol, which gave the best post-thaw fertilization rates for grayling and Danube salmon [9]. Substantial species-specific modifications of semen cryopreservation protocols are needed for different fish species [7,8,9,14,15,16,].

Table 1 - Effects of the type of cryoprotectant on post-thaw sperm percent motility and fertility

Cryoprotectant	Motility (%)		Fertilization rate (%)	
	F. l. salmon	A. salmon	F. l. salmon	A. salmon
DMSO 10%	14.5 ± 0.6^b	17.1 ± 0.7^b	$84.2 \pm 3.4^{b\,1}$	$88.0 \pm 2.2^{b\,1}$
DMAE 10%	9.9 ± 1.0^c	15.4 ± 1.2^b	$72.1 \pm 5.7^{c\,1}$	$83.4 \pm 3.8^{b\,1}$
Methanol 10%	4.6 ± 1.0^d	12.4 ± 1.6^b	$71.1 \pm 2.2^{c\,1}$	$88.0 \pm 5.5^{b\,1}$
Control (Frozen semen)			94.5 ± 2.4^a *	
Control (Fresh semen)	95 #	95.1 ± 1.1^a	96.2 ± 1.4^a **	97.9 ± 1.6^z

Values are means ± standard deviations. N = 4 in all experiments. Values with different superscripts differ significantly (P < 0.05). The diluent was 300 mM glucose. Weight of eggs per experiment was 5 g (about 50 eggs); sperm to egg ratio was 4.2 x 10^6 sperm per egg.
Measured with naked eye.
* Formosan landlocked salmon egg x cryopreserved Formosan landlocked salmon semen.
** Formosan landlocked salmon egg x Formosan landlocked salmon semen.
z Amago salmon egg x Amago salmon semen.
ø Amago salmon egg x cryopreserved Amago salmon semen.
1 Amago salmon egg x cryopreserved Formosan landlocked salmon semen.

4. ACKNOWLEDGMENTS

We are grateful to the staff members of the Wu-Lin Station, Shei-Pa National Park, for providing the fish.

5. REFERENCES

1-Lin, Y.S., Tsao, S.S., and Chang, K.H., Population and distribution of the Formosan landlocked salmon (Oncorhynchus masou formosanus) in Chichiawan Stream, Bull. Inst. Zool. Acad. Sin., 29(3, suppl.),73, 1990.
2-Otha, H., Shinriki, Y., and Honma, M., Sperm motility of masu salmon Oncorhynchus masou in the isotonic solution for egg rinsing, Nippon Suisan Gakkaishi, 52, 4, 609, 1986

3-Ohta, H., Shimma, H., and Hirose, K., Effects of freezing rate and lowest cooling pre-storage temperature on post-thaw fertility of amago and masu salmon spermatozoa, *Fish. Sci.*, 61, 423, 1995.

4-Ohta, H., Shimma, H., and Hirose, K., Relationship between fertility and motility of cryopreserved spermatozoa of the amago salmon *Oncorhynchus masou ishikawae*, *Fish. Sci.*, 61, 886, 1995

5-Yamano, K., Kasahara, E., Yamaha, E., and Yamazaki, F., Cryopreservation of masu salmon sperm by the pellet method., in *Bull. Fac. Fish.*, Hokkaido Univ., 41, 149, 1990.

6-Stoss, J., Fish gamete preservation and spermatozoan physiology, in: *Fish Physiology*, 9, B, Hoar,W.S., Randall, D.J., and Donaldson, E.M., Eds., Academic Press, Orlando, FL, 1993, 305.

7-Gwo, J.C., Strawn, K., Longnecker, M.T., and Arnold, C.R., Cryopreservation of Atlantic croaker spermatozoa, *Aquaculture*, 94, 355, 1991.

8-Lahnsteiner, F. Patzner, R.A., and Weismann, T., Cryopreservation of semen of the grayling (*Thymallus thymallus*) and the Danube salmon (*Hucho hucho*), *Aquaculture*, 144, 265, 1996.

9-Lahnsteiner, F., Berger, B., Weismann, T., and Patzner, R.A., The influence of various cryoprotectants on semen quality of the rainbow trout (*Oncorhynchus mykiss*) before and after cryopreservation, *J. Appl. Ichthyol.*, 12, 99, 1996.

10-Scott, A.P. and Baynes, S.M., A review of the biology, handling and storage of salmonid spermatozoa, *J. Fish Biol.*, 17, 707, 1980.

11-Holtz, W., Schmidt-Baulain, R., and Meiners Gefken, M., A simple saccharide extender for cryopreservation of rainbow trout (*Oncorhynchus mykiss*) sperm, in: *Proc. 4th Int. Symp. Reprod. Physiol. Fish*, Scott, A.P. and Sumpter, P., Kime, D.E. and Rolfe, M.S., Eds., Sheffield University of East Anglia Press.,1991, 250.

12-Erdahl, A.W., Erdahl, D.A., and Graham, E.F., Some factors affecting the preservation of salmonid spermatozoa, *Aquaculture*, 43, 341, 1984.

13-McNiven, M.A., Gallant, R.K., and Richardson, G.F., Dimethyl-acetamide as a cryoprotectant for rainbow trout spermatozoa, *Theriogenology*, 40, 943, 1993.

14-Gwo, J.C., Cryopreservation of yellowfin seabream (*Acanthopagrus latus*) spermatozoa (Teleost, Perciformes, Sparidae), *Theriogenology*, 41, 989, 1994.

15-Gwo, J.C., Kurokura, H., and Hirano, R., Cryopreservation of spermatozoa from rainbow trout, common carp and marine puffer, *Nippon Suisan Gakk*, 59 777, 1993.

16-Piironen, J., Composition and cryopreservation of sperm from some Finnish freshwater teleost fish, *Finnish Fish. Res.*, 15, 65, 1994.

Complete affiliation:

Jin-Chywan Gwo, Department of Aquaculture, Taiwan National Ocean University, Keelung 20224, Taiwan.e-mail: gwonet@ms.16.hinet.net.

PROTOCOLS FOR THE CRYOPRESERVATION OF SALMONIDAE SEMEN, *Lota lota* (GADIDAE) and *Esox lucius* (ESOCIDAE)

F. Lahnsteiner and N. Mansour

1. INTRODUCTION

Salmonidae are traditionally cultured fish in many parts of the world and are of great importance for recreational and commercial fisheries. In these species semen cryopreservation plays a significant role in aquaculture for the synchronization of artificial reproduction, for efficient utilization of semen, maintaining the genetic variability of broodstocks and for interspecific breeding. Nowadays, the method is also of particular importance in setting up gene banks of endangered autochthonous populations for re-stocking purposes.

The Northern pike (*Esox lucius* L.) is a preferred game for recreational fishing and, because of the quality of its meat, is also intensely caught for commerce. Therefore, fry production and restocking is a necessity, but management of artificial fertilization is complicated since mostly single individuals are caught and semen volumes which can be stripped are limited and often insufficient. Cryopreservation of semen is a method of overcoming these problems as semen can be made available throughout the whole reproductive season.

The burbot, *Lota lota,* is a twilight and night active fresh and cold water groundfish. Due to devastation of its habitat, inhibition of spawning migration and intensive fishing pressure many burbot populations are endangered. Today, efforts are being made to restock with adequate fish to maintain genetic diversity. Cryopreservation of semen is an important tool to reach this goal, as it can be used for conservation of biodiversity, efficient and selective fertilization and for synchronization of artificial reproduction. In all mentioned species semen cryopreservation may gain importance for unlimited supply with material for research as e.g., for toxicological tests.

For species of the three families, semen cryopreservation methods have been developed in the last decade and adapted to guarantee easy and reliable handling performances for routine use [1-4].

The present review describes cryopreservation protocols for semen of the Salmonidae (*Oncorhynchus mykiss, Salmo trutta f. lacustris, Salvelinus fontinalis, Salvelinus alpinus, Salmo trutta f. fario, Hucho hucho, Coregonus lavaretus, Thymallus thymallus*), the Northern pike (*Esox lucius*), and the burbot (*Lota lota*).

2. PROTOCOL FOR FREEZING AND THAWING

In Salmonidae the semen cryopreservation method was successfully tested on Coregoninae (*Coregonus lavaretus*), Salmoninae (*Hucho hucho, Oncorhychus mykiss,*

Salvelinus alpinus, Salvelinus fontinalis, Salmo trutta f. lacustris, Salmo trutta f. fario) and Thymalline (*Thymallus thymallus*) [1]. The semen cryopreservation methods for *Lota lota* [3] and *Esox* lucius [2] were not tested on other species of Gadidae and Esocidae, respectively.

2.1 Equipment

- Freezing vessels. For freezing of semen, straws with a volume of 0.5 ml and 1.2 ml are used which are commercially available from several companies. As the Salmonidae, *Esox lucius* and *Lota lota* are species with external fertilization and seasonal spawning times, they have a high egg production. During natural spawning and in artificial insemination several 1000 eggs are fertilized simultaneously. Therefore, for commercial use the cryopreservation methods also have to be adapted for large scale fertilization. Cryopreservation of large biological samples is limited mainly due to lack of homogeneity and low freezing and thawing rates. Therefore, for freezing of large semen quantities straw packages are used [4]. Straw packages must be prepared manually [4]: Plastic foil from commercially available plastic bags with a thickness of 0.45 - 0.55 mm is used which remains flexible even at liquid nitrogen temperatures (Figs. 1a, e). Two plastic ribbons are cut at a width of 1.5 cm and a length depending on the desired number of straws which should be connected (required length per straw 1.3 cm). The ribbons are placed on top of each other (Fig. 1e) and sealed together at their wide side with a commercial plastic bag sealing apparatus in that way that a 6 mm sealed portion is followed by a 7 mm unsealed portion (Figs. 1b, c, d). The straws are tightly fit in the unsealed portions with their plugged side (Figs. 1 b, c). The distance between the single straws should be approximately 0.5 cm (Fig. 1d).
- Freezing apparatus. Freezing is done in liquid nitrogen vapour [1]. The used freezing apparatus is a self constructed insulated box (inner dimensions: base 27 x 18 cm, height 33 cm) which has an overflow trap in a sidewall at a height of 5.5 cm above the bottom (Fig. 2). The overflow trap is necessary to adjust the liquid nitrogen level. The tray can be moved to different distances (0 to 10 cm) above the surface of the liquid nitrogen (Fig. 2) allowing the application of various freezing conditions. Adjustable trays are of advantage when freezing levels have to be changed frequently. Otherwise, floating trays are also useful.
- Thawing equipment. Thawing of straws is performed in a water bath of ambient temperature. Utilization of a thermostat regulated water bath is recommended as it keeps thawing temperature constant even when numerous straws are thawed simultaneously or when the temperature gradient between water and environment is high.

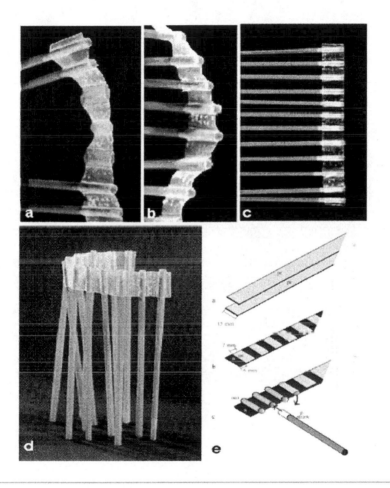

Figure 1 - Straw packages for large scale cryopreservation. (a) Plastic rack during fitting in the straws. (b) Plastic rack with straw. (c) Straw package, placed horizontally as done for freezing. (d) Straw package rolled together as done for storage and for cutting open. (e) Steps in the production of flexible plastic racks: 1) plastic ribbons are placed one above each other, 2) plastic ribbons are sealed together, 3) the unsealed portions are opened and straws are fitted in. (*Source*: Lahnsteiner F., N. Mansour and T. Weismann, *Aquaculture* 209: 359-367, 2002. With permission).

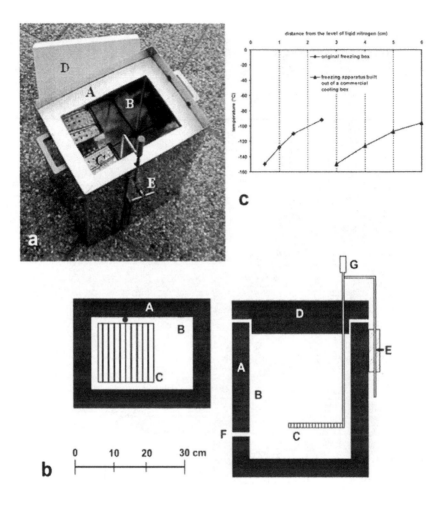

Figure 2 - Freezing box used in the protocol of Lahnsteiner (2000). (a) Picture. (b) Schematic drawing. Left: cross section; right: longitudinal section. A- insulated wall, B- freezing chamber, C-tray for straws, D- cover, E- set screw, F- overflow trap for liquid nitrogen, G- movable rack fixation. *(Source:* Lahnsteiner F., Germ Cell Protocols. Sperm and oocyte analysis. Schatten H. editor Humana Press, New Jersey, volume 2:1-13, 2004. With permission.)

2.2 Solutions

For the preparation of extenders and fertilization solutions chemicals of analytical grade and distilled water are used. Hen egg yolk is prepared freshly and carefully separated from the white which causes agglutination of spermatozoa (Tables 1,2).

Table 1 - Preparation of extenders

	Salmonidae	*Esox lucius*	*Lota lota*
Step 1. Dissolve the following components in 80 ml distilled water:			
NaCl	600 mg	600 mg	580 mg
KCl	315 mg	315 mg	1.5 mg
CaCl$_2$ x 2H$_2$O	15 mg	15 mg	15 mg
MgSO$_4$ x 7H$_2$O	20 mg	20 mg	25 mg
Hepes (sodium salt)	470 mg	470 mg	470 mg
Step 2. Adjust pH to 7.8 with NaOH or HCl, fill up with water to 100 ml			
Step 3. Add the following cryoprotectants:			
Methanol	10% (v/v)	10% (v/v)	10% (v/v)
Sucrose	0.5% (w/v)	0.5% (w/v)	-
Glucose	-	-	1.5% (w/v)
Bovine serum albumin	1.5 % (w/v)	1.5 % (w/v)	-
Hen egg yolk	7% (v/v)	7% (v/v)	7 % (v/v)

Table 2 - Preparation of fertilization solution

	Salmonidae	*Esox lucius*	*Lota lota*
Step 1. Dissolve the following components in 80 ml distilled water:			
NaCl	-	580 mg	145 mg
NaHCO$_3$	500 mg	-	-
Tris base*	600 mg	120 mg	120 mg
Step 2. Adjust pH with NaOH or HCl, fill up with water to 100 ml			
pH	9.0	9.0	8.5

(*2-Amino-2-(hydroxymethyl)-1,3-propanediol)

2.3 Fish handling and semen collection

- In the above described species, semen can be stripped by abdominal massage.
- The genital papilla is dried from adhering water and the semen is collected into reagent tubes by pressure on the abdomen.
- In *Esox lucius,* testicular sperm can also be used when insufficient quantities are obtained by stripping. The males are sacrificed, the testes excised and cleaned from remnants of blood.
- Then the testes are cut into slices and squeezed through a plankton net of 150 µm mesh size.
- The homogenate is collected and diluted in sperm motility inhibiting solution (= extender without cryoprotectants) at a ratio of 1:1 [2]. Semen should be stored not longer than 1 h before freezing.

2.4 Semen dilution and loading into straws

- The semen is diluted in 4°C cold extender whereby the dilution ratio of semen with extender depends on the sperm density. The sperm concentration in the extenders should be $\leq (2.0 - 3.0) \times 10^9$ spermatozoa/ml extender [5]. Higher sperm concentrations decrease post-thaw semen quality [5]. Depending on the sperm density of the semen (Table 2) for *Salvelinus alpinus* a dilution ratio of 1:2 (semen:extender) is recommended, for *Coregonus lavaretus, Hucho hucho, Oncorhynchus mykiss, Salvelinus fontinalis*, and *Thymallus thymallus,* a dilution ratio of 1:3, for *Salmo trutta f. fario* 1:5, and for *Salmo trutta f. lacustris* 1:7 [1]. In *Lota lota* semen is diluted in the extender at a ratio of 1:5 [4], in *Esox lucius* at a ratio of 1:3 (similar for spermatic duct semen and for testicular semen prediluted at a ratio of 1:1 in sperm motility inhibiting solution) [2].
- Sperm density varies with the date of the spawning season, the age of the fishes and may differ between various populations [6]. Therefore sperm density checks are recommended before semen cryopreservation.
- Semen is filled in the straws with micro-pipettes. Before the straws are used they are cooled to 4°C. As the straws have a stopper which avoids liquid penetration the sperm suspension can be sucked in by mouth, too. This is the quickest way to fill the straws in routine work. Equilibration should be ≥ 1 min and ≤ 10 min.

2.5   Freezing and thawing procedures

- Before freezing can be started the freezing box is first cooled with liquid nitrogen and then filled with liquid nitrogen up to a height of 5.5 cm whereby the exact height is adjusted by means of the overflow trap located in one of the side walls.

- Ready for freezing, the box contains a volume of 2.67 l liquid nitrogen. It takes circa 15 and 30 min until the interior of the box is cooled down and stable conditions have been reached.
- Then the tray for the straws is adjusted to the desired freezing level and equilibrated for 5 min to reach the appropriate temperature.
- The straws or straw packages are placed on the tray and frozen for 10 min. Freezing levels and freezing temperatures are species-specific and are shown in Table 3.
- When freezing is terminated the straws are plunged into liquid nitrogen.
- For storage the single straws are transferred in the cans of commercial liquid nitrogen containers. The straw packages remain flexible in liquid nitrogen and can therefore be rolled together and placed in the cans, too (Fig. 1 d).
- For thawing (thawing conditions see Table 1), the straws are taken out of the liquid nitrogen container and transferred immediately into water of the desired temperature, and gently shaken during thawing.
- After thawing the straw stopper is cut away and the sperm suspension released onto the eggs. Straw packages are processed in a similar way. They are taken out of the container, rolled out quickly, and placed in water. Thereafter, they are rolled together again and the plugs of the straws are cut away with scissors (Fig. 2f).

Table 3 - Freezing and thawing conditions for Salmonidae semen. A freezing level of 1.0 cm above the level of liquid nitrogen corresponds to a temperature of $-130 \pm 2°C$, 1.5 cm to $-110 + 2°C$, 2 cm to $-100 + 2°C$, and 2.5 cm to $-92 \pm 2°C$.

Species	sperm density (cells/ml)	sperm to egg ratio	diluted semen for 100g eggs
Coregonus lavaretus	$(4-8) \times 10^9$	$\geq 3.0 \times 10^5 : 1$	4 (10,000 – 15,000)
Esox lucius, testicular semen	$(7-12) \times 10^9$	$\geq 4.5 \times 10^5 : 1$	8 (9,000 – 12,000)
Esox lucius, stripped semen	$(3-7) \times 10^9$	$\geq 4.5 \times 10^5 : 1$	8 (9,000 – 12,000)
Hucho hucho	$4-8) \times 10^9$	$\geq 2.5 \times 10^6 : 1$	4 (1,200 – 3,000)
Lota lota	$(5-8) \times 10^{10}$	$\geq 1.7 \times 10^6 : 1$	100 (115,000 – 130,000)
Oncorhynchus mykiss	$4-8) \times 10^9$	$\geq 2.5 \times 10^6 : 1$	4 (1,200 – 3,000)
Salmo trutta f. fario	$(9-15) \times 10^9$	$\geq 2.5 \times 10^6 : 1$	4 (1,000 – 2,000)
Salmo trutta f. lacustris	$(1-2) \times 10^{10}$	$\geq 2.5 \times 10^6 : 1$	4 (1,000 – 2,000)
Salvelinus fontinalis	$4-8) \times 10^9$	$\geq 2.5 \times 10^6 : 1$	4 (1,200 – 3,000)
Salvelinus alpinus	$(1-3) \times 10^9$	$\geq 2.5 \times 10^6 : 1$	4 (1,200 – 3,000)
Thymallus thymallus	$4-8) \times 10^9$	$\geq 1.2 \times 10^6 : 1$	4 (5,000 – 10,000)

* n.i. - not investigated

2.6 Quality and fertilizing capacity of frozen-thawed semen

Reliable and quick to measure semen quality markers are the sperm motility rate (percentage of progressive moving spermatozoa) and the swimming velocity. These parameters can be exactly determined by analyzing video sequences of the sperm motility in computer programs (e.g., CMA analysis [7], CASA systems [8]). However, these methods need complicated equipment and sophisticated analysis procedures and are therefore difficult to apply in practice. For determination of the sperm motility rate we recommend a subjective method which was originally published by Mansour et al. [9]) (Table 4). One - 2 μl semen is activated on a glass slide with 50 μl of 4°C sperm motility activating saline solution. For the Salmonidae and for *Esox lucius* the sperm motility activating saline solution consists of 60 mmol l^{-1} NaHCO$_3$ and 50 mmol l^{-1} Tris (pH 9.0), for *Lota lota* a 25 mmol l^{-1} NaCl solution (pH 8.5) is used. Alternatively, also water can be used for activation of sperm motility in all investigated species. The activated sperm suspension is covered with a cover slip and the sperm motility rating is performed within 10 sec after activation using a light microscope at 200 x magnification according to the criteria listed in Table 4 [9]. Semen with a motility rating of ≥4 is most suitable for cryopreservation.

In comparison to untreated semen in cryopreserved semen the percentage of immotile spermatozoa is significantly increased, and the rate of motile spermatozoa is decreased [1,4]. At low sperm to egg ratios the fertilization rate (evaluated in the embryo stage before hatching) is also significantly decreased. However, the decrease in fertilizing capacity can be completely compensated by higher sperm to egg ratios and then fertilization rates in the range of fresh semen control are obtained [10]. The percentage of embryonic malformations is similar with cryopreserved and untreated semen [1-3].

Table 4 - Criteria used to determine semen quality by estimation of sperm motility rate. Semen with a motility rating ≥4 is most suitable for cryopreservation.

Motility rating	Criteria
0	No motility (0 %)
1	Almost all sperm cells immotile with slight vibrations; < 10 % progressively motile sperm
2	Most sperm cells immotile with slight vibrations; 10 - 25 % progressively motile sperm
3	Some sperm cells immotile with slight vibrations; 25-50 % progressively motile sperm
4	Only few sperm cells immotile showing vibrations; 50-75 % progressively motile sperm
5	Very few sperm cells immotile showing vibrations and most sperm cells (75-90 %) progressively motile
6	All sperm cells progressively motile (100 %)

2.7 Fertilization with cryopreserved semen

In the Salmonidae fertilization is performed at 4-6°C, in *Esox lucius* at 12°-14°C, and in *Lota lota* at 2°C. Reliable semen doses to fertilize 100 g eggs and the corresponding sperm to egg ratios are shown in Table 5. In the Salmonidae and in *Esox lucius* for fertilization of egg quantities ≤ 50 g wet or dry fertilization can be used, for higher egg quantities and in *Lota lota* dry fertilization is obligatory. The ratio fertilization solution to eggs is 1:2 for Salmonidae and *Esox lucius* and 2:1 for *Lota lota*.

- For wet fertilization eggs are placed in suitable beakers, fertilization solution is added in the required ratio and eggs are distributed in the fertilization solution. Then the thawed semen is mixed with the eggs. Salmonidae and *Esox lucius* eggs have stable quality in the fertilization solution for at least 3 min, and *Lota lota* eggs for 1 min.
- For dry fertilization eggs are placed in suitable beakers, too. The thawed semen is mixed with the eggs and immediately thereafter fertilization solution is added and all components are re-mixed.

Table 5 - Sperm density of fresh semen, recommended sperm to egg ratios, recommended sperm quantities for fertilization of 100 g eggs (values in parenthesis give the number of nonhardened eggs which are represented by 100 g).

Species	sperm density (cells/ml)	sperm to egg ratio	diluted semen for 100g eggs
Coregonus lavaretus	$(4-8) \times 10^9$	$\geq 3.0 \times 10^5$: 1	4 (10,000 – 15,000)
Esox lucius, testicular semen	$(7-12) \times 10^9$	$\geq 4.5 \times 10^5$: 1	8 (9,000 – 12,000)
Esox lucius, stripped semen	$(3-7) \times 10^9$	$\geq 4.5 \times 10^5$: 1	8 (9,000 – 12,000)
Hucho hucho	$4-8) \times 10^9$	$\geq 2.5 \times 10^6$: 1	4 (1,200 – 3,000)
Lota lota	$(5-8) \times 10^{10}$	$\geq 1.7 \times 10^6$: 1	100 (115,000 – 130,000)
Oncorhynchus mykiss	$4-8) \times 10^9$	$\geq 2.5 \times 10^6$: 1	4 (1,200 – 3,000)
Salmo trutta f. fario	$(9-15) \times 10^9$	$\geq 2.5 \times 10^6$: 1	4 (1,000 – 2,000)
Salmo trutta f. lacustris	$(1-2) \times 10^{10}$	$\geq 2.5 \times 10^6$: 1	4 (1,000 – 2,000)
Salvelinus fontinalis	$4-8) \times 10^9$	$\geq 2.5 \times 10^6$: 1	4 (1,200 – 3,000)
Salvelinus alpinus	$(1-3) \times 10^9$	$\geq 2.5 \times 10^6$: 1	4 (1,200 – 3,000)
Thymallus thymallus	$4-8) \times 10^9$	$\geq 1.2 \times 10^6$: 1	4 (5,000 – 10,000)

3. GENERAL CONSIDERATIONS

In the Salmonidae several investigations on improvement of semen cryopreservation protocols have been performed mainly in aspects of extender composition, as the extender used in the original protocol of Lahnsteiner [1] is complex and complicated to prepare. As alternative extender a 0.3 M glucose solution containing 10% methanol as cryoprotectant was successfully used in the Arctic charr (*Salvelius alpinus*) [9]. Sarvi et al.[11] used an extender, which contained beside of 3 M glucose and 10 % methanol also 10% egg yolk. For the glucose based extenders dilution ratio of semen to extender is 1: 3, equilibration time 1 – 2 min.

The self-made freezing apparatus used in the original protocol of Lahnsteiner [1] is also complicated to construct. An alternative is the adaptation of a commercially available cooling box [12] which is shown in Figure 3. Inside this box, the straws are frozen on a freezing rack which can be moved vertically on a scale fixed in a wooden base within the cooling box (Fig. 3a, 3b & 3c). For this freezing box a commercially available freezing rack is used which has been originally designed for freezing of bull semen and can hold up to fifty 0.5 ml straws. Straws can be fixed on the rack by a metal bar to prevent them from floating when immersing the rack into the liquid nitrogen (Fig. 3b, 3c).

Comparison of freezing conditions between the two freezing boxes revealed that the freezing temperature obtained in different distances from the level of liquid nitrogen and subsequently also the freezing rates differ. This seems to depend on different dimensions, material composition, and/or insulation of the freezing boxes. Deviations in freezing conditions between different boxes can be a serious problem in reproducibility of the methods and indicate that standardization and adjustment of freezing conditions is necessary.

The freezing process itself is performed similar to the original method. The rack is equilibrated for 10 min at the desired temperature, and then it is loaded with straws. Freezing is performed for 10 min, thereafter the straws are plunged into the liquid nitrogen picked up and transferred into liquid nitrogen containers for further storage.

4. ACKNOWLEDGMENTS

The authors are grateful to Austrian BMLF and Canadian Coastal Zone Research Institute for financial support and to the Bundesanstalt in Thalheim-Wels, the Bundesamt in Scharfling, and the fish farm Kreuzstein for excellent research cooperations. Also, we are gratefully acknowledging Professor Dr. Gavin F. Richardson and Professor Dr. Mary A. McNiven, Atlantic Veterinary College, University of Prince Edward Island, Canada, for their scientific cooperation.

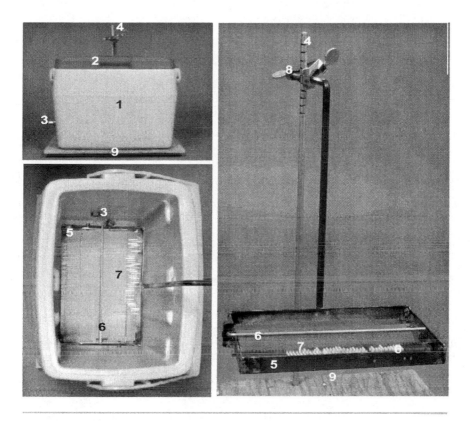

Figure 3 - (top left) Freezing apparatus built out of a commercial freezing box. (bottom left) Open-freezing box showing the freezing rack loaded with 0.5 ml straws. (right) Commercial available freezing rack originally used for freezing of bull semen and loaded with 0.5 ml straws. Rack can be moved up and down on a marked scale. 1: freezing box; 2: cover of the freezing box; 3: liquid nitrogen drainage tap; 4: marked scale; 5: freezing rack; 6: metal bar to prevent the straws from floating when plunging into the liquid nitrogen; 7: 0.5 ml straws; 8: screw clamp for changing the level of the freezing rack above the surface of liquid nitrogen; 9: wooden base. The freezing box has the following dimensions: Inner length: 34 cm, inner width: 31 cm, inner height: 22 cm, drainage hole of liquid nitrogen: 7 cm above bottom, wall thickness: 3 cm.

5. REFERENCES

1-Lahnsteiner, F., Semen cryopreservation in the Salmonidae and in the Northern pike, *Aquac Res*, 31, 245, 2000.

2-Lahnsteiner, F., Weismann, T., and Patzner, R.A., An efficient method for cryopreservation of testicular sperm of the Northern pike, *Esox lucius Linnaeus, Aquac Res*, 29, 341, 1998.

3-Lahnsteiner F., Mansour, N., and Weismann, T., The cryopreservation of spermatozoa of the burbot, *Lota lota* (Gadidae, Teleostei), *Cryobiology*, 45, 195, 2002a.

4-Lahnsteiner, F., Mansour, N., and Weismann, T., A new technique for insemination of large egg batches with cryopreserved semen in the rainbow trout, *Aquaculture*, 209, 359, 2002b.

5-Lahnsteiner F., Weismann, T., and Patzner, R.A., Semen cryopreservation of salmonid fishes. Influence of handling parameters on the postthaw fertilization rate, *Aquac Res*, 27, 659, 1996.

6-Billard, R., Artificial insemination and gamete management in fish, *Marine Behaviour and Physiology*, 14, 3, 1988.

7-Lahnsteiner, F., Berger B., and Weismann, T., Sperm metabolism of the teleost fishes *Oncorhynchus mykiss* and *Chalcalburnus chalcoides* and its relation to motility and viability, *J. Exp. Zool.* 284, 454, 1999.

8-Kime, D.E., Ebrahimi, M., Nysten, K., Roelants, I., Rurangwa, E., Moore, H.D.M., and Ollevier, F., Use of computer assisted sperm analysis (CASA) for monitoring the effects of pollution on sperm quality of fish; application to the effects of heavy metals, *Aquat. Toxicol.* 36, 223, 1996.

9-Mansour, N., Richardson, G.F., and McNiven, M.A., Effect of extender composition and freezing rate on post-thaw motility and fertility of Arctic char, *Salvelinus alpinus* (L.), spermatozoa, *Aquaculture Res.*, 37, 862, 2006.

10-Leung, L.K.B. and Jamieson, B.G.M., Live preservation of fish gametes, in: *Fish evolution and systematics: Evidence from spermatozoa*, Jamieson, B.G.M., Ed., University Press, Cambridge, 1991, 245.

11- Sarvi, K., Hiksirat, H., Mojazi-Amiri, B., Mirtorabi, S.M., Rafiee, G.R., and Bakhtiyari M., Cryopreservation of semen from the endangered Caspian brown trout (*Salmo trutta caspius*), *Aquaculture*, 256, 564, 2006.

12- Mansour, N., Richardson, G.F., and McNiven, M.A., Effect of seminal plasma protein on post-thaw sperm motility, viability and fertility of Arctic char, *Salvelinus alpinus* (L.), Submitted.

Complete affiliation:

Franz Lahnsteiner, Institute of Zoology, University of Salzburg, Hellbrunnerstrasse 34, -5020 Salzburg, Austria. e-mail: Franz.Lahnsteiner@sbg.ac.at.

RAINBOW TROUT (*Oncorhynchus mykiss*) SPERM CRYOPRESERVATION

P. Herráez, V. Robles and E. Cabrita

1. INTRODUCTION

Rainbow trout is one of the most widely farmed fish. By 2002, 64 countries were reporting rainbow trout farming production to the FAO. The primary producing areas are in Europe, North America, Chile, Japan and Australia and total world production reached or even exceeded 500,000 t for the first time in 2003.

Natural spawning occurs annually during autumn/winter, with maturation occurring in response to temperature and photoperiod cues, but environmental manipulation under farm conditions allows breeding all year-round, with apparently minimal effects on milt quality. Trout breeding involves the hand-stripping and dry fertilisation of ripe eggs and sperm from 2-4 year old mature fish held on site. Fertilised eggs are incubated in flow-through containers and hatch after approximately 6 to 9 weeks at temperatures of 5 to 10°C. Survival to hatching is usually up to 80%. Fertilization out of season usually reduces this percentage.

Rainbow trout sperm physiology has been largely analyzed and there is a large amount of information on sperm cytophysiology. Single male milt production in the middle of the reproductive season ranges from 5 to 20 ml/kg, pH varies between 7.5 and 8.5, osmolality between 260 and 320 mOsm/Kg and cell density goes from 10^9 to 10^{11} [1,2]. Motility is prevented in seminal plasma due to the high content in K^+, and is activated in media lacking this ion and with an osmolality lower than 200 mOsm/Kg [3]. Spermatozoa have one mitochondrion and a very limited ability to produce ATP, so the duration of motility is very short (less than 60 s). Rainbow trout eggs are 8-9 mm in diameter and have only one mycropile. Consequently, approximately 10^5 to 10^6 sptz are needed to fertilize a single egg, and this is considered an appropriate rate for fertilization with fresh sperm.

Many studies have been developed on rainbow trout cryopreservation. Obtained results have been very variable between populations, individuals or ejaculates, and often refer to fertilization on an experimental scale [4-7]. Nevertheless, some researchers have reported standardized methods providing good fertilization and hatching rates, even on a commercial scale [8]. Interest in sperm cryopreservation, as in other species that require artificial fertilization, concerns routine sperm handling, the creation of sperm banks for crossbreeding planning, the banking of germplasm from particular individuals or strains, as well as the creation of a sperm stock obtained at the height of the reproductive season for further use all year round. Taking into account the large milt volumes required for fertilization, loading in containers larger than traditional 0.5 ml French straws is recommended.

2. PROTOCOL FOR FREEZING AND THAWING

2.1 Semen collection and selection of samples

- Sperm can be obtained by stripping after cleaning and drying the genital papile, but catheterization (previous anesthesia) is recommended to avoid contamination with urine, mucus or faeces (Fig. 1). Catheterization is performed using a round tip catheter for urine collection 4 mm in diameter.
- Sperm motility is tested by placing a drop of semen, previously diluted 1:100 in a nonactivating solution (i.e., the freezing extender) under a light microscope and activating it with water or another activator such as DIA 532 [9] (0.51 g/l NaCl, 3.75 g/l glycine, 2.42 g/l Tris) at a rate of 1:10. Samples with less than 60% spermatozoa with progressive motility are discarded.
- Tests for cell viability and resistance to hypoosmotic shock are also recommended [4,10,11]. Samples with less than 80% viable cells or with over 20% cells that are extremely sensitive to hypoosmotic shock should be discarded.
- Sperm must be collected during December or January to ensure better fertilization rates after thawing.

2.2 Sperm cryopreservation and thawing

- The extender, the mineral solution #6 from Erdhal and Graham [12] ($CaCl_2 \cdot x\, 2H_2O$ 0.102 g/l, $MgCl_2 \cdot x\, 6H_2O$ 0.22 g/l, Na_2HPO_4 0.26 gr/l, KCl 2.61 g/l, NaCl 5.84 g/l, citric acid 0.10 g/l, glucose 10.00 g/l, KOH 10 ml of a solution 1.27 g/100 ml and bicine 20 ml of a solution 5.3 g/100 ml; 345 mOsm/Kg, pH 7.8), can be prepared in advance and kept at 4°C until use (no longer than 4-5 days). Immediately prior to sperm dilution, cryoprotectant (DMSO 7% v/v) and membrane stabilizers (egg yolk 10% v/v alone or in addition to the soybean protein complex Promine® 7.5 mg/ml) are added to the mineral extender.
- Before freezing, sperm is diluted (1:3 v/v) in the previously prepared cryoprotectant solution. The equilibration time required before freezing is 10-15 min.
- Diluted sperm is loaded into French straws (0.5 ml) (IMV) or 5 ml macrotubes (280 x 5 mm, Minitub) during the equilibration time.
- For freezing, straws are placed in a net rack, floating either 2 cm (0.5 ml straws) or 1 cm (5 ml macrotubes) above the liquid nitrogen (LN_2) surface in a closed styrofoam box for 10 min and then plunged into the liquid nitrogen.
- The straws are stored in LN_2 tanks until used and thawed in a water bath at 25°C (0.5 ml straws) or 60°C (5 ml macrotubes) for 30 s.

2.3 Post-thaw sperm evaluation

- Motility is evaluated as described above, immediately after thawing and activation is by DIA 532 instead of water to reduce hypoosmotic shock, because thawed trout spermatozoa are extremely sensitive.
- Viability of fresh and frozen/thawed spermatozoa is evaluated using propidium Iodide (PI) staining combined with flow cytometry or fluorescence microscopy. Dilute 20 µl of sperm in 1 ml of an nonactivating solution. Add 20 µl of PI working solution (10 µl of stock solution 0.5 mg/ml in 0.1 M PBS- in 100 µl of 0.1 M PBS). Analyse using flow cytometry or fluorescence microscopy. Dead cells appear labelled in red.
- Resistance to hypoosmotic shock is evaluated analing the cell viability in a time course after dilution in hypoosmotic solutions (NaCl, 10-100mOsm/Kg).

2.4 Fertilization

- Immediately after thawing spread the sperm over the eggs at a rate of 1ml thawed sperm per 10 ml eggs (approx. 100 eggs), giving approximately 1.6×10^7 spermatozoa/egg.
- Mix the sperm and eggs rapidly and gently using a bird feather, and add DIA 532.
- After 10 min, wash the eggs gently with water and transfer them to incubation baskets. Keep the eggs incubating in the dark at temperatures below 10°C.

Figure 1 - Sperm extraction after trout canulation.

3. GENERAL CONSIDERATIONS

Sperm quality decreases rapidly after freezing/thawing: motility duration is significantly reduced, the percentage of viable cells diminishes and their resistance to hypoosmotic shock also decreases. Therefore, successful cryopreservation requires immediate use of semen after thawing. Fertility rates achieved in trials with small egg batches (400 eggs) were 84% when sperm was frozen in 0.5 ml straws and 73% when frozen in 5 ml macrotubes [8]. The same protocol provided 71% eyed eggs when larger batches of 150 ml of eggs (1,600-2,000 eggs) were fertilized with 8 ml of sperm frozen in 5 ml macrotubes (equivalent to 2 ml of net sperm) [8]. Spermatozoa of rainbow trout are very sensitive to cryopreservation and special care should be taken in the selection of samples suitable for cryopreservation.

Seasonality clearly affects the success of the cryopreservation process, which should always be carried out with sperm obtained in the natural breeding period, despite the fact that the evaluation of fresh sperm could indicate similar quality of samples.

4. ACKNOWLEDGMENTS

Authors thanks to Las Zayas and Vegas del Condado fishfarms staff and all the students collaborating in this project. Research partially supported by Junta de Castilla y León, Spain.

5. REFERENCES

1-Piironen, J., Composition and cryopreservation of sperm from some Finish freshwater teleost fish, *Fishish Fisheries Research*, 15, 27, 1994.

2-Suquet, M., Billard, R., Cosson, J., Dorange, G., Chauvaud, L., Mugnir, C., and Fauvel, C., Sperm features in turbot: a comparison with other freshwater and marine fish species, *Aquat Living Resour*, 7, 283, 1994.

3-Billard, R., Cosson, J., Crim, L.M., and Suquet, M., Sperm physiology and quality, in *Broodstock management and egg and larval quality*, Bromage and Roberts, Eds., Blackwell Science, 25, 1995.

4-Cabrita, E., Alvarez, R., Anel, R., Rana, K.J., and Herráez, M.P., Sublethal damage during cryopreservation of rainbow trout sperm, *Cryobiology*, 37, 245, 1998.

5-Lahnsteiner, F., Weissman, T., and Patzner, R., A uniform method for cryopreservation of semen of the salmonid fishes *Oncorhyncus mykiss, Salmo trutta fario, Salmo trutta lacustris, Coregonus sp.*, *Aquac Res*, 26, 801, 1995.

6-Lahnsteiner, F., Berger, B., Weismann, T., and Patzner, R., The influence of various cryoprotectant on semen quality of the rainbow trout before and after cryopreservation, *J Appl Ichthyol*, 12, 99, 1996.

7-Conget, P., Fernández, M., Herrera, G., and Minguell, J.J., Cryopreservation of rainbow trout spermatozoa using programmable freezing, *Aquaculture*, 143, 319, 1996.

8-Cabrita, E., Robles, V., Alvarez, R., and Herráez, M.P., Cryopreservation of rainbow trout sperm in large volume straws: application to large scale fertilization, *Aquaculture*, 201, 301, 2001.

9-Billard, R., A new technique or artificial insemination for salmonids using a sperm diluent, *Fisheries*, 2, 24, 1977.

10-Cabrita, E., Alvarez, R., Anel, E., and Herráez, M.P., The hypoosmotic-swelling test performed with coulter counter: a method to assay functional integrity of sperm membrane of rainbow trout, *Anim Reprod Sci*, 55, 279, 1999.

11-Cabrita, E., Martínez, F., Real, M., Alvarez, R., and Herráez, M.P., The use of flow cytometry to assess membrane stability in fresh and cryopreserved trout spermatozoa, *Cryoletters,* 22, 263, 2001.

12-Erdahl, D.A. and Graham, E.F., Cryopreservation of spermatozoa of the brook and rainbow trout, *Cryoletters*, 1, 203, 1980.

Complete affiliation:

Paz Herráez, Ph.D., Department of Molecular Biology, Faculty of Biological and Environmetal Sciences, University of León, 24071 Leon, España. Phone: +34 987291912, Fax: +34 987291917 e-mail: paz.herraez@unilcon.es.

SPERM CRYOPRESERVATION OF SEX-REVERSED RAINBOW TROUT
(*Oncorhynchus mykiss*)

V. Robles, E. Cabrita and P. Herráez

1. INTRODUCTION

The production of sex-reversed trout benefits the aquaculture industry allowing "all female" populations to be obtained. Females become sexually mature 1 year later than males, reaching their marketable size before maturation. Therefore, the production of entire female populations reports important economical benefits in commercial rainbow trout culture [1]. Sex-reversed rainbow trout have similar external morphology to normal males but lack sperm ducts [2], meaning that the animals must be sacrificed to obtain the milt. The peculiarities of the sperm, obtained directly from the testicle, make necessary the development of a specific cryopreservation protocol. Spermatozoa of sex-reversed females are immotile when collected from the testicle and need exogenous maturation before activation. General characteristic of sperm are significantly different from those of ejaculated milt (Table 1). The lack of sperm ducts makes it necessary to maintain a high number of animals in fish farms, since they must be killed to obtain their sperm. Therefore, cryopreserving the sperm of these individuals appears to be of great use since it would make sperm available all year round, reducing the need for a large male stock, and facilitate artificial fertilization using the milt of the same male in different breeding periods.

2. PROTOCOL FOR FREEZING AND THAWING

2.1 Fish handling and semen collection

- Sex-reversed females shall be sacrificed and their testicles surgically extracted and carefully cleaned, removing the main blood vessels (Fig. 1A, B).
- Milt is obtained by making cuts with a scalpel in the testicles and collecting the dripping sperm (Figure 1C). This method reduces sperm contamination and significantly increases fertilization rates with cryopreserved/thawed sperm in comparison with direct homogenization of the testicle, the method usually employed on fish farms.
- Samples must be diluted 1:9 (v/v) in a commercial solution (MATURFISH®, IMV, France) to promote sperm maturation, and kept in that solution 2 h at 4°C with oxygen supply.

Sperm must be collected during December or January to ensure better fertilization rates.

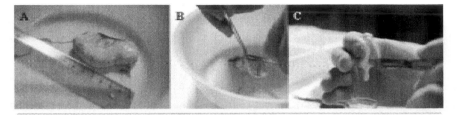

Figure 1 - Sperm extraction. A: testicle from a sex-reversed trout. B: Testicle cleaning removing all main blood vessels. C: sperm extraction making cuts with a scalpel.

Table 1 - Sperm parameters in fresh and cryopreserved samples from normal and sex-reversed trout males.

Parameters	Normal Males	Sex-reversed Males
Viability		
Fresh sperm	98.94±1.23	91.60±0.57
Shock resistance (%)		
(10mOsm, fresh sperm)	96.57±2.20	80.00±0.00
30 s	42.00±0.00	17.27±8.92
15 min		
Motility (%)		
Fresh sperm	57.70±0.00	90.83±7.88
Cryopreserved sperm		24.58±8.54
Fertility (%)		
Fresh sperm	84.00±0.00	82.74±3.36
Cryopreserved sperm		57.73±22.83
ATP concentration (nmol/spz)		
Fresh sperm		$7.99 \times 10^{-8} \pm 4.88\ 10^{-8}$
Cryopreserved sperm		$8.29 \times 10^{-8} \pm 5.25\ 10^{-8}$

2.2 Sperm cryopreservation and thawing

- For sperm dilution, combine the cryoprotectant dimethyl sulfoxide (DMSO, SIGMA) (7%) with two membrane stabilizers, egg yolk (SIGMA) (10%) and Dan Pro S760® (soybean protein complex, Central Soya Potein Group, Denmark) (7.5 mg/ml) in the mineral solution #6 from Erdahl and Graham [3].
- Before freezing dilute the sperm 1:3 (v/v) in the extender and maintain in this solution for 15 min to allow the proper penetration of DMSO into the cells.
- Load the sperm in French straws (0.5 ml) (IMV) during the equilibration time, and then place it 2 cm above the liquid nitrogen (LN_2) surface in a Styrofoam box for 10 min (freezing rate 63°C/min) to finally plung the straws in it.

- Store the straws in LN_2 tanks until use and thaw it in a water bath at 25°C for 30 s (thawing rate 666°C/min).
- To establish freezing and thawing rates, temperatures inside the straw were recorded with a thermocouple placed in the middle of the straws during the freezing and thawing process.

2.3 Post-thaw sperm evaluation

- Motility of fresh and frozen spermatozoa can be tested using DIA532 [4] as activation solution. Place 2 µl of sperm diluted 1:9 in MATURFISH ® on a glass slide under a microscope, and immediately add 12 µl of the activator. For evaluation values from 0 to 5 are attributed to the samples (being 0, 0% of spermatozoa with progressive motility, and 5, 100% of spermatozoa with progressive motility). Three observations must be done per sample and the same evaluator should always perform motility determinations.
- Viability of fresh and frozen/thawed spermatozoa is evaluated using propidium Iodide (PI) staining combined with flow cytometry or fluorescence microscopy. Dilute 20 µl of sperm in 1 ml of an nonactivating solution. Add 20 µl of PI working solution (10 µl of stock solution- 0.5 mg/ml in 0.1 M PBS- in 100 µl of 0.1 M PBS). Analyse using flow cytometry or fluorescence microscopy. Dead cells appear labelled in red.
- Resistance to hypoosmotic shock is evaluated analyzing the cell viability in a time course after dilution in hypoosmotic solutions (NaCl, 10-100 mOsm/Kg).

2.4 Fertilization

- Divide the spawn in batches of 100 eggs and place them in Petri dishes.
- Sperm shall be poured homogeneously over the eggs, and after quick and gentle mixing with a bird feather, sperm activator must be added.
- After 10 min gently wash the eggs with water and transfer them to incubation baskets. Kept at 10°C with a water flow of 2 l/min.
- Fertility rates can be established at eyed stage (Table 2).
- The sperm/egg ratio must be 8.5×10^6 spermatozoa/egg with fresh sperm, and 17×10^6 with frozen (the spermatozoa concentration can be determined using a Neubauer chamber, doing three readings per sample, and for each reading, counting three different fields of the chamber).

3. GENERAL CONSIDERATIONS

Seasonality clearly affects the success of the cryopreservation process, which should always be carried out with sperm obtained in winter, the natural breeding period.

Motility of fresh sperm was, in some cases, higher in spring than in winter, but those differences are not reflected in fertilizing capacity, since cryopreserved sperm collected in winter reported higher fertility rates than that obtained in spring [5].

Table 2 - Fertility rates with fresh and cryopreserved sperm obtained from sex-reversed males by two different methods: homogenization (H) and scalpel (S), and using three different solutions for sperm activation: DIA solution, DIA plus caffeine and DIA plus theophilline. Differences between fresh and cryopreserved sperm are expressed with the symbol %, and between the two extraction methods with the symbol *

	Fresh-H	Cryopreserved-H	Fresh-S	Cryopreserved-S
DIA	80.3±9.8▪	42.54±13.0	82.7±3.4▪	57.7±22.8
caffeine	73.8±21.7▪	41.57±19.0	76.1±11.5	65.2±8.5*
theophilline	84.3±5.5▪	54.26±11.9	81.9±6.3▪	52.1±35.8

It is well known that contaminants reduce seminal quality and aptitude for cryopreservation in salmonid sperm [6]. The development of a clean sperm extraction method significantly improved the fertility rates obtained with sex-reversed trout cryopreserved sperm. The use of a scalpel provided a clean extraction method that avoids most of the sample contamination and reported an increase in motility and fertility rates.

The use of methylxanthines as motility stimulators did not always improve motility and fertility rates, nevertheless it was observed that by adding 5 mM caffeine to DIA to activate scalpel-obtained sperm extracted in winter, a similar fertility rate as the respective fresh milt control was obtained (65.18% and 76.12%, respectively) [5].

4. ACKNOWLEDGMENTS

The authors would like to thank Los Rigales fish farm (Huesca, Spain) staff for their collaboration, in particular Ana Acín.

5. REFERENCES

1-Panadian, T.J. and Sheela, S.G., Hormonal induction of sex reversal in fish, *Aquaculture*, 138, 1, 1995.

2-Geffen, A.J. and Evans, J.P., Sperm traits and fertilization success of male and sex-reversed female rainbow trout (*Onchorhynchus mykiss*), *Aquaculture*, 182, 61, 2000.

3-Erdahl, D.A. and Graham, E.F., Cryopreservation of spermatozoa of the brown, brook and rainbow trout, *Cryo-Lett*, 1, 203, 1980.

4-Billard, R., Utilisation d'un systeme tri-glycocolle pour tamponer le dilueur d'insemination pour truite, *Bull. Fr. Pisc*, 264, 102, 1977.

5-Robles, V., Cabrita, E., Cuñado, S., and Herráez, M.P., Sperm cryopreservation of sex-reversed rainbow trout (*Oncorhynchus mykiss*): parameters that affect its ability for freezing, *Aquaculture*, 224, 212, 2003.

6-Rana, K.J., Gupta, S.D., and McAndrew, B.J., The relevance of collection techniques on the quality of manually stripped Atlantic Salmon (*Salmo salar*) milt, Workshop on Gamete and Embryo Storage and Cryopreservation in Aquatic Organism, Marly le Roy, France, *Aquac News*, 14, 4, 1992.

Complete affiliation

Vanesa Robles, CMR(B)-Center of Regenerative Medicine in Barcelona, c/ Dr. Aiguader, 88, 08003 Barcelona, Spain, Phone: +34933160315, e-mail: vrobles@cmrb.eu.

CRYOPRESERVATION OF TESTICULAR SPERM FROM
EUROPEAN CATFISH (*Silurus glanis*)

B. Ogier de Baulny, G. Maisse and C. Labbé

1.INTRODUCTION

Wild European catfish (*Silurus Glanis*) is distributed from the south of Sweden to the Alps (Morat, Neuchatel, Bienne and Constance lakes) and eastwards to the Black and Caspian Seas. European catfish is a commercially important species in east-European countries, but its rearing is now being developed in western countries. European catfish is a carnivorous fish reared in outdoor ponds. Males become sexually mature at 2-3 years old, females at 3-5 years old. Breeding occurs in the spring (May-July), when water temperature reaches 20°C. Sperm cryopreservation has been very little documented in this species [1,2], contrarily to other catfish species such as African catfish *Clarias gariepinus* [3-11], *Heterobranchus longifilis* [12], channel catfish *Ictalurus punctatus* [13-15], or blue catfish *Ictalurus furcatus* [16]. Males from European catfish are oligospermic, and sperm is always contaminated and activated by urine during stripping [17]. As a consequence, testicular sperm is often preferred to sperm obtained after stripping of the males. The present paper describes a method for cryopreservation of sperm obtained from testis of mature European catfish. The reader is referred to the very informative papers of Legendre et al. [18] and Linhart et al. [19] for thorough data on artificial reproduction of European catfish, since only information related to sperm cryopreservation will be given here.

2. PROTOCOL FOR FREEZING AND THAWING

2.1 Fish handling and semen collection

- Mature males kept at 20°C are injected intraperitoneally with a synthetic analogue of gonadotropin releasing agent (GnRH-A) at a dose of 30 µg/kg body weight. Carp pituitary extracts (3-5 mg/kg of body weight) or Ovaprim (ref 13430, Syndel International Inc. for aquaculture, Vancouver, Canada, 0.5 ml/kg of body weight) can also be used. Dopamine inhibitors do not improve treatment efficiency in this species [19].
- Two days after injection, males are sacrificed by deep anaesthesia, spinal trans-section, or a blow on the head.
- Testes are removed and placed in cold (10°C) Stor-Fish™ (IMV Technologies, L'Aigle, France) at a ratio of 1 g of tissue into 5 ml of Stor-Fish™, an isoosmotic medium allowing European catfish spermatozoa to remain immotile.
- Testis fragments are then minced to allow sperm release.

- Sperm suspension is filtered through a 150 μm mesh, centrifuged at 200 g, 20 min at 4°C.
- One volume pellet is resuspended in one volume of Stor-Fish™, giving a final concentration of about 9.10^9 spermatozoa/ml medium.
- From now, sperm is handled at 4°C.

Tips:
- Sperm exhibiting 90% motility should be preferred for cryopreservation.
- We observed that squeezing sperm out of the dry testis fragment in a mesh will induce more flagella breakage than the Stor-Fish™ procedure will.
- We observed that storage of testicular sperm at 4°C is possible for long periods (5 to 12 days) if sperm is diluted (1 v sperm + 9 v Stor-Fish™) and stored as a thin layer (3 mm) under air (average motility = 51% on day 12). This is important when cryopreservation has to be deferred from sperm collecting.
- Motility of testicular spermatozoa is not depending on the pH of the storage medium (Stor-Fish™), since pH from 4.7 to 9.7 were all suitable for a 24 h storage (same motility as fresh testicular sperm). As a consequence, a step of testicular sperm maturation *in vitro* does not apply for European catfish, contrarily to salmonidae testicular sperm.

2.2 Freezing and thawing procedures

- Mix 1 volume of testicular sperm suspension in Stor-Fish™ with 3 volumes cryoprotecting medium Cryo-Fish™ (ref 017295, IMV technologies, L'aigle, France) at 4°C.
- To reconstitute the cryoprotecting medium, mix 0.8 volume of cryofish buffer with 0.1 volume of egg yolk and 0.1 volume of dimethylacetamide (DMA).
- Load the mixture into 0.5 ml French straws (Ref. AAA 101).
- Leave 5-6 mm air at the free end of the straw to allow liquid expansion upon crystallization.
- Seal the straw and allow it to equilibrate for 10 min at 4°C.
- Place the straws horizontally on a freezing rack (ref 007118, IMV technologies) and lay the rack on a polystyrene frame floating on liquid nitrogen (Fig.1). Adapt the frame thickness so that the straws are 3 to 4 cm above the liquid nitrogen surface, to achieve an appropriate freezing rate (Fig. 2).
- After 10 min, the straws can be plunged in liquid nitrogen for long term storage.
- For thawing, plunge the straws into a 37°C water bath for 10 sec. Under these conditions, the temperature inside the straw will not rise above 4-5°C. A water bath should always be preferred to air, because thermal exchange is facilitated in water.
- After straw opening with scissors, transfer sperm into vials and keep it at 4°C. It is generally recommended to perform fertilization no longer than 2 to 3 min after thawing. There is no need to wash the cells prior to fertilization. With this

method, the motility of thawed testicular sperm is 60% when fresh sperm motility was 90%.

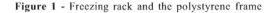

Figure 1 - Freezing rack and the polystyrene frame

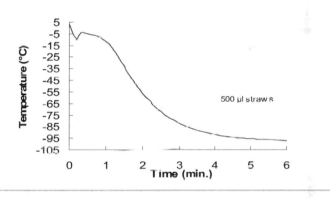

Figure 2 - Temperature decrease in 0.5 ml French straws laid 3 cm above liquid nitrogen. Crystallization occurs at -10°C. No seeding is required.

3. GENERAL CONSIDERATIONS

3.1 Choice of cryoprotectant

Several other cryoprotectants were tested ([2], Fig. 3): DMSO, methanol, glycerol and propylene glycol. DMSO and methanol at a concentration of 10% (v/v) sustained membrane integrity and mitochondrial activity at almost the same range as DMA did, but DMA was the most efficient at sustaining ATP content and motility of the thawed cells.

Interestingly, we observed that incubation of testicular sperm with DMA increased ATP content of spermatozoa from 172 (T0) to 423nmol/10^9spz (after 60s incubation). One consequence is that European catfish sperm cryopreserved with DMA has 2 times more ATP than fresh spermatozoa [2]. This may explain the better cryoprotection provided by DMA.

3.2 Sperm to egg ratio for fertilization

Few data exist on the optimal sperm to egg ratio for fertilization with thawed European catfish spermatozoa. Linhart et al. [19] showed that with fresh ejaculated sperm, 800, 8,000 and 80,000 spermatozoa per egg gave the same hatching success. For fertilization with frozen-thawed spermatozoa, and considering the quite good motility of thawed spermatozoa, it is often recommended to use 10 times more spermatozoa than what is used for fertilization with fresh spermatozoa.

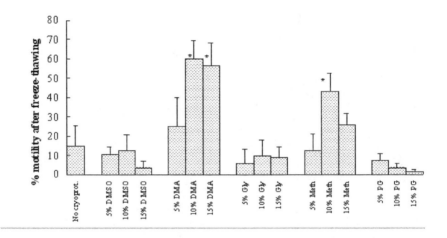

Figure 3 - Effect of the cryoprotectant on testicular sperm motility after cryopreservation. Several concentrations (5 to 15% v/v) were tested. Values are the mean of 6 different males ± SEM. *: significant difference ($p < 0.05$) with sperm frozen without cryoprotectant (only Cryo-Fish™ and egg yolk).

4. REFERENCES

1-Linhart, O., Billard, R., and Proteau, J.P., Cryopreservation of European catfish (*Silurus glanis* L.) spermatozoa., *Aquaculture*, 115, 347, 1993.
2-Ogier de Baulny, B., Labbe, C., and Maisse, G., Membrane integrity, mitochondrial activity, ATP content, and motility of the European catfish (*Silurus glanis*) testicular spermatozoa after freezing with different cryoprotectants, *Cryobiology*, 39, 177, 1999.
3-Steyn, G.J. and Van Vuren, J.H.J. The fertilizing capacity of cryopreserved sharptooth catfish (*Clarias gariepinus*) sperm, *Aquaculture*, 63, 187, 1987.
4-Van der Bank, F.H. and Steyn, G.J., The effect of cryopreservation and various cryodiluents on allozyme variation of glucose phosphate isomerase in the F1 progeny of African catfish (*Clarias gariepinus*), *Comp. Biochem. Physiol.,* 103B, 641, 1992.
5-Steyn, G.J., The effect of freezing rate on the survival of cryopreserved African sharptooth catfish (*Clarias gariepinus*) spermatozoa., *Cryobiology*, 30, 581, 1993.

6-Van der Walt, L.D., Van der Bank, F.H., and Steyn, G.J., The suitability of using cryopreservation of spermatozoa for the conservation of genetic diversity in African catfish (*Clarias garienipus*), *Comp. Biochem. Physiol. 106A*, 313, 1993.

7-Tiersch, T.R., Goudie, C.A., and Carmichel, G.J., Cryopreservation of channel catfish sperm: Storage in cryoprotectants, fertilization trials, and growth of channel catfish produced with cryopreserved sperm, *T Am Fish Soc*, 123, 580, 1994.

8-Horvath, A. and Urbanyi, B., The effect of cryoprotectants on the motility and fertilizing capacity of cryopreserved African catfish *Clarias gariepinus* (Burchell 1822) sperm, *Aquac Res*, 31, 317, 2000.

9-Rurangwa, E., Volckaert, F.A.M., Huyskens, G., Kime, D.E., and Ollevier, F., Quality control of refrigerated and cryopreserved semen using computer-assisted sperm analysis (CASA), viable staining and standardized fertilization in African catfish (*Clarias gariepinus*), *Theriogenology*, 55, 751, 2001.

10-Viveiros, A.T.M., So, N., and Komen, J., Sperm cryopreservation of African catfish, *Clarias gariepinus*: Cryoprotectants, freezing rates and sperm: egg dilution ratio, *Theriogenology*, 54, 1395, 2000.

11-Viveiros, A.T.M., Lock, E.J., Woelders, H., and Komen, J., Influence of cooling rates and plunging temperatures in an interrupted slow-freezing procedure for semen of the African catfish, *Clarias gariepinus*, *Cryobiology*, 43, 276, 2001.

12-Otémé, Z.J., Nunez Rodriguez, J., Kouassi, C.K., Hem, S., and Agnèse, J.F., Testicular structure, spermatogenesis and sperm cryopreservation in the African clariid catfish *Heterobranchus longifilis*, *Aquac. Res.*, 27, 805, 1996.

13-Christensen, J.M. and Tiersch, T.R., Refrigerated storage of channel catfish sperm, *J World Aquacult Soc*, 27, 340, 1996.

14-Christensen, J.M. and Tiersch, T.R., Cryopreservation of channel catfish spermatozoa: Effect of cryoprotectant, straw size, and formulation of extender, *Theriogenology*, 47, 639, 1997.

15-Christensen, J.M. and Tiersch, T.R., Cryopreservation of channel catfish sperm: effects of cryoprotectant exposure time, cooling rate, thawing conditions, and male-to-male variation, *Theriogenology*, 63, 2103, 2005.

16-Bart, A.N., Wolfe, D.F., and Dunham, R.A., Cryopreservation of blue catfish spermatozoa and subsequent fertilization of channel catfish eggs, *T Am Fish Soc*, 127, 819, 1998.

17-Linhart, O and Billard, R., Spermiation and sperm quality of European catfish (*Silorus glanis*, L.) after GnRH implantation and injection of carp pituitary extracts, *J Appl Icthyol*, 10, 182, 1994.

18-Legendre, M., Linhart, O., and Billard, R., Spawning and management of gametes, fertilized eggs and embryos in Siluroidei, *Aquat. Liv. Res.*, 9, 59, 1996.

19-Linhart, O., Gela, D., Rodina, M., and Kocour, M., Optimization of artificial propagation in European catfish, *Silurus glanis* L, *Aquaculture*, 235, 619, 2004.

Complete affiliation:

Catherine Labbé, Fish Reproduction, INRA, UR 1037 SCRIBE, Campus de Beaulieu, F-35000 Rennes. e-mail: Catherine.labbe@rennes.inra.fr

SEMEN CRYOPRESERVATION OF THE AFRICAN CATFISH, *Clarias gariepinus*

A.T.M. Viveiros and J. Komen

1. INTRODUCTION

Catfish are an economically important group of fresh and brackish water fish worldwide. Several species have been successfully introduced in aquaculture. The African catfish, *Clarias gariepinus*, is, together with *Pangasius* sp., the most important species, not only in Africa but also in S-E Asia and in Europe.

In captivity, African catfish gametogenesis is continuous once sexual maturity is reached. Females can be stripped of eggs after hormone treatments. Stripping of semen, however, is impossible due to obstruction of the sperm duct by a structure, typical of most catfish, the seminal vesicle. Thus, for artificial reproduction, male brood fish is sacrificed or part of their testes is surgically removed and the intratesticular semen is squeezed out and mixed with the eggs.

African catfish spermatozoa were first successfully cryopreserved in glucose 5% and glycerol 5%. Post-thawed motility was 40% [1]. Since then, many protocols have been successfully developed, and recently reviewed [2]. For artificial reproduction, 1 ml of fresh semen can be used to effectively fertilize 5 kg of eggs. Cryopreserved semen from one individual male (6-10 ml) can fertilize a total of 8 kg of eggs (*ca.* 4,800,000 eggs) and produce at least 2,400,000 larvae throughout the years [3,4]. Vials should be thawed according as required; the content of one vial is enough to fertilize 20-40 g eggs.

2. PROTOCOL FOR FREEZING AND THAWING

2.1 Materials to collect intra-testicular semen

- 40 l tank to anaesthetise fish
- Tricaine methane sulphonate (MS-222 or TMS)
- Tissue to dry fish skin
- Scissors, blade, scalpel, callipers to open fish and remove testes
- Needles to perforate testes
- 10 ml tubes to collect semen
- Microscope and microscope slides to check sperm motility
- Haemocytometer chamber (Neubauer, Bürker) to estimate sperm concentration
- Polystyrene and rack to keep semen in ice bath during handling
- Crushed ice

2.2 Materials to freeze/thaw semen

- pH meter to adjust semen extender pH
- Ginsburg fish Ringer (semen extender): 123.2 mM NaCl; 3.75 mM KCl; 3.0 mM CaCl$_2$; 2.65 mM NaHCO$_3$, deionised water, pH 7.6, osmolality/osmolarity 244 mOsm
- Methanol as cryoprotectant agent
- Test tubes to prepare the freezing medium in
- Micropipettes: 1 µl, 200 µl and 1000 µl
- Osmometer to check semen extender osmolality/osmolarity
- 1.0 ml cryovials
- Basic laboratory glass material
- Liquid nitrogen vessel and liquid nitrogen
- Programmable freezer
- Water bath
- Tissue to dry the vials

2.3 Materials to fertilize eggs

- Carp pituitary extract (cPE; 4 mg/kg) to induce female ovulation and facilitate stripping of eggs
- Scale to weigh eggs
- Incubation system with 28-30°C water

2.4 Fish handling and semen collection

- Anesthetise catfish male in 8 g of TMS dissolved in 10 l of tap water for 30 min, then sacrifice it by spinal transaction.
- Place the male in a support designed to hold the ventral part upwards to facilitate testes removal. Open the coelomic cavity with scalpel and scissors and remove both testes. Dry testes carefully with tissue, eliminating any blood or water contamination. Perforate each testis using a needle and avoid perforating the blood vessels as blood contamination decreases semen quality. Squeeze intratesticular semen in 10 ml test tubes and store in an ice bath until freezing.

2.5 Pre-freezing sperm evaluation

- Evaluate sperm motility under microscopy 200 x magnification. First, observe 5 µl of semen in a microscope slide. Sperm cells should be immotile, as fish spermatozoa are immotile in seminal plasma. Any sperm movement suggests water or blood contamination and the sample should be discarded. Activate the immotile samples by adding 15 µl of tap water and estimate the percentage of moving cells. Only samples with at least 80% of moving cells should be used for cryopreservation.

- Determine sperm concentration in a haemocytometer chamber. The number of sperm cells per ml will be used to calculate the optimal sperm: egg ratio during artificial fertilization.

2.6 Freezing and thawing procedures

- Prepare semen extender (Ginsburg fish Ringer) at least 24 h in advance. Then adjust pH to 7.6 and check osmolality/osmolarity which should be around 244 mOsm.
- Dilute one part of methanol in eight parts of Ginsburg fish Ringer, add one part of semen and mix gently. The final ratio is 1 semen: 8 extender:1 methanol. Equilibration time is not needed.
- Use a pipette to transfer 0.5 ml of diluted semen into 1.0 ml cryovials. Close the cryovials immediately and place them inside the programmable freezer. The freezer has the capacity to freeze 33-36 cryovials simultaneously.
- Programme the freezer as follows: start chamber temperature +5°C; rate 1: -5°C/min; end chamber temperature: -40°C; rate 2:0°C/min (hold) for 5 min; end chamber temperature: -40°C; end of the programme. This programme freezes semen at a rate of -5°C/min from +5 to -40°C, and holds semen at -40°C for 5 min. The holding period allows sperm cells to dehydrate enough and maintain high post-thaw quality. At the end of the programme, transfer all cryovials to the nitrogen vessel immediately.
- Thaw only 2 or 3 cryovials at a time. Remove the cryovials from the nitrogen vessel, and plunge into a 27°C-water bath. The cryovial cap should not be in contact with water as the cap tends to loosen and allows water contamination which activates sperm motility. To avoid this, hang the vials on a polystyrene surface (5 x 10 x 1 cm height) with the caps up and place in the water bath. After 3 to 5 min, remove the vials from the polystyrene surface and dry with tissue.
- Post-thaw semen should be immediately rediluted so that the cryoprotectant would become less toxic to sperm cells. The dilution used here will depend on the initial sperm concentration. Catfish sperm concentration varies from 1.8 to 21 x 10^9 sperm cells/ml of pure semen. Thus each 0.5 ml cryovial contains from 9 to 105 x 10^7 sperm cells. As 8.5 x 10^6 of post-thaw sperm is enough to fertilize 1 g eggs (*ca* 750 eggs), the contents of one vial should be enough to fertilize from 10 to 123 g of eggs. One simple and practical way to obtain this ratio and still be on the safe side, is diluting 0.5 ml of post-thaw semen in 10-20 ml of Ginsburg fish Ringer and fertilizing 20-40 g eggs.

2.7 Post-thaw sperm evaluation

- To assess post-thaw sperm quality, motility can be estimated using the same subjective procedure described for fresh semen. To quantify sperm motility, the use of a computer-assisted sperm analysis (CASA) system provides objective methods of assessment, using at least a dozen computer-calculated motility characteristics. For further details, refer to Rurangwa et al. [5].

- To assess sperm viability, dilute 5 μl of post-thaw sperm in 10 μl of eosin-nigrosin or 0.4% trypan blue in a microscope slide. Stained (dead) and unstained (live) spermatozoa are counted and converted to percentage live spermatozoa.

2.8 Fertilization

- Even when the most standardized protocol is used, sperm cells lose quality after freezing and thawing processes. Sperm quality continues to decrease after thawing, so post-thaw sperm should be used for fertilization as soon as possible.
- Inject female catfish with cPE 4 mg per kg BW to induce ovulation. At 23.5°C, hand stripping of eggs is possible after 15 h. Females should be anaesthetized in TMS 4 mg dissolved in 10 l water before stripping. Eggs should be collected in a dry plastic beaker (avoid any contact with water).
- Spread post-thaw semen of one vial (already rediluted in 10-20 ml of extender) over 20-40 g of eggs and mix for 5 s. Activate sperm motility with 50-70 ml of tap water, and gently mix for 80 s. Incubate the fertilized eggs at 28-30°C. Larvae will hatch in about 24 h, but should remain inside the incubators until first feeding (3 days after hatching).

3. GENERAL CONSIDERATIONS

The importance of preserving genetic resources for the future is widely recognized, and the conservation of semen would be a major contribution with great potential application in agriculture, biotechnology, species conservation and clinical medicine. As males need to be sacrificed for artificial reproduction in catfish culture, semen cryopreservation offers an indispensable tool for the preservation of genetic variation to be used in breeding programs (back-up and control for genetic progress). Conservation of genetic variability of wild catfish populations is also of importance, as domestication from a too narrow genetic basis could lead to the formation of genetically altered strains with reduced genetic variability in a very short period of time. Van der Walt et al. [6] reported that the use of limited numbers of brood stock without adding genes from unrelated fish to successive generations is responsible for the occurrence of genetic drift and inbreeding in most domesticated African catfish populations.

4. ACKNOWLEDGMENTS

We thank Dr. Henri Woelders for his constructive remarks during the development of this protocol.

5. REFERENCES

1-Steyn, G.J., van Vuren, J.H.J., Schoonbee, and H.J., Chao, N, Preliminary investigations on the cryopreservation of *Clarias gariepinus* (Clariidae: pisces) sperm, *Water SA,* 11, 15, 1985.

2-Viveiros, A.T.M., Semen cryopreservation in catfish species, with particular emphasis on the African catfish, *Journal of Animal Breeding Abstracts,* 73, 1N-9N, 2005.

3-Viveiros, A.T.M., So, N., and Komen, J., Sperm cryopreservation of African catfish, *Clarias gariepinus*: cryoprotectants, freezing rates and sperm:egg dilution ratio, *Theriogenology,* 54, 1395, 2000.

4-Viveiros, A.T.M., Lock, E.-J., Woelders, H., and Komen, J, Influence of cooling rates and plunging temperatures in an interrupted slow freezing procedure for semen of the African catfish, *Clarias gariepinus*, *Cryobiology,* 43, 276, 2001.

5-Rurangwa, E., Volckaert, F.A.M., Huyskens, G., Kime, D.E., and Ollevier, F., Quality control of refrigerated and cryopreserved semen using computer-assisted sperm analysis (CASA), viable staining and standardized fertilization in African catfish (*Clarias gariepinus*), *Theriogenology,* 55, 751, 2001.

6-van der Walt, L.D., van der Bank, F.H., and Steyn, G., The suitability of using cryopreservation of spermatozoa for the conservation of genetic diversity in African catfish (*Clarias gariepinus*), *Comparative Biochemistry and Physiology A,* 106, 313, 1993.

Complete affiliation:

Ana T M Viveiros, Departamento de Zootecnia, Universidade Federal de Lavras, caixa postal 3037, 37200-000, Larvas, MG, Brazil. Phone: +55 35 38291223 Fax: +55 35 38291231, e-mail: ana.viveiros@ufla.br.

CRYOPRESERVATION OF SPERM FROM SPECIES OF THE ORDER ACIPENSERIFORMES

Á. Horváth, B. Urbányi and S.D. Mims

1. INTRODUCTION

Sturgeons and paddlefishes of the order Acipenseriformes are an ancient group of fishes that exclusively inhabit the Northern Hemisphere. Several species are used in aquaculture programs and farmed for their roe sold as caviar and their high quality boneless meat. However, some species are critically threatened or endangered for reasons including destruction of habitat and spawning grounds, overexploitation of stocks and uncontrolled poaching and require restoration efforts. The acipenseriform species that have been studied share a number of characteristics especially concerning their spermatozoa and composition of seminal fluid. Distinctly different from the general morphology of teleost spermatozoa, sturgeon and paddlefish spermatozoa have long cylindrical heads topped with functioning acrosomes, well-defined midpieces with several mitochondria and flagella with fin-like extensions on both sides of the posterior end of the tails [1,2]. Osmolality/osmolarity of sturgeon and paddlefish seminal fluid is 60-100 mOsm/kg which is significantly lower than that of teleost fish (260-300 mOsm/kg) [3]. Lower osmotic shock occurs when released into freshwater and is probably the reason why these spermatozoa swim for a longer period of time (4-6 minutes) than spermatozoa of other freshwater teleosts (less than a minute). Concentration of acipenseriform sperm, typically up to 1-1.5 billion spermatozoa per ml, is also much lower than that of teleosts [2,3,4].

2. PROTOCOL FOR FREEZING AND THAWING

2.1 Fish Handling and Semen Collection

Methods of milt collection from sturgeons and paddlefishes depend on the species and on the technique practiced at a particular farm. This group of fish is generally docile and they do not have to be anaesthetized before milt extraction, though some culturists anaesthetize smaller species to provide better milt collection.

- Spermiation can be induced using an injection of carp, sturgeon or paddlefish pituitary extract (1-6 mg/kg body weight) or different GnRH/LHRH analogues (10-100 µg/kg body weight) administered 24 hours prior to the planned stripping [4].
- Milt is typically collected into dry 15-30 ml syringes with attached elastic tubes (diameter 5 mm, length 5-7 cm) which are inserted into the urogenital pore of the

fish. The volume of milt that can be collected from a fish depends on the species and on the size of the individual, thus smaller species like pallid sturgeon (*Scaphirhynchus albus*) or sterlet (*Acipenser ruthenus*) can give approximately 10-20 ml of milt, while as much as 500-800 ml can be stripped from a beluga (*Huso huso*). Individuals of smaller species need to be handled with care as they tend to release their milt when removed from the water potentially causing a complete loss of milt. Milt does not need to be diluted upon collection as its sperm concentration is typically low.

- Milt can be stored in different containers (i.e., cell culture flasks, 50 ml centri-fuge tubes, zip-lock bags under oxygen) in a refrigerator at 4°C for at least 24 hours without a significant loss in sperm motility ability as well as in its usefulness in cryopreservation.

2.2 Freezing and Thawing Procedures

- Sperm is diluted in a 1:1 ratio with the freezing diluents containing both the basic extender and the cryoprotectant. The extender that has been used successfully for the cryopreservation of sturgeon and paddlefish sperm is called modified Tsvetkova's (mT) extender. Its composition is 23.4 mM sucrose, 0.25 mM KCl, 30 mM Tris and an adjusted pH of 8.0 using hydrochloric acid. Recent experiments on paddlefish sperm indicate that, in some cases, the addition of KCl up to 2.5 mM may be necessary to prevent activation of spermatozoa. The osmolality/osmolarity of this extender is 73 ± 2 mOsm/kg. The cryoprotectant of choice for acipenseriform sperm is methanol with a concentration of 5 or 10% (volume/volume, final concentration). Thus, for one 0.5 ml French straw 250 μl of sperm is added to 225 μl of extender and 25 μl of methanol for 5% methanol or 200 μl of extender and 50 μl of methanol for 10% of cryoprotectant. Recent results indicate that sperm cryopreserved with 5% methanol may result in a higher fertilization rate than that with 10% methanol. The protocol of dilution with extender and cryoprotectant for storing in large straws is similar to that stored in 0.5 ml French straws.

- The mixture is loaded into 0.5 ml French straws (larger straws must have one end closed either by heat sealing or by inserting a cotton plug) and frozen in the vapor of liquid nitrogen. For this cryo-procedure, a styrofoam box is used and filled with liquid nitrogen. A 3 cm high styrofoam frame floats on the surface of the nitrogen. Straws are distributed on this frame and left to freeze for 3 minutes (5 minutes for larger straws), then plunged directly into liquid nitrogen. This results in a freezing rate of approximately 70°C/min (Fig.1). Alternatively, 5-ml straws can be used for the cryopreservation of a larger volume of sturgeon and paddlefish sperm for commercial applications.

- Following storage in liquid nitrogen, straws are thawed in a 40°C water bath for 13 seconds (30 seconds for larger straws).

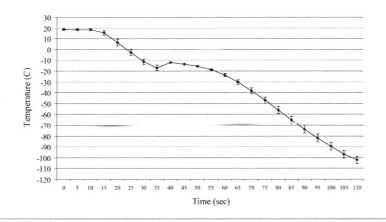

Figure 1 - Freezing curve (mean ± SD, n − 2) with modified Tsvetkova's (mT) extender and 10% methanol measured in 0.5 ml French straws with a thermocouple connected to a datalogger (OM-550, Omega Engineering, Stamford, Connecticut). [5].

2.3 Fertilization

- A semi-dry fertilization method is used for sturgeon and paddlefish sperm.
- Milt is added in a ratio of up to 1:200 (milt:water) to hatchery water and then poured onto the eggs.
- Egg batches of up to 5 g have been fertilized with one 0.5 ml straw of thawed mixture (0.25 ml milt), while 40 g of eggs were fertilized with one 5-ml straw of paddlefish mixture (2.5 ml milt).

3. GENERAL CONSIDERATIONS

Sturgeon and paddlefish sperm cryopreservation was problematic for a long time as the described protocols often resulted in high motility but low fertilization rates. In the literature (review by Mims et al. [2]) dimethyl-sulfoxide (DMSO) has almost exclusively been used as a cryoprotectant. However, we found that methanol resulted in a significantly higher fertilization and hatching rate than DMSO (Table 1). DMSO as cryoprotectant drastically increases the osmolality/osmolarity of the extender while the addition of methanol causes a slight decrease in osmolality/osmolarity (Table 2). When sperm was frozen in a hyperosmotic extender (such as the original Tsvetkova's extender described by Tsvetkova et al. [6]) equally poor fertilization rates were observed with methanol and DMSO. The use of hyperosmotic extenders regardless of the cryoprotectant resulted in lower fertilization and hatching rates compared to isoosmotic extenders (Table 3). Thus, we found that the osmolality/osmolarity of the freezing diluent rather than the type of cryoprotectant defines fertilization success with cryopreserved sturgeon sperm. We found it interesting that when measuring osmolality/osmolarity of a diluent with methanol, correct readings were observed only when a vapor pressure osmometer was used rather than freezing point osmometers.

The above described method has successfully been tested on several species including the sterlet, Siberian sturgeon (*A. baeri*), Russian sturgeon (*A. gueldenstaedti*), European sturgeon (*A. sturio*) [7-9] and beluga as well as North American species such as the shortnose sturgeon (*A. brevirostrum*), pallid sturgeon [5] and paddlefish.

Table 1 - Post-thaw motility, fertilization and hatching rates (mean ± SD) of cryopreserved and fresh shortnose sturgeon sperm (n = 6), DMSO: dimethyl sulfoxide; MeOH: methanol; Control: Fresh sperm. Sperm volume: the volume of sperm used to fertilize 2 ml of eggs. Values sharing a superscript letter within a column were not significantly different (*P* > 0.05) [5].

CPA	CPA concentration (%)	Sperm volume (μl)	Motility (%)	Fertilization (4-cell stage %)	Neurulation rate (%)	Hatching rate (%)
DMSO	5	125	26 ± 13^b	0 ± 0^d	1 ± 1^d	1 ± 1^d
		250		1 ± 2^d	1 ± 1^d	1 ± 1^d
	10	125	17 ± 6^{cb}	2 ± 2^d	0 ± 1^d	0 ± 0^d
		250		1 ± 1^d	1 ± 1^d	0 ± 1^d
	15	125	2 ± 3^d	0 ± 0^d	0 ± 0^d	0 ± 0^d
		250		0 ± 0^d	0 ± 0^d	0 ± 0^d
MeOH	5	125	16 ± 7^{cb}	40 ± 15^b	35 ± 17^b	31 ± 15^b
		250		39 ± 11^b	38 ± 13^b	32 ± 12^b
	10	125	13 ± 8^c	19 ± 13^c	22 ± 17^c	18 ± 16^c
		250		21 ± 12^c	21 ± 11^c	16 ± 8^c
	15	125	10 ± 5^{cd}	3 ± 5^d	2 ± 4^d	2 ± 2^d
		250		8 ± 9^d	7 ± 6^d	5 ± 4^d
Control	--	125	77 ± 8^a	78 ± 26^a	69 ± 24^a	51 ± 22^a
		250		79 ± 23^a	71 ± 20^a	53 ± 21^a

Table 2 - Osmolality/osmolarity of Modified Tsvetkova's and Original Tsvetkova's extenders (mean ± SD) in combination with DMSO or McOH as cryoprotectants in different concentrations (n = 3). DMSO: dimethyl sulfoxide; McOH: methanol, OT: Original Tsvetkova's extender; MT: modified Tsvetkova's extender. Values sharing a superscript letter within a column were not significantly different (P > 0.05). Osmolality/osmolarity of OT extender in combination with 15% DMSO is not shown as it was higher than the measurable range of the used osmometer [5].

Extender	Cryoprotectant	Concentration (%)	Osmolality (mOsmol/kg)
MT	None		73 ± 2[g]
MT	DMSO	5	719 ± 9[e]
MT	DMSO	10	1421 ± 26[c]
MT		15	1985 ± 12[a]
MT	MeOH	5	75 ± 1[g]
MT	MeOH	10	78 ± 15[g]
MT		15	84 ± 14[g]
OT	None		204 ± 8[f]
OT	DMSO	5	840 ± 15[d]
OT	DMSO	10	1481 ± 24[b]
OT		15	>2000
OT	MeOH	5	196 ± 3[f]
OT	MeOH	10	187 ± 3[f]
OT		15	204 ± 8[f]

Table 3 - Post-thaw motility, fertilization and hatching rates (mean ± SD) of cryopreserved and fresh shortnose sturgeon sperm (n = 6). OT: Original Tsvetkova's extender; MT: modified Tsvetkova's extender; mHBSS: modified Hanks' balanced salt solution; McOH: methanol; Control: Fresh sperm. osmolality/osmolarity of mHBSS extender was set to 100 mOsm/kg. Values sharing a superscript letter within a column were not significantly different (P > 0.05).

Extender	Concentration of MeOH (%)	Motility (%)	Fertilization (4-cell stage %)	Hatching rate (%)
OT	5	12 ± 6[bc]	3 ± 7[cd]	2 ± 2[cd]
OT	10	11 ± 2b[cd]	2 ± 3[cd]	3 ± 4[cd]
OT	15	8 ± 3[cd]	0 ± 1[d]	1 ± 1[d]
MT	5	18 ± 10[b]	18 ± 11[b]	17 ± 12[b]
MT	10	8 ± 5[cd]	7 ± 11[cd]	3 ± 4[cd]
MT	15	6 ± 4[cd]	2 ± 2[cd]	1 ± 2[d]
mHBSS	5	11 ± 7[bcd]	12 ± 10[bc]	11 ± 12[bc]
mHBSS	10	8 ± 7[cd]	4 ± 5[cd]	5 ± 6[cd]
mHBSS	15	3 ± 2[d]	1 ± 1[cd]	0 ± 0[d]
Control	--	73 ± 16[a]	40 ± 22[a]	27 ± 13[a]

4. ACKNOWLEDGMENTS

This publication was supported by the Bolyai János Research Scholarship of the Hungarian Academy of Sciences and the post doctoral fellowship of the Hungarian

Ministry of Education number 28345/2003 and USDA 1890 Teaching and Research Capacity Building Grant #KYX-01-11469.

5. REFERENCES

1-Dettlaff, T.A., Ginsburg, A.S., and Schmalhausen, O.I., Sturgeon Fishes, *Developmental Biology and Aquaculture*, Springer-Verlag, Berlin, Heidelberg, 1993.

2-Mims, S.D., Gomelsky, B., Brown, G.G., and Tsvetkova, L.I., Cryopreservation of paddlefish and sturgeon milt, in: *Cryopreservation in Aquatic Species*, Tiersch, T.R. and Mazik, P.M., Eds., World Aquaculture Society, Baton Rouge, LA, USA, 2000.

3-Piros, B., Glogowski, J., Kolman, R., Rzemieniecki, A., Domagala, J., Horváth, Á., Urbanyi, B., and Ciereszko, A., Biochemical characterization of Siberian sturgeon *Acipenser baeri* and sterlet *Acipenser ruthenus* milt plasma and spermatozoa, *Fish Physiol Biochem,* 26, 3, 289, 2002.

4-Linhart, O., Mims, S.D., Gomelsky, B., Hiot, A.E., Shelton, W.L., Cosson, J., Rodina, M., and Gela, D., Spermiation of paddlefish *Polyodon spathula* stimulated with injection of LHRH analogue and carp pituitary extract, *Aquat Living Resour,* 13, 1, 2000.

5-Horváth, Á., Wayman, W.R., Urbányi, B., Ware, K.M., Dean, J.C., and Tiersch, T.R., The relationship of the cryoprotectants methanol and dimethyl sulfoxide and hyperosmotic extenders on sperm cryopreservation of two North American sturgeon species, *Aquaculture*, 247, 1-4, 243, 2005.

6-Tsvetkova, L.I., Cosson, J., Linhart, O., and Billard, R., Motility and fertilizing capacity of fresh and frozen-thawed spermatozoa in sturgeons *Acipenser baeri* and *A. ruthenus, J. App. Ichthyol.,* 12, 107, 1996.

7-Horváth, Á. and Urbányi, B., Cryopreservation of sterlet (*Acipenser ruthenus*) sperm, 6th Proc. *International Symposium on Reproductive Physiology of Fish*, Bergen, Norvégia, 1999, július 4-9, Norberg, B., Kjesbu, O.S., Taranger, G.L., Andersson, E., and Stefansson, S.O., Eds., Bergen, 2000, 441.

8-Glogowski, J., Kolman, R., Szczepkowski, M., Horváth, Á., Urbányi, B., Sieczyñski, P., Rzemieniecki, A., Domaga³a, J., Demianowicz, W., Kowalski, R., and Ciereszko, A., 2002. Fertilization rate of Siberian sturgeon (*Acipenser baeri*, Brandt) milt cryopreserved with methanol, *Aquaculture*, 211, 367, 2002.

9-Urbányi, B., Horváth, Á., and Kovács, B., Successful hybridization of Acipenserid species using cryopreserved sperm, *Aquacult Int,* 12, 47, 2004.

Complete affiliation:

Ákos Horváth Ph.D., Department of Fish Culture, Szent István University, Páter Károly u. 1. Gödöllõ H-2103 Hungary; Phone: +36-28-522-000 ext. 1659; Fax: +36-28-410-804, e-mail: Horvath.Akos@mkk.szie.hu

DIFFERENT PROTOCOLS FOR THE CRYOPRESERVATION
OF EUROPEAN EEL (*Anguilla anguilla*) SPERM

J.F. Asturiano

1. INTRODUCTION

During the last decades, the capture and over-exploitation of eels and elvers has resulted in a decrease in their populations. This has become an ecological and economical problem, making it necessary to develop techniques for the control of reproduction in captivity. Methods for the hormonal induction of gonad maturation in this species have been developed in previous studies, obtaining significant sperm volumes of good quality [1-3] as well as ovarian maturation, spawns, egg fertilisation and even hatching (reviewed by Pedersen [4,5]). Also, hybrids between European eel and Japanese eel (*Anguilla japonica*), using the sperm of European species and Japanese eel oocytes, have recently been achieved [6].

However, methods for the hormonal induction of gonad maturation in this species usually take several weeks [2,3,7,8] and unsynchronized maturation can occur, preventing egg fertilization. In an attempt to improve the unresolved control on the reproduction of this species, sperm cryopreservation media and methods have recently been developed [8-10], based on previous studies in Japanese species ([11], Ohta et al., personal communication).

Hopefully, sperm cryopreservation techniques will be useful for higher flexibility in broodstock management, further genetic improvement programs and the preservation of genetic diversity. However, little research into the effects of cryoprotectants, freezing media and cryopreservation methods on sperm viability has been carried out, and even sperm evaluation techniques are poorly developed.

There is no standard method for the cryopreservation of eel sperm. The present work summarizes our own results in comparison with those obtained by other groups in European and Japanese eels, as well as some ideas to be considered in further experiments to optimize these methods.

2. PROTOCOL FOR FREEZING AND THAWING

2.1 Fish, hormonal treatment and sperm sampling

- Farmed male eels (120-150 g) were gradually acclimatized from freshwater to sea water (salinity: 38%; temperature: 20-21°C) in one week and maintained in 96 l aquaria.
- Once a week, fish were hormonally treated with intraperitoneal injections of hCG (1.5 IU/g BW) for the induction of gonad maturation as described by

Pérez ct al. [2]. Spermiation began in the 7th week of treatment and reached maximum quality (or motility) between the 9 and 11th weeks.

- Sperm was sampled 24 h after hormone administration when the best quality was observed [2]. After cleaning the genital area with fresh water and thorough drying to avoid contamination of samples by faeces, urine and sea water, total expressible milt was collected by applying gentle abdominal pressure to anesthetized males (benzocaine; 60 mg/l). Samples were collected in graduated tubes using a vacuum pump with a retention trap and stored at 4°C.

2.2 Sperm Analysis

Sperm motility was evaluated before cryopreservation and samples showing more than 50% motile cells were selected. Fresh sperm quality was evaluated in triplicate diluting samples 1:1,000 in artificial sea water containing, in mM: 52.4 $MgCl_2$ x $6H_2O$, 28.2 Na_2SO_4, 9.9 $CaCl_2$ x $2H_2O$, 9.4 KCl, and 354.7 NaCl with osmolarity adjusted to 1000 mOsm/kg as an activator solution. Activation must occur very rapidly, in less than 10 seconds, as motility usually drops in a few seconds. Some of the sperm characteristics of fresh samples are displayed in Tables 1-4.

Table 1 - Fresh sperm characteristics [2,13]

Produced volume/sampling	3-4 ml/100 g fish
Concentration	$3-6 \times 10^9$ spermatozoa/ml
Osmolarity (seminal plasma)	325-330 mOsm/kg
pH (seminal plasma)	8.5

2.3 Freezing Media

For the moment, we used four different solutions as the basis for the freezing media: 1) Tanaka´s modified medium [11]: (in mM) 137 NaCl, 76.2 $NaHCO_3$, 20 TAPS, pH 8.2; 2) Ohta´s K30 modified medium [12]: 134.5NaCl, 20 $NaHCO_3$, 30 KCl, 1.6 $MgCl_2$, 1.3 $CaCl_2$, pH 8.1; and two isoionic media with the European eel seminal plasma [13]: 3) P1 medium: 125 NaCl, 20 $NaHCO_3$, 30 KCl, 2.5 $MgCl_2$, 1 $CaCl_2$, pH 8.5, and 4) P2 medium: 70 NaCl, 75 $NaHCO_3$, 30 KCl, 2.5 $MgCl_2$, 1 $CaCl_2$, pH 8.5. All media were supplemented with 10% v/v dimethyl sulphoxide (DMSO) as a cryoprotectant. The addition of 1.4 g L-a-phosphatidylcholine (LEC)/100 ml of freezing media, which showed a clear positive effect in Japanese eel [11], was assayed. Different dilution factors (1:5, 1:20, 1:100) of fresh sperm in the freezing media were evaluated as well [8].

Table 2 - Fresh sperm motility parameters measured by computer assisted sperm analyzer (CASA; *Sperm Class Analyzer*® Microptic, Barcelona, Spain) in hCG-treated fish; n = 21 (Asturiano et al., unpublished results)

Motile spermatozoa (%)	51.96 ± 2.44
Curvilinear velocity (VCL)	78.60 ± 6.30 mm/s
Straight line velocity (VSL)	18.99 ± 1.00 mm/s
Angular path velocity (VAP)	33.92 ± 1.41 mm/s
Beating cross frequency (BCF)	14.20 ± 0.86 Hz

Table 3 - Spermatozoa morphological parameters in fresh sperm: head length, width and perimeter (in i m), and head area (in i m²), evaluated by automated sperm morphology analysis (ASMA; *Sperm Class Analyzer*®, *Morfo Version 1.1*, Microptic, Barcelona, Spain)[17]

	Length	Width	Perimeter	Area
Count (n)	14 898	14 898	14 898	14 898
Average	4.23	1.15	14.55	5.31
Median	4.28	1.12	14.39	5.28
Lower quartile	3.92	1.03	13.00	4.86
Upper quartile	4.59	1.24	15.83	5.73

Table 4 -Ionic composition of European eel seminal plasma in relation to sperm motility categories, where 0 represents no motile sperm, whereas I<25%, II: 25-50%, III: 50-75%, IV: 75-90% and V: 90-100% of the population were vigorously motile. Different letters show significant differences [13]

[mM]	0	I	II	>III
Ca^{2+}	1.00 ± 0.15 c	0.57 ± 0.06 bc	0.41 ± 0.05 ab	0.20 ± 0.02 a
Mg^{2+}	5.74 ± 1.15 b	4.09 ± 0.59 ab	2.37 ± 0.31 ab	1.54 ± 0.28 a
K^+	27.45 ± 2.39 b	34.90 ± 1.83 ab	37.53 ± 2.95 b	36.16 ± 2.94 ab
Na^+	118.18 ± 6.14	116.29 ± 3.28	113.14 ± 4.15	109.34 ± 10.30

2.4 Freezing and Thawing Methods

Freezing was carried out in liquid nitrogen vapour for 10 min, followed by the immersion of 0.25 ml straws in liquid nitrogen for the final freezing. Samples were thawed by immersion in a water bath at 20°C during 45 s. Sperm motility was evaluated after cryopreservation, at a final dilution of 1:1000 in artificial sea water.

3. GENERAL CONSIDERATIONS

The addition of LEC to the freezing media showed a clear positive effect. The highest percentage of post-freezing surviving spermatozoa was obtained with Tanaka's modified medium and the P1 medium plus DMSO and phosphatidylcholino, with 1:5 as the dilution factor (Table 5).

These results suggested DMSO as a useful cryoprotectant for sperm freezing in this species. However, DMSO increases medium osmolarity activating spermatozoa movement, scarcely reducing the time pre-freezing and post-thawing to manipulate the spermatozoa before they become inactive [10]. Trying to avoid spermatozoa activation due to increased osmolarity, first experiments with alternative cryoprotectants are being developed in our laboratory. So far, methanol seems be a useful choice because it has little effect on spermatozoa activation ($1.0 \pm 4.2\%$ with methanol *vs* $50.0 \pm 4.2\%$ with DMSO; Asturiano et al., unpublished results) and the previous results obtained by Ohta in the Japanese eel, who reported 60% post-freezing surviving cells using 15% methanol in the freezing media, as well as the results reported by Muller et al. [9] and Szabó et al. [10], approximately 40% of surviving cells using 10% methanol (Table 5).

As far as we know, dilution factors as low as 1:1 or 1:2 (sperm: freezing medium) give the best post-thawing survival, and allow the maintenance of high concentrated sperm samples in the 250 µl straws. On the other hand, sodium bicarbonate concentrations as high as 75 mM seem to be a critical point due to its role controlling the initiation of spermatozoa motility, as has been demonstrated in the Japanese eel [11,14,15].

The sperm of European species has a density as high as $3\text{-}6 \times 10^9$ spermatozoa ml^{-2}. Moreover, the time of sperm motility is very short (a few seconds), making its manipulation and quality assessment difficult. Specific extenders should be developed for the European species, because trials carried out with extenders developed for Japanese eel [16] showed bad results (Asturiano et al., unpublished results).

4 ACKNOWLEDGMENTS

Supported by the Generalitat Valenciana (GV04A-508), the Universidad Politécnica de Valencia (20030488) and the Spanish Ministry of Science and Technology (AGL2003-05362-C02-01, including FEDER funds). J.F. Asturiano is supported by a Ramón y Cajal research contract, cofinanced by the latter two organizations.

Table 5 - Post-thawed percentage of motile spermatozoa comparing protocols for European and Japanese eels.

European eel

Asturiano et al.,[8].

Medium*	LEC**	Sperm dilution		
(see the text)		1:5		1:100
Tanaka´s modified	-	20.8 ± 12.3	0	16.6 ± 11.1
	+	35.5 ± 14.6	27.5 ± 13.9	28.9 ± 17.7
Ohta´s K30 modified	-	4.5 ± 4.5	0	4.8 ± 4.8
	+	15.2 ± 12.1	19.4 ± 11.3	4.7 ± 4.7
P1 medium	-	14.9 ± 8.7	8.9 ± 8.9	12.9 ± 8.5
	+	36.6 ± 6.8	31.2 ± 18.0	28.4 ± 16.5
P2 medium	-	0	0	0
	+	0	0	0

* Supplemented with 10% v/v DMSO; ** +/- 1.4% LEC in the freezing media.

Muller et al.,[9].

Medium	Cryoprotectant (10% v/v)	Other conditions	Motile cells
Modified Kurokura solution [18]	Methanol	Sperm dilution 1:9 250 μl straws	36 ± 11

Szabó et al., [10].

Medium	Cryoprotectant (10% v/v)	Other conditions	Motile cells
350 mM glucose, 30 mM Tris pH 8.0	DMSO Methanol	Sperm dilution 1:9 250 μl straws	3 ± 6 3 ± 3
Modified Kurokura solution [18]	DMSO Methanol	,,	8 ± 3 12 ± 8
Tanaka medium [11]	DMSO Methanol	,,	47 ± 15 40 ± 10

Japanese eel

Tanaka et al.,[11].

Medium	Cryoprotectant	Other conditions	Motile cells
D#5-D (Tanaka medium) NaHCO₃ 76.2, NaCl 137, TAPS 20 pH 8.2, plus 1.4% LEC	DMSO (10% v/v)	Sperm dilution 1:4 2 ml cryogenic vials	Approx. 40

5. REFERENCES

1-Amin, E.M., Observations on reproduction techniques applicable to the European eel (*Anguilla anguilla* L.), in: *Genetics and breeding of Mediterranean aquaculture species,* Bartley, D. and Basurco, B., Eds., 34, 223, 1997. CIHEAM-IAMZ.

2-Pérez, L., Asturiano, J.F., Tomás, A., Zegrari, S., Barrera, R., Espinós, J.F., Navarro, J.C., and Jover, M., Induction of maturation and spermiation in the male European eel: assessment of sperm quality throughout treatment, *J Fish Biol,* 57, 1488, 2000.

3-Asturiano, J.F., Pérez, L., Pérez-Navarro, F.J., Garzón, D.L., Martínez, S., Tomás, A., Marco-Jiménez, F., Vicente, J.S., and Jover, M., Optimization trial of methods for induction of spermiation in European eel (*Anguilla anguilla*) and test of different sperm activation media, in: *Aquaculture Europe '04 Biotechnologies for quality,* Barcelona ,(Spain), October 2004.

4-Pedersen, B.H., Induced sexual maturation of the European eel *Anguilla anguilla* and fertilisation of the eggs, *Aquaculture,* 224, 323, 2003.

5-Pedersen, B.H., Fertilisation of eggs, rate of embryonic development and hatching following induced maturation of the European eel *Anguilla anguilla, Aquaculture,* 237, 461, 2004.

6-Okamura, A., Zhang, H., Utoh, T., Akazawa, A., Yamada, Y., Horie, N., Mikawa, N., Tanaka, S., and Oka, H.P, Artificial hybrid between *Anguilla anguilla* and *A. japonica, J Fish Biol,* 64, 1450, 2004.

7-Asturiano, J.F., Pérez, L., Tomás, A., Zegrari, S., Espinós, F.J., and Jover, M., Inducción hormonal de la maduración gonadal y puesta en hembras de anguila europea *Anguilla anguilla* L. 1758: cambios morfológicos y desarrollo oocitario, in: *Boletín Instituto Español de Oceanografía,* 18, 1-4, 127, 2002. (In Spanish with abstract in English).

8-Asturiano, J.F., Pérez, L., Marco-Jiménez, F., Olivares, L., Vicente, J.S., and Jover, M., Media and methods for the cryopreservation of European eel (*Anguilla anguilla*) sperm, *Fish Physiol Biochem,* 28, 501, 2003.

9-Müller, T., Urbányi, B., Váradi, B., Binder, T., Horn, P., Bercsényi, M., and Horváth, A., Cryopreservation of sperm of farmed European eel *Anguilla anguilla, J World Aquacult Soc,* 35, 240, 2004.

10-Szabó, G., Müller, T., Bercsényi, M., Urbányi, B., Kucska, B., and Horváth, A., Cryopreservation of European eel (*Anguilla anguilla*) sperm using different extenders and cryoprotectants, *Acta Biol Hung,* 56(1-2), 173, 2005.

11-Tanaka, S., Zhang, H., Horie, N., Yamada, Y., Okamura, A., Utoh, T., Mikawa, N., Oka, H.P., and Kurokura, H., Long-term cryopreservation of sperm of Japanese eel, *J Fish Biol,* 60, 139, 2002a.

12-Otha, H., Kagawa, H., Tanaka, H., and Unuma, T., Control by the environmental concentration of ions of the potential for motility in Japanese eel spermatozoa, *Aquaculture,* 198, 3-4, 339, 2001.

13-Pérez, L., Asturiano, J.F., Martínez, S., Tomás, A., Olivares, L., Mocé, E., Lavara, R., Vicente, J.S., and Jover, M., Ionic composition and physio-chemical parameters of the European eel (*Anguilla anguilla*) seminal plasma, *Fish Physiol Biochem,* 28, 221, 2003.

14-Tanaka, S., Zhang, H., Yamada, Y., Okamura, A., Horie, N., Utoh, T., Mikawa, N., Oka, H.P., and Kurokura, H., Inhibitory effect of sodium bicarbonate on the motility of sperm of Japanese eel, *J Fish Biol,* 60, 1134, 2002b.

15-Tanaka, S., Utoh, T., Yamada, Y., Horie, N., Okamura, A., Akazawa, A., Mikawa, N., Oka, H.P., and Kurokura, H., Role of sodium bicarbonate on the initiation of sperm motility in the Japanese eel, *Fisheries Sci,* 70, 780, 2004.

16-Ohta, H. and Isawa, T., Diluent for cool storage of the Japanese eel (*Anguilla japonica*) spermatozoa, *Aquaculture,* 142, 107, 1996.

17-Marco-Jiménez, F., Asturiano, J.F., Pérez, L., Garzón, D.L., Pérez-Navarro, F.J., Martínez, S., Tomás, A., Vicente, J.S., and Jover, M., Morphometric characterization of the European eel spermatozoa with computer assisted spermatozoa analysis, in: *5th International Symposium on Fish Endocrinology,* Castellón (Spain), 2004.

18-Magyary, I., Urbányi, B., and Horváth, L. Cryopreservation of common carp (*Cyprinus carpio* L.) sperm I. The importance of oxygen supply, *J Appl Ichthyol,* 12, 113, 1996.

Complete affiliation:

Juan F. Asturiano, Grupo de Investigación en Recursos Acuícolas. Departamento de Ciencia Animal. Universidad Politécnica de Valencia. Camino de Vera s/n 46022 Valencia (Spain). Phone: +34 96 387 70 07; ext. 74355; E-mail: jfastu@dca.upv.es

CRYOPRESERVATION OF STRIPED BASS
Morone saxatilis SPERMATOZOA

L. C. Woods III, S. He and K. Jenkins-Keeran

1. INTRODUCTION

Striped bass, of the genus *Morone*, were classified by Johnson [1] into their own family Moronidae, and this classification has been adopted by the American Fisheries Society [2]. Native to North America, they have been transported to numerous other countries including: France, Germany, Israel, Portugal, Russia, Taiwan [3], and most recently, China for aquaculture. Striped bass have an elongated body, silver in color and 7-8 narrow, black stripes running laterally along the sides of the body on alternate rows of scales. Striped bass are iteroparous, and once mature, spawn around vernal equinox annually. Males arrive to spawning areas before females and may remain on the spawning grounds for over a month compared to 7-10 days for females[4]. Testicular maturation is typical of most teleost species as described by Grier [5]. Seminal production, including proliferation and differentiation of spermatocytes as well as filling of testicular efferent ducts with semen has been characterized in wild [6] and domesticated stocks [7], as correlated to rising levels of the androgens testosterone and 11-ketotestosterone. The striped bass aquaculture industry utilizes the conspecific hybrid cross of white bass females and striped bass males [8]. Cryopreservation is of significant interest to the expansion of striped bass culture as it could help minimize spawning problems associated with temporal and latitudinal differences between strains as well as the two species used to make the hybrid cross [9].

2. PROTOCOL FOR FREEZING AND THAWING

2.1 Fish Handling and Semen Collection

- To collect fresh semen, male striped bass are collected during the peak of spawning season and anesthetized in a bath containing 70 mg l^{-1} of buffered MS-222.
- Urine is removed by applying gentle pressure around the urogenital vent.
- Semen was expressed directly into sterile 50 ml graduated, conical bottom tubes (Becton-Dickinson, Falcon®) and placed immediately on ice.

2.2 Freezing and Thawing Procedures

- Based on our lab's previous studies [10-13] the following protocol provides the highest percentage of striped bass spermatozoa post-thaw with both intact plasma membranes and functional mitochondria, and preserves the spermatozoa ultrastructure (Fig. 1).
- Fresh semen is diluted with our striped bass extender (Table 1) in a 1:1 ratio and placed back on ice for 15 minutes to allow the cells to initially acclimate to the extender's high osmolality.
- The extended semen is subsequently diluted 1:1 with the same extender containing 15% DMSO. This provides a final DMSO concentration of 7.5% and a final semen:extender dilution of 1:3.
- Aliquots (150 μl) of the cryoprotected semen mixture should be quickly pipetted into chilled 500 μl cryo-straws (TS Scientific) and sealed.
- Optimal equilibration time of the sperm cells with DMSO, prior to freezing, is 10 min [11]. Cryo-straws containing sperm samples are frozen using a programmable freezer (Planer-Kryosave, Model KS30) with a freezing rate of –40°C min^{-1} until reaching -120°C where straws are placed immediately into liquid nitrogen [14].
- Frozen straws are thawed in a 35°C water bath for 8 seconds [10].

Table 1 - Striped bass extender composition.

NaCl (mg)	KCl (mg)	NaHCO$_3$ (mg)	Glucose (mg)	Glycine (mM)	Water[a] (ml)	pH[b]	Osmolality (mmol/kg)
1400	40	200	100	75	100	7.6	500

[a] Deionized,Ultra-Filtered Water (Fisher Scientific/Catalog No.W2-20); [b] Adjusted with 1 N NaOH

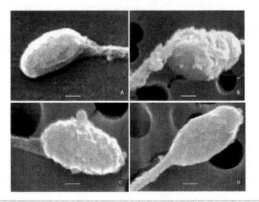

Figure 1 - Post-thaw ultrastructure of striped bass spermatozoa. Bar = 0.5 μm . A) fresh spermatozoa; B) 2.5% DMSO; C) 5% DMSO; D) 10% DMSO.

3. GENERAL CONSIDERATIONS

Our striped bass protocol has been shown in published trials to provide fertilization rates, for striped bass eggs, equivalent to 90% of fresh controls [15]. For fertilization, post-thawed semen is immediately placed directly onto eggs followed by the addition of deionized or fresh, hatchery water (1:5 semen:water). An effective sperm to egg ratio is approximately 4×10^6:1 or two straws (300 µl cryopreserved sperm) per 1,000 eggs. Our striped bass extender (Table 1) containing 7.5% DMSO with 75 mM glycine yielded a significantly higher ($P < 0.05$) percentage of fertilized eggs, at $54 \pm 5.6\%$ or 90% of the fresh semen control (Table 2). No difference ($P > 0.05$) in the egg fertilization percentage was observed between the control and this treatment. Glycine showed a significantly positive ($P < 0.05$) effect on fertilization. Glycine with 7.5% DMSO resulted in a higher ($P < 0.01$) fertilization rate than 7.5% DMSO alone (Figure 2) [15]. Similarly, striped bass spermatozoa cryopreserved under this protocol, stored under liquid nitrogen for one year and shipped in dry dewars to the world's largest commercial hybrid striped bass producer (Kent SeaTech), delivered fertilization rates of white bass eggs equal to that of controls (fresh semen) in multiple trials (Jason Stannard, Kent SeaTech, unpublished data).

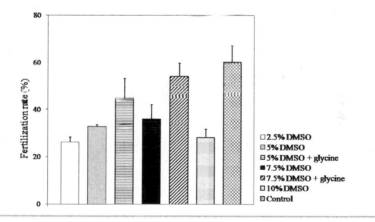

Figure 2 - Fertilization percentage (mean ± SEM) achieved with 300 µl of cryopreserved striped bass sperm with approximately 1000 striped bass eggs. Fresh striped bass semen (1 ml) was used as the control (Woods and He, unpublished data).

Table 2 - Post thaw sperm percentages[a]: motile sperm and fertilization rate

Treatment	Motile Sperm	Fertilization Rate
Fresh Semen	$91 \pm 2.4\%$	$60 \pm 6.9\%$
7.5% DMSO	$7 \pm 1.3\%$	$36 \pm 5.9\%$
7.5% DMSO + 75mM Glycine	$45 \pm 2.0\%$	$54 \pm 5.6\%$

[a]Data taken from He and Woods[15]

4. REFERENCES

1-Johnson, G.D., Percoidei: development and relationships, in: *Ontogeny and systematics of fishes*, Moser, H.G., et al., Eds., American Society of Ichthyologists and Herpetologists, Special Publication No. 1, Miami, Florida, 1984, 464.

2-Kohler, C.C., Striped bass and hybrid striped bass culture, in: *Encyclopedia of aquaculture*, Stickney, R.R., Ed., John Wiley & Sons, New York, 2000, 898.

3-Harrell, R.M. and Webster, D.W., An overview of *Morone* culture, in: *Striped bass and other Morone culture*, Harell, R.M., Ed., Elsevier Science, Amsterdam, 1997, 1.

4-Hocutt, C.H., Seibold, S.E., Harrell, R.M., Jesien, R.V., and Bason, W.H., Behavioral observations of the striped bass on the spawning grounds of the Choptank and Nanticoke Rivers of Maryland, USA, *J Appl Ichthyol*, 6, 21, 1990.

5-Grier, H.J., Cellular organization of the testes and spermatogenesis in fishes, *American Zoologist*, 21, 345, 1981.

6-Sullivan, C.V., Berlinsky, D.L., and Hodson, R.G., Reproduction, in *Striped Bass and Other Morone Culture*, Harrell, R.M., Ed., Elsevier Science, Amsterdam, 1997, 11.

7-Woods III, L.C. and Sullivan, C.V., Reproduction of striped bass (*Morone saxatilis*) brood stock: monitoring maturation and hormonal induction of spawning, *Journal of Aquaculture and Fisheries Management*, 24, 213, 1993.

8-Woods III, L.C., Striped bass and hybrid striped bass culture, in: *Aquaculture in the 21ST Century*, Fish Culture Section, , Kelly, A.M. and Silverstein, J., Eds., American Fisheries Society, Bethesda, MD, 2005, 339.

9-Harrell, R.M., Hybridization and genetics, in: *Striped Bass and Other Morone Culture,* Harrell, R.M., Ed., Elsevier Science, Amsterdam, 1997, 217.

10-He, S. and Woods III, L.C., Effects of glycine and alanine on short-term storage and cryopreservation of striped bass (*Morone saxatilis*) spermatozoa, *Cryobiology*, 46, 17, 2003.

11-He, S. and Woods III, L.C., The effects of osmolality, cryoprotectant and equilibration time on striped bass sperm motility, *J World Aquacult Soc*, 34, 255, 2003.

12-He, S. and Woods III, L.C., Effects of dimethyl sulfoxide and glycine on cryopreservation induced damage of plasma membranes and mitochondria to striped bass (*Morone saxatilis*) sperm, *Cryobiology*, 48, 254, 2004.

13-He, S., Jenkins-Keeran, K., and Woods III, L.C., Activation of sperm motility in striped bass via a cAMP-independent pathway, *Theriogenology*, 61, 1487, 2004.

14-Jenkins-Keeran, K. and Woods III, L.C., An evaluation of extenders for the short-term storage of striped bass milt, *N Am J Aquacult*, 64, 248, 2002.

15-He, S. and Woods III, L.C., Changes in motility, ultrastructure, and fertilization capacity of striped bass *Morone saxatilis* spermatozoa following cryopreservation, *Aquaculture*, 236, 677, 2004.

Complete affiliation:

L. Curry Woods III, Department of Animal and Avian Sciences, University of Maryland, College Park, MD, USA, e-mail: curry@umd.edu.

CRYOPRESERVATION OF SEMEN FROM
STRIPED TRUMPETER (*Latris lineata*)

A.J. Ritar

1. INTRODUCTION

Striped trumpeter (*Latris lineata*) is a demersal perciform found in cool temperate coastal waters of the southern hemisphere, mainly on the continental shelf over rocky bottoms at depths between 5 and 300 m [1,2]. It can grow to 1.2 m in length and weigh up to 25 kg. It is highly regarded as a food fish especially for the exclusive Japanese sashimi market. It forms a small (<100 tonne year[-1]) but valuable commercial fishery in Tasmania, Australia. Wild adults spawn between August and October (late winter to spring) at water temperatures ranging from 10-14°C. The pelagic larvae hatch at a small size and early stage of development and are neustonic [3].

Striped trumpeter is being considered for aquaculture and extensive research has been undertaken on hatchery rearing. Some problems with early larval culture have been overcome and egg incubation and rearing protocols have been established [4]. The mortality peaks associated with first feeding was reduced by using better live feed production techniques and improving water quality particularly at the air/water interface [5]. There are still high mortalities and abnormalities from notochord flexion to metamorphosis caused by an apparent metabolic disorder with a nutritional basis but despite this, in 2004, more than 2,000 juveniles were produced in the hatchery (S. Battaglene, pers. comm.).

Considerable progress has been made in controlling the reproduction of captive animals including the out-of-season spawning by temperature and photoperiod manipulation [6] and semen storage and artificial insemination [7,8]. The cryopreservation of striped trumpeter semen as pellets and straws is described in this paper.

2. PROTOCOL FOR FREEZING AND THAWING

2.1 Fish Handling and Semen Collection

- During the spawning season, adult males weighing 3-4 kg can be hand stripped on consecutive days to consistently produce up to 50 ml semen day[-1] at a density of up to 15×10^9 sperm ml[-1].
- Semen is collected from males anaesthetised with 2-phenoxyethanol. After rinsing and drying the genital area, as much semen as possible is collected into dry syringes or beakers, avoiding contamination with sea water, mucus, urine or faeces that might activate or kill sperm.

- Semen in collection containers is held on ice until assessment for motility and dilution with cryoprotective extender.

2.2 Pre-freezing Evaluation

- Semen, diluent, glassware and equipment coming into contact with semen are kept on ice to avoid temperature shock during handling. Semen may be assessed for motility immediately after fresh collection, dilution or thawing.
- A sample of semen adhering to the tip of a hypodermic needle is transferred onto a microscope slide.
- The droplet is smeared on the slide and 50 µl of filtered sea water is added (dilution rate 1:150 to 1:600) for activation of the sperm and examination of motility. There is little or no sperm movement prior to activation.
- A cover slip is pressed gently onto the diluted semen, rotated lightly to ensure even mixing and dispersal of a thin layer of sperm cells for immediate examination at a magnification of 400 x.
- Sperm motility is defined as the mean percentage of forward-moving cells from at least three fields of view within 15 s after activation.

2.3 Freezing and Thawing Procedures

- Semen is diluted at rates of 1:2 to 1:11 (semen:diluent) with various cryoprotective extenders containing dimethyl sulphoxide (DMSO) or glycerol.
- The diluted semen is held on ice until freezing.
- For freezing as pellets on dry ice (-79°C), diluted semen is rapidly dispensed into small indentations on the surface of dry ice blocks and frozen within 2 min of dilution. For freezing in straws, diluted semen is drawn into 0.25 ml or 0.5 ml clear plastic straws (I.M.V., L'Aigle, France) and the outer surfaces of straws are dried leaving one end unsealed. Straws are suspended on a pre-cooled polystyrene raft in the vapour 4 cm above the level of liquid nitrogen enclosed within a 26 liter polystyrene container (22 cm wide, 33 cm long, 25 cm deep, 1 cm thick). Straws remain in the vapour for 270 s and should reach at least -120°C before plunging into liquid nitrogen where they are held until thawing.
- When frozen pellets have a frosty appearance (2-4 min), they are transferred into liquid nitrogen (-196°C), and stored until thawing.
- Pellets are thawed in dry test tubes (16 mm x 75 mm) held and gently agitated in a waterbath at 20°C. Straws are removed from liquid nitrogen and immersed in a waterbath at 20°C for 30 s.

2.4 Fertilization

- Thawed semen from the pellet and straw methods are used for insemination within 20 s of thawing.

- Eggs for insemination are collected from adult females induced to repeatedly ovulate by treatment with luteinising hormone releasing hormone analogue (LHRHa)[9].
- Implant pellets containing 100 µg LHRHa and prepared according to the me-thod of Lee et al.[10] are admininstered at a rate of 100 µg kg⁻¹ by careful insertion into the dorsal muscle of the fish to a depth of approximately 2 cm.
- Animals ovulate and can be stripped of eggs 3-4 days later after anaesthetisation by applying gentle pressure to the abdomen.
- Eggs from 3-6 females are pooled, placed in dry plastic containers at 12-13°C, inseminated with fresh or cryopreserved semen at a rate of approximately 1 x 10⁶ sperm egg⁻¹ before activation with sea water (0.2 µm filtered, temperature-equilibrated).
- Eggs are incubated at 12-13°C and fertilisation assessed at 4-16 cell stage (within 6 h after insemination) in samples of 100-150 floating eggs.
- Water is exchanged 3 h and 6 h after insemination and then every two days for 6 days when most eggs hatch.
- Dead eggs and larvae are counted in exchange water.
- The percentage of eggs hatching is calculated as the number of live larvae as a proportion of the eggs at insemination.

3. GENERAL CONSIDERATIONS

The patterns of change in temperature during freezing and thawing differed markedly between pellets and straws and were affected by their volume (Figure 1)[8]. The initial decline in temperature was most rapid for small pellets (0.25 and 0.5 ml). Pellets of all sizes plateaued at -77° to -80°C and the time taken to reach the plateau increased with the volume of the pellets. Temperatures in the straws declined to between -133° and -139°C and the time taken to reach the plateau was longer for 0.5 ml than for 0.25 ml straws. After plunging, pellets and straws rapidly reached the temperature of liquid nitrogen. The warming rate during thawing was most rapid for straws, which reached 18°C within 20 s after immersing into the water bath at 20°C. By contrast, individual 2.0 ml pellets held in dry test tubes in the water bath took up to 270 s to reach 18°C, although smaller pellets warmed at a more rapid rate.

The post thawing motilities were higher when a diluent containing DMSO was used compared to the glycerol diluent (61.5 ± 2.8% and 46.0 ± 2.2% motile sperm). The DMSO diluent contained no sea water and did not activate sperm on dilution. It consisted of 2 parts DMSO and 7 parts physiological saline (150 mM NaCl) (equivalent to 2.84 M DMSO and 117 mM NaCl) in distilled water (pH not adjusted). The motilities, fertilisation rates and larval hatch rates were always lower (by 12-30%) for cryopreserved than for fresh semen (Table 1). The best results after cryopreservation was for semen diluted at 1:5 (semen:diluent) and loaded into 0.25 ml straws. The fertilisation rate after insemination was marginally higher for frozen-thawed semen from 0.25 ml straws than from 0.5 ml straws or 0.25-2.0 ml pellets. However, there was no difference when the number of pellets thawed at one time in a dry test tube was 1, 3 or 6.

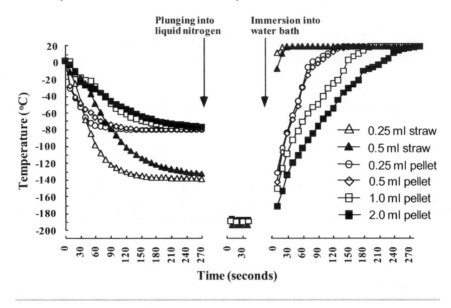

Figure 1 - Temperature during cooling of diluted semen (initially at 2°C, semen diluted 1:5, semen: DMSO diluent) in the centre of straws on styrofoam rafts in the vapour 4 cm above liquid nitrogen and in the centre of pellets on dry ice, after plunging into liquid nitrogen and after immersion (straws transferred directly, pellets in dry test tubes) into a waterbath at 20°C. Adapted from Ritar and Campet [8].

Table 1 - Sperm motility, egg fertilisation and larval hatch (mean ± s.e.m. percentage) of striped trumpeter eggs inseminated with fresh semen or with semen cryopreserved as pellets or straws using DMSO diluent. Values with different superscripts within each parameter differ significantly (P < 0.05). Adapted from Ritar [7].

Parameter	Fresh semen	Semen frozen as:			
		0.25 ml pellets	2.0 ml pellets	0.25 ml straws	0.5 ml straws
Motile sperm (%)	82.7±1.3[a]	53.4±2.4[b]	56.3±3.8[b]	60.0±4.1[b]	58.8±4.3[b]
Fertilisation rate (%)	79.6±2.7[a]	67.3±1.5[b]	69.6±5.3[b]	75.2±1.6[a]	69.5±2.8[b]
Larval hatch (%)	60.5±1.5[a]	29.2±1.7[c]	33.3±3.9[c]	44.3±2.9[b]	44.2±2.0[b]

4. ACKNOWLEDGMENTS

I would like to thank Dr. David Morehead, Mr. Alan Beech, Mr. David Wikely, and Ms. Polly Butler. The research was supported by the Aquaculture CRC.

5. REFERENCES

1-Last, P.R., Scott, E.O.G., and Talbot, F.H., *Fishes of Tasmania*, Tasmanian Fisheries Development Authority, Hobart, Tasmania, 1983, 563.
2-Tracey, S.R. and Lyle, J.M., Age validation, growth modeling, and mortality estimates for striped trumpeter (*Latris lineata*) from southeastern Australia: making the most of patchy data, Fish B-NOAA, 103, 169, 2003.
3-Furlani, D.M. and Ruwald, F.P., Egg and larval development of laboratory-reared striped trumpeter *Latris lineata* (Forster *in* Bolch and Schneider 1801) (Percoidei· Latridiidae) from Tasmanian waters, *New Zeal j Mar fresh*, 33, 153, 1999.
4-Cobcroft, J.M., Pankhurst, P.M., Hart, P.R., and Battaglene, S.C., The effects of light intensity and algae-induced turbidity on feeding behaviour of larval striped trumpeter, *J Fish Biol*, 59, 1181, 2001.
5-Trotter, A.J., Pamkhurst, P.M., and Hart, P.R., Swim bladder malformation in hatchery-reared striped trumpeter *Latris lineata* (Latridae), *Aquaculture*, 198, 41, 2001.
6-Morehead, D.T., Ritar, A.J., and Pankhurst, N.W., Effect of consecutive 9- or 12-month photothermal cycles and handling on sex steroid levels, oocyte development, and reproductive performance in female striped trumpeter *Latris lineata* (Latrididae), *Aquaculture*, 189, 293, 2000.
7-Ritar, A.J., Artificial insemination with cryopreserved semen from striped trumpeter *(Latris lineata)*, *Aquaculture*, 180, 177, 1999.
8-Ritar, A.J. and Campet, M., Sperm survival during short-term storage and cryopreservation of semen from striped trumpeter *(Latris lineata)*, *Theriogenology*, 54, 467, 2000.
9-Morehead, D.T., Pankhurst, N.W., and Ritar, A.J., Effect of treatment with LHRH analogue on oocyte maturation, plasma sex steroid levels and egg production in female striped trumpeter, *Latris lineata* (Latrididae), *Aquaculture*, 169, 315, 1998.
10 Lee, C.S., Tamaru, C.S., and Kelley, C.D., Technique for making chronic-release LHRH-a and 17a-methyltestosterone pellets for intramuscular implantation of fishes, *Aquaculture*, 59, 161, 1986.

Complete affiliation:

Arthur Jeremy Ritar, Tasmanian Aquaculture & Fisheries Institute, University of Tasmania, Marine Research Laboratories, Nubeena Crescent, Taroona Tasmania 7053 AUSTRALIA. E-mail: Arthur.Ritar@utas.edu.au

CRYOPRESERVATION OF ATLANTIC HALIBUT SPERM, *Hippoglossus hippoglossus*

I. Babiak

1. INTRODUCTION

Atlantic halibut, *Hippoglossus hippoglossus* L., is the largest flatfish and one of the largest farmed fish species with body weight exceeding 300 kg [1]. Atlantic halibut show a discontinuous reproductive pattern reaching reproductive readiness in the springtime. Females are batch spawners and males show prenuptial spermatogenesis, with spawning readiness at the peak of gonadal activity [2]. The quantity and quality of sperm decrease toward the end of the reproductive season, likely resulting from low spermiogenesis activity, intensive dehydration and ageing processes to spermatozoa [3-5]. Changes in semen quality towards the end of the season include: lowered fertilization ability [5] and poor resistance to cryopreservation [3]. Asynchrony in gamete quality occurs at the end of the reproductive season when quality eggs are still produced by females [6,4]. One of approaches to overcome the problem of poor quality of sperm late in the season is to cryopreserve the sperm when it is of the best quality. Atlantic halibut semen has been successfully cryopreserved [7,8].

2. PROTOCOL FOR FREEZING AND THAWING

2.1 Fish Handling and Semen Collection

- Milt is collected by hand stripping mature males during the whole reproductive season, normally from January/February to May under natural photoperiod conditions. No additional stimulation of spermiation is needed (e.g., hormonal treatment). Throughout most of the spawning season, the volume of collected milt is high (depending on the size of a male, 5–100 ml) and its viscosity is low making no problems with expressing it into a container (Fig. 1). However, at the end of the reproductive season, milt becomes very dense, its volume decreases and its suitability to cryopreservation drops down [3,8]. After collected, the samples of milt should be refrigerated or stored on crushed ice at 0–2°C.

2.2 Freezing Procedures

- For freezing, the following extender is recommended:
- Diluent: Modified turbot diluent [9]; modified by Vermeirssen et al. [4]: NaCl 70 mM; KCl 1.5 mM; CaCl$_2$ 2.7 mM; NaHCO$_3$ 25 mM; MgCl 6.1 mM; BSA

10 mg/ml; glucose 200 mM; pH 8.1, 400 mOsm/kg. Filtered with 0.22 µm filter and chilled overnight.

Cryoprotectant: 10% (v/v) DMSO, methanol or DMA.

Caution: Adjust the pH after filtering. Keep the diluent cooled (0-2°C). Add the cryoprotectant just prior to freezing the sperm.

- Dilution: 1 part of sperm and 3 parts of the extender (diluent + cryoprotectant). Add extender to the sperm stepwise (within 5–10 sec) swirling the container with diluted sperm vigorously. Make sure that sperm is evenly diluted with the extender.
- No equilibration of sperm with the extender is needed. However, the spermatozoa can remain in the extender safely for at least 30 min after dilution without any negative effects.
- Diluted sperm can be pipetted on a dry ice (-79°C) or it can be frozen in vapours of liquid nitrogen. In the latter case, sperm has to be loaded into cryo-containers such as French straws (0.25–0.5 ml), cryovials (1–5 ml), large straws, Ziploc bags, etc.
- Loaded containers are placed on a metal grid mounted on pieces of polystyrene that float 4 cm above the surface of liquid nitrogen poured into an insulated box. The layer of diluted sperm to be frozen must not exceed 2 mm.
- After sperm is loaded, the containers are placed on the grid horizontally, for 5 (French straws) to 15 min (larger volumes) (Fig. 2).
- Then they are moved directly to the liquid nitrogen.

2.3 Thawing Procedures

- Containers are placed in a Ziploc bag that is put into a water bath (40°C) for 3 s (French straws) or 10 s (bigger containers).
- The contents are then released into a thawing medium. In the case of pellets, they are placed directly into a thawing medium. Hanks' Balanced Salt Solution (Sigma Aldrich Co., H 8264) is the recommended thawing medium. Thawed sperm is not activated in this medium until it is poured into the water. The volume ratio 1:4 to 1:10 of thawed sperm to thawing medium is recommended.
- After thawing, the sperm can be used to fertilize the eggs within 2 hours.

2.4 Fertilization

- The amount of 1 ml of cryopreserved sperm per 1 liter of eggs per 1 liter of water is the safe proportion securing high fertilization success. However, depending on the spermatozoa concentration and the quality of fresh sperm, possibly higher volumes of eggs can be fertilized with 1 ml of cryopreserved sperm. NB: the volume of cryopreserved sperm refers to sperm before any dilution proceeded.

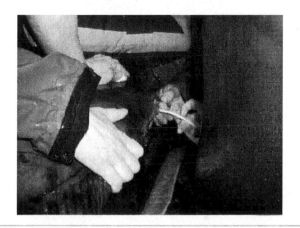

Figure 1 - Stripping the milt from Atlantic halibut.

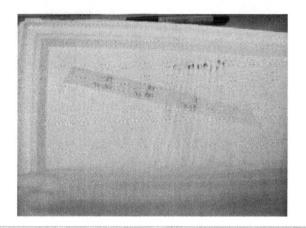

Figure 2 - French straws (0.5 ml, left) and Petri dish lids (5 ml, right) are filled with diluted sperm and placed horizontally on a grid floating on liquid nitrogen to be frozen.

3. GENERAL CONSIDERATIONS

The protocol described above has been proven during the two reproductive seasons at the Atlantic halibut breeding station at the Bodø University College. The method has been used for mass-scale fertilizations with cryopreserved sperm [8].

4. ACKNOWLEDGMENTS

I thank Sylvie Bolla and Oddvar Ottesen, Bodø University College, Norway, for their contribution to experiments on cryopreservation of Atlantic halibut sperm.

5. REFERENCES

1-Mathisen, O.A and Olsen, F. S., Yield isopleths of the halibut, *Hippoglossus hippoglossus*, in *Northern Norway*, FiskDir Skr Ser HavUnders, 1968, 14, 129.

2-Weltzien, F.A., Taranger, G.L., Karlsen, Ø., and Norberg, B, Spermatogenesis and related plasma androgen levels in Atlantic halibut (*Hippoglossus hippoglossus* L.), *Comp Biochem Physio,* 132A, 567, 2002.

3-Billard, R., Cosson, J., and Crim, L.W., Motility of fresh and aged halibut sperm, *Aquat Living Resour*, 6, 67, 1993.

4-Vermeirssen, E.L.M., Mazzora de Quero, C.M., Shields, R.J., Norberg, B., Kime, D.E., and Scott, A.P., Fertility and motility of sperm from Atlantic halibut (*Hippoglossus hippoglossus*) in relation to dose and timing of gonadotropin-releasing hormone agonist implant, *Aquaculture,* 230, 547, 2004.

5-Babiak, I., Ottesen, O., Rudolfsen, G., and Johnsen, S., Quantitative characteristics of Atlantic halibut, *Hippoglossus hippoglossus* L., semen throughout the reproductive season, *Theriogenology,* 65, 1587, 2006.

6-Shields, R., Gara, B., and Gillespie, M.J.S., Effects of stress on fish reproduction, gamete quality, and progeny, *Aquaculture*, 197, 3-24, 1999.

7-Bolla, S., Holmefjord, I., and Refstie, T., Cryogenic preservation of Atlantic halibut sperm, *Aquaculture*, 65, 371, 1987.

8-Babiak, I., Bolla, S., and Ottesen, O., Cryopreservation of Atlantic halibut *Hippoglossus hippoglossus* semen, (in preparation).

9-Suquet, M., Dreanno, C., Dorange, G., Normant, Y., Quemener, L., Gaignon, J.L., and Billard, R., The ageing phenomenon of turbot spermatozoa: effects on morphology, motility and concentration, intracellular ATP content, fertilization, and storage capacities, *J Fish Biol* 52,31, 1998.

Complete affiliation:

Igor Babiak, Department of Fisheries and Natural Sciences, Bodø University College, 8049 Bodø, Norway, e-mail: Igor.Babiak@hibo.no, phone: +47 75517922, fax:+47 755 17349.

SPERM CRYOPRESERVATION: AN ACTUAL TOOL FOR SEABASS (*Dicentrarchus labrax*) REPRODUCTION AND BREEDING CONTROL IN RESEARCH AND PRODUCTION

C. Fauvel and M. Suquet

1. INTRODUCTION

Seabass is widely represented within a large geographic area including the whole Mediterranean, East Atlantic up to Irish waters, and the Channel. Its reproductive season lasts approximately two months starting differentially in the area due to climatic gradient. The reproduction season generally starts in January in the Mediterranean, mid-February in the western Channel and in April in Irish waters. The spermiation period largely covers spawning time but sperm quality decreases as the season progresses [1].

Seabass industrial production presently reaches more than 150,000 tons a year. The necessary fry is all year long produced by hatcheries from broodstocks subjected to shift cycles of photoperiod and temperature. After considerable progress in husbandry and nutrition, most improvements in production are expected from genetics. For this, close control of individual male and female genital activity is necessary, as well as optimized management of gametes allowing special crosses. Ovulation in seabass females can be controlled by heterologous hormones [2] and oviposition prevented in order to obtain unfertilised eggs by stripping the females. These eggs can rapidly become overripe, which is a major cause of reproductive failure (unpublished).

Seabass sperm is also characterised by a reduced capacity to survive long after collection [3]. Due to this short survival of both gametes in seabass, cryopreservation was developed as an indispensable alternative for multiple crosses [4]. This method was applied successfully to perform large scale multiple crosses the progenies of which are all represented in resulting population (unpublished).

2. PROTOCOL FOR FREEZING AND THAWING

2.1 Equipment

- Cooling system: Basically, freezing devices consist of a liquid nitrogen (LN_2) reservoir and a cooling tray placed in LN_2 vapors. The freezing temperature can be adjusted either by controlling the vapour flux at a fixed height over the LN_2 using a computer remote-controlled gas propeller or by moving the tray in the vapor along the temperature gradient. For seabass cryopreservation, the second type was chosen because of its rusticity and efficiency in the aquaculture field (no energy required). The device used to set the method up is a simple 15 cm interior height styrofoam box used as LN_2 reservoir. The cooling tray is placed on a styrofoam raft the height of which can be easily adapted with a simple cutter. The temperature is checked by a thermocouple inserted in a sperm freezing container (straw or vial) filled with a standard sample of sperm. The design and material of the LN_2 reservoir are not of primary importance and some can easily be obtained from cryobiology product traders.

- Reagents: All the reagents used are from Sigma Aldrich, France.

2.2 Semen Collection

- The males are anaesthetized and sperm is collected by applying gentle pressure peristastically to the testis after drying the genital pore area. Only sperm with a dense, whitish appearance is collected with a syringe to avoid the possibility of urine contamination, and the syringes are put on ice until further treatment. Generally, a volume of 2 ml can be collected in very good conditions for males weighing more than 1 kg.

- Sperm concentration assessment: An aliquot of 20 µl of semen is added to 10 ml of distilled seawater in a tube and immediately mixed by vigorous shaking. This dilution to 1/500 can be extemporaneously assessed for absorption at 260 nm wavelength to check the optical density. The cell concentration of the sample is SC= $(0.806 \times OD-0.032) 10^8 \times 500$.

2.3 Freezing and Thawing Procedures

- Dilution extender: Sperm can be adjusted to the desired concentration with a dilution extender before cryopreservation. The use of an isotonic saline at pH 7.7 efficiently prevents sperm motility and, as in the case for other fish, the addition of protein limits the motility decrease induced by dilution. We recommend the use of 10 mg ml^{-1} of protein. Bovine serum albumin was used to set up the method but due to bovine spongiform encephalopathy, other sources of proteins like ovalbumin or milk proteins can successfully replace BSA. The extender contains NaCl, 3.5; KCl, 0.11; $MgCl_2$, 1.23; $CaCl_2$, 0.39; $NaHCO_3$, 1.68; Glucose, 0.08 (in mg ml^{-1} distilled water) the resulting osmolarity is 310 mOsm kg^{-1}.

- Cryoprotectant: Since Mounib medium ($KHCO_3$, 10.1; Reduced glutathione, 1.99; Sucrose, 42.78 (in mg ml^{-1} distilled water); resulting osmolarity (310 mOsm kg^{-1}) complemented with 10 mg ml^{-1} BSA or other protein, and combined with 10% dimethyl sulfoxyde (DMSO) gave interesting results in first trials, no attempts have been made to improve this cryoprotectant.

- Sperm samples are diluted to 1/3 (v/v) in cryoprotectant and subjected to cryopreservation without equilibration delay.

- Sperm loading: Sperm samples are routinely frozen in straws of 250 and 500 µl (Cryobiosystem and IMV France) and in in 1.8 ml cryovials (Nalgene).
- The straws are hand filled using a 1000 µl Gilson micropipette and closed using a sealing machine (Syms, IMV, France) to avoid all types of cross pollution or infection. Now, for large series of cryopreservation, straw filling is automated (IMV, France).
- Depending on the type of loading, two protocols are used:
  1-Straws (either 250 or 500 µl):
- Freezing: the straws are placed immediately after filling on a 6.5 cm high tray floating on an LN_2 surface. They are kept at least 10 min on the tray and then dropped into LN_2.
- Thawing: Straws are removed from LN_2 and immediately shaken in a water bath at 35°C for 10 s. They are then used for motility assessment or fertilisation.
  2-Cryovials (1.8 ml):
- Freezing: the vials are first plunged into LN_2 for 15 s under shaking then placed for at least 10 min on a 2 cm high tray floating on an LN surface before being dropped into LN_2.
- Thawing: the vials are removed from LN_2 and immediately shaken in a water bath at 50°C until thawing (around 1 min). They are then used for motility assessment or fertilisation.

3. GENERAL CONSIDERATIONS

3.1 Method Monitoring

The introduction of a thermocouple in a straw or vial allows monitoring the decrease in temperature to be monitored and, in particular, the seeding point at which intracell crystallisation occurs. Figure 1 describes the curves obtained in straw and in vial under the protocols described above.

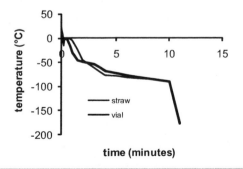

Figure 1 - Record of temperature decrease during the process.

3.2 Results

The results for the setup of the method and particularly the features of treated sperm were reported in Fauvel et al. [4] for both the experimental and scaled up techniques. Cryopreservation slightly decreases sperm fertility but this loss may be compensated by an increase in available sperm cells per egg from 7 x 10^4 (routine for optimized fresh sperm fertilization) to 2 x 10^5 as shown in Table 1. Moreover, the comparison of survival of seabass larvae originating from the same sperm but subjected or not to cryopreservation treatment shows that freezing/thawing may slightly decrease the yield of artificial fertilization and hatching rate (Table 2).

Finally, this protocol was used twice for genetic purposes to perform large-scale combinations of factorial crosses between sires and dams (253 and recently 988 progenies) for which it was impossible to strip the males just prior to fertilization. As a result, DNA microsatellite analysis demonstrated that all the 253 families were represented in the population after 2 years of rearing. [7].

Table 1 - Probability of difference between fresh and frozen sperm fertility at different sperm to egg ratios

	Male 1	Male 2	Male 3	Male 4	Male 5	Male 6
35 10^3 spz ml^{-1}	* p<0.05	** p<0.01	* p<0.05	p=0.21	p=0.15	p=0.74
70 10^3 spz ml^{-1}	* p<0.05	p=0.25	p=0.19	p=0.82	p=0.47	p=0.63
200 10^3 spz ml^{-1}	p=0.4	p=0.25	p=0.45	p=0.34	p=0.11	p=0.16

Table 2 - Effect of cryopreservation on production figures at pilot scale

	Fresh sperm			Frozen/thawed		
	Male A	Male B	Male C	Male A	Male B	Male C
Fertilization rate (%)	71.3	73.6	73.3	66.3	67	46
Hatching rate (%)	84.1	79.5	81.8	76.8	67.1	64.5
Loss compared to fresh				15	21	50
Number of fry from 120,000 eggs	72,000	68,400	72,000	61,200	54,000	35,700
Total fry from 360,000 eggs	212,400			150,900		

3.3 Recommendations

In the present state of the art, the main bottleneck in seabass sperm management is the rapid decrease in sperm quality after collection. When semen is stored on ice or at 4°C, spermatozoa motility and fertility are drastically decreased 24 h after collection (Fauvel and Suquet, unpublished), and their cryopreservation aptitude is already altered 6 hours post sampling [5]. We strongly recommend avoiding long pre-freezing sperm storage in favour of small series of sperm collection and *in situ* freezing process.

Moreover, since seabass sperm is subject to ageing as the season progresses, it is recommended to program sperm preservation at the beginning of the reproductive period.

Effort should be brought to short term preservation of seabass sperm so as to allow *ex situ* gamete freezing and guarantee both efficient sperm collection even from rearing seacages and high quality laboratory freezing processes.

Table 3 - Seabass sperm characteristics

	Fresh sperm	Frozen thawed	
Sperm Concentration	Up to 60 10^9 spz ml^{-1}		
Initial Motility	100%	80%	NS
Initial Flagellar Beat Frequency	60 hz	60 hz	
Initial ATP	12 10^{-8} nmole spz^{-1}		
Initial straightline velocity	100 μm s^{-1}	70 μm s^{-1}	*
Initial curvilinear velocity	120 μm s^{-1}	100 μm s^{-1}	*
Motility Duration	40 s	30 s	NS
Maximum Fertility at 10 s	88%	68%	NS
Fertility Decrease/time	$F=a(t+1)^{1.5}$, $a = f(sperm/egg\ ratio)$		

3.4 PERSPECTIVES

Much progress can be expected from modifications of medium formula and protocols of sample preparation. Sansone et al. [5] showed the influence of diluting medium on sperm motility after short term conservation. Moreover, the same team published the post-thawed sperm motility improvement potentialities by analytical studies of modifications of a reference protocole in seabass [6]. However, research would largely benefit from standardization of analytical protocols such as sperm counting, dilution for motility assessment, *in vitro* fertilization conditions.

4. REFERENCES

1-Dreanno, C., Suquet, M., Fauvel, C., Le Coz, J.R, Dorange, G., Quemener, L., and Billard, R., Effect of the aging process on the quality of sea bass (*Dicentrarchus labrax*) semen, *J. Appl. Ichthyol / Z. Angew. Ichthyol* 15, 6, 176, 1999.

2-Mylonas, C.C, Sigelaki, I., Divanach, P., Mananos, E., Carrillo, M., and Afonso-Polyviou, A., Multiple spawning and egg quality of individual European sea bass (*Dicentrarchus labrax*) females after repeated injections of GnRHa, *Aquaculture*, 221, (1-4), 605, 2003.

3-Fauvel, C., Savoye, O., Dreanno, C., Cosson, J., and Suquet, M., Characteristics of sperm of captive seabass in relation to its fertilization potential, *J. Fish Biol*, 54, (2), 356, 1999.

4-Fauvel, C., Suquet, M., Dreanno, C., Zonno, V., and Menu, B., Cryopreservation of sea bass (*Dicentrarchus labrax*) spermatozoa in experimental and production simulating conditions, *Aquat Living Resour*, 11, (6), 387, 1998.

5-Sansone, G., Fabbrocini, A., Zupa, A., Lavadera, S.L., Rispoli, S., and Matassino, D., Inactivator media of sea bass (*Dicentrarchus labrax* L.) spermatozoa motility, *Aquaculture*, 202, (3-4), 257, 2001.

6-Sansone, G., Fabbrocini, A., Ieropoli, S., Langellotti, A., Occidente, M., and Matassino, D., Effects of extender composition, cooling rate, and freezing on the motility of sea bass (*Dicentrarchus labrax*, L.) spermatozoa after thawing, *Cryobiology*, 44 (3), 229, 2002.

7-Chatain, B., Final report of European project, *CRAFT QCR*-2002-71720, *«Heritabolum»*, 2005.

Complete affiliation:

Christian Fauvel, Ifremer LRPM Chemin de Maguelone, e-mail: Christian.Fauvel@ifremer.fr

NOTES ON *Diplodus puntazzo* SPERM CRYOPRESERVATION

F. Barbato, S. Canese, F. Moretti, A. Taddei, A. Fausto,
L. Abelli, M. Mazzini and K.J. Rana

1. INTRODUCTION

In common with other sparid species, *Diplodus puntazzo* is of commercial interest in Mediterranean fishery. In recent years, the scientific research on this species has resulted in successful captive breeding. The species is fast growing and is well accepted by consumers, but its vulnerability to some diseases is the main obstacle to current development of this species for aquaculture. The reproductive season is at the end of summer, beginning of autumn.

Sexes are, in our relative experience, separated, but there is some hint in the literature of possible hermaphroditism, at least for some individuals. Sperm cryopreservation was attempted by our team for the first time in the late 1990s [1-6]. Cryopreservation protocols proved to be successful with post-thaw motility observations and the creation of hybrids employing *Sparus aurata* oocytes [7].

2. PROTOCOL FOR FREEZING AND THAWING

2.1 Equipment and Reagents

- Polystyrene cryobox
- Metal rack for 0.5 ml straws
- LN$_2$ Cryotank
- Straws Storage Cryotank
- Chronometer
- 0.5 ml straws
- 50 ml test tubes
- Graduated test tubes (various size)
- Various size pipettes
- Straw sealing machine
- Polyvinyl alcohol
- Na citrate (Na$_3$C6H$_5$O$_7$ x 2H$_2$O)
- Me$_2$SO (DMSO)
- Distilled water
- LN$_2$
- Gloves

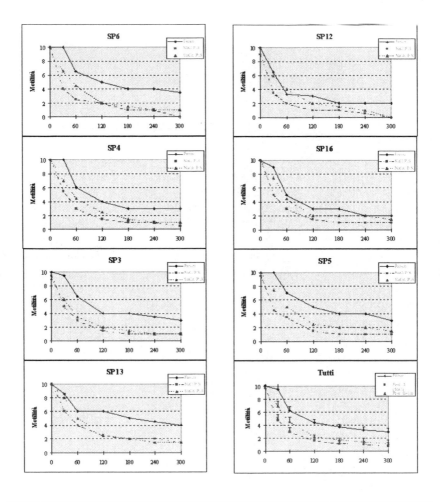

Figure 1 - Motility evaluation (0-10 scores) vs. time in seconds from activation for fresh (Fresco) and post thawed (P.S.) *D. puntazzo* sperm from 7 donors. Prot. A : diluent NaCl, Prot. B : diluent NaCitrate. Tutti = All.

2.2 Freezing and Thawing Procedures

- Procurement of semen: A needless syringe was used to collect sperm from the live, anaesthetized fish.
- Cryosolution: The cryosolution is composed of Me_2SO and Na citrate 0.1 M, with a final proportion v/v/v of 2/1/7, respectively, in semen/ Me_2SO /Na citrate 0.1 M. One volume of semen has been added to four volumes of the cryosolution.

- Sperm loading: Diluted sperm was loaded in 0.5 ml straws with a 1 ml Gilson; sealing was executed with a hot sealing machine or polyvinyl alcohol.
- Freezing procedure: Freezing was carried out by laying the 0.5 ml straws horizontally on a metal rack at a distance of 4.3 cm from LN_2 in a covered polystyrene cryobox, for ten minutes. Following freezing, the straws were plunged in the underlying LN_2.
- Thawing procedure: Straws were thawed in a 28°C water bath, for 20 seconds.

3. GENERAL CONSIDERATIONS

The cooling and warming rates within straw were measured with a copper constance thermocouple inserted into a straw with the cryosolution diluted semen. The straws were cooled at 8°C/min from 4°C down to –80°C, and 360°C/min down to LN_2 temperature. The straws were thawed, for 20 s, at 580°C/min up to –2°C and 133°C/min up to 20°C in a water bath at 28°C.

Motility was evaluated by a skilled operator on a subjective scale (from 0 to 10, 0: no motion, 10: fast movement of 90-100% spermatozoa). Milt samples were placed on a slide under a dark field microscope and activated with filtered and sterilised seawater (39% salinity) to yield a final dilution of 1:100 (milt to other liquids); no coverslip was added to the sample. For the motility profiles, scores were attributed to samples 0, 30, 60, 120, 180, 240 and 300 seconds after activation (Figure 1).

The best fertility data available to date, obtained in 1999, are for the fecundation of *S. aurata* oocytes with thawed *D. puntazzo* sperm, with an average result of 33.7% (best batch 47.1%, worst 11.2%) fertility rate 72 h after from fecundation; egg-sperm ratio was 1/237.000 and contact time 3 min; control with fresh *S.aurata* sperm gave 61.3%.

4. REFERENCES

1-Barbato, F., Canese, S., Moretti, F., and Misiti, S., Sviluppo di metodiche affidabili per la crioconservazione dello sperma di telostei marini di interesse economico.Atti di "Le ricerche sulla pesca e sull'acquacoltura nell'ambito della L.41/82", *Biol. Mar. Medit.*, 5, fasc.3 Parte II, 894, 1988.
2-Barbato, F., Canese, S., Moretti, F., Misiti, S., Laconi, F., and Rana, K., Preliminary experiences for cryopreservation of *Sparus aurata* and *Diplodus puntazzo* semen, in: *Cahiers Options Mediterranennes*, Ciheam, Zaragoza, 1988, 34, 281.
3-Canese, S., Moretti, F., Schino, G., and Barbato F., Effects of different cryopreservation protocols on the motility of *Diplodus puntazzo* thawed sperm, presented at *International Aquaculture Conference*, in Book of abstracts, Verona, February, 1999, 31.
4-Taddei, A.R., Fausto, A. M., Barbato, F., Canese, S., Abelli, L., Baldacci, A., Gambellini, G., and Mazzini, M.,. Modificazioni ultrastrutturali in spermatozoi crioconservati di *Diplodus puntazzo* Cetti (Teleostea, Sparidae), Atti *59° Congr. Naz. Unione Zoologica Italiana*, 1998.
5-Taddei, A.R., Abelli, L., Baldacci, A., Barbato, F., Fausto, A.M., and Mazzini, M., Ultrastructure of spermatozoa of marine teleosts *Umbrina cirrosa* L. and *Diplodus puntazzo* (Cetti) and preliminary observations on effects of cryopreservation, in: *Proc. 33rd Int. Symp. on New Species in Mediterranean Aquaculture*, April 22-24, 1998 Alghero, Elsevier, 1999, 407.

6-Taddei, A.R., Barbato, F., Abelli, L., Canese, S., Moretti, F., Rana, K.J., Fausto, A.M., und Mazzini, M., Is cryopreservation a homogeneous process? Ultrastructure and motility of untreated, prefreezing and postthawed spermatozoa of *Diplodus puntazzo* (Cetti), *Cryobiology*, 42, 244, 2001.
7-Barbato, F., Canese, S. , Barbaro, A., Francescon, A., and Rana, K. J., First results in obtaining hybrids employing *Sparus aurata* fresh oocytes and *Diplodus puntazzo* or *Pagrus major* cryopreserved sperm, in: *Proc. 33rd Int. Symp. on New Species in Mediterranean Aquaculture*, April 22-24,1998, Alghero, Elsevier, 1999, 149.

Complete affiliation:

Dr. Fabio Barbato, Enea Biotec Amb, Centro Ricerche Casaccia s.p. 028, Via Anguillarese 301,00123 Santa Maria di Galeria (Roma) Italy. e-mail: fabio.barbato@casaccia.enea.it

SPERM CRYOPRESERVATION FROM THE
MARINE TELEOST, *Sparus aurata*

E. Cabrita, V. Robles, C. Sarasquete and P. Herráez

1. INTRODUCTION

Gilthead seabream, *Sparus aurata,* is an interesting species in aquaculture. This species has been produced since the 80s in different countries in the Mediterranean area and is marketed mainly in European countries. It is a well-studied species in the fields of nutrition, growth, reproduction and husbandry, which have contributed to solving the main problems of commercial farming. However, the first sperm cryopreservation studies were only published in the 90s by Chambeyron and Zohar [1] and an exhaustive study of milt quality before and after freezing has never been reported. Sperm cryopreservation in this species is particularly interesting because gilthead seabream is a protandric species in which males change into females after the first 2 years of life, making it difficult to plan crossbreeding. Therefore, the male broodstock needs to be constantly increased in order to ensure enough sperm release during fertilization

Sperm can be obtained all year-round from different broodstocks maintained under photoperiod control, thus allowing the freezing period to be extended for several months. Sperm production can be variable between males and during the reproductive season and 7 ml per male can be extracted.

At present, and due to interest in the characterization of *S. aurata* broodstocks, cryopreservation technology is being requested by the fishfarm industry to create sperm banks from males selected according to their phenotypic characteristics or to the presence of specific genetic markers.

A seabream sperm freezing protocol has been designed over the past two years using 0.5 ml and 5 ml straws with the aim of transferring to a commercial scale. The results obtained guarantee the application of this technique on a commercial scale or in the creation of a sperm bank.

2. PROTOCOL FOR FREEZING AND THAWING

2.1 Equipment

- Polystyrene tubes (15 ml) with a rack
- Styrofoam box and rack for keeping straws horizontally
- 5 ml straws (macrotubes)
- Forceps for handling goblets with liquid nitrogen
- 1 ml syringes (without a needle)

- Paper to clean mucus
- Support for holding fish
- 50 litre tank for anesthetising fish
- Micropipettes (1 μl, 200 μl and 1000 μl)
- Basic laboratory equipment (microscope, osmometer, pHmeter)
- Basic laboratory glass material
- Basic cryobiology equipment (N$_2$ tanks, goblets)

2.2 Reagents

- MS-222 (tricaine)
- phenoxyethanol
- DMSO (dimethylsulfoxide)
- BSA (bovine serum albumin)
- NaCl solution at 1% (300 mOsm/kg)
- Sucrose solution at 10% (300 mOsm/kg)
- Artificial seawater (ASW: 30 g/l NaCl, 0.8 g/l KCl, 1.3 g/l CaCl$_2$, 6.6 g/l MgSO$_4$, 0.18 g/l NaHCO$_3$, pH 8.2, 1000 mOsm/Kg^2.
- Distilled water
- Liquid nitrogen

2.3 Fish Handling and Semen Collection

- Anesthetise gilthead seabream males with phenoxyethanol in the broodstock tanks (100 ppt, concentration used only to allow fish capture and to avoid stress) and transport them to a small tank containing 125 mg/l MS-222. Wait 5 to 10 minutes.
- Collect the fish from the anaesthesia tank, wash with sea water to eliminate anaesthesia from skin and put males in a support designed to hold the ventral part upwards for easy handling during sperm extraction.
- Eliminate faeces, urine and mucus by applying gentle pressure in the abdominal region and dry the urogenital pore.
- Extract the sperm by abdominal massage collecting it from the urogenital pore with a 1 ml syringe without a needle. This procedure is repeated several times until all the sperm is collected. The sperm is stored in polystyrene tubes at approximately 7°C until further use. Avoid contact between the tubes and the ice by protecting them in a styrofoam rack.
- Place the fish in a recovery tank endowed with air supply and before the males return to the broodstock tank, open the water flow as much as possible to eliminate phenoxyethanol residues.
- Wait one week until the next stripping.

2.4 Pre-freezing Sperm Evaluation

Before freezing sperm, check motility, pH, osmolarity and cell density. For routine use, evaluate sperm motility under phase contrast light microscopy (magnification 20x), by activating the sperm with seawater (1 μl sperm: 1000 μl SW) and estimating the percentage of cells with progressive movement (0-0%; 1-20%, 2-40%; 3-60%; 4-80%; 5-100%). Samples already activated before contact with seawater or with less than 80% of cells moving after activation, are discarded for cryopreservation. Measure the pH with a pHmeter and osmolarity with a cryo-osmometer. Samples with pH and osmolarity lower than 7.7 and 360 mOsm/Kg, respectively, should be discarded. To determine cell density use a haemocytometer chamber (Neubauer, Bürker, Thoma or Makler). Sperm samples with densities lower than 2.5×10^9 spermatozoa/ml should also be discarded.

A more selective study of sperm samples could be done if equipment is available. In this case motility can be determined using the CASA system and cell viability can also be determined (see Post-thawed sperm evaluation).

2.5 Freezing and Thawing Procedures

- Dilute the sperm (1:6) in the extender containing 1% NaCl (w/v), plus 5% DMSO (v/v) and 10 mg/ml BSA in distilled water. BSA is added to the extender to protect the plasma membrane and avoid sperm aggregation. No sperm equilibration time in the extender medium is needed and straw packaging starts immediately.
- Load the sperm into 5 ml straws (macrotubes) with a micropipette and place the samples in a horizontal rack. The straws are previously sealed (with heat or specific sealing caps) at one end, leaving the other one open during the freezing process.
- Place the rack in the styrofoam box containing liquid nitrogen. Sperm freezing is performed at 1 cm above the surface of liquid nitrogen in vapour phase for 10 min. Before loading the rack with the straws, the nitrogen level should be checked in order to guarantee the correct distance from the straws to the nitrogen surface.
- After 10 minutes (temperature inside the straw: approximately -140°C), immerse the straws into liquid nitrogen and store the samples in a nitrogen container.
- Thaw samples in a controlled temperature water bath with a recirculation system set at 60°C for 30 seconds. Avoid contact between the water and the open end of the straw.
- Dry the straws with paper and cut the closed end, collecting the sperm in eppendorf tubes.
- After thawing, use the sperm immediately to avoid quality loss.

2.6 Post-thaw Sperm Evaluation

Sperm lose quality after freezing/thawing. For routine analysis prior to fertilization, check motility as described before. Immotile samples should be discarded. For detailed studies, post-thawed motility is analysed using the computer-assisted sperm analysis (CASA). For this analysis, dilute the sperm 1:80 (sperm:solution) (v/v) in a nonionic solution (300 mOsm/Kg sucrose prepared in milli-Q water) and maintain it at 20°C. Collect a 0.5 µl droplet and place it in a Makler chamber to obtain a fully stabilised image. Activate the spermatozoa with 20 µl seawater (if possible artificial seawater (ASW), 1000 mOsm/Kg). The loaded chamber should be immediately placed under a microscope. Spermatozoa are observed with a 10x negative phase contrast (Ph-) objective and recorded using a Blaster camera connected to a PC. Microptics software is used to analyse spermatozoa motility characteristics, however, other software are available on the market.

Special attention should be paid to mitochondrial status, cell viability and spermatozoa ATP levels [3]. Membrane viability can be assessed using the live/dead stain SyBr/PI or other fluorescent dyes (Hoechst) and mitochondrial status can be determined using JC-1 or rhodamine. These techniques are well documented and can be performed either in a fluorescent microscope or using flow cytometry. Spermatozoa ATP can be determined using commercial kits from different suppliers.

2.7 Fertilization

- Strip female oocytes into a plastic beaker (avoid any contact with water). For general studies use pooled eggs from different females to avoid quality interference.
- Spread the sperm (50 µl for fresh sperm and 350 µl for frozen sperm) over the 2 ml oocytes and gently mix (3 s) before activation with 4 ml seawater (twice the volume of oocytes). The spermatozoa/oocyte proportion is maintained at approximately 1.3×10^5.
- After 10 minutes, add 50 ml seawater to the eggs (already fertilized).
- Incubate the 2 ml eggs (± 2400 eggs) in small plastic incubators (3 l) until one day after hatching (72 h).

3. GENERAL CONSIDERATIONS

Sperm production in this species enabled us to work with volumes of sperm from each individual ranging between 1 and 7 ml in each extraction. Extraction periods should be at one week intervals to allow further production and total recovery of sperm quality. For selection programs and crossbreeding planning, individual sperm cryopreservation should be used. Nevertheless, for mass production or other purposes, sperm from different males can be pooled before freezing to avoid interference with individual male quality. Sperm extraction is quite simple using the method proposed and urine contamination is avoided if the bladder is emptied. Our freezing protocol was adapted for macrotubes from Fabbrocini et al.[4]. Fertility rates obtained with post-thawed sperm were similar to those with fresh sperm (75.6% and 77.2%, respectively). Fertility tests were done using 1.3×10^5 spermatozoa/oocyte, however, we considered that a lower sperm/oocyte ratio can be used under laboratory conditions, when the volume of oocytes can be reduced and fertilization conditions can be easily controlled. An exhaustive study of sperm quality after freezing/thawing revealed that samples had some damage, which in some cases were not detected by an estimation of motility. Plasma membrane, mitochondria and nucleus DNA could be a target for cryodamage. Thus, depending on the sperm application, other analysis such as cell viability or DNA damage should be determined. The latter parameter is important specially if samples are being used for banking individuals with particular genetic characteristics in selection breeding programs.

4. ACKNOWLEDGMENTS

We thank Dr Jeff Wallace for helpful discussion of this protocol. This work was supported by the Spanish Ministry of Science and Education, project PETRI 95.1026.OP.01.

5. REFERENCES

1-Chambeyron, F. and Zohar, Y., A diluent for sperm cryopreservation of gilthead sea bream, *Sparus aurata*, *Aquaculture*, 90, 345, 1990.
2-Gwo, J.C., Chen, C.W., and Cheng, H.Y., Semen cryopreservation of small abalone (*Haliotis diversicolor supertexa*), *Theriogenology*, 58, 1563, 2002.
3-Cabrita, E., Robles, V., Rebordinos, L., Sarasquete, C., and Herráez, M.P., Evaluation of DNA damage in rainbow trout (*Oncorhynchus mykiss*) and gilthead sea bream (*Sparus aurata*) cryopreserved sperm, *Cryobiology*, 50, 144, 2005.
4-Fabbrocini, A., Lavadera, L., Rispoli, S., and Sansone, G., Cryopreservation of sea bream (*Sparus aurata*) spermatozoa, *Cryobiology*, 40, 46, 2000.

Complete affiliation:
Elsa Cabrita, Center for Marine Sciences-CCMAR, University of Algarve, Campus de Gambelas, 8000 Faro, Portugal. Phone: +351.289800900 ext. 7595 Fax: +351.289800069, e-mail: ecabrita@ualg.pt

CRYOPRESERVATION OF SEABREAM SEMEN

J-C. Gwo

1. INTRODUCTION

The Sparidae family contains a number of economically important species throughout the world. Four seabreams (*Acanthopagms latus, A. scheligeli, Pagrus major* and *Rhabdosargus sarba*) are the most popular for aquaculture in Taiwan. They all are protandrous, being a functional male at an early age (usually one year old) but reversing sex to become a functional female in later years. Self-fertilization is possible for these four species. Such studies could facilitate genetic understanding and quickly lead to the development of inbred lines. The cryopreservation of sperm would also aid broodstock management, gene conservation and genetic selection of the beneficial traits for intensive commercial farming. Kurokura et al. [1] succeeded in using frozen crimson seabream (*Evonnis japonica*) semen to fertilize female red seabream and produced hybrids with increased growth rate and desired body color.

2. PROTOCOL FOR FREEZING AND THAWING

2.1 Fish handling and semen collection

- Four seabreams (*Acanthopagms latus, A. scheligeli, Pagrus major* and *Rhabdosargus sarba*) were obtained from local fish farms in Peng-Hu, Taiwan. Semen was collected during the breeding season (October to Feburary) by abdominal massage from single males weighing between 250 and 500 g. The semen was placed on crushed ice immediately after collection; freezing was accomplished and sperm motility was determined within 30 min.
- The activation of fresh sperm motility, the measurement of semen osmolality, the protocol of freezing and thawing and the fertilization tests were examined as the same as those of grouper semen in this book.

2.2 Freezing and thawing procedures

- Sperm motility was evaluated for 5 cryoprotectants (DMSO, glycerol, propylene glycerol, ethylene glycerol and methanol). The effects of cryoprotectant concentrations of 10, 20 and 30% and an equilibration time of 5, 30 and 60 min were surveyed.
- Five extenders were compared: 150 mM unbuffered NaCI, 300 mM glucose, marine teleost Ringer, 150 mM sodium citrate, and distilled water. Ten percent dimethyl sulfoxide (DMSO; Sigma) was chosen as the sole cryoprotectant.

The semen dilution ratio (volume extender / volume cryoprotectant, volume semen) was 10:1 for all extenders.

- Semen to be frozen was diluted in the described extenders and cryoprotectants, and was drawn up into 0.25 ml straws (Instruments de Medecine Vetennaire-IMV L'Algle, France).
- The straws were frozen either in liquid nitrogen vapor at different distances (1 or 8 cm) to adjust freezing rates or they were buried in dry ice.
- After freezing (about 10 min), the straws were stored in liquid nitrogen. One day later straw samples were thawed in a 20°C water bath, and motility was checked and determined. Other aliquots of straws, frozen by burying in dry ice were saved in liquid nitrogen until fresh ova were available.

2.3 Fertilization

- Both 1 and 2 g ova samples were fertilized with 1 straw (0.25 ml) of either fresh (diluted 10 fold with 150 mM NaCl) or cryopreserved sperm.
- Immediately after addition of semen to ova, 10 ml of sea water were added to activate the sperm. Frozen sperm was used as soon as they were thawed. The percentage of fertilization was determined by the percentage of viable embryos just before hatching.

3. GENERAL CONSIDERATIONS

Sperm of seabream become motile only by contact with NaCl, KCl, glucose, artificial sea water and sea water solutions with osmolalities of 550 mOsm/kg or higher. No motility was observed when methanol was used at concentrations of 10, 20 and 30% prefreezing, and the motility score decreased with increased cryoprotectant concentrations of glycerol, ethylene glycerol, propylene glycerol and DMSO (Table 1).

The highest motility for DMSO was observed with a concentration of 10%; lower rates occurred with DMSO at concentrations of 20 and 30%, respectively. Drastic decreases in motility were observed with increased equilibration times (30 and 60 min) when glycerol, ethylene glycerol, propylene glycerol and DMSO were used at concentrations of 20 and 30%.

Intensive motility was observed as soon as seabream sperm were diluted (1:10) with glycerol concentrations ranging between 10 and 30%. Upon thawing, these sperm were spontaneously motile without further addition of seawater. Using 10% DMSO plus 150 mM NaCl at a semen dilution ratio of 1:10 in 0.25 ml straws, all freezing methods (above liquid nitrogen surface 1.4 and 8 cm or buried in dry ice) proved to be similar in post-thaw sperm motility. Using a constant volume (1 straw = 0.25 ml) of fresh semen (1 part of semen diluted in 9 parts of 1% sodium chloride), the fertilization rates were not significantly different with either 1 or 2 g of oocytes. However, the fertilization capacity of frozen-thawed seabream sperm, frozen 4 cm above liquid nitrogen surface at a estimated freezing rate of -65°C/min, was reduced when the amount of oocytes was increased from 1 to 2 g. Extenders such as 150 mM NaCl, marine teleost Ringer, 300 mM glucose and 150 mM sodium citrate containing 10% DMSO, and 300 mM glucose alone were appropriate for freezing seabream sperm (Table 1). Methanol is not a suitable cryoprotectant. No post-thaw fertilization was obtained in the present study when seabream sperm were frozen in both DMSO and methanol with distilled water as the extender.

Table 1 - Effect of different concentrations of various cryoprotectants at various equilibration times between prefreeze and post-thaw sperm motility of seabream.

Cryoprotectant	Concentration (%)	Equilibration time (minutes)		
		5	30	60
DMSO	10	4/4	4/4	4/4
	20	4/3	3/2	3/1
	30	1/0	0/0	0/0
Glycerol	10	4/2	1/0	0/0
	20	3/1	1/0	0/0
	30	1/0	0/0	0/0
Propylene glycerol	10	4/4	4/4	4/4
	20	4/1	2/0	0/0
	30	1/0	0/0	0/0
Ethylene glycerol	10	4/3	3/2	2/1
	20	4/1	1/0	0/0
	30	1/0	0/0	0/0
Methanol	10	0/0	0/0	0/0
	20	0/0	0/0	0/0
	30	0/0	0/0	0/0

Motility score: 0) no motile sperm, 1) 0-25%, 2) 25-50%, 3) 50-75% , 4) >75% of sperm with progressive movement.

Prolonged equilibration time had a detrimental effect on seabream sperm motility in the present study. Fish sperm do not require any delay in their processing because a delay may adversely affect the subsequent fertilization rate of frozen-thawed sperm; therefore, fish sperm were always frozen directly upon mixing with the diluent. However, the type of extender and cryoprotectant used may influence optimum equilibration time [2]. Extender containing sugar (glucose or sucrose) has proved to be advantageous in freezing fish semen. In addition, sugars are less toxic to sperm than many cryoprotectants, which may prove beneficial for cryopreservation [2]. A synergistic effect between the combination of glucose and DMSO in the diluent was found when freezing seabream semen in the present study. The combination of 300 mM glucose plus DMSO yielded a higher fertilization rate after freezing-thawing than 300 mM glucose only (Table 2). Similar results also have been observed in Atlantic croaker and marine puffer sperm[2].

Table 2 - Fertilization efficacy (using 1g of oocytes) of cryoprotectant (DMSO) and various extenders for freezing seabream sperm

Cryoprotectant	Extender	Fertilization rate (%)
	150 mM NaCl	51.7 ± 10.2 [a]
	marine teleost Ringer	50.5 ± 8.3 [a]
10% DMSO	glucose	48.8 ± 9.7 [a]
	sodium citrate	45.6 ± 7.8 [a]
	distilled water	0 [c]
None	glucose	35.7 ± 8.8 [b]

Values are means ± standard deviation; n = all experiments.
a,b Values with different superscripts differ significantly ($P < 0.05$).

The present study has shown that DMSO is superior to glycerol for seabream sperm. Some indications are available showing that the presence of glycerol lowers the fertilizing ability of semen in several fish species. For example, glycerol appears to be toxic to certain salmonids [3-5] and grouper sperm [2]. To minimize cryoprotectant toxicity, the equilibration time is usually kept to a minimum. Fish sperm are sufficiently small so that cryoprotectant penetration usually occurs very rapidly. Glycerol is osmotically active and was very slow in permeating seabream sperm membranes. In the present study, seabream sperm, diluted in 10% glycerol, would not become motile when equilibration times were 60 min or longer (Table 1). Various mixtures of extender and cryoprotectant have been tested to find a combination that prevents activation of sperm during dilution prior to freezing, thus preserving energy needed for fertilization and for the protection of sperm morphology [2,3,6,7,8,9,10]. Sperm of fish having external fertilization rely entirely on endogenous high motility energy readily available for a short period before they become incapacitated by osmotic shock. Intensive motility was observed before freezing as soon as seabream sperm were diluted with glycerol. Upon thawing, the seabream sperm regained spontaneous motility without further addition of seawater. Because of the short (about 70 s) duration of progressive motility of seabream sperm, glycerol, with its direct osmotic effect on sperm cells, exhausts the energy needed for sperm to be motile for fertilization. Use of high sperm concentration appears to circumvent this problem. In the present study, a reduction of the glycerol concentration increased the survival of frozen seabream semen (Table 1).

Propylene glycerol and ethylene glycerol are used as cryoprotectants for their ability to penetrate cell structure very rapidly and for their limited toxicity, thereby producing less osmotic shock. The present study also verified that both fresh and frozen-thawed seabream sperm motility were higher when 10% glycerol was replaced by either 10% propylene glycerol or 10% ethylene glycerol (Table 1). The molecular weights of both propylene glycerol and ethylene glycerol are much lower than that of glycerol, which may explain the better results achieved using them rather than glycerol for seabream. Methanol was superior to DMSO and glycerol in protecting tilapia and zebra fish sperm to -196°C [2,3]. However, methanol is not a suitable cryoprotectant for seabream sperm in this study. Methanol seems to damage the sperm membrane of seabream (Gwo, unpublished observations).

4. ACKNOWLEDGMENTS

This work was supported by a grant to J.-C. Gwo from the Council for Agricultural Planning and Development, Taiwan, under project number CA80-AD-7.1-F30-17.

5 REFERENCES

1-Kurokura, H., Kurnai, H., and Nakamura, M., Hybridization between female red sea bream and male crimson sea bream by means of sperm cryopreservation, in: *The First Asian Fisheries Forum*, Maclean J.L., Dizon L.B., and Hosillos L.V., Eds., Manila, The Philippines, 1986; 113.
2-Gwo, J.C., Cryopreservation of sperm of some marine fishes, in: *Cryopreservation in Aquatic Species*, Tiersch, T.R. and P.M. Mazik, Ed., Advances in World Aquaculture, 7, World Aquaculture Society, Baton Rouge, LA, USA, 2000, 138.
3-Stoss, J., Fish gamete preservation and sperm physiology, in: *Fish Physiology*, 9B, Hoar, M.S., Randall, D.J., and Donaldson, E.M., Ed., Academic Press, New York, 1983, 305.
4-Gwo, J.C., Strawn, K., Longnecker, M.T., and Arnold, C.R., Cryopreservation of Atlantic croaker sperm, *Aquaculture*, 94, 355, 1991.
5-Gwo, J.C., Kurokura, H., and Hirano, R., Cryopreservation of Spermatozoa from Rainbow Trout, Carp and Marine Puffer, *Nippon Suisan Gakk*, 59, 777, 1993.
6-Morisawa, M., Initiation mechanism of sperm motility at spawning in teleosts, *Zool. Sci.*, 2, 605, 1985.
7-Chambeyron, F. and Zohar, Y., A diluent for sperm cryopreservation of gilthead seabream, *Sparus aurata*, *Aquaculture*, 90, 345, 1990.
8-Fabbrocini, A., Lubrano Lavadera, S., Rispoli, S., and Sansone, G., Cryopreservation of Seabream (*Sparus aurata*) Spermatozoa, *Cryobiology*, 40, 46, 2000.
9-Cabrita, E., Robles, V., Cuñado, S., Wallace, J.C., Sarasquete C., and Herráez, M.P., Evaluation of gilthead sea bream, *Sparus aurata*, sperm quality after cryopreservation in 5 ml macrotubes, *Cryobiology*, 50, 273, 2005.
10- Cabrita, E., Robles, V., Rebordinos, L., Sarasquete, C., and Herráez, M.P., Evaluation of DNA damage in rainbow trout (*Oncorhynchus mykiss*) and gilthead sea bream (*Sparus aurata*) cryopreserved sperm, *Cryobiology*, 50, 144, 2005.

Complete affiliation:
Jin-Chywan Gwo, Department of Aquaculture, Taiwan National Ocean University, Keelung 20224, Taiwan. e-mail: gwonet@ms.16.hinet.net.

CRYOPRESERVATION OF SPERM FROM WINTER FLOUNDER, *Pseudopleuronectes americanus*

R.M. Rideout, I.A.E. Butts and M.K. Litvak

1. INTRODUCTION

Winter flounder, a eurythermal and euryhaline right-eyed flatfish, is an extremely hardy species [1] which possesses antifreeze proteins allowing it to survive in cold environments. This hardiness, coupled with high market prices, make winter flounder an excellent candidate for aquaculture in Atlantic Canada [2]. Because it has been used as a 'model fish' by scientists for many years, much is known about its reproductive biology. Fish can be maintained in captivity at high densities and are able to withstand frequent handling and stripping. Prior to stripping, fish can often be calmed by placing a wet paper towel over the head and anesthetic is often not necessary. Depending on location, winter flounder can spawn anywhere between March and June. Timing of reproduction in the lab can be controlled through photo-thermal manipulation and, if necessary, carp pituitary extract or GnRh-A. Eggs (0.55-0.86 mm) are negatively buoyant and adhesive [3,4]. Spermatozoa average 0.030-0.035 mm total length [5] with a spermatocrit of 65%, sperm volume of 49 ml kg^{-1} and production of 3.22 x 10^{14} sperm kg^{-1} [6].

Sperm cryopreservation for this species will allow researchers and aquaculturists to store sperm from males with selectively-bred traits, minimize the number of males required to be held in captivity and ensure availability of sperm when females are ready to spawn. Attempts to cryopreserve sperm on a small scale have been very successful [7]. This technique has not yet been modified for commercial-sized scale.

2. PROTOCOL FOR FREEZING AND THAWING

2.1 Equipment

- 3 ml syringes (without needles)
- Paper to clean mucus
- Styrofoam box and 5.5 cm high Styrofoam raft for freezing straws
- 0.25 ml cryogenic straws (minitubes)
- Cryogloves
- Long forceps for retrieving frozen samples
- Micropipettes (1 µl, 200 µl and 1000 µl)
- Basic laboratory equipment (microscope, pH-meter) and glass material
- Basic cryobiology equipment (Dewar Flasks, etc.)
- Refrigerator and crushed ice

- Heated water bath
- Clay or beads to seal cryogenic straws

2.2 Reagents

- Propylene glycol
- Modified Mounib's medium (glutathione, sucrose, potassium bicarbonate)
- Distilled water
- Liquid nitrogen

2.3 Fish Handling and Semen Collection

- Gently remove fish from holding tank
- Completely dry the entire fish and the hands of the collector using paper towel and be especially cautious that all water, faeces, urine, and mucus are removed from around the urogenital pore. No anaesthetic is required while drying the fish.
- To extract semen start by holding the flounder with one hand near the head and the other hand near the caudal peduncle. Strip semen by applying gentle pressure to the testes from the bottom of the testicular lobe and maintaining this pressure while moving up to the urogenital pore. Collect sperm into a 3 ml syringe. This procedure can be repeated several times until the desired amount of semen is obtained. Be aware that urine, and faeces can also be extracted from the urogenital area. If this does occur make sure they are eliminated before semen collection continues. Collecting semen is much easier with two people (one person can strip the semen while the other collects it with the syringe).
- Immediately after collection semen can either be placed in a refrigerator or into a beaker (glass or plastic) that is surrounded by crushed ice inside a cooler. If the sperm is not going to be immediately cryopreserved, oxygen can be injected into the syringe containing sperm to prolong viability. This procedure has proven to be useful for transporting semen, as well as holding it for up to 48 hours prior to use.

2.4 Pre-freezing Sperm Evaluation

- Before semen is cryogenetically frozen, sperm motility is observed. The sperm is activated by placing 10 µl of milt on a clean microscope slide and activating with 10 µl of seawater [7]. A coverslip is then added, and the percentage of sperm cells with   forward movement are assessed using an arbitrary numerical scale of six scores where: 0 is no motility; 1, 1-19%; 2, 20-39%; 3, 40-59%; 4, 60-79%; and 5, 80-100% [7]. Samples with less than 80% of cells moving after activation are not cryopreserved.

2.5 Freezing/Thawing Procedure

- Dilute the semen (1:3) in the modified Mounib's medium (0.125 M sucrose, 0.100 M $KHCO_3$, 0.0065 M reduced glutathione) and then add 10% (v/v) propylene glycol (PG). Mix by swirling or use a vortex.
- Load semen solution into 0.25 ml cryogenic straws (Minitube Canada) using a cryogenic straw pipetter (Minitube Canada). If this instrument is not available the sperm solution can be loaded into the cryogenic straws in two ways. Either (i) a syringe can be used to transfer the solution to the straw or (ii) the straw can be inserted directly into the opening of a syringe (with needle removed) and the syringe plunger pulled up to draw the solution directly into the straw. Once the straw has been completely filled, the unsealed end is filled with a metal bead or clay.
- Place the straws on a Styrofoam rack and then lower into a Styrofoam box containing one inch or less of liquid nitrogen. The rack should float the samples exactly 5.5 cm above the surface of the liquid nitrogen in order to obtain the proper freezing rate. After 12 min the samples have reached -90°C and are ready to be placed directly into the liquid nitrogen. Results have shown that sperm can still maintain motility rates of ~80% after being submerged in liquid nitrogen (within a 34 l dewar) for 3 years.
- Thaw samples for 7 s in a heated water bath set at 30°C. Often the metal bead or clay becomes dislodged from one end of the straw during freezing. Do not allow this end of the straw to become submerged in the water.
- Dry the straws and use scissors to cut off both ends. This permits the sperm to run out freely.
- Use the thawed sperm as soon as possible to guarantee viability.

2.6 Post-thaw Sperm Evaluation

- Sperm motility is tested in the same manner as it was prior to freezing.

2.7 Fertilization

- Once the female's egg pore has swelled outside the body cavity, a light pressure can be applied to the abdomen of the fish and any eggs expelled can be collected into plastic or glass beakers. Egg containers should preferably be held at the temperature upon which the fertilization is to occur.
- There are a number of approaches for rearing eggs [8]. We prefer to fertilize and incubate eggs in Petri dishes. High fertilization (> 90%) and hatch success (>80%) has occurred when incubating at 0.1-0.2 ml of eggs (450-900 eggs) per Petri dish (100 x 15 mm and 150 x 15 mm, respectively).

- Approximately, 40-60 µl of semen solution is added to the Petri dish, the gametes are then quickly swirled together followed by adding 28-30 ppt sterilized seawater (containing 13 mg/l of streptomycin sulfate and 13 mg/l of penicillin G) to the dish.
- The sperm-egg-seawater combination is left for 20 seconds followed by one seawater rinse. Longer gamete contact times can be used when larger volumes of eggs are being fertilized and should be followed by multiple seawater rinses. After rinsing the eggs, the Petri dishes are immediately refilled (¾ full to allow air exchange) with sterilized seawater antibiotic mixture, and covered with the lid. Every second day, 75% of the water in each Petri dish is replaced.

3. GENERAL CONSIDERATIONS

Sperm of this species was very amenable to being cryogenically stored using a number of procedures. The details given here only represent the technique that was most successful and the one that is currently used. Some success was also achieved using a saline-based extender and other cryoprotectants such as DMSO and glycerol. Work has shown that high fertilization success can be achieved using any combination of these extenders and cryoprotectants if the amount of sperm used is sufficient and/or the gamete contact time is long. It is important that sperm and eggs be rinsed thoroughly after fertilization since the reagents used can be toxic and result in low hatching success.

4. REFERENCES

1-Bigelow, H.B. and Schroeder, W.C., Fishes of the Gulf of Maine, *Bulletin of the Fish and Wildlife Service*, 53, 276, 1953.
2-Litvak, M K., The development of winter flounder (*Pleuronectes americanus*) for aquaculture in Atlantic Canada: current status and future prospects, *Aquaculture*, 176, 55, 1999.
3-Smigielski, A.A., Hormonal-induced ovulation of the winter flounder, *Pseudopleuronectes americanus*, *Fish B.-NOA*, 76, 431, 1975.
4-Butts, I.A.E., Parental and stock effects on embryo and early larval development of winter flounder (*Pseudopleuronectes americanus*), M.Sc. thesis, University of New Brunswick, 2006.
5-Breder, C.M., Some embryonic and larval stages of the winter flounder, *Bulletin of the United States Bureau of Commercial Fisheries*, 38, 311, 1924.
6-Shangguan, B. and Crim, L.W., Seasonal variations in sperm production and sperm quality in male winter flounder, *Pleuronectes americanus*: the effects of hypophysectomy, pituitary replacement therapy, and GnRH-A treatment, *Mar Biol*, 134, 19, 1999.
7-Rideout, R.M., Litvak, M.K., and Trippel, E.A., The development of a sperm cryopreservation protocol for winter flounder *Pseudopleuronectes americanus* (Walbaum): evaluation of cryoprotectants and diluents, *Aquac Res*, 34, 653, 2003.
8-Howell, W.H. and Litvak, M.K., Winter flounder culture, in: *Aquaculture Encyclopedia*, Stickney, R., John Wiley & Sons, Inc., Eds., New York, 2000, 98.

Complete affiliation:

Rick Rideout, Fisheries and Oceans Canada, Northwest Atlantic Fisheries Centre, PO Box 5667, St. John's, NL, A1C 5X1, Canada, e-mail: rideoutr@dfo-mpo.gc.ca.

CRYOPRESERVATION OF SPERM IN TURBOT (*Psetta maxima*)

M. Suquet, O. Chereguini and C. Fauvel

1. INTRODUCTION

Turbot is distributed from Norway to Morocco but also in the Mediterranean and Black seas. Depending on the geographic area, its spawning period lasts from February to April (Mediterranean sea), April to August (North sea) and May to July (Channel and North Atlantic) [1]. Turbot farming is a recent industry with a total production of 5,500 tons in 2000 [2].

Spontaneous spawns were collected in captivity but both egg quality and quantity were variable. Thus, stripping females and artificial insemination were recommended for industrial regular production, requiring a good knowledge of biology and management of turbot gametes. In the female, this includes the settlement of a stripping protocol preventing ova overripening and increasing larvae production by 300% [3] but also the availability of analytical tools such as pH measurements of ovarian fluid used to estimate ova quality [4]. In the male, different features of sperm biology such as motility characteristics, energetic metabolism, spermatozoon morphology, seminal fluid composition and changes with time of sperm quality were also described [5].

Furthermore, since turbot was considered as a poor sperm producer in terms of spermatozoa concentration and semen volume collected [6], the improvement of sperm handling techniques was a prerequisite for artificial reproduction improvement in aquaculture.

Based on increased knowledge of gametes biology, the development of management procedures such as short term storage and artificial insemination supported the settlement of a cryopreservation method of turbot sperm.

2. PROTOCOL FOR FREEZING AND THAWING

2.1 Fish Handling and Semen Collection

- Before freezing, particular attention was paid to sperm sampling quality. Because of the close vicinity of sperm and urinary ducts, turbot sperm samples are frequently contaminated with urine (mean contamination rate: 15.3%, urine volume/sperm volume; [7]). The presence of urine in sperm significantly modified the composition of seminal fluid, induced a delay in the initiation of spermatozoa motility and decreased ATP content, sperm velocity, percentage of motile spermatozoa, short term storage and fertilization capabilities.

- Empty the urinary bladder with a catheter inserted through the gonopore prior to sperm collection in turbot.
- Sperm was then sucked into a syringe and immediately stored at 4°C, since the percentage of motile cells significantly decreased after 5 minutes at 15°C.

2.2 Pre-freezing Sperm Evaluation

- Ageing of sperm resulted in a lower cryopreservation capacity of turbot sperm at the end of the spawning period [8]. Before processing, sperm sample quality was checked by measuring the percentage of motile cells using a two step dilution in order to initiate a synchronous activation of spermatozoa [9].
- Only sperm samples presenting a high motility (>70%) were used for freezing.

2.3 Freezing and Thawing Procedures

- Semen was diluted at a 1:2 ratio (sperm:diluent) with a modified Mounib extender (KHCO$_3$ 100 mM, sucrose 125 mM, reduced gluthatione 6.5 mM, BSA 10mg/ml, pH 7.8), supplemented with 10% DMSO.
- Samples were sucked into 200 µl straws, placed on a floating tray in nitrogen vapour and after a 15 minute period, plunged into liquid nitrogen.
- The straws were thawed in a water bath at 30°C for 5s.

3. GENERAL CONSIDERATIONS

The Mounib medium gave the highest post-thaw survival [10,11]. DMSO was proved to be more efficacious than other cryoprotectants (Fig. 1).

Best results have also been recorded using DMSO, compared to propylene glycol and dimethylformamide [11]. The height of the floating tray, determining the kinetics of temperature decrease, influence sperm quality after freezing (Fig. 2).

On the other hand, the nature of proteins added to the diluent to protect the plasma membrane (10% BSA or egg yolk), the dilution rate of sperm in Mounib (from 1:1 to 1:9) and the thawing temperature (from 20 to 40°C) did not influence the post thaw sperm motility [12].

The quality of thawed sperm was described using a panel of bio-tests (Table 1). Compared to fresh sperm, 60 to 80% of the thawed spermatozoa could be reactivated. Cryopreservation decreased intracellular ATP content by 40%, showing this technique was energy-consuming. On the other hand, neither sperm velocity nor respiratory rate were affected by freezing [12]. Flow cytometric analysis revealed that 93 to 96% of thawed turbot spermatozoa had an undamaged plasma membrane [13]. Then, the fertilization capacity of thawed sperm was significantly decreased for low sperm to egg ratios, compared to fresh sperm (Fig. 3). However, no differences were observed using 20,000 sperm per egg [12].

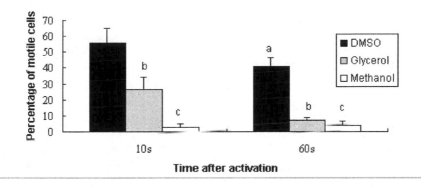

Figure 1 -. Effect of cryoprotectant (concentration: 10%) on thawed sperm motility (for each time after activation, different letters show significantly different results: P<0.05).

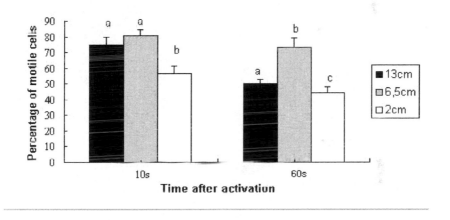

Figure 2 - Effect of the tray height on thawed sperm motility (for each time after activation, different letters show significantly different results: P<0.05).

Table 1 - Turbot fresh and frozen/thawed sperm characteristics

Sperm	Fresh	Frozen/thawed	Difference
Sperm concentration (x 10^9 spz ml^{-1})	Up to 11	-	
Motility (%)	85	75	*
Flagellar beat frequency (Hz)	50	50	NS
ATP (nM.10^{-9} spz)	260	160	*
Respiratory rate (μl O$_2$.min^{-1}.10^{-9} spz)	1.61	1.45	NS
Straight-line velocity (μm s^{-1})	180	200	NS
Motility duration (min.)	5	5	NS
Maximum fertility* (%)	70	68	NS

No significant differences were observed in hatching and survival rates of larvae produced using thawed sperm (Figure 3), but also in length and weight of 7, 10 and 14 day old larvae [14,15]. Furthermore, the survival and changes in weight and length with time of turbot juveniles, produced using thawed sperm, were not significantly different from those of fresh sperm [16]. This method was successfully adapted to seabass (*Dicentrarchus labrax*; Fauvel and Suquet, protocol 19 in this volume). However, cryopreservation of cod (*Gadus morhua*) sperm using the same technique, showed low motility of thawed samples, ranging from 4 to 6% of the value recorded in fresh samples (Rouxel, unpublished results). The cryopreservation process of turbot sperm was up-scaled: no significantly different fertilization rates were observed using two different-sized containers (0.5 ml straws and 2 ml cryotubes). Furthermore, fertilization and hatching rates were not significantly different between fresh and cryopreserved sperm collected in 49 males [17]. Forty millilitres ova batches were successfully fertilized, showing fertilization and hatching rates similar to those of fresh sperm [11].

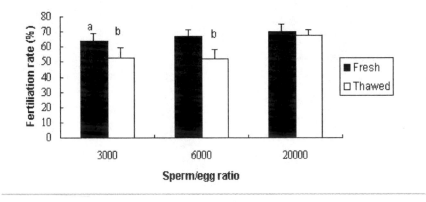

Figure 3 - Effect of sperm to egg ratio on the fertilization rate of fresh and thawed sperm.

In conclusion, a simple and practical method of cryopreservation of turbot spermatozoa was settled. It gives a high survival and fertilization capacity of spermatozoa. This method was extrapolated to production scale showing similar rearing performances between offspring produced using cryopreserved or fresh sperm. This observation suggests the absence of genome alterations and supports the use of this technique for gene banking. These techniques contribute to the improvement of turbot hatchery management since industrial production of turbot depends on mastering artificial fertilization, cryopreservation, short term storage and artificial insemination.

4. REFERENCES

1-Quéro, J.C. and Vayne, J.J., *Les poissons de mer des pêches françaises*, Delachaux and Niestlé, Eds., Lausanne, 1997, 304.

2-Person-Le Ruyet, J., Turbot (*Scophthalmus maximus*) grow-out in Europe: practices, results, and prospects, *Turkish Journal of Fisheries and Aquatic Sciences*, 2, 29, 2002.

3-Fauvel, C., Omnes, M.H., Suquet, M., and Normant, Y., Enhancement of the production of turbot, *Scophthalmus maximus* (L.), larvae by controlling overripening in mature females, *Aquaculture and Fisheries Management*, 23, 209, 1992.

4-Fauvel, C., Omnes, M.H., Suquet, M., and Normant, Y., Reliable assessment of overripening in turbot (*Scophthalmus maximus*) by a simple pH measurement, *Aquaculture*, 117, 107, 1993.

5-Dreanno, C., Régulation de la mobilité des spermatozoïdes de turbot (*Psetta maxima*) et de bar (*Dicentrarchus labrax*): Etude du métabolisme énergétique, du contrôleionique, de la morphologie et du pouvoir fécondant, Thesis, Rennes, France, 1998, 81.

6-Suquet, M., Billard, R., Cosson, J., Dorange, G., Chauvaud, L., Mugnier, C., and Fauvel, C., Sperm features in turbot (*Scophthalmus maximus*): a comparison with other freshwater and marine fish species, *Aquat Living Resour*, 7, 283, 1994.

7-Dreanno, C., Suquet, M., Desbruyères, E., Cosson, J., Le Delliou, H., and Billard, R., Effect of urine on semen quality in turbot (*Psetta maxima*), *Aquaculture*, 169, 247, 1998.

8-Suquet, M., Dreanno, C., Fauvel, C., Cosson, J., and Billard, R., Cryopreservation of sperm in marine fish, *Aquac Res*, 31, 231, 2000.

9-Chauvaud, L., Cosson, J., Suquet, M., and Billard, R., Sperm motility in turbot, *Scophthalmus maximus*: initiation of movement and changes with time of swimming characteristics, *Environl Biol Fish*, 43, 341, 1995.

10-Chereguini, O., Maria Cal, R., Dreanno, C., Ogier de Baulny, B., Suquet, M., and Maisse, G., Short-term storage and cryopreservation of turbot (*Scophthalmus maximus*) sperm, *Aquat Living Resour*, 10, 251, 1997.

11-Chen, S.L., Ji, X.S., Yu, G.C., Tian, Y.S., and Sha, Z.X., Cryopreservation of sperm from turbot (*Scophthalmus maximus*) and application to large scale fertilization, *Aquaculture*, 236, 547, 2004.

12-Dreanno, C., Suquet, M., Quemener, L., Cosson, J., Fierville, F., Normant, Y., and Billard, R., Cryopreservation of turbot (*Scophthalmus maximus*) spermatozoa, *Theriogenology*, 48, 589, 1997.

13-Ogier de Baulny, B., Le Vern, Y., Kerboeuf, D., Heydorff, M., and Maisse, G., Flow cytometric analysis of plasma membrane damages of rainbow trout and turbot frozen sperm, in: *Proceedings of the Commission C2, Refrigeration and production, International Symposium "Froid et Aquaculture"*, Bordeaux, France, 1997, 65.

14-Suquet, M., Dreanno, C., Petton, B., Normant, Y., Omnes, M.H., and Billard, R., Long-term effects of the cryopreservation of turbot (*Psetta maxima*) spermatozoa, *Aquat Living Resour*, 11, 45, 1998.

15-Chereguini, O., Garcia de la Banda, I., Rasines, I., and Fernandez, A., Larval growth of turbot, *Scophthalmus maximus* (L.) produced with fresh and cryopreserved sperm, *Aquac Res*, 32, 133, 2001.

16-Chereguini, O., Garcia de la Banda, I., Rasines, I., and Fernandez, A., Growth and survival of young turbot (*Scophthalmus maximus* L.) produced with cryopreserved sperm, *Aquac Res*, 33, 637, 2002.

17-Chereguini, O., Garcia de la Banda, I., Herrera, M., Martinez, C., and De la Hera, M., Cryopreservation of turbot *Scophthalmus maximus* (L.) sperm: fertilization and hatching rates, *Aquac Res*, 34, 739, 2003.

Complete affiliation:

Marc Suquet, Ifremer, Laboratoire ARN, BP 70, 29280 Plouzané, France. Tel: 0298895755, e-mail: Marc.Suquet@ifremer.fr

CRYOPRESERVATION OF GROUPER SEMEN

J-C. Gwo and H. Ohta

1. INTRODUCTION

The grouper is a delicious and expensive food in Japan, Taiwan and southeast Asia; among groupers the malaba grouper (*Epinephelus malabaricus*) and kelp grouper (*E. bruneus*) are protogynous hermaphrodite fish, and functional males are reported to mature at 5 to 10 years of age, when reaching a weight of 19 kg or more. Due to their large size, the source of male broodstock is limited; thus, collecting semen is always a problem. Cryopreservation of grouper sperm would reduce the number of males needed, minimize handling stress through less frequent stripping, facilitate artificial propagation when ova are available, promote genetic and breeding studies, and enhance seed production on a commercial scale.

2. PROTOCOL FOR FREEZING AND THAWING

2.1 Fish Handling and Semen Collection

- Male malaba grouper (30~40 kg body weight) and kelp grouper (21~28 Kg) were obtained from Taiwan and Japan, respectively. Spermiation was induced by intramuscular injection of human chorionic gonadotropin (hCG; China Chemical & Pharmaceutic Co.Ltd., Taipei, Taiwan) at a dose of 500~600 IU/ kg of body weight. Spermiation started 24 to 36 hours following hormone administration.
- Semen was collected during the breeding season (March to December) by abdominal massage from single males. Semen was placed on crushed ice immediately after collection; freezing was accomplished and sperm motility was determined within 30 minutes.

2.2 Freezing and Thawing Procedures

- Sperm motility was evaluated for 3 cryoprotectants (DMSO, glycerol and methanol). The effects of cryoprotectant concentrations of 10, 20 and 30% and an equilibration time of 5 minutes were surveyed.
- Five extenders were compared: 150 mM unbuffered NaCl, 300 mM glucose, marine teleost Ringer, 150 mM sodium citrate, and distilled water. Twenty percent dimethyl sulfoxide (DMSO; Sigma) was chosen as the sole cryoprotectant. The semen dilution ratio (volume extender + volume cryoprotectant: volume semen) was 10:1 for all extenders. When 150 mM NaCl

was used as the extender, two semen dilution ratios (10 and 100) and two equilibration times (5 and 60 minutes) were tested.

- Semen to be frozen was diluted in the described extenders and cryoprotectants, and was drawn up into 0.25 ml straws (Instruments de Medecine Vetennaire-IMV L'Algle, France).
- The straws were frozen either in liquid nitrogen vapor at different distances (1 or 8 cm) to adjust freezing rates or they were buried in dry ice.
- After freezing (about 10 minutes), the straws were stored in liquid nitrogen. One day later straw samples were thawed in a 20°C water bath, and motility was checked and determined. Other aliquots of straws, frozen by burying in dry ice were saved in liquid nitrogen until fresh ova were available.

2.3 Fertilization

- A sample of about 100 oocytes was sampled from each female via intraovarian biopsy, and their diameters were measured. Females, with oocytes > 0.4 mm in diameter, were injected with hCG at a dose of 0.5 IU/g of body weight. Ovulation occurred about 24 to 30 hours after a single injection at 25°C water temperature, and ovulated oocytes were hand-stripped and inseminated within 20 minutes with a similar concentration of either fresh or cryopreserved sperm.
- Both 1 and 2 g ova samples were fertilized with 1 straw (0.25 ml) of either fresh (diluted 10 fold with 150 mM NaCl) or cryopreserved sperm. One-gram samples of ova comprised about 2,000 oocytes.
- Immediately after addition of semen to ova, 10 ml of sea water were added to activate the sperm. Frozen sperm was used as soon as they were thawed. The percentage of fertilization was determined by the percentage of viable embryos just before hatching, as described by Gwo et al. [1].

3. GENERAL CONSIDERATIONS

3.1 Selection of Diluents and Cryoprotectants

The concentration of malaba sperm in the 3 males ranged from 4.2-8.6 x 10^9/ml. Sperm were immotile in electrolyte (150 mOsm/kg (75 mM) NaCl, 150 mOsm/kg (75 mM) KCl and 300 mOsm/kg artificial sea water) and nonelectrolyte (300 mOsm/kg glucose) solutions (Table I). The osmolarity of the seminal plasma was 340 mOsm/kg. Sperm become motile only by contact with NaCl, KCl, artificial sea water, and glucose solutions with osmolalities of 600 mOsm/kg or higher. Sperm maintained full (Score 4) motility in NaCl, KCl, artificial sea water, sea water and glucose (Table 2) among the semen dilution ratios tested (10, 20 and 100). The duration of sperm motility decreased from approximately 40 minutes to about 2 minutes with an increase of semen dilution ratio from 10 to 100 (Table 2).

Table 1 - The effect of NaCI, KCl, artificial sea water (ASW) and glucose on sperm motility of malaba grouper at a semen dilution ratio of 100

Chemical	Osmolality (mOsm/Kg)				
	150	300	450	600	1,100
NaCl, KCl, ASW and glucose	0	0	0	4	4

Motility score 0: no motile sperm. Motility score 1: >0-25% of sperm with progressive movement. Motility score 2: >25-50% of sperm with progressive movement. Motility score 3: >50-75% of sperm with progressive movement. Motility score 4: >75% of sperm with progressive movement.

Table 2- The effect of semen dilution ratio on malaba grouper sperm motility in NaCl, KCl, artificial sea water, sea water and glucose solutions

	Semen dilution ratio	Duration (min) of motility	Motility score
NaCl, KCl, ASW,	10	40[a]	4
sea water	20	10[b]	4
glucose	100	2[c]	4

Column values with different superscripts differ significantly (P<0.05).

Table 3 - The effect of different concentrations of DMSO, glycerol and methanol on post-thaw sperm motility of malaba grouper

Cryoprotectant	Concentration (%)	Motility score
DMSO	10	3
	20	4
	30	1
Glycerol	10	0
	20	0
Methanol	30	0

Motility score see Table 1.

Table 4 - The effect of ova quantity on post-thaw grouper sperm fertilization capacity. Fertilization rate (%), are means ± standard deviation

Type of semen	Ova quantity (grams)				
	1	2	2.5	10	20
Malaba grouper semen					
Fresh	60.8 ± 2.4[a]	55.4 ± 4.2[a]			
Frozen-thawed	56.7 ± 3.3[a]	33.1 ± 2.3[b]			
Kelp grouper semen[1]					
Fresh			91.9 ± 1.6[c]	90.3±1.5[c]	88.3 ± 2.8[c]
Frozen-thawed			86.8 ± 4.8[c]	83.7±7.9[c]	77.2 ± 7.4[d]

Values with different superscripts differ significantly. [1]Using 1.0 ml of a mixture (DMSO:300mM glucose:semen = 5:93:2) of either fresh or frozen-thawed kelp grouper semen.

Because 150 mM NaCl yielded score 4 along with marine teleost Ringer extender for malaba grouper sperm and because of its simplicity, it was chosen for use in the following experiments. A semen dilution ratio of 10 was superior (sperm motility score 4) to that of 100 (score 1; Table 3). The post-thaw motility score decreased with increasing equilibration time. The short equilibration time (5 min) produced higher post-thaw motility than that obtained with an equilibration time of 60 min. On the basis of the above results, semen dilution ratio of 10 and equilibration time of 5 minutes were used. The highest motility for DMSO was observed with a concentration of 20%; lower levels occurred at DMSO concentrations of 10 and 30%, respectively (Table 3). No motility was obtained when both glycerol and methanol were used at concentrations of 10, 20 and 30% (Table 3). Using 20% DMSO plus 150 mM NaCl at a semen dilution ratio of 10 in a 0.25 ml straw, all freezing methods (above liquid nitrogen surface 1 or 8 cm above a liquid nitrogen surface or buried in dry ice) proved to be similar in post-thaw sperm motility scores. Using a constant volume (1 straw = 0.25 ml) of fresh malaba grouper semen (1 part semen diluted in 9 parts of 150 mM NaCl), the fertilization rates were not significantly ($P < 0.05$) different with 1 and 2 g of ova (Table 4) for malaba grouper. However, the fertilization capacity of cryopreserved grouper sperm was significantly ($P < 0.05$) reduced when the number of oocytes was increased (Table 4).

Sperm of oviparous fish are immotile in seminal plasma [2-4]. Vigorous motility of good quality sperm occurs after dilution in a proper medium. It is important that sperm motility is inhibited before freezing and during thawing, because this motility will exhaust the energy reserved for sperm motility to fertilize the oocytes. In marine teleosts, sperm motility is usually initiated when osmolality increases beyond 300 mOsm/kg [5]. However, the lowest osmolality (600 mOsm/kg) which triggered the initiation of motility in malaba grouper sperm in the present study was much higher than those reported in other marine fish species.

The decrease in duration of sperm motility following the increase in the semen dilution ratio in sea water, artificial sea water, NaCl, KCl and glucose seems to be due to a dilution effect. A minimum concentration of the essential material(s) contained in the seminal plasma may be necessary for sperm motility [1,6,7].

3.2 Factors Affecting Post-thawed Quality

High semen dilution ratio with 150 mM NaCl could be responsible for the reduced motility score after the freezing-thawing of malaba grouper sperm. Withler and Lim [8] indicated a better post-thaw motility score at a semen dilution ratio of 9 than of 24 in sperm of E. tauvina another grouper species. Post-thaw motilities of sperm frozen either at distances of 1 and 8 cm above the surface of liquid nitrogen, or those buried in dry ice, were similar.

There have been no reports suggesting that fish sperm are sensitive to cold shock, and fish semen is usually stored on ice prior to processing. The difference in susceptibility of sperm to cold shock suggests that the membrane composition of fish sperm may differ from those of mammalian sperm. It was found that DMSO was a successful cryoprotectant while glycerol was not for the cryopreservation of *E. tauvina* [8], *E. malabaricus* and *E. bruneus* sperm in the present study. Furthermore, while distilled water provides some protection for rainbow trout sperm undergoing freezing [3], distilled water is not effective for either Atlantic croaker [1], seabream [4] or for grouper sperm as shown in the present study. Both egg yolk and milk powder have been claimed to protect rainbow trout semen [3], but they have not provided benefit for Atlantic croaker [1]. Sodium citrate plus DMSO save good post-thaw fertilization rates for the marine puffer [9] and several seabreams [4]; however, no post-thaw fertilization was reported in the Atlantic croaker and common carp [9]. The increasing evidence of a species-specific requirement of extenders and cryoprotectant for freezing implies specific variations in the composition of seminal plasma as well as sperm membranes and sperm structure [4]. The study of lipid composition and its distribution and arrangement in the fish sperm membrane may contribute toward better understanding of membrane biochemistry, the underlying molecular mechanism responsible for membrane damage at reduced temperatures. It would also promote the development of more satisfactory cryopreservation techniques.

In summary, malaba and kelp grouper sperm equilibrated for 5 minutes in a 300 mM glucose extender containing 10~20% DMSO with a freezing rate of 32.5°C/min can be successfully cryopreserved, while retaining a fertilization capacity similar to that of fresh sperm, when a proper amount of ova were inseminated.

4. ACKNOWLEDGMENTS

This work was supported by a grant to J.-C. Gwo from the Council for Agricultural Planning and Development, Taiwan, under project number CA80-AD-7.1-F30-17. We also thank the staff members of Komame Station and Amami Station, National Center for Stock Enhancement, Fisheries Research Agency, Japan.

5. REFERENCES

1-Gwo, J.C., Strawn, K., Longnecker, M.T., and Arnold, C.R, Cryopreservation of Atlantic croaker sperm, *Aquaculture*, 94, 355, 1991.
2-Sin, S. and Hirano, R., Study on cryopreservation of fish sperm, in *:Proceedings Japanese Society of Scientific Fisheries, abstr.* 1972, 115.
3-Stoss, J., Fish gamete preservation and sperm physiology, in: *Fish Physiology*, 9B, Hoar, M.S., Randall, D.J., and Donaldson, E.M., Eds, Academic Press, New York, 1983, 305.
4-Gwo, J.C., Cryopreservation of sperm of some marine fishes, in: *Advances in World Aquaculture*, 7, Cryopreservation in Aquatic Species, Tiersch, T.R. and P.M Mazik, Eds., World Aquaculture Society Baton Rouge, LA, USA, 2000,138.
5-Morisawa, M. and Morisawa, S., Acquisition and initiation of sperm motility, in: *Controls of Sperm Motility: biological and clinical aspects*, Gagnon, C., Ed., CRC Press, Boca Raton, FL, 1990, 137.
6-Scott, A.P. and Baynes, S.M., A review of the biology, handling and storage of salmonid spermatozoa, *J. Fish Biol.*, 17, 707, 1980.

7-Billard, R., Effects of coelomic and seminal fluids and various saline diluents on the fertilizing ability of sperm in the rainbow trout *Salmo gairdneri*, *J Repord. Fertil.*, 68, 77, 1983
8-Withler, F.C. and Lim, L.C., Preliminary observations of chilled and deep-frozen storage of grouper (*Epinephelus tauvina*) sperm, *Aquaculture*, 27, 389, 1982.
9-Gwo, J.C., Kurokura, H., and Hirano, R., Cryopreservation of Spermatozoa from Rainbow Trout, Carp and Marine Puffer, *Nippon Suisan Gakk*, 59, 777, 1993.

Complete affiliation:

Jin-Chywan Gwo, Ph.D., Department of Aquaculture, Taiwan National Ocean University, Keelung 20224, Taiwan. Phone: +886 2 246 22 150 Fax: +886 2 250 93 985. e-mail: gwonets.16@hinet.net

CRYOPRESERVATION OF SPERM FROM ATLANTIC COD (*Gadus morhua*) AND HADDOCK (*Melanogrammus aeglefinus*)

R.M. Rideout and D.L. Berlinsky

1. INTRODUCTION

Atlantic cod (*Gadus morhua*) and haddock (*Melanogrammus aeglefinus*) are being examined as potential cold water aquaculture species, due in part to worldwide decreases in landings from wild fisheries. Both species are batch spawners with individual females releasing numerous batches of eggs over several weeks [1,2]. Fish spawn readily in captivity but gametes can also be manually stripped for selective breeding and experimental purposes.

Like many potential aquaculture species, the integration of biotechnology, including the cryopreservation of sperm, is being explored as a means to ensure success of the industry. Primary interest in cryopreservation lies in the ability to bank sperm from males with superior genetic traits (high growth rate, survivorship, etc.), as well as storage from disease-free fish and prevention of vertically transmitted diseases such as nodavirus.

A procedure developed many years ago for cryopreserving cod sperm [3,4] has been improved upon in recent years and has also served as a guide for the development of techniques for the successful cryopreservation of haddock sperm [5,6]. Current procedures use small cryogenic straws. Larger scale freezing procedures, suitable for commercial application, have not yet been developed.

2. PROTOCOL FOR FREEZING AND THAWING

2.1 Equipment

- 3-5 ml syringes (without needles)
- Paper to clean mucus
- Flat, clean surface to support fish
- Programmable cryogenic freezer or Styrofoam box and 5.5 cm high Styrofoam raft for freezing straws
- 0.25 – 0.5 ml cryogenic straws (minitubes)
- Cryogloves
- Long forceps for retrieving frozen samples
- Micropipettes (1 µl, 200 µl and 1000 µl)
- Basic laboratory equipment (microscope, pH-meter)
- Basic laboratory glassware
- Basic cryobiology equipment (Dewar Flasks, etc.)

- Refrigerator and crushed ice
- Heated water bath
- Clay, sealing powder or beads to seal cryogenic straws

2.2 Reagents

- MS-222 (tricaine methanesulfonate) or metomidate
- Propylene glycol or DMSO
- Modified Mounib's medium (glutathione, sucrose, potassium bicarbonate)
- Distilled water
- Liquid nitrogen

2.3 Fish handling and collection of sperm

- Transfer fish one at a time from holding tank to water bath containing anaesthetic.
- Eliminate faeces and urine by applying gentle pressure to the abdominal region and carefully dry urogenital area with paper towels or lab wipes.
- Lay fish on its side or back on a smooth flat surface. Massage the abdomen with one hand and use the other to collect the expressed sperm into 3-5 ml syringes (without needle). Multiple syringes may be needed to collect all the sperm from a single male. Syringes filled with sperm are placed in a refrigerator or container on crushed ice.

2.4 Pre-freezing sperm evaluation

- Sperm quality can be tested via (i) motility scores or (ii) differential staining. Motility is tested by diluting sperm in seawater and assessing the proportion of sperm moving against a stationary grid. Dilution can be approximate (e.g., touch a pipette tip that was dipped in sperm onto a clean microscope slide and dilute the resulting 'dot' of sperm with a single drop of water) or exact (e.g., 1:20).
- The grid used to assess motility can be on a hemacytometer or drawn on a clear plastic sheet covering a computer viewing screen. Sperm motility can be assessed as percent motility or as an arbitrary motility score (1 = very slow vibrating, 5 = rapidly swimming). The technology for computer-assisted sperm analysis (velocity and motility) has also been developed (Hamilton Thorne Biosciences, Beverly, MA; CEROS® V12.2 g; CASA).
- Sperm viability can also be assessed using a LIVE/DEAD® sperm viability kit (Molecular Probes, Eugene, OR) which differentially stains live and dead cells.

2.5 Freezing/Thawing procedure

- Dilute the semen (1:3) in modified Mounib's medium (0.125 M sucrose, 0.1 M KHCO$_3$, 0.0065 M reduced glutathione) and then add cryoprotectant. Mix by swirling or use a vortex. Both dimethyl sulphoxide (DMSO) and propylene glycol (PG) can be used as cryoprotectants. DeGraaf and Berlinsky [5] reported greater success with 10% than 5% DMSO and Rideout et al. [6] reported greater success with 10% PG than 10% DMSO.
- Semen can be loaded into 0.25-0.5 ml cryogenic straws using either (i) a cryogenic straw pipetter (Minitube Canada) or (ii) a syringe with a needle attached to transfer the semen solution to the straw. If 0.25 ml straws are used, the straw can be inserted directly into the opening of a syringe (with needle removed) and the syringe plunger pulled up to draw the solution directly into the straw. Once the straw has been completely filled, the unsealed end is filled with a metal bead, clay or sealing powder.
- Cod and haddock sperm can be frozen using a programmable freezer or by floating samples over liquid nitrogen. In the case of the programmable freezer, samples should be frozen at a rate of -5°C min^{-1} until they reach -150°C. Then they are plunged directly into liquid nitrogen for storage. In the absence of a programmable freezer, samples should be floated 5.5 cm above liquid nitrogen on a Styrofoam rack until they reach at least -90°C. This takes approximately 12 minutes with 0.25 ml cryogenic straws. Then the straws are plunged directly into the liquid nitrogen for storage.
- The effect of thawing rate on post-thaw viability has not been thoroughly evaluated but two thawing procedures have been used successfully. Rideout et al. [6] thawed samples in 30°C water for 7 s and DeGraaf and Berlinsky [5] thawed samples in 20°C water for 30 s. Often the sealing material becomes dislodged from one end of the straw during freezing. Do not allow this end of the straw to become submerged in the water.
- Dry the straws and use scissors to cut off both ends. This permits the sperm to run out freely.
- Use the thawed sperm as soon as possible to guarantee viability.

2.6 Post-thaw sperm evaluation

- Sperm viability is tested in the same manner as it was prior to freezing.

2.7 Fertilization

- Both cod and haddock females can spawn freely in captivity. Analogues of gonadotropin releasing hormone (e.g., salmon, mammalian) have also be used to induce ovulation.

- Eggs are stripped into large plastic or stainless steel collection containers. The optimal sperm to egg ratio has not been determined. In experimental trials, Degraaf and Berlinsky [5] fertilized 2 ml aliquots of eggs with 500 μl of post-thawed extended semen, while Rideout et al. [6] used 200 μl of thawed semen solution to fertilize 10 ml aliquots of eggs. Fertilization success is determined 4-48 h later.

3. GENERAL CONSIDERATIONS

The relationship between spermatocrit and sperm cryopreservation success has not been directly tested. Spermatocrit is known to increase throughout the spawning season for haddock, however, and motility is reduced at very high spermatocrits [7]. It is therefore suggested to avoid cryopreserving sperm from males sampled very late in the spawning season or those that have 'leftover' semen when females have completed spawning.

Although both cod and haddock can and have been stripped successfully, cod appear to be more amenable to such manipulations. It is not uncommon for male haddock to die after stripping. Mortality can be reduced by only collecting gametes from those fish that readily release sperm with only light abdominal pressure and by minimizing the time that fish are held out of water.

4. REFERENCES

1-Chambers, R.C. and Waiwood, K.G., Maternal and seasonal differences in egg sizes and spawning characteristics of captive Atlantic cod, *Gadus morhua, Can J Fish Aquat Sci*, 53, 1986, 1996.
2-Rideout, R.M., Trippel, E.A., and Litvak, M.K., Effects of egg size, food supply and spawning time on early life history success of haddock *Melanogrammus aeglefinus, Marine Ecology Progress Series*, 285: 169, 2005.
3-Mounib, M.S., Hwang, P.C., and Idler, D.R., Cryogenic preservation of Atlantic cod (*Gadus morhua*) sperm, *Journal of the Fisheries Research Board of Canada*, 25, 2623, 1968.
4-Mounib, M.S., Cryogenic preservation of fish and mammalian spermatozoa, *Journal* of *Reproduction and Fertility*, 53, 13, 1978.
5-DeGraaf, J.D. and Berlinsky, D.L., Cryogenic and refrigerated storage of Atlantic cod (*Gadus morhua*) and haddock (*Melanogrammus aeglefinus*) spermatozoa, *Aquaculture*, 234, 527, 2004.
6-Rideout, R.M., Trippel, E.A., and Litvak, M.K., The development of haddock and Atlantic cod sperm cryopreservation techniques and the effect of sperm age on cryopreservation success, *J Fish Biol*, 65, 299, 2004.
7-Rideout, R.M., Trippel, E.A., and Litvak, M.K., Relationship between sperm density, spermatocrit, sperm motility and spawning date in wild and cultured haddock, *J Fish Biol*, 65, 319, 2004.

Complete affiliation:
Rick Rideout, Fisheries and Oceans Canada, Northwest Atlantic Fisheries Centre, PO Box 5667, St. John's, NL, A1C 5X1, Canada. Phone: + 1 709 772 6975, Fax: + 1 709 772 4105, e-mail: rideoutr@dfo-mpo.gc.ca.

CRYOPRESERVATION OF SMALL ABALONE
(*Haliotis diversicolor supertexa*) SEMEN

J-C. Gwo

1. INTRODUCTION

As the name indicates, the shell length of small abalone (*Haliotis diversicolor supertexa*) reaches only about 12 cm and it weighs 100 g or so. Due to the slow growth rate and body size limit of this species, developing new strains using hybridization with other species to enhanced growth rate and better tolerance to adverse environments are worth studying. Here, I describe a simple, reliable and economic cryopreservation technique for small abalone semen.

2. PROTOCOL FOR FREEZING AND THAWING

2.1 Fish Handling and Semen Collection

- The sex of the animals was determined by gamete external examination on the gonad index (male gonad milky white, female gonad dark green). Animals were collected (n = 50) during the spawning seasons (November to December) and kept out of water for 100 minutes before placing them back into the aquarium and sequentially repeated temperature shock (4-5°C above ambient room temperature 20°C for 3-4 hours) 2 to 3 times to induce animals spawning.
- The sperm-seawater suspension was centrifuged at 3,000 rpm for 10 min at 20°C to obtain concentrated sperm.
- The concentrated sperm was placed on crushed ice after collection; sperm motility was determined and freezing was accomplished within 10 min. Unless otherwise stated, the experiments were performed with natural filtered seawater at salinity 25 (pH 8.0) and incubated at room temperature (20°C).

2.2 Freezing and Thawing Procedures

- Only sperm with almost 100% vigorous forward movement were used in this study.
- Sperm was diluted in artificial seawater (ASW, consisting of 30 g/l NaCl, 0.8 g/l KCl, 1.3 g/l $CaCl_2$, 6.6 g/l $MgSO_4$, 0.18 g/l $NaHCO_3$ with an osmolality of 1,100 mOsm/kg and a pH of 8.2) extender with 10% DMSO as cryoprotectant, filled into 1.5 ml microtubes (39 x 10 mm ø; Sarstedt, Aktiengesells chaft & Co.,

German) and immediately frozen with liquid nitrogen vapor in an insulated styrofoam box (inner diameter 14 x 14 cm ø) within 5 min of dilution.

- A two-step freezing method was used to define the best freezing rate and transition temperature. Once reaching the target transition temperature (the temperature at which the sperm sample was put into the liquid nitrogen; first step), sperm samples were transferred directly to liquid nitrogen (second step).

- To examine the effect of cooling rate, these microtubes in each treatment were cooled from 20°C to a transition temperature of –30°C at various cooling rates (-3.5, -5, -7.5, -12, -15, and –20°C/min) and plunged into liquid nitrogen immediately. The cooling rates were adjusted by varying the height of the microtubes above the surface of liquid nitrogen.

- To examine the effect of transition temperature on post-thaw sperm motility, sperm was diluted in ASW extender with 10% DMSO as cryoprotectant and frozen in 1.5 ml microtubes in the above styrofoam box at 4 cm above the surface of liquid nitrogen (the final temperature was –90°C with a cooling rate of –15°C/min) from 20°C to various transition temperatures (0,-30,-60,-90, and –120°C). Upon reaching the selected transition temperature, samples were placed directly into liquid nitrogen.

- Frozen samples were thawed by immersing tubes in a gently stirred water bath at various ttemperatures (25, 50, 70 and 90°C). One portion of each thawed samples was evaluated using motility score; the other part was utilized to fertilize eggs and assess sperm viability.

- In brief, the semen was diluted 1:1 with the extender, inserted into 1.5 ml microtubes and cooled to –90°C at a cooling rate of –12°C/min and then immersed in liquid nitrogen. The microtubes were thawed at 70°C for 1 min and immediately mixed with freshly spawned eggs at various sperm concentrations of 10 to 10^5.

3. GENERAL CONSIDERATIONS

The concentration of gametes affected fertilization success. The sperm concentrations of 10, 10^2 and 10^3 cells/ml had the maximal fertilization rates (Figure 1). Fertilization rates were higher for gametes fertilized within 2 h post-spawning and decreased after 3 h (Figures 2 and 3). There was a significant influence of sperm-egg contact time on fertilization rates (Figure 4). Fertilization rates were high and deformation rates low when fertilized eggs were washed right after eggs were inseminated at sperm concentrations of 10^3 cells/ml. Washing the fertilized eggs 3 times resulted in better fertilization rates (Figure 5). The highest post-thaw sperm motility was obtained when sperm was frozen at a rate of -12 or –15°C/min to the transition temperature –30°C (Table 1). The transition temperature of –90°C seemed to be more beneficial than other tested temperatures (Table 2). The optimal thawing condition for the sperm of small abalone was in a water bath at 70°C for 1 min. Small abalone sperm cryopreserved in DMSO resulted in fertilization rates between 7 to

48%, depending on the sperm concentration (Figure 6). The fertilization capacity of the frozen sperm is about half that of the fresh sperm using the same sperm concentration of 10^3 (Figs. 1 and 6).

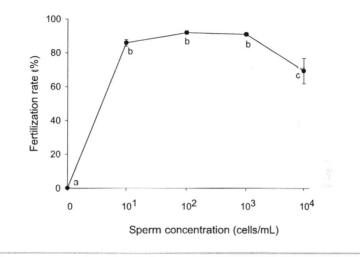

Figure 1 - Fertilization rates achieved with different sperm concentration.

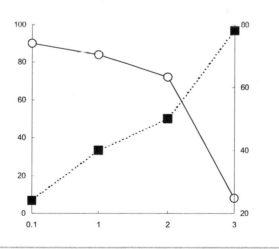

Figure 2 - Fertilization (O) and deformation rates (■) obtained with eggs at different time after spawning (0.1-3 h).

A sperm concentration of between 10^1 and 10^3 is required for maximal fertilization and normal embryo development in small abalone. Lower fertilization and normal development rates at higher sperm concentrations are attributed to polyspermy, demonstrated in various *Haliotis* species, while no fertilizations were reported below concentration of 10^3 sperm ml^{-1} [1-3]. The gametes (sperm and eggs) of small abalone in the present study were still viable 3 h post-spawning with both fertilization and normal development rates significantly decreasing with time. A gradual decline in the hatching rates as the fresh milt of small abalone was maintained at room temperature was also reported.

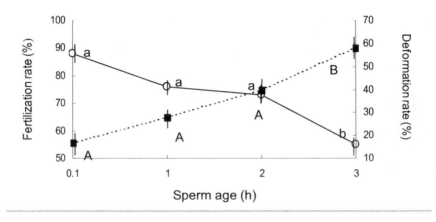

Figure 3 - Fertilization (O) and deformation rates (■) obtained with sperm at different times after collection.

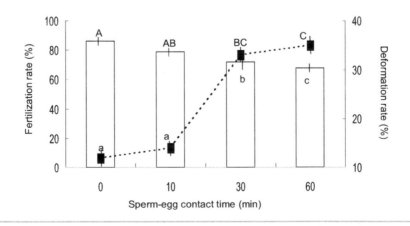

Figure 4 - Fertilization (bars) and deformation rates (■) obtained after different sperm-egg contact time (0-60 min).

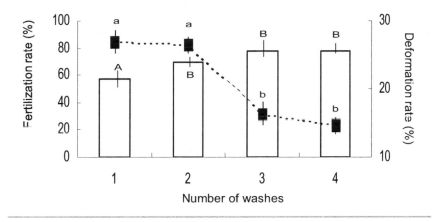

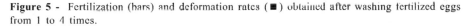

Figure 5 - Fertilization (bars) and deformation rates (■) obtained after washing fertilized eggs from 1 to 4 times.

Table 1 - Effect of cooling rate on the post-thaw sperm motility scores[a] of small abalone sperm after two step freezing at transition temperature -30°C.

Cooling rate (°C/min)	Motility score
-3.5	1
-5	2
-7.5	2
-12	3
-15	3
-20	2

[a]Motility score 0: no motile sperm. Motility score 1: 0-25% of sperm with progressive movement. Motility score 2: 25-50% of sperm with progressive movement. Motility score 3: 50-75% of sperm with progressive movement. Motility score 4: >75% of sperm with progressive movement

Table 2 - Effect of transition temperature on the post-thaw sperm motility scores[a] of small abalone sperm after two-step freezing with various transition temperatures and a cooling rate of -15°C/min.

Transition temperature (a)	Motility score
0	0
-30	2
-60	2
-90	3
-120	2

[a] See Table 1

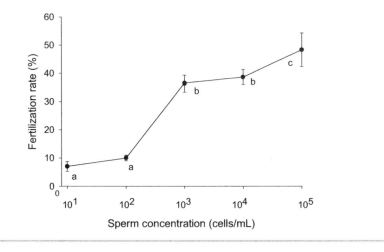

Figure 6 - Fertilization rates achieved with cryopreserved sperm at different concentrations.

Diluted fresh or artificial seawater has been widely used for cryopreservation of mollusk sperm [1,2]. However, successful cryopreservation of small abalone does not correlate with the complexity of the extenders. The simple sugar solutions (glucose and sucrose) with osmolality of 800 mOsm/kg or similar to seawater have been successfully used to freeze oyster and small abalone sperm [1-3]. The choice of cryoprotectant seems species-specific in aquatic invertebrate [1,2]. Most cryoprotectants were toxic to aquatic invertebrate sperm [1,2]. Motility of small abalone sperm was extensively decreased with equilibration time extended to 30 min. These results indicated that to preserve motility better, sperm of small abalone should be exposed to cryoprotectant for a short time periods not longer than 10 minutes. A short equilibration time (< 3 min) in the cryopreservation extender is strongly recommended for aquatic invertebrate sperm [1,2].

The spermatozoa of aquatic invertebrates make poor use of external energy sources [1,2]. Few studies have examined factors influencing motility of small abalone sperm. The sperm of small abalone motile were in 10% DMSO-ASW extender with an osmolality of 3,000 mOsm/kg before freezing (Gwo, J-C., unpublished data). Sperm of small abalone had diminished motility when placed in DW or ASW with osmotic pressure below 400 mOsm/kg (Gwo, J.-C., unpublished data). Other factors besides osmotic pressure for controlling sperm motility intensity and duration of small abalone should be also investigated. To conserve the finite endogenous energy reserves of small abalone sperm, developing an extender tailored to maintaining the sperm in complete quiescence might be worth trying.

4. ACKNOWLEDGMENTS

I sincerely thank Mr. Jeng, C.-H.(Bytoijeau Gao-Karn Nursery) for his help in generously providing the incredible broodstocks.

5. REFERENCES

1-Gwo, J.C., Cryopreservation of aquatic invertebrate semen: a review, *Aquac Res,* 31, 259, 2000.
2-Gwo, J.C., Cryopreservation of sperm of some marine fishes, in: *Cryopreservation in aquatic species,* Tiersch, T.R. and Mazik, P.M., Eds., Baton Rouge, World Aquaculture Society, LA, USA, 2000, 138.
3-Gwo, J.C., Chen, C.W., and Cheng, H.-Y., Semen cryopreservation of small abalone (*Haliotis diversicolor supertexa*), *Theriogenology*, 58, 1563, 2002.

Complete affiliation:
Jin-Chywan Gwo, Department of Aquaculture, Taiwan National Ocean University, Keelung 20224, Taiwan. e-mail: gwonet@ms.16.hinet.net

CRYOPRESERVATION OF PACIFIC OYSTER SPERM

Q. Dong, C. Huang, B. Eudeline and T. R. Tiersch

1. INTRODUCTION

The Pacific oyster, *Crassostrea gigas* (Thunberg, 1793), also referred to as the Japanese oyster or giant oyster, is one of the most important species of bivalves cultured worldwide. The basic life history of the Pacific oyster is similar to that of the eastern oyster, *C. virginica* [1] with both functioning as protandrous hermaphrodites. The young are functionally male during their first spawning, while adults function as separate male or female animals in any given reproduction season [2]. Sexual maturity is reached during the first year, and spawning generally occurs in the summer months (June to August in the west coast of United States). Oysters are highly prolific, and males generally produce 5×10^{10} sperm per gram of gonad wet weight [3].

The first report of sperm cryopreservation in oysters was published [4] with *C. virginica*. Since then some 27 reports have been published to address oyster sperm cryopreservation, of which ~80% have addressed sperm from the Pacific oyster. Despite that, procedural standardization is lacking in the cryopreservation of oyster sperm [5]. Inconsistency of various components of cryopreservation technology have been observed among and within studies, such as initial sperm quality, gamete collection methods, extender formulation, cryoprotectant choice, cooling rate and method, thawing rate and method, insemination protocols, and evaluation of post-thaw sperm quality [5]. The procedures outlined below are mainly derived from our 4 years of research on this species.

2. PROTOCOL FOR FREEZING AND THAWING

2.1 Equipment

- Sperm collection: gloves, oyster knife, tweezers, scissors, scalpel, spatula, 15 ml centrifuge tubes, permanent marker, Kimwipes, and 40 μm cell strainer (BD Biosciences Discovery Labware, Bedford, Massachusetts).
- Solution preparation: top-loading balance, spatula, stir plate (Barnstead/ Thermolyne, Dubuque, Iowa), stir bar, 0.45 μm CA (cellulose acetate) filter (Corning Incorporated, Corning, New York), and vapor pressure osmometer (model 5500, Wescor Inc., Logan Utah).
- Motility estimation: micropipette and tips (10 and 100 μl), two-well glass microscope slide, and microscope with darkfield and 20 × objectives.

Straw filling: 0.25 or 0.5 ml straws, 3 ml syringe without needle, goblets, cane holders, PVC powder, and paper towels.
- Freezing: a controlled-rate freezer with accompanying low-pressure liquid nitrogen cylinder, liquid nitrogen storage dewar, and cryogloves.
- Sample shipping: shipping Dewar (CP35, Taylor-Wharton, Theodore, Alabama), plastic cable tie wraps, and tape.
- Thawing: Safety glasses, water bath with temperature controls (Model 1141, VWR Scientific, Niles, Illinois), long tweezers, scissors, paper towels, and 1.5 ml microcentrifuge tubes.
- Fertilization: Constant supply of filtered seawater at 34 ppt salinity and 25°C, 25 μm and 60 μm meshes, 1 l beakers, and 100 l tanks.

2.2 Reagents

- C-F HBSS at 1000 mOsm/kg (calcium free Hanks' balanced salt solution: 26.32 g N aCl; 1.32 g KCl; 0.65 g $MgSO_4$ x $7H_2O$; 0.18 g N a_2HPO_4 x $7H_2O$; 0.18 g KH_2PO_4; 1.15 g $NaHCO_3$; 3.30 g $C_6H_{12}O_6$ (glucose) in sufficient distilled water to yield 1000 ml, pH 7.0-7.4)[5].
- Cryoprotectants: methanol, dimethyl sulfoxide, propylene glycol, ethylene glycol, and polyethylene glycol (formula weight of 200).

2.3 Sperm Collection

- Carefully remove the top shell by use of an oyster knife (wear gloves, at least on the hand holding the oyster).
- Use tweezers and scissors to carefully peel off the mantle, gills, labial palps and other tissues, but leave gonad intact and attached to the abductor muscle, and use tweezers to remove the heart and associated tissues.
- Use filtered sea water or C-F HBSS to rinse the gonad 3 times, and use Kimwipes to dry the gonad and clean the inside shell.
- Use the scalpel to cut openings on the gonad, and use the spatula to collect the gonad tissue into the 15 ml centrifuge tube. Avoid contamination with digestive gland (yellowish-green material).
- Label the tubes, and store the undiluted sperm samples at 4°C before shipping or use.
- After suspension with C-F HBSS, filter samples through a 40 μm cell strainer prior to adjusting cell concentration.

2.4 Sperm Motility Estimation

- Take 10 ml C-F HBSS from a stock solution stored at -20°C and allow it warm to room temperature (it is recommended to use the same activation solution throughout a single working season. C-F HBSS can be aliquotted into small volumes (such as 10 ml) and stored at -20°C) . Evaluate the C-F HBSS with microscopy for presence of bacterial contamination.

- Place 30 μl of C-F HBSS inside the wells on the two-well glass microscope slide.
- Determine initial motility by adding undiluted nonmotile sperm [6] or 1 μl of sperm suspension (after suspending in C-F HBSS at 1000 mOsm/kg) and gently mix with the tip of the pipette.
- Evaluate post-thaw motility by adding 2 μl of sperm suspension and gently mixing with the tip of the pipette. Equilibrate sperm suspensions inside the wells on the glass slide at 23°C for 2 min before motility estimation.
- Estimate the percent of sperm that are actively moving in a forward direction at 200 × magnification using darkfield microscopy. Estimate the percent motility in increments of 5%. Samples with motility below 5% that have motile sperm are recorded as 1%.
- Identify the treatments (to avoid bias associated with observer, it is especially important to estimate sperm motility without knowledge of the treatment of the straw).

2.5 Freezing and Thawing Procedure

- Prepare a double-strength cryoprotectant solution within 2 h of use with C-F HBSS at 1000 mOsm/kg, and store at 4°C.
- Adjust sperm concentration to 2×10^9 cells/ml with a spectrophotometer using standard curves, and add an equal volume of the cryoprotectant solution in a stepwise addition within 1 h (e.g., 10 or more increments).
- Selected cryoprotectants (final concentration) include 6% methanol, 5% dimethyl sulfoxide, 5% propylene glycol, 5% ethylene glycol, and combined cryoprotectants of 4% methanol and 2% polyethylene glycol at a formula weight of 200.
- Load sperm suspensions into 0.25 ml or 0.5 ml French straws (IMV International, Minneapolis) (for detailed instructions see Appendix A in [5]). Place eight 0.25 ml straws or five 0.5 ml straws into a 10 mm plastic goblet, and attach two goblets to a 10 mm aluminum cane. Insert the canes into the rack inside the freezing chamber.
- Equilibrate samples at 5°C for 5 min before cooling at the desired rate.
- Cool samples in two steps, initially to -30°C at 5°C per min, followed by cooling at 45°C per min from –30°C to –80°C. Hold straws at –80°C for 5 min.
- Remove samples swiftly from the freezing chamber and immediately plunge them into liquid nitrogen in a storage dewar.
- Thaw samples after a minimum of 12 h of storage in liquid nitrogen.
- Remove individual straws from the 10 mm goblets with the long tweezers.
- Thaw in a 40°C water bath for 7 s for 0.5 ml straws, or 6 s for 0.25 ml straws.

- Wipe the straws with paper towel and cut the PVC powder sealed end. Empty the straw into 1.5 ml microcentrifuge tubes by cutting the cotton plug end to release the contents into the tubes.
- In case of sperm agglutination, disrupt the agglutination before motility estimation or fertilization trials.

2.6 Fertilization

- Use the shipping dewar to ship frozen sperm samples to the hatchery.
- Use diploid females for fertilization trials. Eggs from individual females are obtained by dissection, and are sieved, washed through 60 μm mesh, retained on 25 μm mesh, and suspended in filtered seawater (34 ppt) at 25°C.
- Pool unfertilized eggs (fresh) from three females and determine the number of eggs per ml (by Coulter Counter if possible). Hold the eggs in seawater at 25°C for at least 30 min to observe germinal vesicle breakdown at 100× magnification. Separate eggs into 1 liter beakers with each beaker containing 500,000 eggs (fresh) in 250 ml of seawater.
- Thaw ten straws of each treatment as described above.
- Conduct fertilization trials by mixing 5 ml of thawed sperm suspension (the pooled contents of ten 0.5 ml straws) with 500,000 eggs in 250 ml of seawater.
- Incubate the gametes at 25°C and calculate percent fertilization by counting developing embryos at 2 h after insemination (n = 100).
- Hold treatments for further evaluation of percent hatch by transferring to 100 liter tanks filled with fresh seawater. Drain the tanks 24 h after fertilization through a 45 μm mesh and concentrate larvae into 1 liter of seawater. After mixing, calculate percent hatch in five 1 ml subsamples by counting normal straight-hinge larvae with a dissecting scope. If the samples cannot be evaluated immediately, fix the subsamples with Lugol's solution (mix 100 ml acetic acid with 1 liter of Lugol solution, Sigma Chemical Corporation., St. Louis, Missouri) and analyze later.
- For a negative control, monitor eggs after treatment as described above without addition of sperm.
- For the evaluation of egg quality, collect fresh (nonfrozen) sperm from diploid males using the techniques described above, wash the sperm through a 70 μm mesh, and add to fresh eggs to obtain about 100 spermatozoa per egg. Perform sperm counts with a spectrophotometer using standard curve [3].
- For the evaluation of cryoprotectant toxicity, expose fresh sperm at the same concentration as thawed sperm samples to the same treatments (concentration, equilibration time, batch of eggs), and estimate percent fertilization.
- To avoid contamination of gametes among individuals, handle the animals with care and wash all surfaces with 0.01% bleach. Hold the sexes separately in different containers to avoid unintended fertilization.

3. GENERAL CONSIDERATIONS

The cryoprotectants suggested above were found to yield percent fertilization above 90% in different trials. Sperm samples frozen at the suggested concentration can lead to sperm agglutination, which was found to have no effect on fertilization success [5]. However, crushing of the sperm agglutination after thawing is required for motility estimation and fertilization trials.

4. ACKNOWLEDGMENTS

This project was supported in part by funding from the USDA-Small Business Innovation Research program, 4Cs Breeding Technologies, Inc., and the Louisiana Sea Grant College Program. We thank S. K. Allen for assistance and discussion. Approved for publication by the Director of the Louisiana Agricultural Experiment Station as manuscript number 05-11-2007.

5. REFERENCES

1-Galtsoff, P.S., The American oyster, *Crassostrea virginica* Gmelin, in: *Fisheries Bulletin of the United States Fish and Wildlife Service*, 64, 1964, 480.

2-Pauley,G.B., Van Der Raay, B., and Troutt, D., Species profiles: life histories and environmental requirements of coastal fishes and invertebrates (Pacific Northwest)-Pacific oyster, in: *U.S. Fish and Wildlife Service Biological Report*, 82 (11.85), U.S. Army Corps of Engineers, TR EL-82.4., 1988, 28.

3-Dong, Q., Eudeline, B., Huang, C., and Tiersch,T.R., Standardization of photometric measurement of sperm concentration from diploid and tetraploid Pacific oysters, *Crassostrea gigas* (Thunberg), *Aquac Res*. 36, 86, 2005.

4-Lannan, J.E., Experimental self-fertilization of the Pacific oyster, *Crassostrea gigas*, utilizing cryopreserved sperm., *Genetics*, 68, 599, 1971.

5-Dong, Q., Comparative studies of sperm cryopreservation of diploid and tetraploid Pacific oysters, in: *Dissertation*, Baton Rouge, LA, Louisiana State University, 2005, 271.

6-Dong, Q., Eudeline, B., Huang, C., Allen, S.K Jr., and Tiersch, T.R., Commercial-scale sperm cryopreservation of diploid and tetraploid Pacific oysters, *Crassostrea gigas*, *Cryobiology*, 50, 1, 2005.

Complete affiliation:

Terrence R. Tiersch, Aquaculture Research Station, Louisiana Agricultural Experiment Station, Louisiana State University Agricultural Center, Baton Rouge, Louisiana 70820 USA. ttiersch@agcenter.lsu.edu.

A SIMPLE METHOD FOR FREEZING PACIFIC OYSTER (*Crassostrea gigas*) SPERM IN QUANTITIES SUITABLE FOR COMMERCIAL HATCHERIES

S.L. Adams, J.F. Smith, R.D. Roberts, A.R. Janke, H.F. Kaspar,
H.R. Tervit, P.A. Pugh, S.C. Webb and N.G. King

1. INTRODUCTION

The Pacific oyster (*Crassostrea gigas* Thunberg) is the most commonly farmed shellfish species worldwide with an annual production of 4,200,000 tonnes [1]. It is also one of the few shellfish species for which culture is based substantially on hatchery-reared juveniles. Oyster broodstock can be easily conditioned in hatcheries by manipulating temperature [2]. In natural populations, Pacific oysters spawn during summer months [3]. The ability to cryopreserve sperm of this species enables hatcheries to store sperm from individual males for selective breeding programmes [4,5]. Since adult oysters conditioned with ample food are often predominantly female [6], cryopreservation also provides a guaranteed supply of sperm for spat production outside the natural breeding season and reduces the costs associated with conditioning broodstock.

Cryopreservation of oyster sperm was achieved decades ago [7,8] but is still not used by oyster hatcheries. Most past work has been conducted at small scale, often with controlled rate freezers. Dimethyl sulphoxide (DMSO) has been found to be more effective than other cryoprotectants (CPA) in maintaining post-thaw fertility [5,9-12] and the addition of trehalose to the base medium has also been found to be beneficial [5,13]. The cooling rates used to freeze oyster sperm have ranged from 4.7°C/min [12] to immediate plunging in liquid nitrogen [9], suggesting that the sperm are able to survive a wide range of cooling rates. The method outlined here is a commercial scale method which is affordable and reliable and will enable hatcheries to use cryopreserved sperm routinely.

2. PROTOCOL FOR FREEZING AND THAWING

- Prepare a stock trehalose solution by adding 15.2 g of trehalose to 40 ml of distilled water. Combine 30 ml of trehalose stock solution, 7.8 ml of distilled water and 2.2 ml of DMSO to make the CPA solution and cool on ice.
- Prepare a methanol/dry ice bath for sperm to be frozen in by placing several litres of methanol and dry ice in a polystyrene box. The bath should be prepared in advance to ensure it has cooled below ~-70°C before using. As a guide, the bath is cold enough if there is plenty of dry ice left, but bubbling and gas release has reduced to a minimum.

- Strip sperm from mature males by opening the oyster, lacerating the gonad wall and gently scraping without any addition of seawater. Avoid contaminating the sperm with any gut contents. Hold sperm "dry" on ice until ready to be used.
- Determine the concentration of the collected sperm by taking a sub-sample, diluting with a known volume of seawater and Lugol's iodine and counting the sperm on a Neubauer haemocytometer slide. In general, a large mature male will produce around 7 ml of dry sperm at a concentration of $\sim 2 \times 10^{10}$ ml^{-1} [5] and at 5°C, the sperm will retain its fertility for several days [14].
- Eggs can be obtained in similar fashion by gently washing the lacerated gonad with seawater into a beaker.
- Dilute sperm 1:10 with CPA solution. The solution should be added to the sperm in 10 steps, 10 to 20 s apart with mixing between each step, to avoid osmotic injury.
- Load the sperm mixture into chilled 4.5 ml cryotube vials with internal thread (Nalgene Nunc International, Denmark) and attach to aluminium canes (for ease of handling and storage).
- Place the canes with the loaded cryovials directly into the methanol/dry ice bath. After 10 min, transfer the canes quickly from the bath into liquid nitrogen.
- Thaw the cryovials in a lidded container of seawater at room temp until contents become liquid (5 to 8 min), then use for fertilization (N.B. Nalgene Nunc warn that on rare occasions, liquid nitrogen may leak into the cryotube vials, potentially causing the vials to explode when thawing, hence the need to use a lidded bath during thawing).

3. GENERAL CONSIDERATIONS

The motility of cryopreserved sperm is poor, but fertility remains adequate. Typically ~10 to 100 fold higher sperm to egg ratio is required with cryopreserved sperm for the equivalent fertilization of fresh sperm (Figure 1).

For large commercial scale fertilizations, thawed sperm are combined with freshly spawned or stripped eggs at a ratio of ~2000 sperm per egg, and an egg density of ~100,000 eggs ml^{-1}. After a contact time of 10 min, the eggs are diluted into larval hatching tanks and reared following normal practices. Assuming: (1) good freezing success; (2) a dilution ratio of 1+10; (3) an initial raw sperm density of 2×10^{10} ml^{-1}; then > 50% of 1 million eggs can be fertilized with about 0.55 ml of frozen sperm mixture. As a commercial example, we obtained 81% fertilization of 30 million eggs using 2,000 frozen sperm per egg in a total volume of ~300 ml. Normal commercial rearing procedures yielded 3.7 million settled spat.

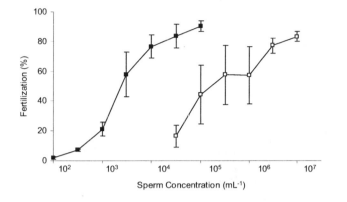

Figure 1 - Fertilization (mean ± SE, n = 3 pools) of eggs fertilized with unfrozen (closed squares) and cryopreserved (open squares) Pacific oyster sperm. The pools of sperm were collected, cryopreserved and assayed independently. The same batches of eggs were used with both the unfrozen and cryopreserved sperm. (modified from Adams et al. [13]).

4. ACKNOWLEDGMENTS

We thank staff at the Cawthron Institute for technical assistance. This research was funded by the New Zealand Foundation for Research Science and Technology (contracts CAW801, CAWX0004 and CAWX0304).

5. REFERENCES

1-Food and Agriculture Organization (FAO), Yearbook of Fisheries Statistics: Summary tables. World aquaculture production of fish, crustaceans mollusks, etc., by principal species in 2002, 2004.http://www.fao.org/fi/statist/summtab/default.asp

2-Muranaka, M.S. and Lannan, J.E., Broodstock management of *Crassostrea gigas*: environmental influences on broodstock conditioning, *Aquaculture*, 39, 217, 1984.

3-Tanaka, Y., Spawning season of important bivalves in Ariake Bay – *Ostrea rivularis* Gould and *O. gigas* Thunberg, *Bulletin of the Japanese Society of Scientific Fisheries* 19, 1161, 1954.

4-McFadzen, I.R.B., Cryopreservation of sperm of the Pacific oyster *Crassostrea gigas*, in: *Cryopreservation and Freeze-Drying Protocols*, Humana Press Inc., Day, J.G. and McLellan, M.R., Eds, Totowa, 1995, 145.

5-Smith, J.F., Pugh, P.A., Tervit, H.R., Roberts, R.D., Janke, A.R., Kaspar, H.F., and Adams, S.L., Cryopreservation of shellfish sperm, eggs and embryos, *Proc. N. Z. Soc. Anim. Prod.*, 61, 31, 2001.

6-Thompson, R.J., Newell, R.I.E., Kennedy, V.S., and Mann, R., Reproductive processes and early development, in: *The Eastern Oyster, Crassostrea virginica*, Kennedy, V.S., Newell, R.I.E. and Eble, A.F., Eds., Maryland Sea Grant College, Maryland, 1998.

7-Lannan, J.E., Experimental self-fertilisation of the pacific oyster *Crassostrea gigas*, utilising cryopreserved sperm, *Genetics*, 6, 599, 1971.

8-Staeger, W.H., Cryobiological Investigations of the Gametes of the Pacific Oyster, *Crassostrea gigas*, Department of Fisheries and Wildlife, Oregon State University, Corvallis, 1974.

9-Hwang, S.W. and Chen, H.P., Fertility of male oyster gametes after freezing and thawing, *Chinese-American Joint Commission on Rural Reconstruction Fisheries Series*, 15, 1, 1973.

10-Bougrier, S. and Rabenomanana, L.D., Cryopreservation of spermatozoa of the Japanese oyster, *Crassostrea gigas*, *Aquaculture*, 58, 277, 1986.

11-Iwata, N., Kurokura, H., and Hirano, R., Cryopreservation of Pacific oyster, *Crassostrea gigas*, sperm, *Suisanzoshoku*, 37, 163, 1989.

12-Yankson, K. and Moyse, J., Cryopreservation of the spermatozoa of *Crassostrea tulipa* and three other oysters, *Aquaculture*, 97, 259, 1991.

13-Adams, S.L., Smith, J.F., Roberts, R.D., Janke, A.R., Kaspar, H.F., Tervit, H.R., Pugh, P.A., Webb, S.C., and King, N.G., Cryopreservation of sperm of the Pacific oyster (*Crassostrea gigas*): development of a practical method for commercial spat production, *Aquaculture*, 242, 271, 2004.

14-Roberts, R., Smith, J., Adams, S., Janke, A., King, N., Kaspar, H., Tervit, R., and Webb, S., Cryopreservation and cool storage of Pacific oyster sperm for selective breeding and commercial spat production, in: *Australasian Aquaculture Conference*, September, 2004, 248.

Complete affiliation:

Serean L. Adams, Cawthron Institute, Private Bag 2, Nelson, New Zealand and AgResearch, Ruakura Research Centre, Private Bag 3123, Hamilton, New Zealand. e-mail: Serean.Adams@cawthron.org.nz.

CRYOPRESERVATION OF EASTERN OYSTER SPERM

C.G. Paniagua-Chavez and T.R.Tiersch

1. INTRODUCTION

Crassostrea virginica (formerly the "American oyster") was designated in 1985 as the "eastern oyster" by the Committee on Scientific and Vernacular Names of Mollusks of the Council of Systematic Malacologists [1]. The eastern oyster is a lamellibranch with pronounced bilateral asymmetry [2] and usually spawns as a male in the first year, a condition called protandry. Fecundity has been estimated to be between 500,000 and 66 million eggs per female depending upon body size [3,4]. Within 8 to 12 h (depending on temperature) fertilized eggs develop into free-swimming larvae or trochophores (50 to 60 μm in width). After ~24 h, the trochophore larvae develop into veliger or D-stage larvae (70 to 125 μm). After metamorphosis, the settled larvae develop into adults within 1 to 2 years. Adult oysters mature and spawn in the summer months (April to September in the Gulf of Mexico) for the next cycle [5,6]. The eastern oyster is the most important bivalve species in the United States [7]. However, along the Atlantic and Gulf coasts, oyster production has declined over the past century due to reasons including a lack of consistent seed supply, excessive harvest, loss of suitable habitat, disease, and natural predation [8]. The use of cryopreserved gametes and larvae can improve hatchery production of seedstock to increase production for the oyster industry. Cryopreservation of oyster sperm and larvae has been tested at the laboratory level, but given the benefit that this technique offers, cryopreservation of oyster sperm and larvae should be developed for commercial application at the hatchery level. The procedures outlined below are suitable for application at the hatchery scale, and could be scaled up for commercial application.

2. PROTOCOL FOR FREEZING AND THAWING

2.1 Equipment

- Oyster knife
- Dissection kit
- Capillary tubes
- Microscope and slides
- Osmometer
- Nitex screens (15 μm, and 70 μm)
- Micropipettes (20 μl, 1000 μl, and 5000 μl)
- 50 ml beakers
- Hematocytometer

- 5 ml macrotubes and sealing balls
- 30 ml syringes without needles
- Controlled-rate freezer
- Water bath
- Dewars
- Basic laboratory glassware

2.2 Reagents

- Distilled water
- Natural seawater, artificial seawater, or calcium-free Hanks' balanced salt solution FHBSS: 24 g/l NaCl, 1.20 g/l KCl, 0.60 g/l $MgSO_4$ x $7H_2O$, 0.36 g/l Na_2HPO_4 x $7H_2O$, 0.18 g/l KH_2PO_4, 1.05 $NaHCO_3$, 3 g/l $C_6H_{12}O_6$ adjusted to 600 mOsm/kg
- Propylene glycol
- Liquid nitrogen

2.3 Semen Collection

- Gamete preparation: A gonad sample is collected with a capillary tube and smeared on a glass microscope slide for examination at 200 x magnification. Sex is identified based on the presence of eggs or sperm. Gamete samples are removed from each oyster by the dry stripping method [9]. The gonad is gently disrupted and gonadal material is collected with a Pasteur pipette. A 10 µl sample is removed from the gonad to measure osmolality with a vapor pressure osmometer (model 5500, Wescor Inc., Logan, Utah).
- Sperm samples are placed in 50 ml beakers until suspension in an extender. After suspension, the sperm samples are washed through 70 µm and 15 µm Nitex screens (Aquacenter, Leland, Mississippi), with artificial seawater (ASW), natural seawater, or calcium-free Hanks' balanced salt solution (C-F HBSS) adjusted to 600 mOsm/kg [10].

2.4 Sperm Motility Estimation

- Unlike sperm of most fishes, oyster sperm can remain continuously motile for hours or days. A 10 µl sample is removed from the sperm suspension to estimate motility. The percentage of sperm exhibiting vigorous forward movement is estimated at 200 x magnification using dark-field microscopy. Sperm vibrating in place are not considered to be motile. Only males with actively swimming sperm (> 90%) are selected for experimentation [10,11].

2.5 Description of the Protocol for Refrigerated Storage of Sperm

- After microscopic identification, undiluted sperm samples can be placed in sealed tubes and be stored in a refrigerator at 4°C for as long as 4 days. Sperm should maintain >50% motility in this condition. Extended sperm (1:1, v:v) may be refrigerated for as long as 2 days if diluted in C-F HBSS. Sperm can maintain 50% motility in this condition [10].

2.6 Freezing and Thawing Procedures

- Sperm may be suspended at 1 x 10⁹ cells per ml in ASW, natural seawater, or C-F HBSS. The cryoprotectant can be prepared in these extenders to yield a final concentration of 10% or 15% propylene glycol (PG).
- After suspension, sperm are equilibrated for 20 min, and 5 ml aliquots are used to fill 5 ml macrotubes (Minitube of America, Inc., Madison, Wisconsin).
- The macrotubes are cooled in a controlled-rate freezer (Kryo 10 series II; Planer Products, Sunbury-on-Thames, UK). The initial temperature is 15°C, and the samples are cooled at a rate of 2.5°C per min until reaching a final temperature of –30°C, which is held for 5 min. Macrotubes are plunged into liquid nitrogen and stored in a dewar until use.
- A water bath is used to thaw the samples at 70°C for 15 s. After thawing, the samples are resuspended in 5 ml of fresh ASW, natural seawater or C-F HBSS. Sperm should be used immediately after thawing to fertilize eggs [11,12,13].

3. ACKNOWLEDGMENTS

This project was supported in part by the Louisiana Sea Grant College Program. We thank J. Arias for technical assistance. Approved for publication by the Director of the Louisiana Agricultural Experiment Station as manuscript number 05-11-2007.

4. REFERENCES

1-Turgeon, D.D., Bogan, A.E., Coan, E.V., Emerson, W.K., Lyons, W.G., Pratt, W.L., Roper, C.F.E., Scheltema, Thompson, F.G., and Williams, J.D., Common and Scientific Names of Aquatic Invertebrates from the United States and Canada: *Mollusks*, American Fisheries Society Special Publication, 16, 28, 1988.

2-Seed, R., Structural organization, adaptive radiation, and classification of mollusks, in: *The Mollusca*, Hochachka, P.W., Ed., Academic Press, San Diego, California, 1983, 1.

3-Galtsoff, P.S., Fecundity of the oyster, *Science*, 72, 97, 1930.

4-Cox, C. and Mann, R., Temporal and spatial changes in fecundity of eastern oyster, *Crassostrea virginica* (Gmelin 1791) in James River, Virginia, *J Shellfish Res*, 11, 49, 1992.

5-Galtsoff, P. S., The American oyster, *Crassostrea virginica* Gmelin, *Fisheries Bulletin of the United States Fish and Wildlife Service*, 64, 480, 1964.

6-Thompson, R.J., Newell, R.I.E., Kennedy, V.S., and Mann, R., Reproductive Process and Early Development, in: *The eastern oyster Crassostrea virginica*, Kennedy, V.S., Newell, R.I.E., and Eble, A.F., Eds., Maryland Sea Grant College, Maryland University, Maryland MD, 1996, 335.

7-Wendell, J.L. and Malone, S., The cultivation of American oyster (*Crassostrea virginica*), *Southern Regional Aquaculture Center Publication*, 432, 8, 1994

8-Supan, J. and Wilson,C., Oyster seed alternatives for Louisiana, *World Aquaculture*, 24, 79, 1993.

9-Allen, S.K. and 'Bushek, D., Large scale production of triploid *Crassostrea virginica* (Gmelin) using "stripped" gametes, *Aquaculture*, 103, 241, 1992.

10-Paniagua-Chavez, C.G., Buchanan, J.T., and Tiersch, T.R., Effect of extender solutions and dilution on motility and fertilizing ability of eastern oyster sperm, *J Shellfish Res*, 17, 1, 231, 1998.

11-Paniagua-Chavez, C.G. and Tiersch, T.R., Laboratory studies of cryopreservation of sperm and trochophore larvae of the eastern oyster, *Cryobiology*, 43, 211, 2001.

12-Paniagua-Chavez, C.G., Buchanan, J.T., Supan, J.E., and Tiersch, T.R.,. Settlement and growth of eastern oysters produced from cryopreserved larvae, *Cryoletters*, 19, 283, 1998.

13-Paniagua-Chavez, C., Buchanan, J.T., Supan, J.E., and Tiersch, T.R., Cryopreservation of sperm and larvae of the eastern oyster, in: *Cryopreservation in Aquatic Species*, Tiersch, T.R. and. Mazik, P.M., Eds., World Aquaculture Society, Baton Rouge, Louisiana, USA, 2000, 230.

Complete affiliation:

Carmen G. Paniagua-Chavez, Centro de investigación Científica y de Educación Superior de Ensenada-CICESE, Department of Aquaculture, Km 107 Carretera Tijuana-Ensenada, Apartado Postal 2732, 22800, Ensenada, Baja California México. e-mail: cpaniagu@cicese.mx.

CRYOPRESERVATION OF SEA URCHIN
(*Evechinus Chloroticus*) SPERM

S.L. Adams, P.A. Hessian and P.V. Mladenov

1. INTRODUCTION

Sea urchins are used as model systems to study fertilization and embryo development [1]. The sperm, embryos and adults are also used in bioassays as indicators of water pollution [2,3] and some species are commercially fished and farmed [4,5]. *Evechinus chloroticus*, is common throughout New Zealand waters and spawns during the summer months (Nov to Feb) [6]. There are reports of this species being used in developmental and embryological studies [7,8] and there is a small scale commercial fishery in operation.

The ability to cryopreserve sea urchin sperm has a number of practical benefits. For example, cryopreservation facilitates the study of hybrids produced between sea urchin species with different breeding seasons [9]. In bioassays, cryopreservation enables sperm from the same animals to be be used which reduces any influence of genetic heterogeneity between tests. Furthermore, excess sperm from experiments can be stored for later use and for research outside the natural breeding season, reducing costs associated with broodstock conditioning.

Cryopreservation of sperm from a number of other sea urchin species has been reported: *Anthocidaris crassispina* [9]; *Hemicentrotus pulcherrimus* [10]; *Loxechinus albus* [11,12]; *Strongylocentrotus droebachiensis* [13]; *Strongylocentrotus intermedius* [14,15]; *Pseudocentrotus depressus* [16]; *Sphaerechinus granularis* [17]; and *Tetrapigus niger* [11,18]. In these studies, fertilization rates for cryopreserved sperm have varied, ranging from 6% to 96%. Here, a method for cryopreserving sperm of *Evechinus chloroticus* is described.

2. PROTOCOL FOR FREEZING AND THAWING

2.1 Solutions

- Prepare a stock cryoprotectant solution of 5.25% (v/v) dimethyl sulphoxide (DMSO) in filtered seawater and place on ice.

2.2 Sea Urchin Handling and Gamete Collection

- Select ripe adult sea urchins and induce spawning by injecting 1.5 to 3 ml of 0.5 M KCl through the peristomal membrane into the coelomic cavity of each urchin (Strathmann, 1987). Spawning typically starts within 1 min of injection.

Sea urchin sperm are quiescent in the gonad and activated upon contact with seawater (Chia and Bickell, 1983). To minimise any effects of handling and storage on sperm activation, motility and viability, collect sperm as "dry" as possible by aspirating from the surface of each male.

- Place collected sperm into test tubes on ice. Each male should yield ~100 to 500 μl of dry sperm.
- Female sea urchins should also be induced to spawn by KCl injection and are then placed upside down over a beaker containing seawater.

2.3 Freezing and Thawing Procedures

- Aliquot sperm into test tubes and dilute 1:20 with cryoprotectant solution so that the final DMSO concentration is 5%.
- Load the diluted sperm into 0.25 ml plastic straws (IMV, l'Aigle, France) and seal with PVC powder.
- Place the straws into a controlled rate freezer and cool from to -75°C at 50°C min⁻¹ (Kryo-10 Series II, Planer Products Ltd, Sunbury-on-Thames, England). The time between the addition of DMSO and the start of cooling should be kept as short as possible to avoid any possible toxic effects of the DMSO.
- Once -75°C has been attained, hold the straws at -75°C for a further 10 min before removing from the freezer and plunging into liquid nitrogen.
- For thawing, remove the straws and quickly place into a water bath at 15°C for 30 s. Cut the ends off each straw then use for fertilization.

3 GENERAL CONSIDERATIONS

Cryopreservation markedly reduces sperm fertilization ability (Fig. 1). In general, maximum fertilization is attained with unfrozen sperm at a concentration of ~10^5 ml⁻¹ (equivalent to a egg: sperm ratio of 1:1000) whereas a concentration of at least 10^6 ml⁻¹ is required to achieve maximum fertilization with cryopreserved sperm (1:10000 egg:sperm). The reduction in fertilization ability may in part be caused by damage to the sperm mitochondria and plasma membrane during cryopreservation (Table 1).

Concentrations of DMSO ranging from 2.5 to 12.5% (final) have been found to be equally effective in preserving sperm fertilization ability [21]. If storage space is an issue, sperm can be diluted with CPA solution at ratios lower than 1:20 (e.g., 1:5) although there is generally a further reduction in fertility so more sperm may need to be added for fertilization.

4. ACKNOWLEDGMENTS

We would like to thank Dr. Rodney Roberts at the Cawthron Institute and Dr. John Smith at AgResearch for helpful discussions. Funding for this research was provided by the New Zealand Foundation for Research Science and Technology.

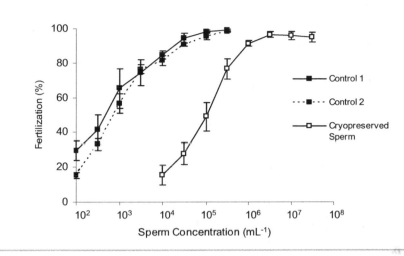

Figure 1 - Fertilization (mean ± SE, n = 3 pools) of eggs fertilized with unfrozen and cryopreserved sperm. Control 1 is assays that were carried out to assess the fertilization ability of the sperm before it was frozen. Control 2 is assays that were carried out using a separate pool of unfrozen sperm to assess the fertility of eggs used in assays with cryopreserved sperm. (modified from Adams et al. [21]).

Table 1 - Mitochondrial function and membrane integrity of unfrozen and cryopreserved sea urchin sperm. Flow cytometry was used in combination with R123- and PI-staining to determine the functional status of the mitochondria and plasma membrane integrity of sperm. Data presented are the mean and S.E. (in parentheses) of three replicates of pooled sperm. (modified from Adams et al. [21]).

	% Live (sperm with functioning mitochondria and intact plasma membranes)	% Dying (sperm with non-functioning mitochondria and intact plasma membranes)	%Dead (sperm with non-functioning mitochondria and damaged plasma membranes)
Unfrozen Sperm	84.7 (0.9)	4.2 (0.2)	10.6 (1.0)
Cryopreserved Sperm	62.6 (5.8)	17.7 (3.4)	19.0 (2.8)

5. REFERENCES

1-Foltz, K.R., Gamete recognition and egg activation in sea urchins, *Am. Zool.*, 35, 381, 1995.

2-Pagano, G., Cipollaro, M., Corsale, G., Esposito, A., Ragucci, E., Giordano, G.G., and Trief, N.M., The sea urchin: bioassay for the assessment of damage from environmental contaminants, in: *Community Toxicity Testing*, Cairns, J., Ed., ASRM STP 920, American Society for Testing and Materials, Philadelphia, 1986, 66.

3-Bay, S., Burgess, R., and Nacci, D., Status and applications of echinoid (Phylum Echinodermata) toxicity test methods., in: *Environmental Toxicology and Risk Assessment*, Landis, W.G., Highes, J.S. and Lewis, M.A., Eds., ASTM STP 1179, American Society for Testing and Materials, Philadelphia, 1993, 281.

4-Hagen, N.T., Echinoculture: from fishery enhancement to closed cycle cultivation, *World Aquacult.*, 27(4), 6, 1996.

5-Keesing, J.K. and Hall, K.C., Review of harvests and status of world sea urchin fisheries points to opportunities for aquaculture, *J. Shellfish Res.*, 17, 1597, 1998.

6-Barker, M.F., The ecology of *Evechinus chloroticus*, in: *Edible Sea Urchins: Biology and Ecology*, Lawrence, J.M., Ed., Elsevier Science B. V., Amsterdam, 2001, 245.

7-Johnson, L.G., Stage-dependent thyroxine effects on sea urchin development, *N. Z. J. Mar Freshwat Res*, 32, 531, 1998.

8-Knapp, G.F. and Barker, M.F., The effect of ultraviolet radiation on development and time to cleavage for embryos of the sea urchin *Evechinus chloroticus*, in: *Echinoderms 2000*, Barker, M.F, Ed., Swets & Zeitlinger, Lisse, 2001, 469.

9-Wu, J., Kurokura, H., and Hirano, R., Hybridization of *Pseudocentrotus depressus* egg and cryopreserved sperm of *Anthocidaris crassispina* and the morphology of hybrid larva, *Nippon Suisan Gakkaishi*, 56, 749, 1990.

10-Asahina, E. and Takahashi, T., Survival of sea urchin spermatozoa and embryos at very low temperatures, *Cryobiology*, 14, 703, 1977.

11-Barros, C., Garrido, C., Muller, A., Roth, A., and Whittingham, D.G., Cryopreservation of gametes and embryos of the sea urchins *Tetrapigus niger* and *Loxechinus albus*, in: *Improvement of the Commercial Production of Marine Aquaculture Species*, Gajardo, G. and Coutteau, P., Eds, Alfabeta, Santiago, 1996, 215.

12-Barros, C., Muller, A., and Wood, M.J., High survival of spermatozoa and pluteus larvae of sea urchins frozen in Me$_2$SO, *Cryobiology*, 35, 341, 1997.

13-Dunn, R.S. and McLachlan, J., Cryopreservation of echinoderm sperm, *Can. J. Zool.*,51, 666, 1973.

14-Asahina, E. and Takahashi, T., Freezing tolerance in embryos and spermatozoa of the sea urchin, *Cryobiology*, 15, 122, 1978.

15-Asahina, E. and Takahashi, T., Cryopreservation of sea urchin embryos and sperm, *Dev. Growth Diff.*, 21, 423, 1979.

16-Kurokura, H., Yagi, N., and Hirano, R., Studies on cryopreservation of sea urchin sperm, *Suisanzoshoku*, 37, 215, 1989.

17-Anthony, P., Ausseil, J., Bechler, B., Benguria, A., Blackhall, N., Briarty, L.G., Cogoli, A., Davey, M.R., Garesse, R., Hager, R., Loddenkemper, R., Marchant, R., Marco, R., Marthy, H.J., Perry, M., Power, J.B., Schiller, P., Ugalde, C., Volkmann, D., and Wardrop, J., Preservation of viable biological samples for experiments in space laboratories, *J. Biotechnol.*, 47, 377, 1996.

18-Barros, C., Muller, A., Wood, M.J., and Whittingham, D.G., High survival of sea urchin semen (*Tetrapigus niger*) pluteus larvae (*Loxechinus albus*) frozen in 1.0 M Me$_2$SO, *Cryobiology*, 33, 646, 1996b.

19-Strathmann, M.F., Reproduction and Development of Marine Invertebrates of the Northern Pacific Coast, in: *University of Washington Press*, Seattle, 1997.

20-Chia, F.S. and Bickell, L.R., Echinodermata, in: *Reproductive Biology of Invertebrates*, Spermatogenesis and Sperm Function, Adiyodi K.G. and Adiyodi, R.G., Eds., John Wiley & Sons Ltd., Chichester, 1983, II, 545.

21-Adams, S.L., Hessian, P.A., and Mladenov, P.V., Cryopreservation of sea urchin (*Evechinus chloroticus*) sperm, *Cryoletters*, 25,(4), 287, 2004.

Complete affiliation:

Serean L. Adams, Cawthron Institute, Private Bag 2, Nelson, New Zealand and Departments of Physiology and Marine Science, University of Otago, P. O. Box 56, Dunedin, New Zealand. e-mail: Serean.Adams@cawthron.org.nz.

PROTOCOL FOR CRYOPRESERVATION OF *Penaeus vannamei* SPERM CELLS

M. Salazar, M. Lezcano and C. Granja

1. INTRODUCTION

While vertebrate gamete cryopreservation has become a well-established technology in assisted reproduction, it is not a routine procedure for invertebrates [1], although the cryopreservation of penaeid prawn sperm or embryos has definite applications for the aquaculture industry.

One of the main drawbacks for Penaeid sperm cryopreservation is the lack of a reliable method for determining post-thawing cell viability. In many species, visualization of sperm motility and fertilization rate is widely used for the quantification of cell viability [2]. However, *Penaeid* sperm cells are flagellated and nonmotile therefore it is not possible to use motility as an indicator of cell viability. Fertilization rate methods are very difficult to use due to the fact that spermatophore attachment and spawning are not simultaneous and although thawed cells could be viable they have to remain attached to the female until spawning. In *Penaeus vannamei*, as in all decapods, sperm cells are produced in the testes. As they merge into the proximal *vas deferens,* spermatozoa are embedded in seminal fluids of testicular origin creating clusters of sperm. These sperm masses become surrounded by a coat of secretions that generates the spermatophore layer. During the copula spermatophores are deposited on the ventral surface of the female called thelyca, in order to fertilize the eggs during spawning. *In vitro* fertilization of eggs using cryopreserved sperm has not yet been possible. Therefore, evaluation of cell viability is mainly done by morphotype analysis under optical microscopy. More recently, viability determined by flow cytometry using propidium iodide as a marker of membrane integrity was compared to morphotype analysis [3]. Although cell viability determined by flow cytometry was lower than by optical microscopy, there was a significant positive correlation between both methods. These findings indicate that it is feasible to perform an initial screening using morphotype analysis, a simpler and faster technique, and to reserve flow cytometry for confirmation of results.

2. PROTOCOL FOR FREEZING AND THAWING

2.1 Equipment and materials

- Microcentrifuge at variable speed
- Micropipettes
- 2.0 ml microcentrifuge tubes

- 0.5 ml French Straws
- Programmable freezer Cryologic CL2000 (CryoBiosystem, Paris, France)
- Water bath

2.2 Reagents

- Methanol (Mallinckrodt Baker S.A., Mexico, Mexico)
- Sterile sea water Salinity 35 ppt
- Sucrose (Sigma, S-1888)
- Extender solution (10% egg yolk and 0.2 M sucrose in Sterile Sea Water)

2.3 Freezing procedure

- Collect spermatophores from mature *Penaeus vannamei* males by applying slight pressure on the lateral region of the genital pore. Select spermatophores without signs of necrosis.
- To obtain spermatic mass gently press the posterior region of the spermatophore.
- Transfer spermatic mass to a micro centrifuge tube with 500 µl cold (4°C) freezing solution (10% egg yolk, 0.2 M sucrose in filtered sea water).
- Add 25 µl of methanol (final concentration 5%), let the solution stabilize for 10 minutes keeping it at 4°C. Add another 25 µl of methanol to reach the 10% final concentration.
- Transfer the freezing medium with the spermatic mass into 0.5 ml French straws. Seal them with polyvinyl alcohol (Sigma) and transfer them to a programmable Cryologic CL2000 freezer (CryoBiosystem, Paris, France).
- Adjust freezing rate to –0.5°C/min down to –6°C. Perform manual seeding at –6°C. Continue freezing at –0.5°C per minute up to –32°C. At this point transfer the vials to liquid nitrogen.

Note: Freezing rate is also a critical factor in sperm cryopreservation of *P. vannamei*. Cells are very susceptible to fast cooling or direct plunging into liquid nitrogen. Slow freezing protocols should be used.

2.4 Thawing procedure

- Withdraw the straws from the liquid nitrogen.
- Transfer them to a water bath at 20°C for 10 seconds.
- Remove the contents and transfer them to 1ml sterile sea water with 0.2 M sucrose at room temperature.
- Analyze viability by optical microscopy as described below.

2.5 Viability evaluation by optical microscopy

- To obtain a homogenous cell suspension, gently disrupt the spermatic mass by mechanical agitation.
- Take an aliquot of 100 µl of cell suspension and dilute it 1:5 in sterile sea water with 0.2 M sucrose.
- Analyze cell viability in a Neubauer chamber.
- Count at least 100 cells, at 400x magnification and classify them into three morphological groups: a) spiked cells, b) nonspiked cells and c) everted cells.
- Express viability as the percentage of cells with spike over the total number of evaluated cells.

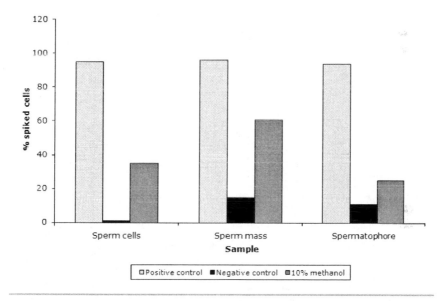

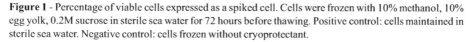

Figure 1 - Percentage of viable cells expressed as a spiked cell. Cells were frozen with 10% methanol, 10% egg yolk, 0.2M sucrose in sterile sea water for 72 hours before thawing. Positive control: cells maintained in sterile sea water. Negative control: cells frozen without cryoprotectant.

3. GENERAL CONSIDERATIONS

Penaeus vannamei sperm cells are highly sensitive to osmolarity changes. Osmolarity of solutions should be similar to that of sea water (1,048 mOsmKg^{-1}). Incubation of cells in calcium free artificial sea water with an osmolarity of 700 mOsmKg^{-1} induced 100% cell mortality in less than 2 minutes. Regarding the cryoprotectants, they are not very toxic to sperm cells. As shown in Table 1, cell survival is very high, even after 2 hour incubation at 10% concentration of cryoptotectants.

Cryopreservation of sperm mass is more reliable and produces highest survival post-thawing when compared to sperm cells or complete spermatophore. It is possible that the components of sperm mass might protect the cells against direct freezing injuries and changes in osmolarity. On the other hand, freezing complete spermatophore produced the highest variability rates among samples. This result can be attributed to the wall of the spermatophore, which acts as a barrier, preventing the effect of cryoprotectant upon the cells. Figure 1 shows the percentage of viable cells (spiked cells) after freezing and thawing sperm cells, sperm mass and complete spermatophore. The highest percentage of spiked cells was found using sperm mass as the biological material.

Table 1. *In vitro* survival of sperm cells from *P. vannamei* exposed to cryoprotectants (10% final concentration), at room temperature, for up to 120 minutes

Treatment	Percentage survival (Mean SD) Time (minutes)				
	0	30	60	90	120
SSW Me$_2$SO 10%	84.8 ± 8.1	65.8 ± 22	67.8 ± 14.9	67.0 ± 16.6	66.0 ± 17
SSW glycerol 10%	89.0 ± 3.4	88.0 ± 4.1	81.6 ± 4.4	86.0 ± 3.5	86.2 ± 4.5
SSW methanol 10%	90.6 ± 1.5	88.4 ± 2.8	87.8 ± 4.0	86.6 ± 5.8	85.6 ± 3.6
SSW ethylene glycol 10%	93.2 ± 2.6	89.2 ± 1.3	87.4 ± 1.5	86.6 ± 1.1	85.2 ± 0.8
Control SSW	91.0 ± 5.0	88.0 ± 5.6	89.0 ± 4.7	87.0 ± 6.4	87.0 ± 6.0

SSW: Sterile sea water 35 ppm salinity. Each mean (±SD) represents the data from five replicates of 100 cells each. Me$_2$SO: dimethyl sulfoxide.

4. REFERENCES

1-Gwo, J.C., Cryopreservation of aquatic invertebrate semen; a review, *Aquac Res*, 31, 259, 2000.
2-Kronenberger, K., Brandis, D., Turkay, M., and Storch, V., Functional Morphology of the Reproductive System of *Galathea intermedia* (Decapoda: Anomura), *J morphol*, 262, 500, 2004.
3-Lezcano, M., Granja, C., and Salazar, M., The use of flow cytometry in the evaluation of cell viability of cryopreserved sperm of the marine shrimp (*Litopenaeus vannamei*), *Cryobiology*, 48, 349, 2004.

Complete affiliation:
Marcela Salazar, Centro de Investigaciones de la Acuacultura en Colombia CENIACUA. Carrera 8A # 96-60 Bogota Colombia. Phone number 571 6369770 e-mail: msalazar@ceniacua.org

TECHNIQUES FOR CRYOPRESERVATION OF SPERMATOPHORES OF THE GIANT FRESHWATER PRAWN, *Macrobrachium rosenbergii* (DE MAN)

P. Damrongphol and K. Akarasanon

1. INTRODUCTION

The giant freshwater prawn, *Macrobrachium rosenbergi,* de Man, is an economically important crustacean species belonging to the family Palaemonidae. It is found in freshwater and brackish water of tropical and subtropical areas. It generally becomes reproductively mature within 5 months. Mature males and females are able to mate throughout the year. Mating occurs after the female experiences a premating-molt. The male deposits spermatophores on the medial surface between the pereiopods (walking legs) of the female. Eggs are normally spawned at night within 24 h after mating. Hatching commences about 19 days after spawning.

The male reproductive system consists of a pair of testes, *vas deferenses* and genital pores. The testes are elongate structures lying above the hepatopancreas and below the heart. The vas deferens arises from the mid-lateral region of each testis and extends posterior-ventrally to the gonopore. The gonopores are located on the medial surface of the coxa of the fifth pereiopods. The sperm mass is ejaculated as a form of spermatophore [1,2]. The spermatophores are white in color; they are rod-shaped, with varying sizes. The sperm are enclosed in gelatinous matrices secreted by the epithelium of the *vas deferenses* [3]. The sperm of *M. rosenbergii* are immotile, the head containing the nucleus is convex shaped, resembling an inverted umbrella with a long spike extending from the center of the convex head [4].

Cryopreservation of spermatophores of *M. rosenbergii* has important applications for genetic conservation, artificial reproduction, selective breeding and for availability of gametes on demand. Storage of spermatophores of *M. rosenbergii* has been reported. Studies involved a short-term cold storage at 2°C [5], preservation in liquid nitrogen (LN$_2$) to a maximum of 31 days [6] and a satisfactory long-term storage of up to 150 days at -196°C [7].

2. PROTOCOL FOR FREEZING AND THAWING

2.1 Equipment and Reagents

- 2 ml cryogenic vial (Corning, Inc., New York, USA)
- glycerol (Gly) (Sigma Chemicals Co. Ltd., St. Louis, Mo, USA)
- ethylene glycol (EG) (Sigma Chemicals Co. Ltd., St. Louis, Mo, USA)
- 0.4% trypan blue

- thermoprobe (type PC, Shinko Electric Industries Co., Ltd. Osaka, Japan)
- a quick-set cyanoacrylate adhesive (Alteco Inc., Osaka, Japan)

2.2 Collection and Assessment of the Spermatophores

- Collection of spermatophores is achieved by electrical stimulation. A pair of electrical probes are placed on each side at the base of the coxa of the fifth pereiopods just below the opening of the gonopore. (Figure 1). A 9 volt-electrical stimulus is applied three to four times. The extruded spermatophore (Figure 2) is picked up by a pair of fine forceps and placed in a cryogenic vial.
- Assessment of the spermatophores is performed by determining the normality of sperm morphology and sperm survival. A small piece of the spermatophore is cut and placed on a glass slide. After a drop of deionized water (DW) is added, the specimen is covered with a cover slip and pressed gently to disperse the sperm aggregates in the spermatophore. The sperm morphology (Figure 3) is evaluated under a phase-contrast microscope.

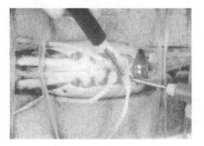

Figure 1 - Electro-ejaculation.

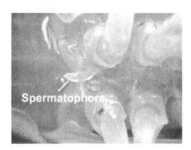

Figure 2 - Extruding spermatophore.

- Sperm of *M. rosenbergii* are nonmotile; their survival can be determined by vital-dye staining such as eosin-nigrosin [8] or trypan blue [9] stain. Live sperm exclude the stain while dead sperm incorporate the stain. The survival of *M.*

rosenbergii sperm can be evaluated using trypan-blue staining. Small pieces of spermatophores are cut and stained with 0.4% trypan blue for 15 min at room temperature. The specimen is then placed on a glass slide, covered with a cover slip and examined under a bright-field microscope. The percentage of live sperm is then determined from a random sperm count.

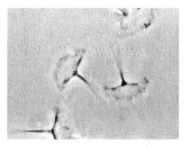

Figure 3 - Sperm.

2.3 Short-term Storage of the Spermatophores

- Cryopreservation at -20°C in the presence of a cryoprotective agent (CPA) is a simple and efficient method for short-term storage of up to 10 days. Gly and EG at 10%-20% (v/v) serve satisfactory as CPAs. One ml of 10%-20% Gly or EG is added slowly to vials containing spermatophores to avoid sudden osmotic changes. The vials are then equilibrated for 15 min at room temperature before storing immediately at -20°C.
- A long storage period at -20°C reduces sperm quality both in term of sperm survival rate and fertilizing ability (Table 1). Storage at -20°C without CPA can be done by simply wrapping spermatophores with aluminum foil, placing in a sealed plastic bag and immediately storing at -20°C. Storage of up to 5 days using this method maintains a moderate level of sperm survival but does not retain a satisfactory fertilizing ability. An extended storage period of up to 10 days reduces the sperm quality markedly (Table 1).

2.4 Long-term Storage of the Spermatophores

- Storage of the spermatophores at -196°C in the presence of 20% EG provides a satisfactory protocol for long-term cryopreservation of up to 150 days; the sperm survival rate and fertilizing ability are relatively high. A decline in fertilizing ability occurs if the storage period is extended.
- One ml of 20% EG is gradually added into vials containing spermatophores to avoid sudden osmotic changes as described in the previous section; the vials are equilibrated at room temperature for 15 min.

- The specimens are then cooled to -70°C at a slow cooling rate of 1.5-2.5°C min⁻¹ by placing the vials in a cooling-container filled with 95% ethanol; dry-ice cubes (1-2 cm³) are slowly dropped in the 95% ethanol. The temperature is monitored using a thermoprobe (type PC, Shinko Electric Industries Co., Ltd. Osaka, Japan).
- After cooling to about -70°C, the vials are exposed to LN_2 vapor (-110°C to -130°C) for 1-2 min and are then plunged directly into a container filled with LN_2 (-196°C) for storage.
- A simple method of cryopreservation at -20°C in the presence of CPA by storing immediately after equilibration may be performed but the sperm quality, the sperm survival rate and fertilizing ability, is significantly lowered (Table 1).

Table 1- The quality of cryopreserved sperm from various different protocols.

Type of storage	Temp. (°C)	Storage period (days)	CPA	Sperm survival (%)	Fertilizing ability (%)
Direct freezing	-20	5	-	75	40
		10	-	60	20
		10	10% Gly	85	60
			20% Gly	85	60
			10% EG	85	65
			20% EG	85	65
		60	20% EG	75	40
		150	20% EG	40	<10
		300	20% EG	<10	0
In LN_2	-196	60	20% EG	85	70
		150	20% EG	85	70
		300	20% EG	80	60

2.5 Thawing and Assessment of the Cryopreserved Sperm Quality

- The cryopreserved spermatophores are thawed with a rapid thawing procedure by immersing the vial in a 30°C water bath for 5 min. After thawing, the CPAs are removed by washing with DW.
- Cryopreserved sperm quality is evaluated for sperm survival rate and fertilizing ability. The sperm survival rate can be evaluated by staining with 0.4% trypan-blue as described in the previous section. The percentage of live sperm is then determined. The fertilizing ability is assessed using an artificial insemination technique [5,10].
- The frozen-thawed cryopreserved spermatophores are applied to the sternum close to the third pair of pereiopods of the female prawn which has experienced a premating-molt with a quick-set cyanoacrylate adhesive (Alteco Inc., Osaka, Japan). After spawning, fertilizing ability is determined by assessing the survival rate of the early developing embryos and/or the rate of spermatophores produced embryos survived to hatching. During the early developing stage of embryos

(1-5 days after spawning), the unfertilized eggs and the degenerated embryos remain in the brood chamber. Determination of the survival rate of early developing embryos is possible. Degenerated eggs or embryos are gradually discarded at later stages of development; the rate cannot be determined.

3. GENERAL CONSIDERATIONS

The presence of CPA is essential for efficient cryopreservation of giant freshwater prawn spermatophores. Moreover, the efficiency depends on the type of CPA, temperature and the storage periods. A change in any parameter alters the preservation efficiency. Selection of an appropriate method to serve a specific purpose should be considered. Direct freezing at -20°C with CPA is simple and economic producing relatively well-preserved sperm. This may be applied to short-term storage for transport of spermatophores from one location to another. Preservation in LN_2 requires a more substantial procedure. This can serve as long-term storage for future applications.

Successful cryopreservation of the spermatophores means the preservation of fertilizing ability of the sperm. A high sperm survival rate does not necessarily indicate a high fertilizing ability; alterations at a cellular or subcellular level of the sperm cell may affect fertilization success without a significant change in sperm survival, which in the present case is detected by a staining method not by a conventional sperm motility method.

4. ACKNOWLEDGMENTS

The authors would like to thank S. Pornirai for assistance in preparation of the manuscript. This work was funded by a grant from the Institute of Science and Technology for Research and Development, Mahidol University.

5. REFERENCES

1-Ra'anan, Z. and Sagi, A., Alternative mating strategies in male morphotypes of the freshwater prawn, *Macrobrachium rosenbergii (de Man), Biol Bull,* 169, 592, 1985.

2-Sagi, A., Milner, U., and Cohen, D., Spermatogenesis and sperm storage in the testes of the behaviorally distinctive male morphotypes of *Macrobrachium rosenbergii* (decapoda, palaemonae), *Biol Bull-us,* 174, 330,1988.

3-Dougherty, W.J., Dougherty, M. M., and Harris, S. G., Ultrastructural and histochemical observation on electroejaculated spermatophores of the palaemonid shrimps, *Macrobrachium rosenbergii, Tissue and Cell,* 18, 709, 1986.

4-Lynn, J. W. and Clark, W. H., The fine structure of the mature sperm of the freshwater prawn, *Biol Bull-us,* 164, 459, 1983.

5-Chow, S., Artificial insemination using preserved spermatophores in the palaemonid shrimp Macrobrachium rosenbergii, *Bulletin of the Japanese Society of Scientific Fisheries,* 48, 1693, 1982.

6-Chow, S., Taki, Y., and Ogasawara, Y, Cryopreservation of spermatophore of the freshwater shrimp, *Macrobrachium rosenbergii, Biol Bull-us,* 168, 471, 1985.

7-Akaiasanon, K., Damrongpol, P, and Poolsanguan, W., Long-term cryopreservation of spermatophore of the giant freshwater prawn, *Macrobrachium rosenbergii (de Man)*, *Aquac res*, 35, 1415, 2004.

8-Jeyalectumie, C. and Subramoniam, T., Cryopreservation of spermatophores and seminal plasma of the edible crab *Scylla serrata*, *Biol Bull-us*, 177, 247, 1989.

9-Bhavanishankar, S. and Subramoniam, T., Cryopreservation of spermatozoa of the edible mud crab *Scylla serrata* (Forskal), *J Exp Zool*, 277, 326, 1997.

10-Sandifer, P.A. and Smith, T.I.J., A method for artificial insemination of Macrobrachium prawns and its potential use in inheritance and hybridization studies, *Proceedings World Mariculture Society*, 10, 403, 1979.

Complete affiliation:

Praneet Damrongphol, Department of Biology, Faculty of Science, Mahidol University, Rama VI. Rd., Bangkok 10400, Thailand. e-mail: scpdr@mahidol.ac.th.

PROTOCOL DEVELOPMENT FOR THE CRYOPRESERVATION OF SPERMATOZOA AND SPERMATOPHORES OF THE MUD CRAB *Scylla serrata*

S. Bhavanishankar and T. Subramoniam

1. INTRODUCTION

Scylla serrata (Forskal) of the family Portunidae, commonly known as the red mud crab has a wide distribution throughout the Indo-Pacific region. They abound shallow estuarine and brackish waters and are commercially fished in maritime countries like Indonesia, Australia, Fiji, Philippines, Thailand, Malaysia, Vietnam, Japan, Sri Lanka and India. Much as the class Crustacea comprises animals which have long been fished for their delectable flesh, research in crustaceans has failed to entice adequate attention on gamete preservation as the pride of place in cryobiological research has always been given to mammals. Understandably, no serious attempt has been made to extend the low-temperature preservation techniques of mammalian gametes to those of invertebrates. However, recent years have witnessed a steady increase in scientific interest in low-temperature storage of gametes of cultivable invertebrates due chiefly to their need in commercial gains in animal production. Sadly, literature on gamete preservation exposes the lacuna where crustaceans are concerned, albeit the few reports on successful storage of male gametes [1-12]. It is evident from previous studies that a cryoprotectant more effective for one biological system is less effective for another, underlining the importance of testing the cryoprotective ability of different cryoprotectants for hitherto unstudied material like the male gamete of crabs. More so, questions pertaining to the toxicity of individual cryoprotective agents for gametes, freezing damage, osmotic response to various addition and dilution procedures and to the freeze-thaw cycle are yet to be answered. Problems related to successful cryopreservation are further inflated in the case of atypical sperm of *Scylla serrata*, which are nonmotile and delivered during mating as sperm bundles or spermatophores. The assessment of post-thaw viability of free sperm accordingly poses problems, making mandatory parallel evolution of viability assessment techniques [8,11].

In this context, the present contribution is a holistic approach towards providing a sequence of methods for developing protocols for the cryopreservation of spermatophores and free spermatozoa of *S. serrata*.

2. PROTOCOL FOR FREEZING AND THAWING

2.1. Animal Handling and Semen Collection

- In the current study, owing to failure in our efforts to achieve expulsion of seminal substances by electrical stimulation or extract semen by using catheters, we resorted to killing experimental animals for collection of spermatophores and spermatozoa. Adult intermolt crabs (*S. serrata*) with a carapace width of 10 cm to 14.5 cm were used.

- The live crabs were either sacrificed immediately after being transported to the laboratory or were maintained in sea water (salinity ~20 ppt; 100% exchange every 24 h) in well aerated 150 liter fiber-reinforced plastic tanks with a feed of locally available bivalves and sand crabs. All samples were prepared in an extender medium (in most cases, calcium free artificial sea water - $Ca^{++}FASW$).

- Spermathecae (8-10) from the female crabs were cut open, and the contents gently shaken in the extender medium to release the free spermatozoa.

- The luminal contents of the mid *vas deferens* from adult male crabs (3-4) were centrifuged at 200 x g for 5 min to pellet the spermatophores, which were either subjected to enzyme treatment or mechanical shearing to release the enclosed spermatozoa. Enzyme treatment involved digestion of the spermatophore wall using 1% w/v pronase (in $Ca^{++}FASW$). Mechanical shearing involved gentle homogenization of spermatophore pellet in $Ca^{++}FASW$. In each case 3 to 5 min treatment was given with gentle shaking, followed by centrifugation at 440 x g for 3 min.

- Spermatozoa were collected from the pellet and washed repeatedly in small volumes of extender. The isolated spermatozoa were diluted to a concentration of ~10^6 cells/ml in the extender [8].

- The normal spermatozoa, as released from the spermathecae and from the spermatophores, appeared uniformly spherical in shape measuring 3-4 μm in diameter, with several slender radiating nuclear arms and a prominent acrosome overlying the nuclear mass (Figure 1). The recovery of normal sperm, retaining their morphological integrity, was higher when released from the spermathecae (~93%) and spermatophores by enzymatic digestion (~90%), as compared to those liberated from the spermatophores by mechanical shearing (~73%) (Table 1). However, free sperm released from the spermatophores after enzyme treatment were highly sensitive to induced acrosome reaction and cryopreservation procedures apart from showing wide fluctuations among different samples (Table 2). Therefore, only the spermatozoa obtained from spermathecae and those released from spermatophores by mechanical shearing were considered for most experiments.

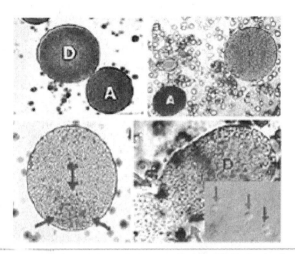

Figure 1 - Viability assessment by staining techniques:a. Trypan Blue b. Eosin nigrosin c. Incomplete dye exclusion using trypan blue (Note area indicated by large arrows showing intense blue color suggesting incorporation of dye by a small fraction of the enclosed sperm) d. Ineffectiveness of dye exclusion methods. Notice wall leakage in eosin nigrosin stained "dead" spermatophore (sperm oozing out); arrows indicate region of spermatophore wall breakage. Inset at a higher magnification (x1000) demonstrating dye exclusion by a few spermatozoa. (A – "alive"; D – "dead").

Table 1 - Percentage "normal" spermatozoa (sptz retaining morphological integrity) upon release from the spermatophores (sph) by different means as compared to those released from the spermathecae (spth). Values expressed are mean ± Standard Error (SE); (n = 6; * n = 10).

SAMPLE		% 'NORMAL' SPERM	REMARKS
SPTH		92.71 ± 0.79	pooled sample from 8 -10 spth
SPH	1% pronase	90.05 ± 1.20*	used for some experiments
SPH	0.75 % pronase	88.29 ± 0.32	
SPH	0.5 % pronase	90.01 ± 0.83	digestion incomplete
SPH	1% trypsin	89.86 ± 0.44	digestion incomplete
SPH	2% trypsin	77.31 ± 2.35	
SPH	mechanical shearing	73.20 ± 0.47	used for cryo-experiments

Table 2 - Inherent differences in (A) the sensitivity of the spermatophores among different individuals to enzyme treatment and (B) the spermathecal spermatozoa. (n = 5, decimal values rounded off to whole numbers)

SOURCE OF FREE SPERM	% MORPHOLOGICALLY INTACT SPERM
SPH 1%P (A)	81, 95, 86, 79, 70
SPH 2%T (A)	89, 62, 72, 83, 69
SPTH (B)	81, 96, 83, 90, 81

2.3 Pre-freezing and Post-thawing Quality Assessment

Assessment of quality poses a difficult problem in the atypical spermatozoa of decapod crustaceans, which are conspicuous by their packaging into sperm sacs or spermatophores, assuming bizarre shapes, lacking motility altogether. This has necessitated the introduction of parallel methods for viability assessment alongside protocols for cryopreservation. Fresh and frozen-thawed spermatophores were assessed for quality using the following techniques:

- Staining techniques or "dye exclusion" methods [5,13,14]:
  1. Trypan blue-Twenty microliters of 1% (w/v) trypan blue was mixed with 180 μl spermatophore suspension, allowed to stand for 10 min and scored for dye incorporation. The number of spermatophores stained blue "dead" and those unstained "live" were counted in 10 random fields.
  2. Eosin/Nigrosin-To 100 μl of spermatophore suspension, 100 μl of 0.5% (w/v) eosin were added, followed by 200 μl of 10% (w/v) nigrosin, allowed to stand for 2 - 5 min and scored for dye incorporation. The number of spermatophores that stained pink "dead" and those unstained "live" were counted in 10 random fields.
- All the stains were prepared in $Ca^{++}FASW$ and the pH adjusted to 7.4. Dye exclusion was taken as an index for 'survival.'
- The percentage of live spermatophores was calculated as: % live spermatophores = 100 x live (unstained) units/total units counted. Each score involved examination of spermatophores in the order of 4-6 x 10^2. For convenience, the spermatophores demonstrating incomplete dye exclusion were grouped under 'live' as most of the sperm enclosed within excluded the dyes.
- Sperm viability assessment using staining techniques with eosin-nigrosin [5] and trypan blue [13] yielded unsatisfactory results. Appreciable differences (11.01 ± 1.12 SE, n = 9) were noticed in stain incorporation in unfrozen controls when eosin-nigrosin and trypan blue were used (Table 3). Differences were also noticed with a single stain on different aliquots (up to 17% in trypan blue and 16% in eosin-nigrosin). Moreover, dye incorporation into the spermatophores was incomplete in some cases (1% in eosin-nigrosin and 8% in trypan blue) whereas in other completely stained spermatophores, a few enclosed spermatozoa excluded the dyes altogether (40% in eosin-nigrosin and 15% in trypan blue). The photographic evidence exposing the ineffectiveness of dye exclusion techniques are represented in Fig. 1.

Table 3 - Differences in staining using trypan blue /eosin nigrosin. Values expressed are mean ± SE (n = 8)

SOURCE	TRYPAN BLUE	EOSIN NIGROSIN
SPTH	97.20 ± 1.70	85.28 ± 0.49
SPH (MS)	92.03 ± 1.10	80.70 ± 2.29
SPH (PD)	82.07 ± 1.66	77.03 ± 2.41

- Hypo/hyper-osmotic sensitivity tests with modification of the procedures followed by Hammerstedt et al. [15], Lomeo and Giambersio [16] and Chan et al [17]:

1. To 1.0 ml of spermatophore suspension 1.0 ml of hypo-osmotic extender medium $Ca^{++}FASW$ (< 760 mOsm) was added drop wise. The osmotic response, as indicated by swelling of spermatophoric mass, was taken as an index for membrane integrity.

2. Additionally, 1% (w/v) trypan blue was mixed with the hypo-osmotic extender medium. Spermatophores incubated in this medium were scored for hypo-osmotic viability by observing those which remained unstained and swollen. A 40% (w/v) sucrose solution in $Ca^{++}FASW$ was used as a stock hyperosmotic solution and used after adjusting osmolality to fall in the range 1.5 to 1.8 Osm with $Ca^{++}FASW$. Volume changes as indicated by shrinkage were recorded microscopically. For the purpose of assessing the viability of free spermatozoa, they were released from the pelleted spermatophores [8].

- Induced acrosome reaction:

1. Acrosome reaction was induced artificially in the spermatozoa by treatment with the divalent cation ionophore A23187. Spermatozoa were incubated for varying periods of time (2 to 10 min) in 2 mM $CaCl_2$ in $Ca^{++}FASW$ and subjected to ionophore treatment.

2. A 1 mg/ml stock solution of A23187 (Sigma Chemical Co., St. Louis, USA) was prepared by mixing in Me_2SO and diluted in $Ca^{++}FASW$ to concentrations of 1 - 20 µM. Aliquots of 100 µl were added to sperm suspensions of varying volumes (100-400 µl), agitated and incubated for 2 - 5 min.

3. Acrosome reaction was scored by light microscopic observations with DIC optics. Each score involved examination of spermatozoa in the order of 10^3. All observations were made under Carl Zeiss Axioplan Universal Microscope, Germany, at magnifications ranging from 1000 x to 2500 x. Adequate care was taken to curtail the risk of damage to the cells due to the weight of the coverslips. Bisected coverslips (No. 00/13 mm diameter/800 µg) were used for observations. Alternatively, transparent acetyl sheets of 80 µm thickness/6 mm diameter/200 µg were used.

4. The treatment involved a twofold swelling of nuclear mass (spermatozoa measured 6.5 to 7.0 µm in diameter) accompanied by retraction of nuclear arms until they were completely drawn into the nucleus. This was followed by the complete eversion of the sub-acrosomal material through the apical acrosomal cap, culminating in the formation of a slender acrosomal filament (acrosomal process) (Figure 2).

5. In order to check the existence of nuclear material in the extensions of the nucleus, and to trace these during induced acrosome reaction, free spermatozoa were incubated in fluorochrome acridine orange 0.1% (w/v) in 0.2 M PBS and observed for fluorescence. This staining revealed an interesting feature, the DNA extending through the protruding anterior acrosomal filament, towards the completion of the reaction. During the process of swelling the nuclear mass,

fluorescent signals weakened. The initial stages of eversion of subacrosomal material did not carry the nuclear material within. However, the onset of filament formation triggered the extension of nuclear material into the acrosomal filament (Fig. 3).

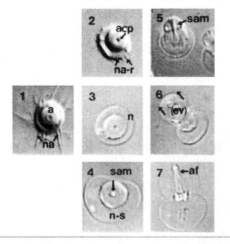

Figure. 2 - Acrosome reaction in the spermatozoa of *S. serrata*. 1. Normal spermatozoon. Retraction of nuclear arms 2-3 swelling of nuclear mass 3-4, eversion of the sub-acrosomal material through apical acrosomal cap 5-6 (arrows in 6 indicate the direction of progressive eversion), formation of acrosomal filament 7. Bar – 4 µm. a- acrosome; acp-acrosomal cap; af-acrosomal filament; (ev) eversion; n-nucleus; na-nuclear arms; na-r-nuclear arms (retracted); n-s-nucleus (swollen); sub-subacrosomal material.

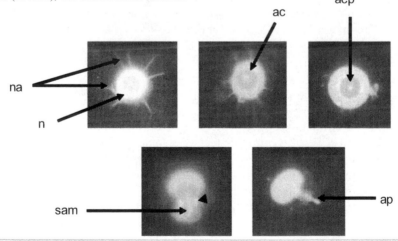

Figure 3 - Fluorescence micrographs of the pronase sensitized spermatozoa *Scylla serrata* stained with acridine orange. Demonstration of the extension of nuclear material into the acrosomal filament (acrosomal process); na-nuclear arms; n-nucleus; ac-acrosome; acp-acrosomal cap; sam-sub-acrosomal material; ap-acrosomal process. Blunt triangle indicates the position of the acrosomal cap.

2.4 Selection of Diluents and Cryoprotectants

- Several extenders were used in order to identify the most suitable extender (Table 4): filtered sea water, artificial sea water [18], calcium free artificial sea water (Ca^{++}FASW) [18] and PBS-glycine-NaCl-Cryoprotectant [5]. Ca^{++}FASW was the only medium in which most spermatozoa retained their morphological integrity for over 90 min and therefore, was considered as the extender medium for cryopreservation experiments. Calcium free artificial sea water was prepared following the procedure of Leung-Trujillo and Lawrence [13]. A one liter solution in double distilled water (ddH_2O) consisted of NaCl (21.63 g), KCl (1.12 g), H_3BO_3 (0.53 g), NaOH (0.19 g) and $MgSO_4 \cdot 7H_2O$ (4.93 gm), pH 7.4, osmolality 890-920 mOsm/Kg. To remove Ca^{++} impurities 0.1 mM EGTA was added as a chelator as recommended by Gesteira and Halcrow [19].

Table 4 - Maintenance of morphological integrity of spermatozoa of *S serrata* in different extender media. Time dependent effect on spontaneous exocytosis. Values denote percentage loss in morphological integrity due to exocytosis. Sperm samples pooled from spermathecae of four animals for each experiment. Values represented are mean ± SE (n = 4).

Time in min.	Ca^{++}FASW	FSW	ASW	PBS-GLY-NaCl-5%G
0 (Control)	7.95 ± 0.06	7.95 ± 0.06	7.95 ± 0.06	7.95 ± 0.06
1-5	8.16 ± 0.05	18.2 ± 0.38	16.78 ± 0.52	9.13 ± 0.09
5-10	8.38 ± 0.07	25.43 ± 0.62	21.6 ± 0.37	9.93 ± 0.06
15-20	8.69 ± 0.1	27.93 ± 0.57	27.98 + 0.33	11.38 ± 0.22
25-30	9.20 ± 0.04	30.27 ± 0.33	31.68 ± 0.53	13.2 ± 0.24
40-45	9.77 ± 0.06	34.08 ± 0.43	34.58 ± 0.34	13.85 ± .12
55-60	9.99 ± 0.08	36.93 ± 0.47	37.3 ± 0.27	14.23 ± 0.21
85-90	10.25 ± .09	39.28 ± 0.43	40.5 ± 0.25	15.15 ± 0.19

- Permeating cryoprotectants like dimethyl sulfoxide (Me_2SO), ethylene glycol (EG), glycerol and methanol (MeOH) were prepared in Ca^{++}FASW to different concentrations ranging from 5 to 30% v/v.
- Effective cryoprotectant concentrations of 2.5% to 15% v/v were obtained by mixing aliquots of equal volumes of cryoprotectants to sperm/spermatophore suspension. Step mode or drop-mode addition of cryoprotectants was employed to minimize damage to the spermatophores and/or spermatozoa.
- Spermathecae: Equilibration studies at different temperatures (30°C, 23°C, 12°C and 5°C) revealed that volume changes, due to the addition of cryoprotectants, were rapid and tonic equilibrium could be achieved minutes after introduction of cryoadditives. The cryoprotectants tested could permeate through the

spermatophore wull well within 3-5 min of introduction. Therefore, equilibration was extended to 90 min to determine the toxic effects of the cryoprotectants, if any. However, for freezing experiments, equilibration of 10 to 15 min was found to be optimal. Lowering the equilibration temperature to 5°C did not enhance the viability of spermatophores, as evidenced by dye exclusion methods. It was noticed that the spermatophores could tolerate cryoprotectant concentrations of up to 15% (v/v) (Table 5). Toxicity of the cryoprotective agents to spermatophore was in the order Me_2SO > glycerol > MeOH and the temperature dependency for addition was most pronounced in the case of Me_2SO.

Table 5 - Loss of morphological integrity in spermatophore samples subjected to cryoprotectant treatment at different temperatures. Values are rounded off to the nearest whole number (pooled samples from 8 animals).

| | | DAMAGE (% NORMAL SPERM) | | | | | | | | | | | |
| | | 30°C | | | 23°C | | | 12°C | | | 5°C | | |
CPA	Eq. time min.	5	10%	15%	5%	10%	15%	5%	10%	15%	5%	10%	15%
Me_2SO	5	-	2	4	-	2	3	-	2	4	-	-	4
	15	2	3	9	2	2	8	3	2	7		1	5
	30	3	3	9	3	3	8	3	2	8		2	6
	60	3	4	9	3	3	7	4	3	8		3	8
	90	3	4	9	3	4	9	4	3	8		4	8
Glycerol	5	-	-	7	-	-	7	-	-	6	-	-	5
	15	-	1	7	-	-	7	-	-	8		-	7
	30	-	2	8	-	2	8	-	1	8		-	7
	60	1	3	7	1	4	7	1	2	9		2	8
	90	1	3	7	1	4	9	2	2	9		2	9
MeOH	5	-	-	-	-	-	-	-	-	-	-	-	-
	15		-	2			-		1	-			-
	30		-	2			-		1	2			1
	60		-	2			2		1	2			2
	90		1	2			2		1	2			2

- Spermatozoa: Equilibration studies carried out at different temperatures (30°C, 23°C, 15°C) revealed that most cryoprotectants tested could be used in the concentration range of 5% to 12.5% v/v. However, loss of morphological integrity in the sperm was greater at concentrations over 12.5% v/v. Lowering the addition temperatures considerably reduced the cryoprotectant-induced damage, specially with higher concentrations of Me_2SO. Although 17.8% of spermatozoa lost morphological integrity at 30°C when equilibrated in 15% v/v Me_2SO, the damage could be reduced at 15°C to 13.3%.
- Treatment with the divalent cation ionophore A23187 reveals that it could trigger acrosome reaction in the free spermatozoa of *S. serrata* in the presence of calcium as $CaCl_2$ in the extender medium. With the progress of time, a sizeable percentage of the sperm samples released in FSW and ASW (42% and 46%

respectively) underwent precocious eversion, resembling acrosome reaction. This reaction could be enhanced further in the presence of A23187. However, the samples released in Ca^{++}FASW failed to undergo spontaneous acrosome reaction. Differences, both in the duration of acrosome reaction and the percentage of completely reacted sperm could be noticed for different aliquots (Table 6).

Table 6 - Percentage of acrosome reacted sperm in fresh and frozen-thawed aliquots at various concentrations of ionophore A23187. Values represented are mean ± SE (n = 4).

		DAMAGE (% NORMAL SPERM)											
		30°C			23°C			12°C			5°C		
CPA	Eq. time min.	5	10%	15%	5%	10%	15%	5%	10%	15%	5%	10%	15%
Me$_2$SO	5	-	2	4	-	2	3	-	2	4	-	-	4
	15	2	3	9	2	2	8	3	2	7		1	5
	30	3	3	9	3	3	8	3	2	8		2	6
	60	3	4	9	3	3	7	4	3	8		3	8
	90	3	4	9	3	4	9	4	3	8		4	8
Glycerol	5	-	-	7	-	-	7	-	-	6	-	-	5
	15	-	1	7	-	-	7	-	-	8		-	7
	30	-	2	8	-	2	8	-	1	8		-	7
	60	1	3	7	1	4	7	1	2	9		2	8
	90	1	3	7	1	4	9	2	2	9		2	9
MeOH	5	-	-	-	-	-	-	-	-	-	-	-	-
	15	-	2			-		1	-				-
	30	-	2			-		1	2				1
	60	-	2			2		1	2				2
	90	1	2			2		1	2				2

- The hyper-osmotic sensitivity test, reveals that the spermatophores and spermatozoa exhibited volume changes as a response to osmotic changes in the extender medium. After initial shrinkage upon treatment with 5% - 10% (w/v) sucrose solution, they regained their original shape when the medium was replaced with Ca^{++}FASW. However, in hypo-osmotic sensitivity tests, 43% of spermatophores permitted dye entry indicating membrane leakage. The osmotic changes were complete and irreversible in the spermatozoa enclosed in the spermatophores. When the tonicity of the extender medium was reduced (Ca^{++}FASW diluted in ddH$_2$O in the ratios 2:1), swelling of the nuclear mass was noticed and complete eversion of sub-acrosomal material through the acrosomal cap took place, mimicking the morphological changes evidenced during the ionophore-induced acrosome reaction.

2.5 Cryopreservation Protocol Development

- The samples for short-term preservation and cryopreservation were aspirated into 0.5 ml straws 1.8 ml polypropylene vials and equilibrated at physiological temperatures (30°C, 23°C, 15°C) for 10 min and assessed for damage owing to cryoprotectant toxicity. The volume of the aspirated sperm suspension or spermatophores was 300 μl/straw and ~1.5 ml in the polypropylene vials. The straws were sealed with the help of artery forceps.

- The equilibrated samples were conventionally frozen, either by direct plunging into liquid nitrogen after lowering the samples slowly (t<10 min) in the vapor phase over liquid nitrogen or in a rate-controlled programmable freezer, Kryo 10/1.7 (Planer Biomed, U.K.). Programmed freezing involved subjecting the samples at physiological temperatures (start temperatures) to cooling at rates varying from -0.5°C/min to -15°C/min. The samples were frozen to different subzero temperatures (-30°C, -40°C, -50°C and -80°C) (final temperatures).

- During freezing, the samples were nucleated exogenously at -7°C, by dipping a pair of tongs in liquid nitrogen and touching the outer wall of the sample holder to avoid supercooling and to initiate ice crystal growth.

- The frozen samples were either subjected to isothermal holds at final temperature for 1 h and thawed to room temperature (~23°C) or plunged into liquid nitrogen.

- After 8 h of storage the samples were thawed to room temperature. Thawing was effected by immersing the samples in a water bath maintained at 55°C for 10 s to 15 s.

- Dilution of the sperm suspension upon thawing involved drop-mode addition of 2 ml of extender medium to 2 ml of the frozen-thawed sample. The diluting medium was cooled to 4°C before addition to the frozen-thawed samples.

- On completion of the dilution procedure the samples were observed microscopically to assess survival. These diluted samples were centrifuged at 440 x g for 3 min and the pellet re-suspended in 2 ml of the extender medium.

- Direct plunging of equilibrated spermatophore samples (2 - 4 x 10^2/ml) in straws/vials into liquid nitrogen from physiological temperatures was seen to be damaging to their survival, causing irreversible damage to more than 60% of spermatophore samples. Leakage of enclosed material and wall fracture were evident in most spermatophores. However, a marginal increase in the recovery of post-thaw samples could be noticed when several serial holds were given through the vapor phase over LN_2, accomplished by slow lowering at time intervals of 5 min each with phased lowering down every 1.0 cm column length.

- With spermathecae, programmed freezing, at cooling rates <-15°C/min to various subzero temperatures (-30°C to -80°C) and subsequent storage at -196°C was successful with most concentrations of the cryoprotectants tested. Using eosin-nigrosin, a post-thaw recovery of 57.2% was noticed in the samples plunged into LN_2 from a pre-storage temperature of -80°C in a 4 ramp freezing program with 10% (v/v) glycerol. The same aliquot projected a recovery of 68.1% when trypan blue staining was employed. Post-thaw survival was greater (72.3% with trypan blue staining and 67% with eosin-nigrosin) in samples frozen to a pre-storage temperature of -30°C, and thawed to room temperature. Thawing was done in a water bath at 60°C and transferred to a water bath at 30°C, in one quick change (Table 7).

- Direct plunging of equilibrated spermatozoa samples into LN_2 from physiological temperatures was seen to be detrimental to their survival. However, programmed freezing using a rate-controlled freezer at cooling rates <-15°C/min to various subzero temperatures (-30°C to -50°C) and subsequent storage at -196°C was successful with most concentrations of the cryoprotectants tested (Table VIII).

- Cryosurvival, as evidenced by induced acrosome reaction, was maximum in the spermatozoa released from the spermathecae and in those obtained by mechanical shearing of the spermatophores. Me_2SO, EG and glycerol (5% to 12.5% v/v) offered cryoprotection with a maximum survival of 52% of the control using 12.5% glycerol at a cooling rate of -5°C/min to a final temperature of -40°C with manual seeding at -7°C and subsequent plunging into liquid nitrogen. Although Me_2SO provided effective cryoprotection at 10% v/v (48% survival), increasing the concentration of Me_2SO proved to be damaging to the spermatozoa. Methanol, which exhibited minimal toxicity, did not provide effective cryoprotection with the concentrations tested. Cryoinjuries were evident in all the samples, the damage being prominent in the nuclear and acrosomal regions (Figure 4).

Table 7 - "Freezability" of spermatophores by direct plunging in liquid nitrogen as compared to programmed freezing. "Viability" scored by dye exclusion methods. Number in parentheses indicates number of observations. Start temperature 15°C; Cooling rate -1°C/min to 0°C; -5°C/min to -30°C, -80°C; hold time 60 min; seeding temperature (optional) 7°C. Trypan blue staining was incomplete in 14.2% of rapidly plunged spermatophores; 11.27% in programmed frozen spermatophores.

MODE OF FREEZING	EOSIN-NIGROSIN	TRYPAN BLUE*
Rapid plunging	36.67 ± 2.91 SE (6)	41.17 ± 4.57 SE (9)
Rapid plunging (serial holds in vapor phase)	39.23 ± 1.71 SE (5)	41.83 ± 1.23 SE (5)
1.Programmed freezing (seeding at -7°C)	67.00 ± 2.2 SE (6)	72.3 ± 3.4 SE (9)
1.Programmed freezing (-30°C) No seeding	64.9 ± 1.74 SE (5)	71.8 ± 2.59 SE (5)
2.Programmed freezing (-80°C)	57.16 ± 2.32 SE (4)	68.14 ± 1.89 SE (4)

2.6 Short-term storage of spermatophores

- Short-term storage of spermatophores was attempted at 2 ± 1.8°C, in a cold room.
- Equal volumes of 100 µl aliquots of spermatophores in Ca^{++}FASW and cryo-protective agents (Me_2SO, glycerol, EG or MeOH) were loaded in a 96-well ELISA plate, sealed with parafilm and maintained for varying periods of time (8 hr to 96 hr).
- Microscopic observations were carried out at 8 h intervals to assess damage, if any, to the stored samples. Prior to observation, each aliquot was diluted to 1.0

ml in Ca^{++}FASW in 4 steps, pelleted at 440 x g for 5 min and re-suspended in 200 μl Ca^{++}FASW to remove any trace of cryoadditives.

- Spermatophores subjected to storage at low temperatures (2 ± 1.8°C) exhibited microbial contamination with progress of time. Close to 23% of the samples revealed spermatophore wall leakage when stored at 2-3°C after 96 h storage. A remarkable improvement in viability (82% of control excluded trypan blue) was noticed when samples were equilibrated in 10% glycerol for 30 min. Storage at 2 ± 1.8°C rendered the free sperm more susceptible to ionophore-induced acrosome reaction. Moreover, a small fraction of the sperm lost morphological integrity with the progress of time.

Table 8 - Programmed freezing of the spermatozoa of *S serrata*. Summary of protocols.

SAMPLE	EQBN TEMP (°C)	MODE OF ADDN.	START TEMP. (°C)	COOLING (°C/min)	TEMP. -1 (°C)[1]	COOLING (°C/min)	SEEDING TEMP. (°C)	FINAL TEMP. (°C)[2]	HOLD TIME (min)	CPA USED (% v/v) AND DAMAGE (%)[3]
SPTH	15	I STEP	15	-5	-	-	-7	-40	60	2.5Me₂SO (70), 2.5G (69), 2.5EG (80), 2.5MeOH (100)
SPTH	15	DROP	15	-5	-	-	-7	-40	60	2.5Me₂SO (70), 2.5G (70), 2.5EG (81), 2.5MeOH (100)
SPH-P	15	DROP	15	-5	-	-	-7	-40	60	2.5Me₂SO (67), 2.5G (71), 2.5EG (90), 2.5MeOH (100)
SPTH	15	DROP	15	-5	-	-	-7	-40	60	5Me₂SO (57), 5G (54), 5EG (61), 5MeOH (87)
SPH -MS	15	DROP	15	-5	-	-	-7	-40	60	5Me₂SO (59), 5G (47), 5 EG (62), 5MeOH (94)
SPTH	15	DROP	15	-5	-	-	-7	-30	15	10Me₂SO (40), 10G (46), 10EG (39), 10MeOH (79)
SPH-MS	15	DROP	15	-5	-	-	-7	-40	60	10Me₂SO (40), 10G (47), 10EG (40), 10MeOH (79)
SPH-MS	15	DROP	15	-5	-	-	-7	-30	30	10Me₂SO (42), 10G (39)
SPH-P	15	DROP	15	-5	-	-	-7	-30	30	10Me₂SO (71), 10G (83)
SPTH	15	DROP	15	-5	-	-	-7	-40	60	12.5Me₂SO (59), 12.5G (32), 12.5EG (42), 12.5MeOH (86)
SPH -P	15	DROP	15	-5	-	-	-7	-30	10	12.5Me₂SO (76), 12.5G (81), 12.5EG (76), 12.5MeOH (82)
SPTH	15	DROP	15	-5	-	-	-7	-30	30	15Me₂SO (80), 15G (56), 15EG (61), 15MeOH (92)
SPTH	15	DROP	15	-5	-	-	-7	-40	60	15Me₂SO (81), 15G (54), 15EG (62), 15MeOH (54)
SPTH	15	I STEP	15	-1	4	-6	-7	-30	5	10Me₂SO (49), 10G (53)
SPTH	15	DROP	15	-1	4	-6	-7	-30	5	*10Me₂SO (31), 10G (22)
SPTH	15	DROP	15	-0.5	4	-5	-7	-40	10	*12.5Me₂SO (49), 12.5G (29)
SPTH	15	DROP	13	-0.5	4	-20	-	-40	15	10Me₂SO (91), 10G (97)
SPH-P	15	DROP	13	-0.5	4	-20	-	-40	15	10Me₂SO (100), 10G (100)

Differences in post-thaw sperm recovery based on the samples, concentration of the cryoadditives and the freezing rates are represented. Values from 2 experiments.

[1]Intermediary temperature in a 3 - ramp cooling program where from cooling rates change.

[2]Samples from final temperatures plunged into liquid nitrogen for storage.

[3]Damage denotes loss of morphological integrity.

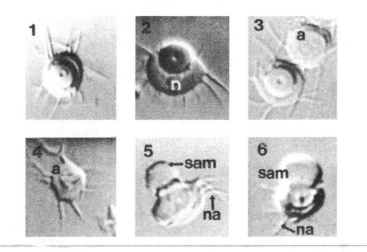

Figure 4 - Freezing damage in spermatozoa of *S. serrata*. 1. Normal spermatozoon 2-6. Morphological damage in the regions of nucleus and acrosome due to injuries during freezing. Bar 4 μm. a- acrosome; n- nucleus; na- nuclear arms; sam- sub-acrosomal material.

3. GENERAL CONSIDERATIONS

Spermatozoa of *Scylla serrata* are packaged into a spermatophoric container during transfer into the female reproductive tract. The inherent differences in the behavior of male gametes obtained from different individuals entails the importance of extending a detailed study on their responses to experimental treatment at physiological temperatures for the purpose of cryopreservation.

The selection of a suitable extender is as important as the selection of cryoprotectants. Ideally the extender for a marine crustacean which predominates estuarine and brackish waters, should mimic the immediate environment, taking into account factors like osmolality, pH and ionic strength. Sea water has been the best extender for most marine species. After studying the effects of several extenders, $Ca^{++}FASW$ was selected for use during most cryopreservation experiments as it maintained the morphological integrity of male gametes over extended periods of time.

Lengthy equilibration at physiological temperatures affected the survival of spermatophores and spermatozoa, probably indicative of the biochemical toxicity of added cryoadditives. Microscopic observations on volume changes in spermatophores revealed that the entry of cryoprotectants was rapid. While in the spermatophores of *S. serrata*, entry of cryoprotectants was permitted within minutes of introduction, an equilibration time of 10 min was found to be insufficient for the spermatophores of *M. rosenbergii* [2,10] but not for *Penaeus monodon* [12], indicating apparent variations in spermatophore wall permeability among different crustacean groups. Interestingly, in the current study, lowering addition temperatures to 12°C did not enhance the 'survival' of spermatophores considerably, indicating that the spermatophore wall permeability to cryoadditives remains unaltered at 12°C.

The toxicity of cryoprotectants used could be osmotic at most lower concentrations, as slowing addition rates and/or dilution of cryoprotectant with a view to limiting osmotic damage increased survival. However, the fact that male gametes did not tolerate cryoprotectant concentrations over 15% over a definite period of time suggests the possibility of biochemical toxicity at higher concentrations. Interestingly, the spermatophores of *Macrobrachium rosenbergii* have been shown to be tolerant to concentrations up to 20% v/v[10]. Fahy [20,21] discussed that increasing the speed of addition/dilution of cryoprotectant will help in reducing biochemical toxicity, whereas decreasing the speed of addition/dilution will favor reduction of osmotic toxicity.

Spermatophores of *S. serrata* displayed limited tolerance to direct plunging in LN_2. As assessed by dye exclusion techniques using eosin-nigrosin, ~37% of spermatophores subjected to direct plunging were 'viable' upon thawing when Ca^{++}FASW was used as an extender and ~41%, when TG-PBS was used. It is interesting to note that Jeyalectumie [22] and Jeyalectumie and Subramoniam[5] recorded remarkable recovery of spermatophores (~95%) upon thawing, when frozen by direct immersion in LN_2. On the contrary, attempts at programming freezing of spermatophores have shown better recovery (~73% of control) in terms of post-thaw 'survival.' Notwithstanding the success, the use of spermatophores for cryopreservation has its limitations due to the fact that injury, if any, to the enclosed spermatozoa cannot be ascribed to the cryoprotectant-induced or to freeze-thaw induced impairment alone, as methods employed in releasing the same from spermatophores for viability scores also resulted in loss of morphological integrity.

Generally, the temperatures beyond which perceivable biochemical reactions do not occur can be considered as 'pre-storage' temperatures. Such temperatures have been shown to have intense effects on sperm survival [23,24]. Present findings indicate that over a cooling rate of -1°C to -10°C/min the final temperatures of -40°C to -80°C did not have any appreciable impact on sperm survival, although higher temperatures -20°C, -30°C, were not suitable as final pre-storage temperatures. This might suggest the possibility of extension of sub-optimal extracellular freezing between the temperature range of -20°C to -30°C. Interestingly, lowering the pre-storage temperatures to -90°C in the abalone *Haliotis diversicolor* sperm yielded greater post-thaw viability [25].

As in our earlier study, tolerance to glycerol was greater when compared to Me_2SO, EG or MeOH. However, the temperature-dependent toxicity of Me_2SO was not as pronounced as in the case of free spermatozoa. This can be due to the additional protective covering offered in the case of the spermatophores.

The reduced survival in higher concentrations of Me_2SO may be attributed to its effect on pH prevailing at sub-zero temperatures [26]. Anchordoguy et al. [27,28] have reported the ineffectiveness of Me_2SO in preventing damage to liposomes subjected to freeze/thaw experiments. Moreover, the occurrence of damage at the region of the acrosome has been attributed to the toxicity of Me_2SO at physiological temperatures [14]. Nonetheless, the observations reveal that Me_2SO-induced toxicity can be reduced to some extent if addition temperatures are lowered.

Although ethylene glycol has been used successfully in the cryopreservation of sea urchin sperm [29] and penaeid shrimp larvae [30,31] their use in male gamete cryopreservation in marine invertebrates is limited. In the current study, its use has not yielded consistent results. Methanol was least toxic at physiological temperatures, but failed to provide appreciable cryoprotection.

In the current study, seeding at -6°C or -7°C did not have much impact on post-thaw sperm membrane integrity if the freezing rate was <-5°C/min. However, increasing the freezing rates to over -7°C/min necessitated sample nucleation. Increase in acrosomal damage was pronounced in both seeded and unseeded sperm aliquots released by enzymatic means, but unseeded samples showed better recovery over the entire range of freezing velocities tested. Methods of providing isothermal holds at temperatures of -5°C and -8°C have marginally improved the survival of unseeded sperm samples obtained from spermatophores by enzyme-aided digestion. However, such holds were of little help when extended to spermatozoa obtained from spermathecae and from spermatophores by mechanical shearing. As with the spermatozoa of *S. ingentis* [4], the spermatophores of *S. serrata* did not tolerate fast freezing rates. Cooling at rates greater than -7°C/min reduced post-thaw survival considerably.

Unlike the spermatozoa of *S. ingentis,* the frozen gametes of *S. serrata* are not amenable to slow thawing procedures. A major limiting factor during thawing is osmotic stress to the cells. The present study shows that the sperm of *S. serrata,* when inside the spermatophores, are amenable to fast thawing procedures. Thawing could be successfully achieved by immersing the straws/vials in water baths maintained at different temperatures ranging from 23°C to 75°C or in air at 30°C.

After thawing, the removal of cryoprotectants becomes important and is directly related to sperm membrane permeabilities to water and cryoprotectants. In an atypical model like the crab sperm, enclosed in sperm bundles or spermatophores, which does not tolerate hypo-osmotic conditions as the human sperm, osmotic swelling due to the exit of cryoprotectants should not far exceed the iso-osmotic volume. This issue gains magnitude as progressively increasing hyper-osmotic conditions during freezing are reversed during thawing, directing the cells through increasing rehydration. Logically therefore, step-wise dilution upon thawing is presumably most suited for optimal cryoprotectant removal. In general, a highly membrane bound system like sperm, with little protoplasmic water, does not require protracted methods for cryoprotectant addition or removal. However, the limited tolerance exhibited by spermatophores to wide cell volume excursions in the current study, seemingly restricts the use of methods involving steep changes in osmolality of the medium.

In brachyuran crabs, as fertilization occurs internally utilizing the stored spermatozoa from the spermathecae [32], it is not possible to obtain unfertilized eggs for the purpose of in-vitro fertilization experiments. Furthermore, the fertilized eggs, immediately upon release from the ovipores, get attached to the pleopodal hairs making egg collection difficult. The nonavailability of ovigerous and freely spawning females in brackish waters, and their in-sea spawning migrations [33-35], in no way aid egg collection. Hence, the possibility of inclusion of "fertilizability" as an index for post-thaw survival of the spermatozoa is ruled out. Although several attempts were made to attach the free spermatozoa to mature oocytes obtained from females, they have not proven successful. The study did not proceed towards artificial insemination as it was difficult to ascertain whether the female crabs had already mated. Attempts towards extraction of spermathecal fluid using catheters were unsuccessful. In order to advance methods to assess post-thaw viability of cryopreserved spermatozoa in the absence of an 'absolute' test, techniques like dye exclusion methods, hypo-/hyper-osmotic response and ionophore-induced acrosome reaction were attempted as pointers for "survival," membrane integrity and simulated physiological response, respectively. Parallel assessment methods have evolved in recent years and it is of interest to note that flow cytometry, has been used effectively to assess freezing damage in the sperm of the shrimp *Litopenaeus vannamei* [11]. The inherent difficulties in establishing reliable viability assays for crustacean sperm have been reviewed by Gwo [24].

In this study, staining techniques used for testing the viability of spermatophores using eosin-nigrosin [5], and spermatozoa using eosin-nigrosin and trypan blue [13,14], have not proven successful. This highlights the inequality existing in membrane permeability in the spermatophores and spermatozoa and selective permeability variations for the different stains used. It was also evident that incorporation of the dyes into the spermatophores did not include all the spermatozoa, indicating retention of membrane properties by the spermatozoa, the differences existing between the membrane properties of the spermatophores and spermatozoa [36,37] and the possibility of the inclusion of noncellular material during the formation of one or more layers of spermatophore wall. Hinsch [38] and Bamba [39], using eosin-nigrosin staining, found difficulty in classifying boar spermatozoa as 'live' or 'dead' as some spermatozoa allowed incorporation of dye only in the post-acrosomal region. The gross variations among the different stains to permeate the sperm plasma membrane highlights the inefficiency of a single staining method as an index of sperm viability.

Considering the disparity existing between the composition of spermatophore wall and the sperm plasma membrane, the incorporation of stains into, or exclusion of stains from spermatophores cannot be taken as indices for survival of the enclosed spermatozoa. In addition, viability of male gametes based on staining techniques should take into account selective variance in the staining characteristics of the different stains and the distinct dissimilarities in the permeation properties of sperm membranes that exist between individuals in a population.

Volume changes that occur when cells are exposed to hyper/hypo-osmotic media have been used as indices for judging membrane integrity [40]. Such protocols have been routinely used for mammalian spermatozoa [15,16,41,42]. The present experiments have shown that volume changes in sperm, exposed to hypo-osmotic media are irreversible, resembling the acrosome reaction. This is in agreement with

Pochon-Masson [43], who has suggested that tonicity of the medium bathing the sperm, might be an extraneous factor in triggering acrosome reaction in decapods.

As a prerequisite to fertilization, most sperm must undergo acrosome reaction. The reaction is initiated on contact of the sperm with an artificial inducer like free calcium ions. Induced acrosome reaction has been shown to be the best alternative to fertilization in cryopreserved sperm of other crustaceans [4]. The current investigation suggests that the ionophore A23187 is responsible for the induction of acrosome reaction probably by its dual function, aiding in opening membrane channels and effecting calcium transport into the sperm cell as postulated by Szabo [44]. In most studies reported to date, the sperm of decapod crustaceans have been noticed to undergo an ionophore-induced acrosome reaction in the presence of extracellular free Ca^{++} [45,46]. The calcium dependency of the acrosome reaction in the lobster *Homarus americanus* has been confirmed [15] using Ca^{++}FASW. Increased concentration of calcium ions in the medium initiated spontaneous acrosome reaction. This was true for the sperm of *S. ingentis*, where Clark et al. [47] have reported a Ca^{++} influx dependent A23187 mediated acrosome reaction. Furthermore, the dependence of ionophore-induced acrosome reaction on the availability of Ca^{++} in the sperm extender has been verified using the ionophore A23187 in the spermatozoa of the lobster *H. Americanus* [48], in the sperm of the sand lobster *Thenus orientalis* [49] and in the fiddler crab *Uca tangeri* [50]. Accordingly, the use of an extender medium devoid of Ca^{++} ions for maintaining morphological integrity of the free sperm of *S. serrata* in the present study is justified.

Our results also indicated that on exposure to proteolytic enzymes the spermatozoa are sensitized and the percentage of sperm undergoing total acrosome reaction is elevated. Griffin and Clak [51] have suggested that proteases might bind to membrane-bound substrates and initiate ionic changes that accompany filament elongation in *S. ingentis* spermatozoa. Reduced cryosurvival may also be due to increased sensitivity to osmotic changes as a result of enzyme treatment. It has also been reported that freezing alters the sensitivity of spermatozoa to osmotic stress in *C. virginica* [52]. The reduced rates of overall success in the cryopreservation protocols employed in the current study can be assigned to this freezing damage, probably caused by membrane leakage, which is more pronounced in the case of spermatozoa released from the spermatophores by enzymatic means.

It has been noticed in the current study that a percentage of free spermatozoa undergo spontaneous exocytosis resembling acrosome reaction when released in FSW and ASW, but not in an artificial medium devoid of Ca^{++} ions. This is in agreement with the earlier reports 1. that free sperm of *H. americanus* [48] and *S. ingentis* [53] undergo spontaneous acrosome reaction in SW and ASW, respectively, strengthening the dependence of acrosome reaction associated events to Ca^{++} entry.

The practical problems of optimizing a freezing technique for the sperm of any given species are complicated because there are so many variables to consider. If the fundamental causes and effects of cryoinjury to male gametes were clearly identified, logical steps could be introduced to minimize their impact. Direct visualization of freezing in action would aid in understanding the actual process of freezing damage. However, it has to be recognized that cryomicroscopy is only an extension of the "trial and error" method for semen freezing. In the current study, cryomicroscopic observations extended to the spermatophores revealed damage which was predominantly osmotic. Even in the absence of a cryoadditive, the spermatophores demonstrated effective shrinkage at temperatures between -5°C and -12°C and regained the original volume upon thawing. However, dye exclusion techniques using trypan blue showed incorporation of dye after the spermatophoral volume increased several fold. It can be inferred that the process of thawing might cause membrane leakage which magnifies after completion of the rewarming phase. Addition of 15% EG depressed the freezing point by ~10°C. Recovery after thawing, nevertheless, was not comparable with that obtained by programmed freezing in Kryo 10/1.7. These preliminary observations might suggest inherent experimental errors and demand further investigation.

It can be concluded from the current observations that an atypical reptantian sperm demands development of accurate methods for viability assessment alongside methods for developing cryopreservation protocols. Glycerol might be the cryoprotectant of choice owing to its dual effect of reduced toxicity at physiological temperatures and optimal protective ability, although the use of other low molecular weight glycols cannot be ruled out. To offset the wide individual fluctuations in freezability, pooling of sperm samples from different individuals would afford optimization of procedures. Freezing rates varying between -3°C/min to -7°C/min from ambient temperature to final sub-zero temperatures of -40°C to -80°C and subsequent storage in LN_2 can be attempted. Thawing procedures should ideally be rapid and precise in order to minimize osmotic and high temperature-induced damages.

An extension of the procedures followed in the current study to the successful cryopreservation of male gametes in other aquaculture-important crustaceans should take into consideration the inherent differences in the freezability of sperm existing between individuals. The design of a protocol should examine all the fundamental cryobiological properties while determining the nature of injuries during cryopreservation [54,55]. The fact that a sperm population contains a heterogeneous mixture of unit cells varying in maturity and functional status has to be appreciated. Designing methods to homogenize the population would certainly be of help. Viability assessments should proceed in the direction of devising methods for assessing fertilizability rather than inferring data from inaccurate methods like dye exclusion.

4. ACKNOWLEDGMENTS

Financial support from Government of India - Department of Science and Technology (Grant SP/SO/C33/88); Department of Biotechnology (Grant BT/AA/03/53/93) and UGC (SAP- Department of Zoology, University of Madras) are gratefully acknowledged.

5. REFERENCES

1-Chow, S., Artificial insemination using preserved spermatophores in the palaemonid shrimp, *Macrobrachium rosenbergii, Bull. Japan Soc. Sci. Fish.,* 48, 1693, 1982.

2-Chow, S., Taki, Y., and Ogasawara, Y., Cryopreservation of spermatophore of the fresh water shrimp, *Macrobrachium rosenbergii, Biol. Bull.,* 168, 471, 1985.

3-Ishida, T., Talbot, P., and Kooda-Cisco, M., Technique for the long-term storage of lobster (*Homarus*) spermatophores, *Gamete Res.,* 14, 183, 1986.

4-Anchordoguy, T.J., Crowe, J.H., Griffin, F.J., and Clark, W.H., Jr., Cryopreservation of sperm from the marine shrimp *Sicyonia ingentis, Cryobiology,* 25, 238, 1988.

5-Jeyalectumie, C. and Subramoniam, T., Cryopreservation of spermatophores and seminal plasma of the edible crab *Scylla serrata, Biol. Bull.,* 177, 247, 1989.

6-Dumont, P., Levy, P., Simon, C., Diter, A., and Francois, S., Freezing of sperm ball of the marine shrimp *Penaeus vannamei.,* in: *Abstract from the workshop on gamete and embryo storage and cryopreservation in aquatic organisms.,* Marly Le Roi, France., 1992, 54.

7-Ke, Y. and Cai, N., Cryopreservation of spermatozoa from the marine shrimp *Penaeus chinensis, Oceanol. Limnol. Sin. Haiyang-Yu-Huzhao,* 27, 187, 1996.

8-Bhavanishankar, S. and Subramoniam, T., Cryopreservation of spermatozoa of the edible mud crab *Scylla serrata* (Forskal), *J. Exp. Zool.,* 277, 326, 1977.

9-Bhavanishankar, S. and Browdy, C.L., Cryopreservation of male gametes of the pacific white shrimp *Litopenaeus vannamei, Aquaculture,* Abstract, 619, 2001.

10-Akarasanon, K., Damrogphol, P., and Poolsanguan, W., Long-Term cryopreservation of spermatophore of the giant freshwater prawn, *Macrobranchium rosembergii (de Man), Aquac Res,* 35, 1, 6, 2004.

11-Lezcano, M., Granja, C., and Salazar, M., The use of flow cytometry in the evaluation of cell viability of cryopreserved sperm of the marine shrimp (*Litopenaeus vannamei*), *Cryobiology,* 48, 349, 2004.

12-Nimrat, S., Sangnawakij, T., and Vuthiphandchai, V., Preservation of black tiger shrimp *Penaeus monodon* spermatophores by chilled storage, *J. World Aquacult. Soc.,* 36, 76, 2005.

13-Leung-Trujillo, J.R. and Lawrence, A.L., Observations on the decline of sperm quality of *Penaeus setiferus* under laboratory conditions, *Aquaculture,* 65, 363, 1987.

4-Kurokura, H., Namba, K., and Ishikawa, T., Lesions of spermatozoa by cryopreservation in oyster *Crassostrea gigas, Nippon Suisan Gakkaishi,* 56, 1803, 1990.

15-Hammerstedt, R.H., Keith, A.D., Snipes, W., Amann, R.P., Arruda, D., and Griel, L.C., Jr., Use of spin labels to evaluate effects of cold shock and osmolality on sperm, *Biol. Reprod.,* 18, 686, 1978.

16-Lomeo, A.M. and Giambersio, A.M., "Water – test": a simple method to assess sperm - membrane integrity, *Int. J. Androl.,* 14, 278, 1991.

17-Chan, P.J., Tredway, D.R., Pang, S.C., Corselli, J., and Su, B.C., Assessment of sperm for cryopreservation using the hypo-osmotic viability test,. *Fertil. Steril.,* 58, 841, 1992.

18-Cavanaugh, G.M., *Formulae and Methods VI of the Marine Biological Laboratory Chemical Room,* Marine Biological Laboratory, Woods Hole, Massachusetts, 1956.

19-Gesteira, T.C.V. and Halcrow, K., Influence of some external factors on the acrosome in the spermatozoa of *Homarus americanus,* Milne Edwards, 1837, *J. Crust. Biol.,* 8, 317, 1988.

20-Fahy, G.M., Cryoprotectant toxicity: biochemical or osmotic, *Cryo-Lett,* 5, 79, 1984.

21-Fahy, G.M., Cryoprotectant toxicity reduction: specific or non-specific., *Cryo-Lett.,* 5, 287, 1984.

22-Jeyalectumie, C., Biochemical investigations on the reproductive tissues and cryopreservation of seminal secretions of a brachyuran crab *Scylla serrata* (Forskal) (Decapoda:Portunidae), in: Doctoral thesis, Department of Zoology, University of Madras, India, 1989, 170.

23-McAndrew, B.J., Rana, K.J., and Penman, D.J., Conservation and preservation of genetic variation in aquatic organisms, in: *Recent Advances in Aquaculture IV,* Muir, J.F. and Roberts R.J., Ed.,. Blackwell Scientific Publications, NY., 1993, 295.

24-Gwo, J.C., Cryopreservation of aquatic invertebrate semen: a review, *Aquacult. Res.*, 31, 259, 2000.

25-Gwo, J.C., Chen, C-W., and Cheng, H.Y., Semen cryopreservation of small abalone (*Haliotis diversicolor supertexa*), *Theriogenology*, 58, 1563, 2002.

26-van den Berg, L. and Soliman, F.S., Effects of glycerol and dimethyl sulphoxide on changes in composition and pH of buffer salt solutions during freezing, *Cryobiology*, 6, 93, 1969.

27-Anchordoguy, T.J., Rudolph, A.S., Carpenter, J.F,. and Crowe, J.H., Modes of interaction of cryoprotectants with membrane phospholipids during freezing, *Cryobiology*, 24, 324, 1987.

28-Anchordoguy, T.J., Cecchini, C.A., Crowe, J.H., and Crowe, L.M., Insights into the cryoprotective mechanism of dimethyl sulfoxide for phospholipid bilayers, *Cryobiology*, 28, 467, 1991.

29-Asahina, É. and Takahashi, T, Freezing tolerance in embryos and spermatozoa of the sea urchin.,*Cryobiology*, 15, 122, 1978.

30-Subramoniam, T. and Newton, S.S., Cryopreservation of penaeid prawn embryos, *Curr. Sci.*, 65, 176, 1993.

31-Newton, S.S. and Subramaniam, T., Cryoprotectant toxicity studies on the embryos of a penaeid prawn *Penaeus indicus*, *Cryobiology*, 33, 172, 1996.

32-Subramoniam, T., Spermatophores and sperm transfer in marine crustaceans, in: *Advances in marine biology*, 29, Blaxter, J.H.S. and Southward, A.J., Eds., Academic Press, New York. 1993, 129.

33-Burke, J.B., Gillespie, N.C., Hill, B.J., Hyland, S.J., and Williams, M.J., The Queensland mud crab fishery, in: *Information Ser*, QI 84024, Hill, B.J., Ed., Queensland Department of Primary Industries, Incl by Bs. Chkd, 1984, 54.

34-Hyland, S., Mud crab fisheries in Australia, *Fins*, 20, 3, 1987.

35-Prasad, P.N. and Neelakantan, B., Maturity and breeding of the mud crab, *Scylla serrata* (Forskal) (Decapoda: Brachyura: Portunidae), *Proc. Ind. Acad. Sci. (Anim. Sci.)*, 98, 341,1989.

36-Subramoniam, T., Chemical composition of the spermatophores in decapod crustaceans, in: *Crustacean Sexual Biology*, Bauer, R.T. and Martin, J.W., Eds., Columbia University Press, New York, 1991, 308.

37-Hinsch, G.W., Structure and chemical composition of spermatophores and seminal fluid of reptantian decapods, in: *Crustacean Sexual Biology*, Bauer, R.T. and Martin, J.W., Eds., Columbia University Press, New York., 1991, 290.

38-Hinsch, G.W., A comparison of sperm morphologies, transfer and sperm mass storage between two species of crab, *Ovalipes ocellatus* and *Libinia emarginata*, *Int. J. Invert. Reprod. Dev.*, 10, 79, 1986.

39-Bamba, K., Evaluation of acrosomal integrity of boar spermatozoa by bright field microscopy using an eosin-nigrosin stain, *Theriogenology*, 29,: 1245, 1988.

40-Curry, M.R. and Watson, P.F., Osmotic effects on ram and human sperm membranes in relation to thawing injury, *Cryobiology*, 31, 39, 1994.

41-Chan, S. Y. W., Wang, C., Ng, M., So, W.W.K., and Ho, P.C., Multivariate disscriminant analysis of the relationship between the hypo-osmotic swelling test and the *in vitro* fertililizing capacity of human sperm,. *Int. J. Androl.*, 11, 369, 1988.

42-Schweisguth, D.C. and Hammerstedt, R.H., Evaluation of plasma membrane stability by detergent-induced rupture of osmotically swollen sperm., *J. Biochem. Biophys. Methods*, 24, 81, 1992.

43-Pochon-Masson, J., L'ultrastructure des spermatozoïdes vesiculaires chez les crustacés décapodes avant et au cours de leur dévagination expérimentale. I. Brachyoures et anomoures, *Ann. Sci. Nat. Zool. Biol. Anim. Ser.*, 12, 1, 1968.

44-Szabo, G., Structural aspects of ionophore function, *Fed. Proc.*, 40, 2196, 1981.

45-Du, N. and Xue, L., Induction of acrosome reaction of spermatozoa in the decapoda *Eriocheir sinensis*, *Chin. J. Oceanol. Limnol.*, 5, 118, 1987.

46-Du, N., Lai, W. and Xue, L., Acrosome reaction of the sperm in the Chinese mittenhanded crab, *Eriocheir sinensis* (Crustacea, Decapoda), *Acta Zool. Sin.*, 33, 9, 1987.

47-Clark, W.H., Jr., Kleeve, M.G., and Yudin, A.I., An acrosome reaction in natantian sperm, *J Exp. Zool.*, 218, 279, 1981.

48-Talbot, P. and Chanmanon, P., Morphological features of the acrosome reaction and the role of the reaction in generating forward sperm movement, *J. Ultrastruct. Res.*, 70, 287, 1980.

49-Silas, M.R., Studies on the spermatophores of the sand lobster, *Thenus orientalis* (Lund) (Decapoda, Scyllaridae) from the Madras coast, Ph.D. thesis, University of Madras, India, 1991.

50-Medina, A. and Rodriguez, A., Structural changes in sperm from the fiddler crab *Uca tangeri* (Crustacea, Brachyura), during the acrosome reaction, *Mol. Reprod. Dev.*, 33, 195,1992.

51-Griffin, F.J. and Clark, W.H., Jr., Induction of acrosomal filament formation in the sperm of *Sicyonia ingentis.*, *J. Exp. Zool.*, 254, 296, 1990.

52-Zell, S.R., Bamford, M.H., and Hidu, H., Cryopreservation of spermatozoa of the American oyster *Crassostrea virginica* (Gmelin), *Cryobiology*, 16, 448, 1979.

53-Griffin, F.J., Clark, W.H., Jr., Crowe, J.H., and Crowe, L.M., Intracellular pH decreases during the *in vitro* induction of the acrosome reaction in the sperm of *Sicyonia ingentis*, *Biol. Bull.*, 173, 311, 1987.

54-Holt, W.V., Fundamental aspects of sperm cryobiology: the importance of species and individual differences, *Theriogenology*, 53, 911, 2000.

55-Woods, E.J., Benson, J.D., Agca, Y., and Critser, J.K., Fundamental cryobiology of reproductive cells and tissues, *Cryobiology*, 48, 146, 2004.

Complete affiliation.

S. Bhavanishankar, Department of Biotechnology, KRMM College of Arts and Science, University of Madras, #4, Crescent Avenue Road, Gandhi Nagar, Chennai 600 020. India, e-mail: shankarsubra@yahoo.co.uk

INDEX

A

Spermiation,9-13,81-83,94 133, 151,
185-187, 201-205, 220-226, 237-248,
251, 297, 339, 345-348, 351, 358, 363,
377, 385, 391, 397, 403, 409, 417,
421-424, 433, 440, 445, 449, 459, 463,
484, 490, 505, 515-532
Spermiation, 4, 9-12, 19, 37, 40, 54-62,
82, 101, 345, 361, 409, 416, 433, 437,
469
Sphaerechinus granularis, 501
Sphoeroides annulatus, 43, 58
Spotted grouper, 51
Spotted sea trout, 26, 50
Starry flounder, 61
Stellate sturgeon, 210
Sterilization, 116
Sterlet, 191, 192, 194, 410, 412
Steroids, 17-19, 25-27
Stickleback, 24, 119, 120, 200
Straws, 244-248, 258, 270, 297-312,
339, 346, 352, 362, 374, 385, 392,
398, 410, 417, 422, 427, 434, 439,
443, 447, 454, 459, 464, 470, 475,
488, 502, 506, 523
Streptomycin, 224-226, 462
Stress responsiveness, 134
Striped bass, 6, 26, 97, 115, 134, 168,
223, 226, 421-424
Striped snakehead, 47
Striped trumpeter, 427-430
Stripping , 6, 29, 37-42, 60-62, 82, 111,
154, 189-192, 266, 343, 345, 351, 361,
378, 385, 397, 403, 409, 433, 437,
448, 459, 463, 469, 478, 498
Strongylocentrotus droebachiensis, 501
Strongylocentrotus intermedius, 501
Sturgeons, 59, 95, 112, 187-190, 206,
352, 409-413
Summer flounder, 60, 61
Supercooling, 239, 524
SYBR, 108, 450
Synchronous, 12, 24, 36, 39, 46, 49, 50,
58, 60, 188, 252, 464

T

Tail moment, 127
Takifugu niphobles, 96, 98
Teleost, 12, 20, 94-100, 105, 122, 149-
173, 185-188, 196, 207, 224, 227, 238,
266, 274-279, 339, 409, 421, 447, 453,
469
Tench, 199, 223
Testes dissection, 333
Testicular sperm, 12,26,83,84,221-
223,226,378,397-400
Testosterone, 17,19,25,58,421
Tetrapigus niger, 501
Tetraploidization, 183,196,199,201
Theca, 6, 7, 8, 25, 26, 94, 251, 516,
517, 521,524, 525, 529, 530
Thelycum, 85, 94
Thenus orientalis, 531
Thermal conductivity, 301
Thermal cycle, 32,44,46,48,50,60
Thermal mass, 296, 297
Thermocouple, 239, 297, 299, 325, 393,
411, 438, 439, 445
Thoracoabdominal segment, 85
Three-spined stickleback, 200
Thunnus albacares, 52
Thunnus maccoyii, 52
Thunnus orientalis, 51
Thunnus thynnus, 43, 51
Thymallus thymallus, 120, 374, 378
Thyroid hormones, 170, 171, 200
Tiger barb, 208
Tinca tinca, 198, 223
Torpedo-shaped catfish, 56
Trachurus japonicus, 49
Transgenesis, 253, 284
Transmembrane pore formation, 283
Transplantation, 198, 252, 253, 280
Trehalose, 255, 282, 283, 285, 493
Tricarboxylic acid cycle, 122, 124
Trichogaster trichopterus, 24